Selected Papers from the Sixth International Symposium on Marine Propulsors

Selected Papers from the Sixth International Symposium on Marine Propulsors

Special Issue Editors

Kourosh Koushan

Sverre Steen

MDPI • Basel • Beijing • Wuhan • Barcelona • Belgrade • Manchester • Tokyo • Cluj • Tianjin

Special Issue Editors
Kourosh Koushan
SINTEF Ocean and Norwegian University of Science and Technology
Norway

Sverre Steen
SINTEF Ocean and Norwegian University of Science and Technology
Norway

Editorial Office
MDPI
St. Alban-Anlage 66
4052 Basel, Switzerland

This is a reprint of articles from the Special Issue published online in the open access journal *Journal of Marine Science and Engineering* (ISSN 2077-1312) (available at: https://www.mdpi.com/journal/jmse/special_issues/symposium_marine_propulsors).

For citation purposes, cite each article independently as indicated on the article page online and as indicated below:

LastName, A.A.; LastName, B.B.; LastName, C.C. Article Title. *Journal Name* **Year**, *Article Number*, Page Range.

ISBN 978-3-03936-248-6 (Hbk)
ISBN 978-3-03936-249-3 (PDF)

Contents

About the Special Issue Editors . . . . **vii**

Preface to "Selected Papers from the Sixth International Symposium on Marine Propulsors" . **ix**

Kourosh Koushan and Sverre Steen
Selected Papers from the Sixth International Symposium on Marine Propulsors
Reprinted from: *J. Mar. Sci. Eng.* **2020**, *8*, 319, doi:10.3390/jmse8050319 . . . . **1**

Kourosh Koushan, Vladimir Krasilnikov, Marco Nataletti, Lucia Sileo and Silas Spence
Experimental and Numerical Study of Pre-Swirl Stators PSS
Reprinted from: *J. Mar. Sci. Eng.* **2020**, *8*, 47, doi:10.3390/jmse8010047 . . . . **3**

Jin Gu Kang, Moon Chan Kim, Hyeon Ung Kim and I. Rok Shin
Study on Propulsion Performance by Varying Rake Distribution at the Propeller Tip
Reprinted from: *J. Mar. Sci. Eng.* **2019**, *7*, 386, doi:10.3390/jmse7110386 . . . . **27**

Chaosheng Zheng, Dengcheng Liu and Hongbo Huang
The Numerical Prediction and Analysis of Propeller Cavitation Benchmark Tests of YUPENG Ship Model
Reprinted from: *J. Mar. Sci. Eng.* **2019**, *7*, 387, doi:10.3390/jmse7110387 . . . . **39**

Mohammed Arab Fatiha, Benoît Augier, François Deniset, Pascal Casari and Jacques André Astolfi
Morphing Hydrofoil Model Driven by Compliant Composite Structure and Internal Pressure
Reprinted from: *J. Mar. Sci. Eng.* **2019**, *7*, 423, doi:10.3390/jmse7120423 . . . . **55**

Da-Qing Li, Per Lindell and Sofia Werner
Transitional Flow on Model Propellers and Their Influence on Relative Rotative Efficiency
Reprinted from: *J. Mar. Sci. Eng.* **2019**, *7*, 427, doi:10.3390/jmse7120427 . . . . **75**

Gisu Song, Hyounggil Park and Taegoo Lee
The Effect of Rudder Existence on Propeller Eccentric Force
Reprinted from: *J. Mar. Sci. Eng.* **2019**, *7*, 455, doi:10.3390/jmse7120455 . . . . **97**

Nachum E. Eisen and Alon Gany
Theoretical Performance Evaluation of a Marine Solid Propellant Water-Breathing Ramjet Propulsor
Reprinted from: *J. Mar. Sci. Eng.* **2020**, *8*, 8, doi:10.3390/jmse8010008 . . . . **117**

Luca Savio, Lucia Sileo and Sigmund Kyrre Ås
A Comparison of Physical and Numerical Modeling of Homogenous Isotropic Propeller Blades
Reprinted from: *J. Mar. Sci. Eng.* **2020**, *8*, 21, doi:10.3390/jmse8010021 . . . . **129**

Wenyu Sun, Qiong Hu, Shiliang Hu, Jia Su, Jie Xu, Jinfang Wei and Guofu Huang
Numerical Analysis of Full-Scale Ship Self-Propulsion Performance with Direct Comparison to Statistical Sea Trail Results
Reprinted from: *J. Mar. Sci. Eng.* **2020**, *8*, 24, doi:10.3390/jmse8010024 . . . . **151**

Lotan Arad Ludar and Alon Gany
Experimental Study of Supercavitation Bubble Development over Bodies in a Duct Flow
Reprinted from: *J. Mar. Sci. Eng.* **2020**, *8*, 28, doi:10.3390/jmse8010028 . . . . **173**

Noriyuki Sasaki, S. Kuribayashi, M. Fukazawa and Mehmet Atlar
Towards a Realistic Estimation of the Powering Performance of a Ship with a Gate Rudder System
Reprinted from: *J. Mar. Sci. Eng.* **2020**, *8*, 43, doi:10.3390/jmse8010043 . . . . **185**

Liang Li, Bin Zhou, Dengcheng Liu and Chao Wang
Numerical Analysis of Influence of the Hull Couple Motion on the Propeller Exciting Force Characteristics
Reprinted from: *J. Mar. Sci. Eng.* **2019**, *7*, 330, doi:10.3390/jmse7100330 . . . . **201**

Bradford Knight Kevin Maki
Multi-Degree of Freedom Propeller Force Models Based on a Neural Network and Regression
Reprinted from: *J. Mar. Sci. Eng.* **2020**, *8*, 89, doi:10.3390/jmse8020089 . . . . **225**

Seungnam Kim, and Spyros A. Kinnas
Prediction of Unsteady Developed Tip Vortex Cavitation and Its Effect on the Induced Hull Pressures
Reprinted from: *J. Mar. Sci. Eng.* **2020**, *8*, 114, doi:10.3390/jmse8020114 . . . . **247**

Ville Viitanen, Timo Siikonen, Antonio Sánchez-Caja
Cavitation on Model- andFull-Scale Marine Propellers: Steady and Transient Viscous Flow Simulations at Different Reynolds Numbers
Reprinted from: *J. Mar. Sci. Eng.* **2020**, *8*, 141, doi:10.3390/jmse8020141 . . . . **291**

Batuhan Aktas, Naz Yilmaz, Mehmet Atlar, Noriyuki Sasaki, Patrick Fitzsimmons and David Taylor
Suppression of Tip Vortex Cavitation Noise of Propellers using PressurePores^{TM} Technology
Reprinted from: *J. Mar. Sci. Eng.* **2020**, *8*, 158, doi:10.3390/jmse8030158 . . . . **325**

Stephan Helma
Surprising Behaviour of the Wageningen B-Screw Series Polynomials
Reprinted from: *J. Mar. Sci. Eng.* **2020**, *8*, 211, doi:10.3390/jmse8030211 . . . . **347**

Anna Maria Kozlowska, Øyvind Øksnes Dalheim, Luca Savio and Sverre Steen
Time Domain Modeling of Propeller Forces due to Ventilation in Static and Dynamic Conditions
Reprinted from: *J. Mar. Sci. Eng.* **2020**, *8*, 31, doi:10.3390/jmse8010031 . . . . **411**

Spyros A. Kinnas
VIScous Vorticity Equation (VISVE) for Turbulent 2-D Flows with Variable Density and Viscosity
Reprinted from: *J. Mar. Sci. Eng.* **2020**, *8*, 191, doi:10.3390/jmse8030191 . . . . **431**

About the Special Issue Editors

Kourosh Koushan is Senior Advisor at SINTEF Ocean since 2016. He is also Adjunct Professor at the Norwegian University of Science and Technology (NTNU) since 2015. He received his PhD from and worked at the Technical University of Berlin (1992–1997). He joined SINTEF Ocean (formerly MARINTEK) in 1997 as researcher, was promoted to senior researcher in 2000, to principal researcher in 2005 and research director in 2008. He is a member of the Executive Committee (since 2014) and Advisory Council (since 2011) of the ITTC (International Towing Tank Conference). Dr. Koushan, together with Prof. Steen, founded the International Symposium on Marine Propulsors in 2009.

Sverre Steen specializes in marine hydrodynamics, focusing on propulsion and experimental methods. He has been a professor at the Department of Marine Technology at the Norwegian University of Science and Technology since 2004, and Head of Department at the same institute since 2016. In 2009, he founded the Symposium of Marine Propulsors together with Dr. Kourosh Koushan.

Preface to "Selected Papers from the Sixth International Symposium on Marine Propulsors"

Marine propulsors are key components of the many thousands of ships operating in oceans, lakes, and rivers around the world. The performance of propulsors is a vital component in assessing the efficiency, environmental impact (including impact on marine fauna), and safety of ships. Propulsor performance is also important for crew and passenger comfort. New types of propulsors with electric drives, flexible blades, and multi-stage propellers require new knowledge and improved tools. Innovative main or auxiliary propulsor types, using renewable energy from waves or winds, are also being commercialized. The improvement of computers and computational fluid dynamics creates new opportunities for advanced design and performance prediction, and new instrumentation and data collection techniques enable more advanced experimental techniques. This book is devoted to bringing the latest developments in research and technical developments regarding the hydrodynamic aspects of marine propulsors, to the benefit of both academics and industry. It consists of 19 peer-reviewed scientific papers previously published in the Journal of Marine Science and Engineering. Seventeen of these papers are improved and extended versions of papers presented at the Sixth Symposium on Marine Propulsors in May 2019 in Rome, Italy, while the remaining two are completely new submissions. The Symposia on Marine Propulsors are held bi-annually; the next one will take place in Wuxi, China in May 2021.

Kourosh Koushan, Sverre Steen

Special Issue Editors

Journal of
Marine Science and Engineering

Editorial

Selected Papers from the Sixth International Symposium on Marine Propulsors

Kourosh Koushan [1,*] and Sverre Steen [2]

1 SINTEF Ocean and Department of Marine Technology, Faculty of Engineering Science and Technology, Norwegian University of Science and Technology, Otto Nielsens vei 10, N-7491 Trondheim, Norway
2 Department of Marine Technology, Faculty of Engineering Science and Technology, Norwegian University of Science and Technology, Otto Nielsens vei 10, N-7491 Trondheim, Norway; sverre.steen@ntnu.no
* Correspondence: Kourosh.Koushan@sintef.no; Tel.: +47-411-05-297

Received: 14 April 2020; Accepted: 14 April 2020; Published: 1 May 2020

Keywords: propellers; waterjets; unconventional propulsors; cavitation; noise and vibration; numerical methods in propulsion; propulsor dynamics; propulsion in seaways; propulsion in off-design conditions; energy saving devices

---

This Special Issue is following up the success of the latest Symposium on Marine Propulsors (www.marinepropulsors.com, smp'19) by publishing extended or improved versions of the selected papers presented at the symposium. This issue also includes new original contributions. The symposium smp'19 was the sixth in a series of international symposiums dedicated to the hydrodynamics of all types of marine propulsors. The next symposium in this series will be held in China in May 2021. This Special Issue comprises 17 excellent papers originating from the symposium [1–17] and two outstanding new papers [18,19]. The papers disseminate state-of-the-art numerical and experimental research results on marine propulsors and marine renewable devices.

Marine propulsors are key components of the many thousands of ships operating in oceans, lakes, and rivers around the world. The performance of propulsors is vital for the efficiency, environmental impact, including the impact on marine fauna, and safety of ships. Propulsor performance is also important for crew and passenger comfort. New types of propulsors with electric drives, flexible blades, and rim driven propellers require new knowledge and improved tools. Innovative main or auxiliary propulsor types, using renewable energy from waves or winds, are also being commercialized. The improvement of computers and computational fluid dynamics creates new opportunities for advanced design and performance predictions, and new instrumentation and data collection techniques enable more advanced experimental techniques. This Special Issue of the *Journal of Marine Science and Engineering* is devoted to bringing the latest developments in research and technical developments regarding hydrodynamic aspects of marine propulsors, to the benefit of both academics and the industry.

Prof. Dr. Kourosh Koushan and Prof. Dr. Sverre Steen

Guest Editors of "Selected Papers from the Sixth International Symposium on Marine Propulsors"

Author Contributions: K.K. wrote the editorial. S.S. contributed to and reviewed the editorial. All authors have read and agreed to the published version of the manuscript.

Funding: This research received no external funding.

Conflicts of Interest: The authors declare no conflict of interest.

References

1. Li, L.; Zhou, B.; Liu, D.; Wang, C. Numerical Analysis of Influence of the Hull Couple Motion on the Propeller Exciting Force Characteristics. *J. Mar. Sci. Eng.* **2019**, *7*, 330. [CrossRef]
2. Kang, J.; Kim, M.; Kim, H.; Shin, I. Study on Propulsion Performance by Varying Rake Distribution at the Propeller Tip. *J. Mar. Sci. Eng.* **2019**, *7*, 386. [CrossRef]
3. Zheng, C.; Liu, D.; Huang, H. The Numerical Prediction and Analysis of Propeller Cavitation Benchmark Tests of YUPENG Ship Model. *J. Mar. Sci. Eng.* **2019**, *7*, 387. [CrossRef]
4. Fatiha, M.; Augier, B.; Deniset, F.; Casari, P.; Astolfi, J. Morphing Hydrofoil Model Driven by Compliant Composite Structure and Internal Pressure. *J. Mar. Sci. Eng.* **2019**, *7*, 423.
5. Li, D.; Lindell, P.; Werner, S. Transitional Flow on Model Propellers and Their Influence on Relative Rotative Efficiency. *J. Mar. Sci. Eng.* **2019**, *7*, 427. [CrossRef]
6. Song, G.; Park, H.; Lee, T. The Effect of Rudder Existence on Propeller Eccentric Force. *J. Mar. Sci. Eng.* **2019**, *7*, 455. [CrossRef]
7. Eisen, N.; Gany, A. Theoretical Performance Evaluation of a Marine Solid Propellant Water-Breathing Ramjet Propulsor. *J. Mar. Sci. Eng.* **2020**, *8*, 8. [CrossRef]
8. Savio, L.; Sileo, L.; Kyrre Ås, S. A Comparison of Physical and Numerical Modeling of Homogenous Isotropic Propeller Blades. *J. Mar. Sci. Eng.* **2020**, *8*, 21. [CrossRef]
9. Sun, W.; Hu, Q.; Hu, S.; Su, J.; Xu, J.; Wei, J.; Huang, G. Numerical Analysis of Full-Scale Ship Self-Propulsion Performance with Direct Comparison to Statistical Sea Trail Results. *J. Mar. Sci. Eng.* **2020**, *8*, 24. [CrossRef]
10. Arad Ludar, L.; Gany, A. Experimental Study of Supercavitation Bubble Development over Bodies in a Duct Flow. *J. Mar. Sci. Eng.* **2020**, *8*, 28. [CrossRef]
11. Sasaki, N.; Kuribayashi, S.; Fukazawa, M.; Atlar, M. Towards a Realistic Estimation of the Powering Performance of a Ship with a Gate Rudder System. *J. Mar. Sci. Eng.* **2020**, *8*, 43. [CrossRef]
12. Koushan, K.; Krasilnikov, V.; Nataletti, M.; Sileo, L.; Spence, S. Experimental and Numerical Study of Pre-Swirl Stators PSS. *J. Mar. Sci. Eng.* **2020**, *8*, 47. [CrossRef]
13. Knight, B.; Maki, K. Multi-Degree of Freedom Propeller Force Models Based on a Neural Network and Regression. *J. Mar. Sci. Eng.* **2020**, *8*, 89. [CrossRef]
14. Kim, S.; Kinnas, S. Prediction of Unsteady Developed Tip Vortex Cavitation and Its Effect on the Induced Hull Pressures. *J. Mar. Sci. Eng.* **2020**, *8*, 114. [CrossRef]
15. Viitanen, V.; Siikonen, T.; Sánchez-Caja, A. Cavitation on Model- and Full-Scale Marine Propellers: Steady and Transient Viscous Flow Simulations at Different Reynolds Numbers. *J. Mar. Sci. Eng.* **2020**, *8*, 141. [CrossRef]
16. Aktas, B.; Yilmaz, N.; Atlar, M.; Sasaki, N.; Fitzsimmons, P.; Taylor, D. Suppression of Tip Vortex Cavitation Noise of Propellers using PressurePoresTM Technology. *J. Mar. Sci. Eng.* **2020**, *8*, 158. [CrossRef]
17. Helma, S. Surprising Behaviour of the Wageningen B-Screw Series Polynomials. *J. Mar. Sci. Eng.* **2020**, *8*, 211. [CrossRef]
18. Kozlowska, A.; Dalheim, Ø.; Savio, L.; Steen, S. Time Domain Modeling of Propeller Forces due to Ventilation in Static and Dynamic Conditions. *J. Mar. Sci. Eng.* **2020**, *8*, 31. [CrossRef]
19. Kinnas, S. VIScous Vorticity Equation (VISVE) for Turbulent 2-D Flows with Variable Density and Viscosity. *J. Mar. Sci. Eng.* **2020**, *8*, 191. [CrossRef]

Journal of
Marine Science and Engineering

Article

Experimental and Numerical Study of Pre-Swirl Stators PSS

Kourosh Koushan [1,*], Vladimir Krasilnikov [1], Marco Nataletti [1], Lucia Sileo [1] and Silas Spence [2]

1 Department of Ship and Ocean Structures, SINTEF Ocean AS, 7465 Trondheim, Norway; Vladimir.Krasilnikov@sintef.no (V.K.); Marco.Nataletti@sintef.no (M.N.); Lucia.Sileo@sintef.no (L.S.)
2 SkaMik AS, 7940 Ottersøy, Norway; silas@skamik.no
* Correspondence: Kourosh.Koushan@sintef.no; Tel.: +47-41105297

Received: 18 December 2019; Accepted: 13 January 2020; Published: 16 January 2020

Abstract: Energy saving within shipping is gaining more attention due to environmental awareness, financial incentives, and, most importantly, new regional and international rules, which limit the acceptable emission from the ships considerably. One of the measures is installation of energy saving devices (ESD). One type of such a device, known as pre-swirl stator (PSS), consists of a number (usually 3 to 5) of fins, which are mounted right in front of the propeller. By modifying the inflow and swirl into the propeller, the fins of a PSS have the possibility to increase the total propulsion efficiency. However, at the same time, they may introduce additional resistance either due to changes in pressure distribution over the aft ship or due to its own resistance of fins. In this paper, the authors present experimental and numerical investigation of a PSS for a chemical tanker. Numerical analysis of the vessel with and without PSS is performed in the model and full scale. Model testing is performed with and without PSS to verify the power savings predicted numerically. Among other quantities, 3D wake field behind the hull is densely measured at different planes, starting from the PSS plane to the rudder stock plane. 3D wake measurements are also conducted with a running propeller. The measurements show considerable improvement in the performance of the vessel fitted with PSS. On the numerical side, analyses show that scale effect plays an important role in the ESD performance. Investigation of the scale effect on the vessel equipped with an ESD provides new insight for the community, which is investing more into the development of energy saving devices, and it offers valuable information for the elaboration of scaling procedures for such vessels.

Keywords: energy saving devices (esd); pre-swirl stator (pss); scale effect; 3D wake measurements with a working propeller

1. Introduction

Energy Saving Devices (ESDs) represent the technology equally suitable for both the new and retrofit ships. The Propulsion Improving Devices occupy the greatest place on today's ESD market, and they show well-documented power savings over a large range of vessels (ABS [1]), (Mewis [2]). These ESDs are normally classified by their working principle regarding the mechanism that they use to modify the flow around the propeller operating in the wake behind hull, including Wake Equalizing and Flow Separation Alleviating Devices, Pre-Swirl Devices, Post-Swirl Devices, and High-Efficiency Propellers and Ducts. More than one device can be employed as components of an ESD installation, and some devices obtain the benefit of several flow modification mechanisms, such as the Mewis Duct® (Guiard et al. [3]), (Mewis [2]).

The present study is focused on the experimental and numerical investigation of the ESD type known as Pre-Swirl Stator (PSS) as a possible solution for retrofitting of a chemical tanker. A PSS may consist of a number (usually 3 to 5) fins, which are mounted upstream of the propeller. By modifying the axial and tangential velocity components of the inflow on the propeller, a properly designed

PSS should recover part of the kinetic energy losses in the system of the ship hull with an operating propeller. The greatest part of energy saving by PSS comes from utilizing the kinetic energy associated with tangential velocities, which is due to the effect of pre-swirl. The PSS fins may also be used in combination with a duct or a shroud. Ducted configurations known as Pre-Ducts with Stator (PDS) are widely employed (Dang et al. [4]), (Mewis [2]), (Kim et al. [5]). They combine the effect of pre-swirl with the effect of flow equalization. Since the duct may accelerate the inflow on the propeller, which counteracts the effect of flow retardation by the fins, PDSs show usually higher power savings compared to PSS except the range of light propeller loading, where approximately CTh < 1.25. The benefits of duct installation increase with the rise of loading. Shrouds connecting the stator fins may be employed to improve structural integrity, and to reduce tip losses (Voermans [6]). Voermans investigated the determination of hydrodynamic loads on a PSS experimentally and validated a prediction methodology for design loads (Voremans [7]). Zhai studied hull pressure fluctuations numerically with an energy saving device known as the Pre-Shrouded Vane (Zhai et al. [8]). Recently, Controllable Pre-Swirl Fins (CPSF) were introduced. These are mounted in front of the propeller in such a way that they can be adapted to different operating conditions by optimizing their pitch and flap settings (Nielsen et al. [9]). In the present work, only the PSS without duct/shroud is investigated. For the design and performance optimization of the PSS, a Computational Fluid Dynamics CFD-based approach is employed, where a number of design parameters defining configuration of the fins are systematically investigated through a self-propulsion calculation. The objective is to maximize power savings at the true vessel self-propulsion point due to the installation of ESD with the given propeller design. The design is performed for full-scale conditions at the chosen design speed and draught, and the final design is checked in a relevant speed range around the design point. Model tests are performed for the validation of the numerical method and verification of ESD performance. In particular, extensive 3D wake measurements are conducted at different planes starting from the PSS plane in front of the propeller to rudder stock plane aft of the propeller. These measurements are conducted with a running propeller.

2. General Considerations Regarding ESD Design

When designing a PSS to retrofit a single-screw displacement vessel, one needs to take into consideration the following aspects.

The energy losses due to flow rotation induced by the propeller operating behind the hull are greater than energy losses due to flow acceleration, which is opposite of the case of the propeller operating in open water conditions. Hence, one needs to work on the modification of a tangential velocity component.

Depending on the direction of the propeller rotation, the PSS fins should be installed either on the portside or starboard, where the prevailing direction of transverse flow is in the direction of the propeller rotation.

If large vortices develop on the ship hull as a result of flow separation in the stern region, the PSS fins may break those vortices and, to a certain degree, serve the purposes of flow redirection and equalization. If this is of importance, one may consider placing additional fins on the opposite side of the ship as well, even though they may not give any considerable contribution to pre-swirl, or they may even reduce the pre-swirl effect already existing in a non-modified wake.

The shape of fins and their location with respect to hull and propeller should be chosen to maximize the desired effect of pre-swirl, and, at the same time, to minimize separation, vortex shedding, and strength of the tip vortex on the fins. The said phenomena led to energy losses that cannot be regained. This prompts the distributions of pitch (twist angle) and camber along the span, which should, in theory, be derived individually for each fin, which makes the fins adapt to local flow features.

The rudder installed downstream of the propeller utilizes a certain amount of rotational energy in a propeller slipstream, which would reduce the overall positive effect of pre-swirl by PSS. At the same time, since PSS recovers part of the rotational energy and reduces the tangential velocity in the slipstream, one may expect an increase in rudder resistance compared to the case without PSS.

The PSS may be designed to produce an additional thrust. However, this thrust would result in an increase of the thrust deduction force on the hull, so that one cannot expect significant gains in the net thrust. Designing a neutral PSS (regarding the axial force it develops) may, therefore, be the preferred strategy as such a stator would also be less likely to suffer from flow separation, and it would pose less risk to have a negative effect on the flow over stern.

The previously mentioned points indicate the importance of using a system approach to the ESD design, considering the whole system consists of a ship hull, ESD, propeller, and rudder. The choice of a correct objective function for the ESD design is equally important. The maximum achievable reduction of the shaft delivered power, PD, or a maximum increase of propulsive efficiency are the criteria adequate for the purpose. These quantities are representative of the efficiency of an ESD designed for a given ship, when they are defined at the ship self-propulsion point. For comparing ESDs installed on different ships and working under different conditions, additional criteria such as the ESD Quality Index suggested in Voermans [6] may be employed.

Since, for single-screw ships, the scale effect on wake field is considerable, full-scale effective inflow on the propeller needs to be accounted for in the ESD design process.

While the actual design is normally performed for one design condition (ship speed and draft), one needs to verify the performance of the vessel equipped with the designed ESD at other speeds and drafts, since these parameters may have influence on the inflow experienced by the ESD and propeller.

The above considerations, and, in particular, those related to a system design approach and realistic full-scale conditions, suggest that a CFD-based design may be an adequate method of choice. Verification by model tests is, however, important in the light of known deficiencies of the CFD RANS methods related to capturing an anisotropic pattern of turbulence in the wake field and dependency of results on mesh resolution.

The discussion presented in this section is focused on the hydrodynamic aspects of ESD design. Before the design is finalized, one needs to check the structural characteristics of the ESD such as fin stiffness, stress levels, avoidance of resonance, fatigue loads, and connections with the ship hull.

3. Numerical Method and its Validation

The numerical method employed as the main solver in the design exploration procedure is based on the unsteady RANS approach, using the Sliding Mesh technique to fully account for the interaction between the ship hull, ESD, propeller, and rudder. The commercial CFD software STAR-CCM+ (version 12.04) is used for this purpose. STAR-CCM+ provides a powerful design exploration and optimization engine called Design Manager where setups with a different degree of automation can be realized.

When modeling the geometry and generating mesh for the CFD solution, attention is paid to accurate representation of all the main components of the system. The examples of computation mesh in the vessel stern region are shown in Figures 1 and 2. The mesh in the main fluid region containing the ship hull, rudder, and, eventually, ESD, is produced using Trimmer, a predominantly hexahedral mesh where the cells next to the surfaces are trimmed to accommodate the input geometries.

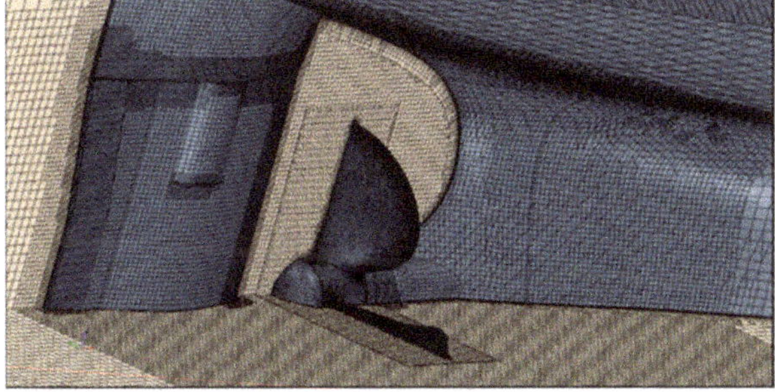

Figure 1. Fragment of computation mesh in the vessel stern region.

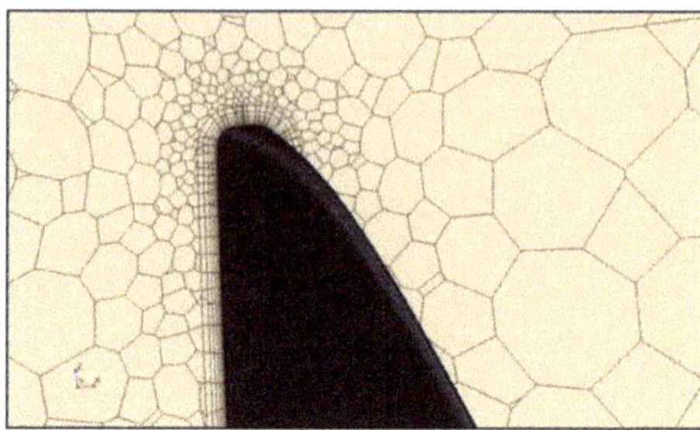

Figure 2. Details of computation mesh around the propeller blade tip.

In the rotating propeller region, a polyhedral mesh is generated. Along all the wall boundaries, a prism layer mesh is generated to resolve velocity profiles in the boundary layer and meet the desired Wall Y+ levels. In the full-scale calculations, 12 prism layers are arranged along the hull, rudder, and ESD surface. Additionally, 10 prism layers are arranged on the propeller blades and hub. The averaged target Y+ values are 150 on the hull surface, and 90 on the rudder and ESD surfaces. The Y+ on propeller blades ranges between 30 and 80, with most of the blade area being close to the lower boundary of this range. Thus, a high-Re near-wall resolution is aimed for on the wall boundaries to reduce computational effort, which is essential in design exploration studies where many variants need to be evaluated. The main turbulence model used in the RANS solution is k-ω SST. However, other turbulence models such as realizable k-ε and RST (Linear Pressure Strain formulation) models were also investigated. The numerical solution for the free surface is obtained using the Volume of Fluid VOF method, using the Flat VOF Wave model and HRIC interface capturing scheme. Considering the importance of sufficiently accurate resolution of the flow past ship stern, around the propeller, ESD, and rudder, the mesh in that region is refined to provide an average cell size (in both the fluid region and propeller region) about 1.35% of the propeller diameter Dp. The target surface cell size on propeller blades, rudder, and ESD fins is twice smaller than 0.675% of Dp. Further refinement of 0.084375% of Dp is applied along the edges and tips of respective parts. With these settings, the cell count in the propeller region is about 4 million (1 million cells per blade), and the cell count of the fluid region is about 11.6 million without ESD and 14.5 with ESD. A grid sensitivity study was conducted prior to the aforementioned mesh recommendations where the volume cell size in the mesh control surrounding the ship stern, ESD, propeller, and rudder (this control is seen in Figure 1 as the zone of refined mesh) varied as 2%, 1.5%, and 1.0% of Dp. The surface cell size on the geometry parts was scaled accordingly. The changes in integral characteristics of propeller due to mesh refinement were found to be within 0.5%. The final cell size value of 1.35% of Dp was assigned to optimize total cell count due to the feature of the Trimmer meshing template where cell size changes by multiples of two from the base size, which, in this case, was equal to the vessel LPP. The influence of hull surface roughness is included in the CFD model by means of modified wall functions. The assumed equivalent height of sand roughness grains is equal to 60 μm. The studied vessel has one bow tunnel thruster and bilge keels. These design features as well as the superstructure are not included explicitly in the CFD model. Their influence on resistance is accounted for by means of standard corrections, which are used in the experimental predictions. The present setup has earlier been validated against the benchmark full-scale dataset of the Lloyd's Register 2016 CFD Workshop (Ponkratov [10]). Additional validation exercises have been conducted in the present work with the vessel used in the ESD optimization study. Table 1 presents a comparison between the CFD predictions, full-scale prognosis based on model tests, and full-scale trial data. In this table and below, n is the propeller rate of revolution, TB is the propeller thrust behind the hull (at the self-propulsion point), Vs is the ship speed, PD is the shaft delivered power, and PB is the brake power. The propulsion coefficients (wake fraction WT, thrust deduction factor t, relative rotative efficiency η_r, and quasi-propulsive efficiency η_D) are compared between the experimental and numerical predictions. Good agreement between the prognosis based on model tests and full-scale trial data is achieved. The CFD prediction also agree well with the experimental prognosis, especially the results obtained with the anisotropic (seven equations) turbulence model RST. This result is mainly explained by the fact that the RST model allows for a more accurate and detailed resolution of the hull wake features. Regarding nominal wake in model scale conditions, this is illustrated by Figures 3 and 4 for the sections at the propeller plane and ESD location, respectively.

Table 1. Comparison of propulsion characteristics between the numerical and experimental predictions and full-scale trial data for the vessel without ESD.

(**a**) Numerical (CFD) vs. experimental (EFD) predictions, 14 (kn), without a sea margin.

	EFD Prediction	**CFD kwSST**	**Δ, %**	**CFD RST**	**Δ, %**
n (rps)	1.44	1.472	2.22	1.453	0.90
TB (kN)	690.72	695.57	0.70	689.72	−0.15
PD (kW)	5596.68	5820.38	4.00	5671.31	1.33
t	0.195	0.197	1.04	0.168	−13.77
WT	0.312	0.294	−5.77	0.311	−0.32
η_r	1.006	0.991	−1.49	0.990	−1.59
η_D	0.715	0.708	−1.04	0.724	1.14

(**b**) Experimental predictions vs. sea trial data with a 15% sea margin.

	Sea Trials	**EFD Prediction**	**Δ, %**
PB (kW)	10,170	10,170	0.00
n (rps)	1.745	1.717	−1.60
Vs (kn)	15.95	15.69	−1.63

Since the RST model solves seven equations for turbulence transport instead of two used by the k-ω SST, it carries considerable computational overhead. In addition to that, the solution with the RST model may experience stability issues, and it is, therefore, desired to initialize it with a converged k-ω SST solution. For these reasons, the RST model becomes rather expensive in the design exploration studies involving many repetitive runs. Therefore, in spite of its limitations and somewhat lower accuracy, the k-ω SST was used in the ESD design studies conducted in the present work. In order to reduce the computational effort even further, it was decided to neglect the presence of free surface in self-propulsion calculations.

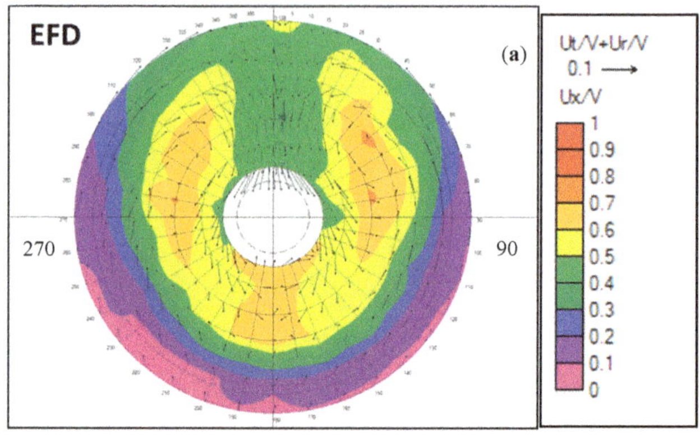

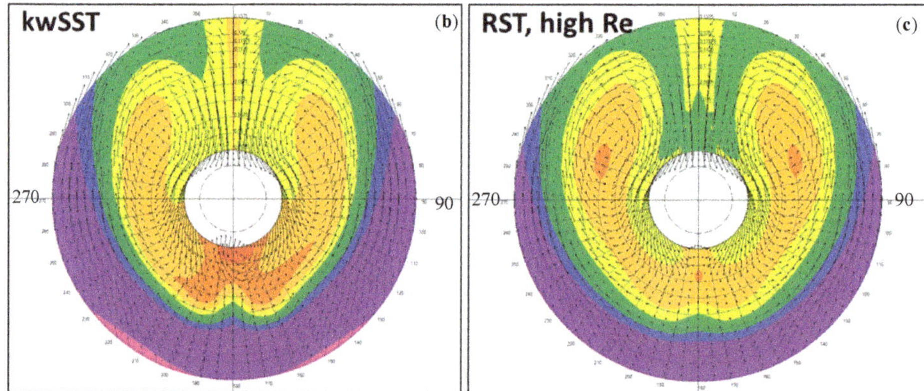

Figure 3. Measured and computed nominal wake field on the propeller plane. (Model scale, Vs = 14 kn, without ESD). (**a**) measured results. (**b**) CFD results using kwSST. (**c**) CFD results using RST, high Re.

The computation domain was restricted by the symmetry plane at the level of design waterline, while the resistance components due to wave making and dynamic vessel sinkage and trim were introduced, along with other additional resistance corrections, into the force balance equation as follows.

$$|T_{BX} + R_{ts} + \Delta R| \leq \varepsilon) \quad (1)$$

where T_{BX} is the axial component of the propeller thrust (accounts for vessel trim and, if present, propeller shaft inclination), R_{ts} is the ship resistance with an operating propeller (includes contributions from hull, rudder, appendages, and ESD). ΔR is the correction accounting for resistance contributions that are not modelled explicitly in the CFD model (in the present case, tunnel thruster, bilge keels, aerodynamic resistance of superstructure, and resistance due to wave making), and ε is the tolerance with which the self-propulsion condition is satisfied (in the present calculations, it was set to 0.5% of propeller thrust T_B).

The validity of the assumption about negligible effect of the free surface on propeller inflow is dependent on simulated conditions. In the present case of design speed and draught, the influence of a free surface was very small, as illustrated by Figure 5, which shows the wake fields on the propeller plane computed with and without a free surface. In this case, the relative difference in averaged axial coefficient of nominal wake, WTN, amounted to about 1.5%. For shallower sitting vessels and higher Froude numbers, this may not be the case, so that the said assumption needs to be checked in each case.

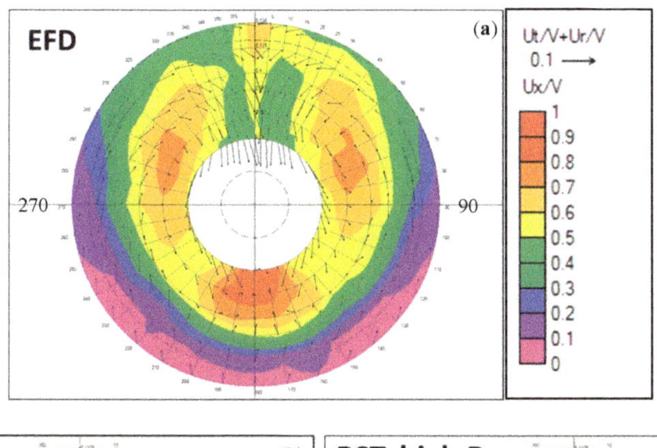

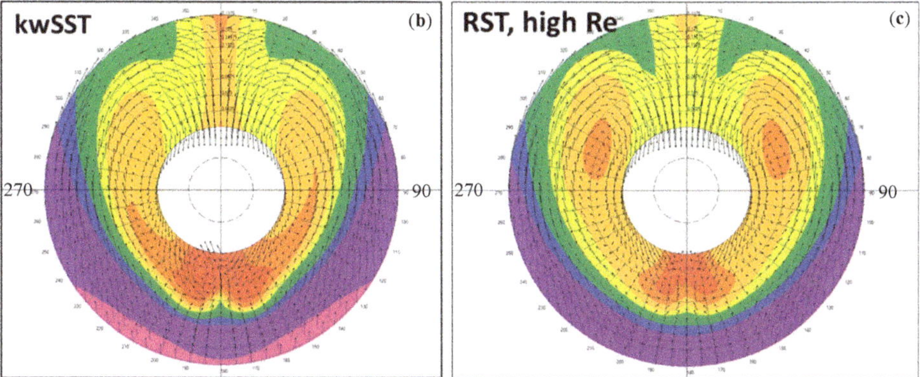

Figure 4. Measured and computed nominal wake field around the ESD location. (Model scale, Vs = 14 kn, without ESD). (**a**) measured results. (**b**) CFD results using kwSST. (**c**) CFD results using RST, high Re.

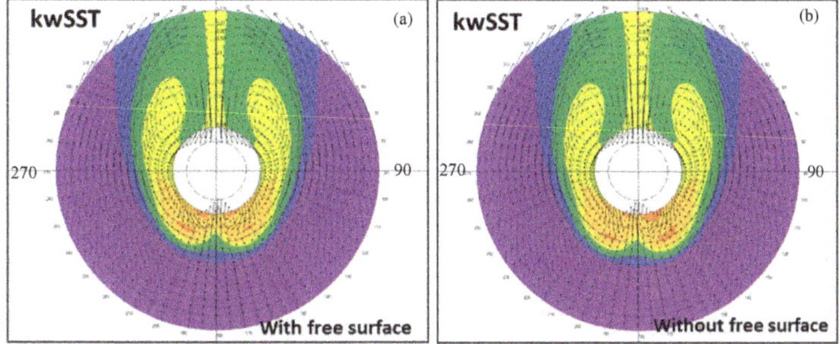

Figure 5. Nominal wake field on the propeller plane, (**a**) computed with and (**b**) without a free surface. (Full scale, Vs = 14 kn, w/o ESD). (same color scale as in Figure 4).

Comparing the wake field computed in model scale (Figure 3) and full scale (Figure 5), one can notice a significant effect of Reynolds number (scale effect). Thus, an ESD installed behind the ship hull would operate in very different conditions on a full scale and on a model scale. Since the ultimate goal is to design an ESD for the full-scale ship, it is natural to use the full-scale CFD setup. At the same time, it needs to be cautioned that uncertainties associated with a full-scale wake prediction may, however, affect the numerical results (Guiard et al. [3]). They are primarily related to the resolution of wake turbulence and influence of hull roughness. Separate tests were performed in the present study to assess the influence of the turbulence model on the predicted magnitude of the scale effect on a nominal wake field. Calculations were done on a model scale and full scale using the k-ω SST model, the RST high-Re model, and the RST 2-Layer (All Y+) model. The relative differences in a nominal wake fraction WTN between full scale and model scale cases amounted to −31%, −30%, and −29%, respectively. The patterns of predicted full-scale wakes were also found quite comparable between the turbulence models. Nevertheless, in view of the known uncertainties and lack of direct validation data for full-scale wake fields, experimental verification of the numerical ESD design is still very important.

4. Design Exploration

The approach used for the design of PSS implies systematic variation of selected geometrical parameters to study their effect on vessel propulsion efficiency. These parameters include:

Fin geometry parameters—fin radius at the tip (determines fin length), distributions of chord length, thickness, pitch, and camber of fin cylindrical sections (chord length and thickness distributions are assumed identical for all fins, while pitch and camber distributions are individual for each fin), and the fin section profile (the same normalized profile applies along the fin span, and for all fins);

Fin installation parameters—position of PSS plane, number of fins, angular positions of fins, and fin installation angles (angles about the fin reference line that passes through a fin root section and is perpendicular to the propeller shaft axis).

The main idea behind the present design workflow is to provide user-driven design exploration with partial automation of some optimization steps. Clearly, the initial number of design variables is very large, and some of their values, or ranges of interest, can be defined from the experience. A limited number of preliminary calculations can also narrow down the design space considerably, so that only the most important parameters and most promising ranges are left for the formal optimization exercise.

For the present study, the PSS configuration with three fins have been selected. The fin geometry was produced using the SINTEF Ocean propeller design/analysis software AKPA, where fin geometry was defined in a parametric form, after the fashion of the conventional propeller blade definition. The Wageningen Duct 7 profile with thickness increased toward the trailing edge used as the basis fin section. This profile would also eventually be suitable to design a shroud connecting the tips of PSS fins. Thickness and camber distributions along the chord were presented in normalized form, so that one can apply the desired maximum thickness-to-chord and camber-to-chord ratios. The initial value of fin radius was given equal to Rf/Rp = 0.923. The initial distributions of chord length, pitch, and camber along the span were assigned as constant functions. This preliminary configuration was meant to eventually serve the purpose of both the PSS (open pre-swirl stator) and PSD (ducted pre-swirl stator) designs. Thus, produced fins were converted to CAD geometry parts in the 3D-CAD tool of STAR-CCM+. As mentioned in the Introduction, only the open PSS configuration was explored further in this study. In 3D-CAD, a number of design parameters were defined, including the position of the PSS plane, the angular positions of fins, and fin installation angles. These parameters were exposed to the STAR-CCM+ solver, and they were used in the design exploration studies with the initial fin geometries. CFD calculations to determine self-propulsion performance of the ship equipped with ESD were done in full-scale conditions. Once the best values for each parameter were established, which resulted in maximum predicted power savings by the PSS (PSS-Var-1), we returned back to the AKPA software to modify chord length, pitch, and camber distributions, in order to align each fin with local flow and minimize separation, while still producing the required amount of pre-swirl on the propeller. The length of fins was increased to Rf/Rp = 1.077, while fins were tapered toward the tip, and tip rounding was applied. With the modified geometry of the fins, a new round of design exploration exercises has been undertaken, varying only the design parameters responsible for the fin installation angle, in a much narrower range than at the first step. The described loop was repeated several times, before the final geometry of the PSS fins was established (PSS-Var-5), which satisfies the criteria of desired power saving (about 4%) and alignment of fins with the flow. This allowed avoiding significant flow separation. After the preliminary design (PSS-Var-1), the fin installation angles remained nearly unchanged with the following fin geometry iterations, which indicates that reliable values of these parameters were found. An overall flowchart illustrating the described design workflow is presented in Figure 6.

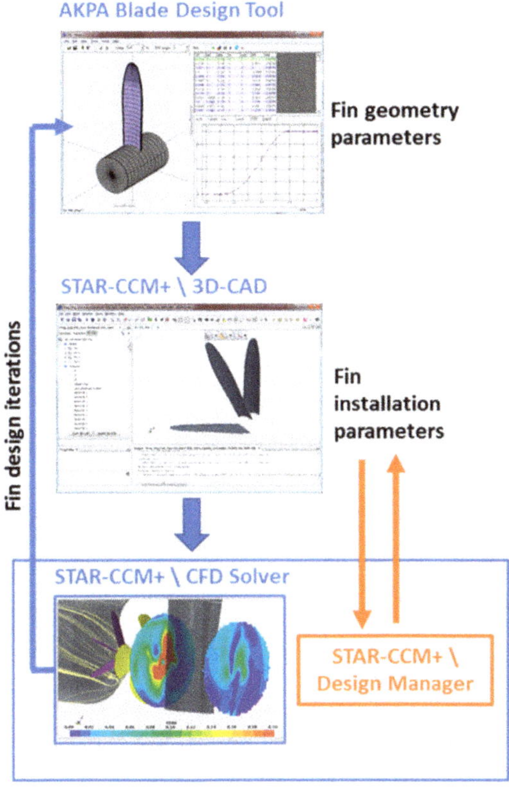

Figure 6. Overall flow chart of the ESD design workflow.

Table 2 shows the main particulars of the ship and propeller for which the retrofit PSS design was carried out. Table 3 presents the design conditions. Figure 7 illustrates the evolution of PSS design in the process of design exploration, according to the described method. Only the "final" designs of the three design families (groups of designs with similar design properties) are presented in this figure. The main properties of these designs are reduced in Table 4. In total, about 70 design alternatives grouped in six design families have been evaluated. As mentioned above, the distributions of pitch and camber are variable along the fin span, and they are derived individually for each of the three fins. Table 5 shows the results of the CFD evaluation of the three selected PSS designs. It can be seen that all three designs offer power savings of more than 3%. However, the variants Var-3 and Var-5, featuring longer and thinner fins with tapering towards the tip, allow for greater power savings, which is about 4%. The pre-swirl action of the PSS can be judged from the values of tangential velocity Vt/V, which are presented for the three slipstream sections in the last three rows of Table 5. The section "Sect 1" is located immediately upstream of the propeller (X/Dp = 0.1 from propeller plane), section "Sect 2" is located in the middle between the propeller and rudder (X/Dp = −0.246 from propeller plane), and section "Sect 3" is located downstream of the rudder (X/Dp = −1.308 from the propeller plane). The tangential velocity is averaged over the slipstream section area. Positive tangential velocity is in the direction of propeller rotation. One can see that, at section "Sect 1," in the wake of the ship without ESD, the averaged amount of swirl is quite close to zero, since the tangential velocities of an opposite direction on the two sides of the ship tend to cancel each other out.

Table 2. Main particulars of the ship and propeller.

Ship		Propeller	
Length between PP, L_{PP} (m)	175.6	Diameter, Dp (m)	6.5
Breadth moulded, B (m)	32.229	Number of blades, Z	4
Draught at $L_{PP}/2$, T (m)	11.846	Blade area ratio, AE/A0	0.52
Displacement, Δ (t)	56,147.7	Pitch, P (0.7)/Dp	0.821
Block coefficient, C_B	0.8123	Hub ratio, dH/Dp	0.205
Prismatic coefficient, C_P	0.8096	Direction of rotation	Right-handed

Table 3. Conditions for PSS design.

Ship Speed, Vs (kn)	14.0
Froude number, Fr	0.1735
Draught at $L_{PP}/2$, T (m)	11.846
Propeller loading, C_{Th} (without ESD)	1.57

Figure 7. Evolution of PSS design in the process of design exploration.

Table 4. Conditions for PSS design.

	PSS-Var-1	PSS-Var-3	PSS-Var-5
Number of Fins	3	3	3
Profile	Duct 7	Duct 7	Duct 7
Fin radius, Rf/Rp	0.923	1.077	1.077
Maximum thickness, t0/c	0.12	0.130	0.135
Chord length, c/Rf, its distribution	0.3, constant, no rounding	0.257, taper from 0.685Rf, tip rounding	0.1, taper from 0.55Rf, tip rounding
ESD plane from AP, X (m)	6.1	6.1	6.1
Angular positions of fins 1/2/3 (deg), from the Y-axis	−22/23.6/69	−22.5/23/68	−22.5/23/68
Installation angles of fins 1/2/3 (deg)	65/72/88	68/77.5/91	68/77.5/91

Table 5. Performance characteristics of the ship equipped with the three PSS designs. CFD predictions at design speed (14 knots) on a full scale.

	No ESD	PSS-Var-1	PSS-Var-3	PSS-Var-5
n (rps)	1.472	1.434	1.422	1.4255
PD (kW)	5820.38	5632.57	5595.35	5589.33
		(−3.23%)	(−3.87%)	(−3.97%)
KT_B	0.1753	0.18725	0.19113	0.18951
KQ_B	0.0244	0.02554	0.02602	0.02580
Ct, ship	0.002976	0.003003	0.003032	0.003023
Ct, rud	0.000133	0.000136	0.000141	0.000138
Ct, ESD	—	0.000012	−0.000008	−0.000008
WT	0.294	0.350	0.366	0.357
Vt/V (Sect 1)	−0.000237	−0.02157	−0.032959	−0.02899
Vt/V (Sect 2)	0.089347	0.071418	0.061025	0.064009
Vt/V (Sect 3)	0.026384	0.022593	0.017588	0.018427

The tangential velocity induced at this section by the propeller is very small. At the section "Sect 2," the slipstream is swirled in the direction of propeller rotation by the action of the propeller. Positive averaged amount of swirl remains at section "Sect 3" while it is reduced by 70% by the presence of a rudder. Installation of PSS results in an increase of negative tangential velocity (pre-swirl) in front of the propeller, and reduction of positive tangential velocity at the sections downstream of the propeller and downstream of the rudder.

The assessment of swirl alone is representative of the recovery of kinetic energy associated with tangential flow, but it is not sufficient for the analysis of ESD performance, since an ESD acts on the axial flow. For example, one can see that all ESD designs presented in Table 5 result in an increase of the axial wake fraction WT, i.e., they slow down the inflow on the propeller. Therefore, the operation point of the propeller ("effective" advance coefficient, J_A) moves toward heavier loading. The increase of propeller thrust and torque coefficients for the cases with PSS is, thus, due to both the pre-swirl and heavier axial wake produced by the ESD. A better adaptation of PSS fins to the local flow implemented in the design variant Var-5 allows for a slightly greater power savings than in the variant Var-3 due to the reduction of axial wake on the propeller, which moves the propeller operation point in the direction of higher efficiency. This is in spite of the fact that the design Var-5 produces less pre-swirl, and, hence, recovers less tangential flow energy compared to the design Var-3. The above result illustrates the importance of careful adaptation of the ESD to the flow in the system of the ship hull and propeller.

The averaged flow characteristics presented in Table 5 are representative of the integral performance of an ESD. More detailed insight into the influence of ESD on the flow around the propeller can be obtained from the distributions of the tangential velocity presented as colored vectors in Figure 8, and distributions of the axial velocity presented as contours in Figure 9. These are instantaneous images corresponding to the propeller key blade position at 0 degrees (12 o'clock). One can notice the reduction of a positive swirl produced by the PSS on the portside of the vessel, where the PSS is installed. While the PSS is installed on the portside, it also has a favorable effect on the pre-swirl on the starboard, which is seen from the reduction of a positive swirl near the hub, and from the increase of a negative swirl at the outer sections. The axial velocity component undergoes smaller changes by the action of PSS compared to the tangential component. Its overall distribution remains similar to the case without PSS, but an increase of flow retardation (lower axial velocity) in front of the propeller is clearly visible.

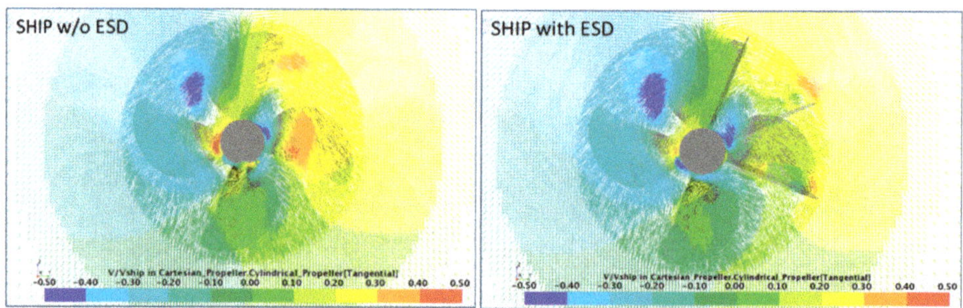

Figure 8. Distribution of the tangential velocity on section "Sect 1" in front of the propeller, seen from upstream (PSS-Var-5, full scale, Vs = 14 kn).

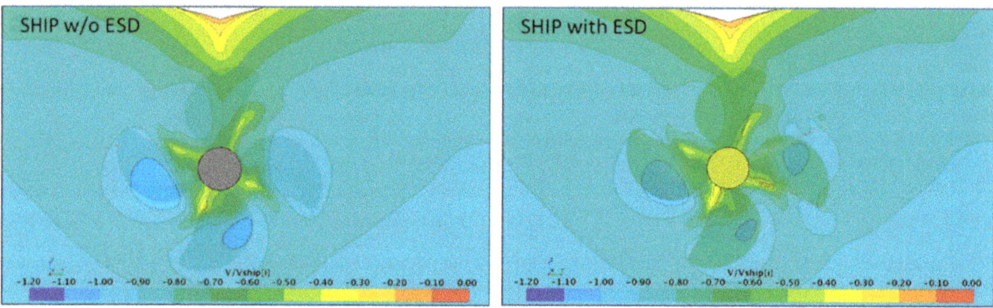

Figure 9. Distribution of the axial velocity on section "Sect 1" in front of the propeller, seen from upstream (PSS-Var-5, full scale, Vs = 14 kn).

In order to present a more complete and formal analysis of energy savings, one can consider the distributions of kinetic energy associated with the transverse and axial flows. Such distributions are presented in dimensionless form for the downstream sections "Sect 2" and "Sect 3" in Figures 10 and 11. The corresponding averaged values of kinetic energy are given in dimensionless form in Table 6. Both the transverse flow energy and axial flow energy are reduced by the action of the PSS, which results in power savings in the whole system, as presented above. The largest portion of energy saving comes from the utilization of tangential flow energy, in particular, energy contained in the portside of the flow. However, the contribution from the axial kinetic energy is also important, as its amount is about one-fourth of total energy saved in the section between the propeller and rudder, and more than one-third of energy saved in the section downstream of the rudder.

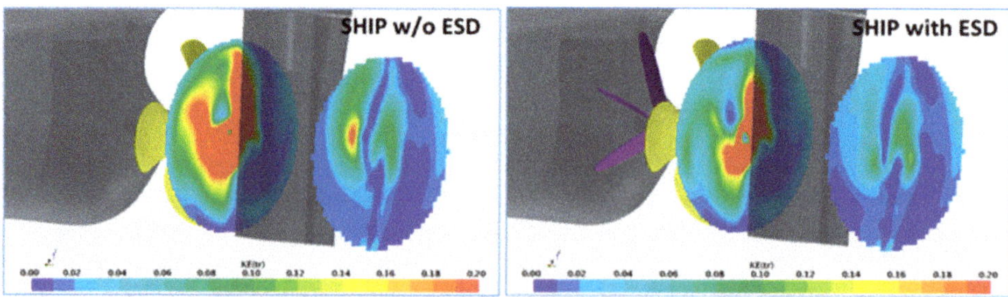

Figure 10. Distribution of the kinetic energy of transverse flow on sections "Sect 2" and "Sect 3" downstream of the propeller (PSS-Var-5, full scale, Vs = 14 kn).

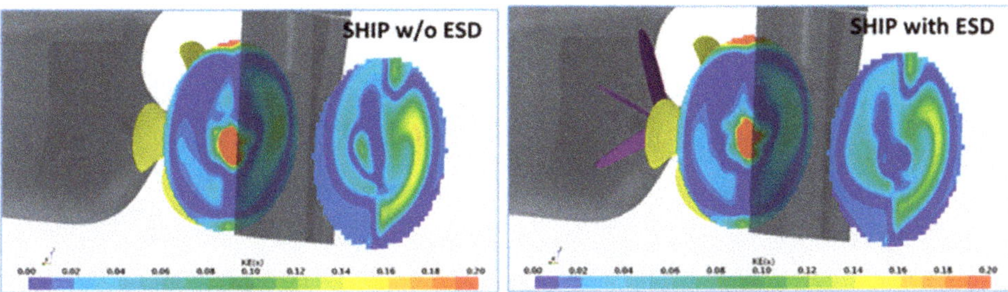

Figure 11. Distribution of the kinetic energy of axial flow on sections "Sect 2" and "Sect 3" downstream of the propeller (PSS-Var-5, full scale, Vs = 14 kn).

Table 6. Averaged magnitudes of kinetic energy on the sections downstream of the propeller ("Sect 2") and downstream of the rudder ("Sect 3"). Full scale, Vs = 14 kn.

	Section "Sect 2"			Section "Sect 3"		
	without ESD	**PSS-Var-5**	**Δ, %**	**without ESD**	**PSS-Var-5**	**Δ, %**
KE(x)	0.0442	0.0408	−7.69	0.0395	0.0368	−6.84
KE(tr)	0.1120	0.0708	−36.79	0.0325	0.0238	−26.77
KE(tot)	0.1562	0.1116	−28.55	0.0720	0.0606	−15.83

The ratio between KE(x) and KE(tr) is different at these two sections. At section "Sect 2," which is downstream of the propeller, the amount of transverse flow energy is greater than the amount of axial flow energy, but at section "Sect 2," the amount of axial flow energy is slightly larger.

This is explained by the action of rudder that destroys the swirl of propeller slipstream and utilizes a significant portion of tangential flow energy. In the presence of PSS, the amount of swirl coming on the rudder is reduced, and, thus, the rudder can utilize less rotation energy. It results in an additional small penalty seen in the increase of rudder resistance (see Table 5). From the kinematics standpoint, it is explained by the reduction of local angles of attack of the rudder section, and, hence, reduction of section lift, which counteracts drag. In the present case, with the PSS-Var-5, the increase of rudder resistance amounted to about 3.75% compared to the case without PSS. The proportion of rudder resistance in the total resistance of the ship does not change significantly, and it amounts to approximately 4.5% in the present case. From the analysis of flow kinetic energy, one can conclude that there is still potential for further energy savings through a better utilization of both the axial and tangential flows. In particular, the area downstream of the propeller hub holds a considerable amount of both KE(x) and KE(tr). It can be utilized by the implementation of post-swirl measures such as Propeller Boss Cap Fins (PBCF) or modification of the rudder design.

Using properly designed PBCF, one can expect additional power savings of an order of 1%–1.5%. These estimated figures are lower than power savings achieved reportedly with PBCF on propellers without pre-swirl ESDs (Mewis [2]) since part of the available energy is already utilized by the PSS. Figure 12 shows pressure distribution over the aftship, constrained streamlines on the hull, and flow streamlines released from the propeller hub (colored by vorticity magnitude) with and without PSS. In particular, one can notice that, in presence of the PSS, the strength of the hub vortex is reduced, and the trajectory of the vortex is more aligned with the shaft axis. The analysis of pressure distribution on the hub reveals a more uniform pattern when the PSS is installed. The pressure reduction in the hub vortex core is also considerably lower than in the case without ESD, even though propeller loading is heavier (see Figure 13). This is a favorable result as far as the rudder operation is concerned.

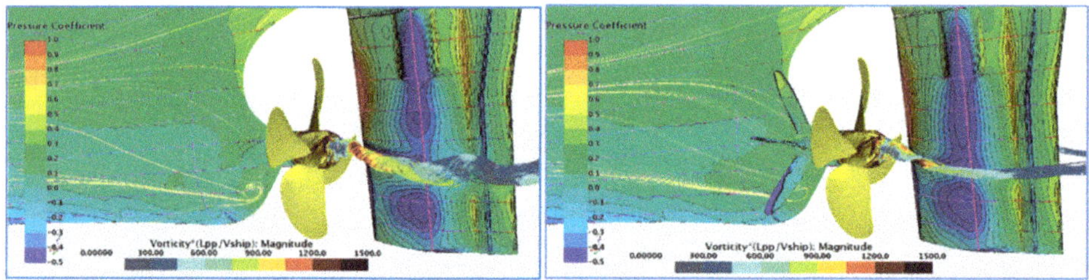

Figure 12. Pressure distribution and flow streamlines past the aft ship. (PSS-Var-5, full scale, Vs = 14 kn).

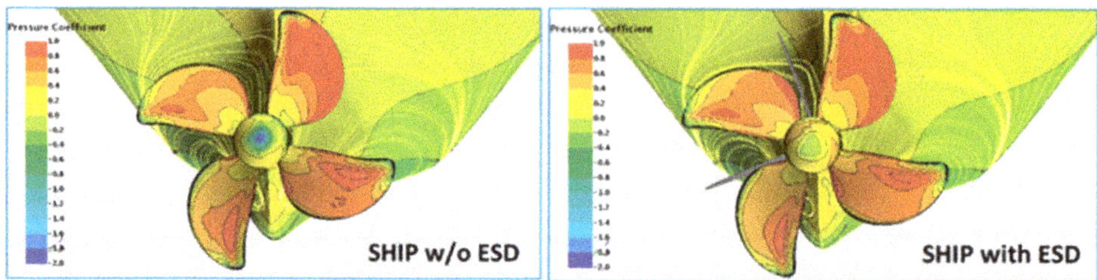

Figure 13. Pressure distribution on the hub and pressure side of propeller blades (PSS-Var-5, full scale, Vs = 14 kn).

The axial wake fraction, WT, is derived based on propeller open water characteristics using the thrust identity method.

$$WT = 1 - J_A/J_V \tag{2}$$

where J_V is the advance coefficient defined based on ship speed, and J_A is the advance coefficient where the thrust identity KT_0 (J) = KT_B (J) is met, with KT_0 being the propeller thrust coefficient in open water and KT_B being the propeller thrust coefficient behind the hull. The thrust deduction factor, t, presented in Table 7 is calculated by considering the ESD as a part of the propulsor.

Table 7. Computed propulsion coefficients of the vessel without ESD and with the designed PSS. (PSS-Var-5, full scale).

	Without ESD				With PSS			
Vs (kn)	**WT**	**t**	**η_r**	**η_D**	**WT**	**t**	**η_r**	**η_D**
12	0.297	0.193	0.992	0.712	0.359	0.203	0.991	0.742
14	0.294	0.197	0.991	0.708	0.357	0.209	0.990	0.737
16	0.295	0.2085	0.993	0.703	0.357	0.218	0.991	0.733

Table 7 presents the propulsion coefficients of the vessel without ESD and with the designed PSS derived from self-propulsion simulations at the three ship speeds around the design condition.

$$t = 1 - R_{tow}/T_{tot} \tag{3}$$

where R_{tow} is the towing resistance of the hull without ESD, and T_{tot} is the total thrust produced by the propeller and ESD at the self-propulsion point. The relative rotative efficiency, η_r, represents the ratio between the propeller torque coefficient in open water, KQ_o, and propeller torque coefficient behind the hull, KQ_B, calculated at the advance coefficient J_A.

$$\eta_r = KQ_o/KQ_B \tag{4}$$

The propulsion efficiency, η_D, represents the ratio between the effective power, P_E, and shaft delivered power, P_D, at the vessel self-propulsion point.

$$\eta_D = P_E/P_D \tag{5}$$

It is observed that the installation of PSS results in an increase of axial wake fraction of the order of 18%–21%, and an increase of thrust deduction of the order of 5%–6%. The relative rotative efficiency remains almost unchanged. The increase of propeller loading due to heavier wake and effect of pre-swirl, as discussed above, results in a larger thrust deduction factor. The predicted trends in changes of propulsion factors due to the installation of ESD correlate well with the findings from the studies by other authors done with similar installations, e.g., (Dang et al. [4]), (Kim et al. [5]).

Table 8 presents relative changes due to the installation of PSS in the propeller rate of revolution, shaft delivered power, and propulsive efficiency for the same speed range. These figures confirm that comparable energy savings are achieved by the designed PSS for all vessel speeds, in the range of interest around the design condition.

Table 8. Relative changes in propeller RPM, shaft delivered power, and propulsive efficiency due to the installation of PSS. (PSS-Var-5, full scale).

Vs (kn)	Δn, %	ΔPD, %	$\Delta\eta_D$, %
12	−3.21	−4.06	4.21
14	−3.16	−3.97	4.10
16	−3.20	−4.14	4.27

In order to evaluate the influence of PSS on the cavitation characteristics of the propeller, additional comparisons were performed for the design speed of 14 kn as shown in Figure 14 and a higher speed of 16 kn as presented in Figure 15. Since propeller cavitation is not evident at the design draught, these comparisons were done for the ballast draught where the cavitation number was estimated based on propeller submergence at even keel, without free surface deformation.

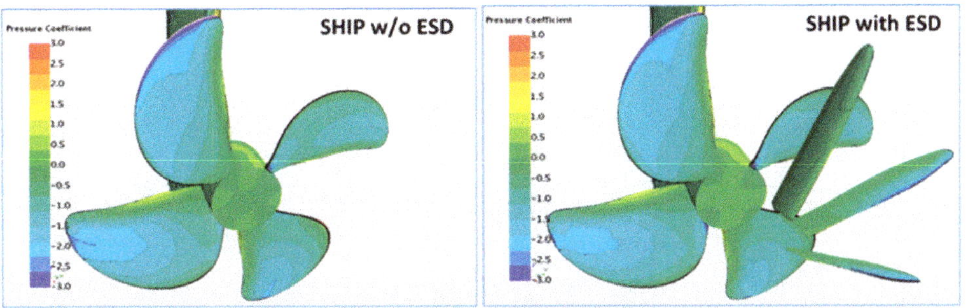

Figure 14. Cavitation domains on propeller and ESD. (PSS-Var-5, full scale, Vs = 14 kn, ballast draught).

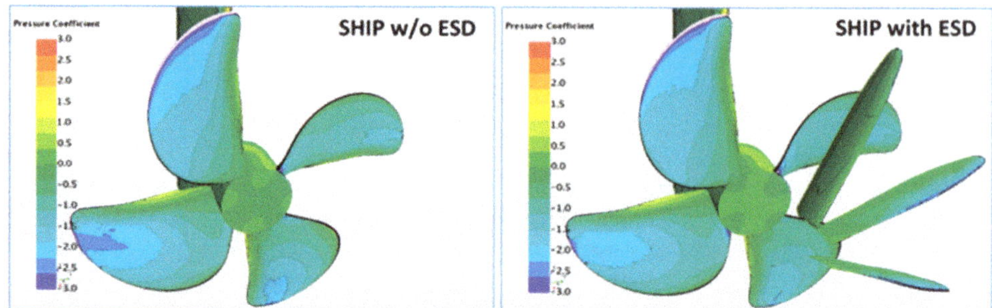

Figure 15. Cavitation domains on the propeller and ESD. (PSS-Var-5, full scale, Vs = 16 kn, ballast draught).

5. Experimental Investigation

In the wake survey tests, the setup shown in Figure 16 was used. The rudder and rudder headbox were removed from the setup. In the measurements with an operating propeller, the model was self-propelled, according to the conditions defined from the propulsion test at the vessel speed and draft corresponding to the design point: Vs = 14 (kn), T = 11.846 (m). The data acquired from the first phase of the test campaign were used in the validation studies with the numerical method. Selected validation results are presented in Section 3 of this paper. After the ESD design exploration studies were completed, and the design variant PSS-Var-5 was chosen, the second phase of the test campaign was realized to verify power savings predicted from the numerical investigations. Figure 17 shows the propulsion test setup with the designed PSS. In addition to the resistance and propulsion tests, 3D wake measurements were repeated with the PSS installed, for the same locations as in the first phase, except the ESD plane. As an example, Figure 18 shows wave profile photographed during resistance tests with PSS mounted at 14 kn.

Figure 16. 3D wake measurements setup.

Figure 17. Propulsion test setup with the designed PSS.

Figure 18. Wave profile during the resistance test with PSS at 14 kn.

Figures 19 and 20 show 3D nominal wake survey at the propeller plane without and with PSS, respectively. Fins installation angles in this angular coordinate system are 247.5°, 293°, and 338°. Fins are clearly affecting the axial, tangential, and radial velocities as intended, which are discussed earlier in this publication.

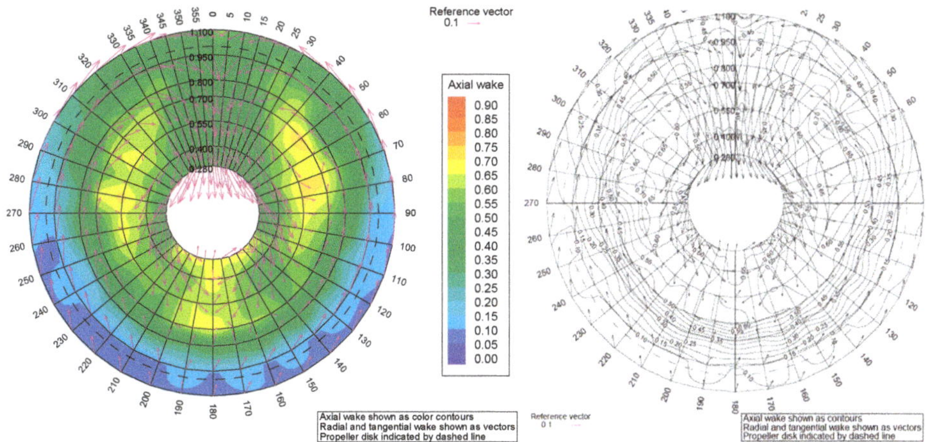

Figure 19. 3D wake measurements at the propeller plane, without PSS, at 14 kn.

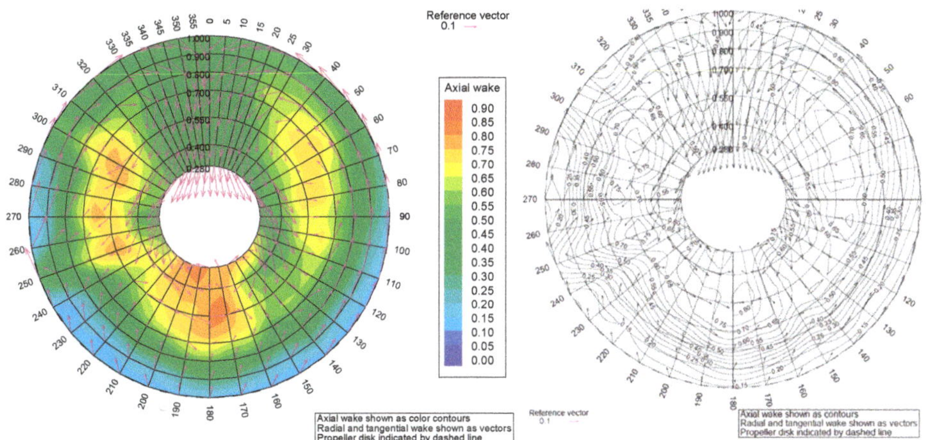

Figure 20. 3D wake measurements at the propeller plane, with PSS, at 14 kn.

Figures 21 and 22 show effective 3D wake measurements behind the working propeller, midway between the propeller and rudder leading edge, without and with PSS, respectively. Additionally, in this case, the presence of fins affect axial, tangential, and radial velocities. The propeller stream results in strong tangential flow.

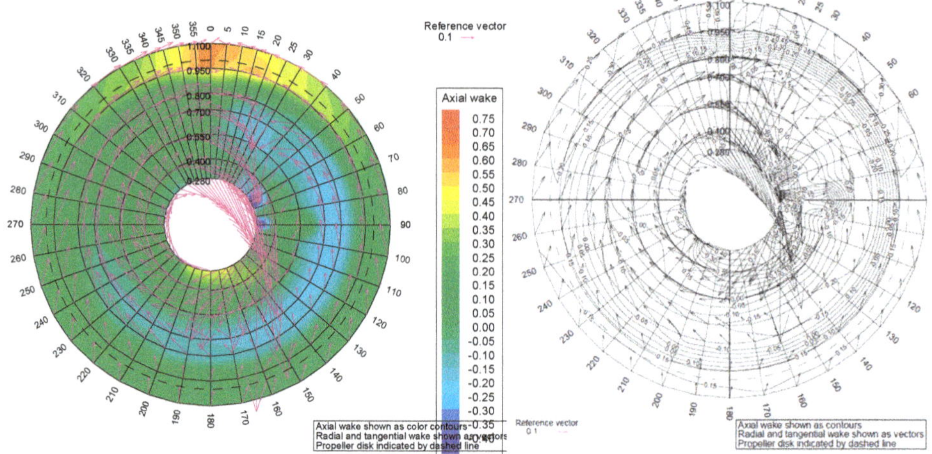

Figure 21. 3D wake measurements at a plane midway between the propeller and rudder, with a working propeller, without PSS, at 14 kn.

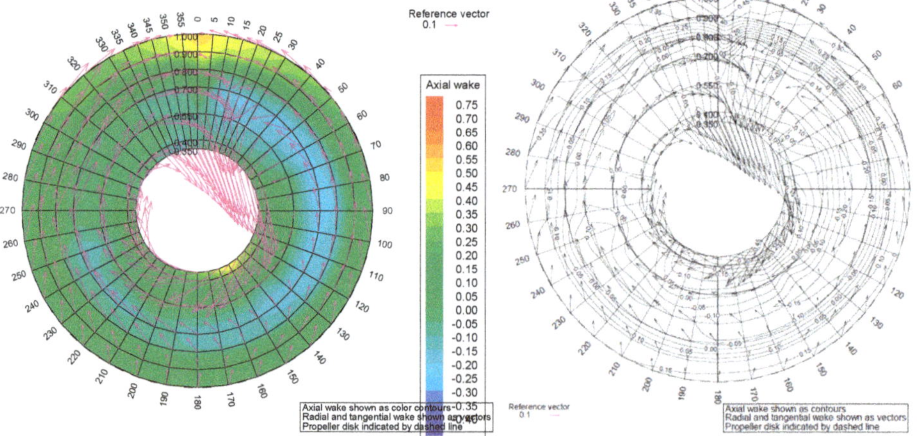

Figure 22. 3D wake measurements at a plane midway between the propeller and rudder, with a working propeller, with PSS, at 14 kn.

Results of 3D wake survey without PSS at a rudder stock plane with working propeller are presented in Figure 23. Figure 24 shows results of the 3D wake survey at a rudder stock plane with a working propeller and mounted PSS. In this survey, measurements included the area close to the propeller centerline, where strong radial and tangential velocities are present. This indicates a strong hub vortex. This leads to the conclusion that, even with presence of PSS, there can be some improvement potential by installing propeller boss cap fins (PBCF). The bilge keel vortex can still be traced on the starboard side, even with a working propeller. A comparison of results without PSS mounted (Figure 23) and with PSS (Figure 24) shows clearly the effect of PSS fins on axial, radial, and tangential components of the wake while the propeller is running.

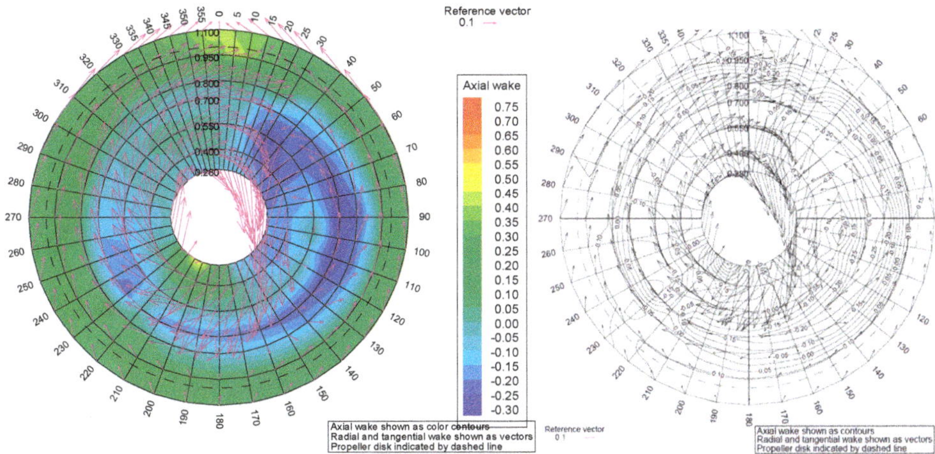

Figure 23. 3D wake measurements at a rudder stock plane (AP), with a working propeller, without PSS, at 14 kn.

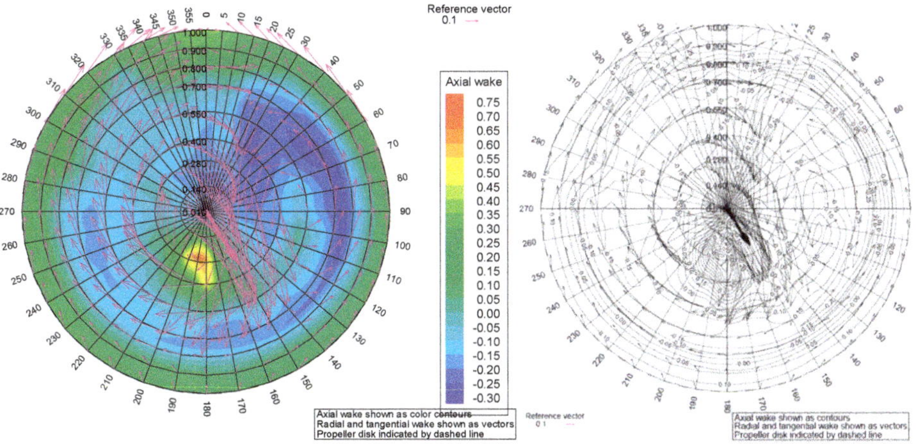

Figure 24. 3D wake measurements at a rudder stock plane (AP), with a working propeller, with PSS, at 14 kn.

6. Verification of PSS Design by Model Tests

Figure 25 shows the power savings achieved with the PSS installation on a model scale (CFD and test results) and on a full scale (CFD results). For comparison, statistical data regarding the full-scale performance of PSS and PSD (Mewis Duct®) installations are also plotted using the data presented in Mewis [2].

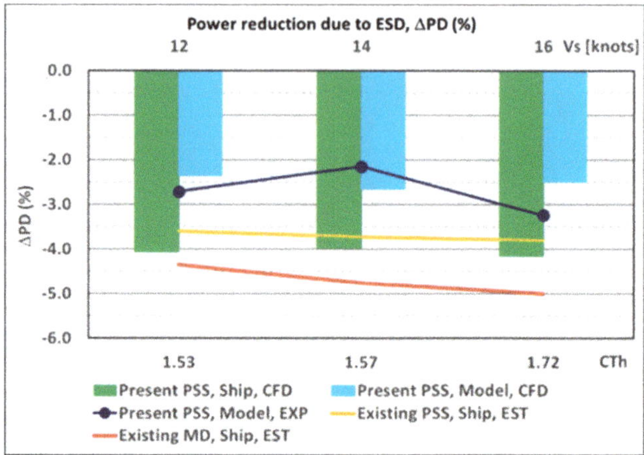

Figure 25. Comparison of power savings due to ESD installation.

It can be seen that, in the model scale, CFD calculations with the designed PSS predict 1.5% lower power savings than on a full scale. On average, the results are 2.5% in the model scale vs. 4% in the full scale. Power savings of the same level (2.2%–3.2%) are reported from model tests. Similar conclusions about better performance of a PSS-type of ESD on a full scale was reported in Park et al. [11]. However, the mentioned work did not present the details about how the PSS was designed. With regard to the present study, a better performance of the PSS on a full scale is explained by the fact that it was designed for full-scale conditions. From the images of the vorticity field presented in Figure 26, one can see that, on a full scale, a smaller part of the ESD fin appears in the hull boundary layer compared to the model scale. The flow separation is delayed on a full scale, and, as a result of overall wake contraction, the large-scale vortices detached from bilges in the stern area shift upwards, above the shaft line, and closer to the center plane. These changes in the vorticity pattern cause differences in the distribution of tangential velocity between the model scale and full-scale conditions. The fins of the PSS designed for full-scale conditions perform better work utilizing the tangential kinetic energy of a full-scale flow. In general, the model scale flow past the ship hull without ESD is less favorable from the standpoint of propeller efficiency, and, hence, it holds a greater potential for energy saving. However, in order to utilize it, the fins need to be adapted to the model scale wake field.

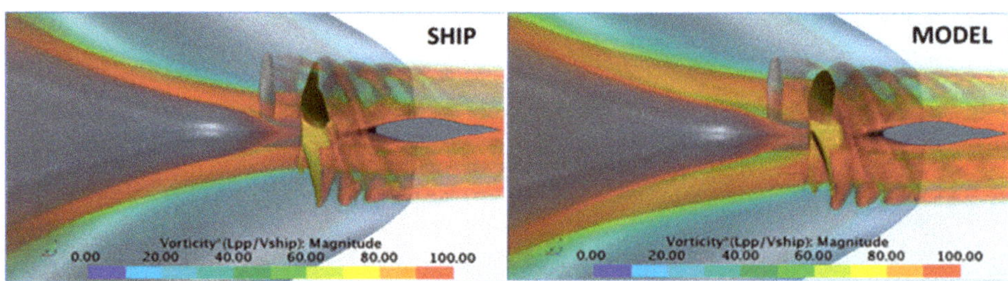

Figure 26. Vorticity field in the ship wake around ESD and propeller location on a full scale and a model scale. (Vs = 14 kn).

Table 9 shows the variation with the scale of propulsion coefficients for the vessel with and without ESD. In the case of the ship with ESD, the values found from the experiment are presented only for the wake fraction, WT, and relative rotative efficiency, η_r, since the thrust deduction factor, t, was defined differently in the CFD calculations (ESD is part of the propulsor, Equation (3)) and in the tests (ESD is the hull appendage). The agreement between the values of WT derived from the model test and CFD calculations in the model scale is very close. In addition, the tests and CFD calculations predict a comparable increase of WT due to the PSS installation. However, larger differences are noticed in full-scale predictions of WT. The approach employed for scaling of WT in the experimental predictions is based on the "Method B" proposed in ITTC [12] for performance prediction of ships with PSS. Since scaling of WT is an essential part of the scaling procedure, this question deserves further investigation, where CFD analyses can be of great help. CFD predictions on a full scale require more validation material. Acquiring reliable full-scale data regarding the wake field with an operating propeller is, therefore, of paramount importance.

Table 9. Relative changes in propeller RPM, shaft delivered power, and propulsive efficiency due to the installation of PSS. (PSS-Var-5, full scale).

WT		without ESD	With PSS	Δ, %
EXP	Model	0.375	0.426	13.60
	Ship (ITTC scaling)	0.312	0.363	16.35
CFD	Model	0.38	0.425	11.84
	Ship	0.294	0.357	21.43
t		**without ESD**	**With PSS**	**Δ, %**
EXP	Model	0.195	—	—
	Ship	0.195	—	—
CFD	Model	0.191	0.2	4.71
	Ship	0.197	0.209	6.09
η_r		**without ESD**	**With PSS**	**Δ, %**
EXP	Model	1.006	1.002	−0.40
	Ship	1.006	1.002	−0.40
CFD	Model	0.992	0.989	−0.30
	Ship	0.991	0.99	−0.10

The thrust deduction coefficients derived from the tests and CFD calculations are found to be at comparable levels. A minor increase of t on a full scale is predicted by the calculations. The installation of ESD results in an increase of t as a penalty for thrust increase. The relative rotative efficiency predicted by CFD is only about 1.3% lower compared to the experimental values. This quantity does not vary significantly with scale, and it shows a slight decrease when the PPS is installed. The presented findings from the CFD analyses justify the assumption of the scaling procedure about t and η_r being weakly influenced by the scale effect.

7. Conclusions

The interaction phenomena in the system hull-ESD-propeller-rudder and the realistic operation conditions on a full scale are essential for successful design of ship ESDs. The CFD-based approach to ESD design is, therefore, found to be an efficient and robust tool. The design workflow can be automated considerably using the design exploration tools such as the Design Manager of STAR-CCM+. It can also be used conveniently in connection with external tools such as potential propeller codes for quick geometry generation or design of wake-adapted propellers. The use of adequate optimization criteria is very important. Power savings, or increase in propulsive efficiency, predicted at the ship self-propulsion point are recommended as such criteria. The analyses of kinetic flow energy in the propeller slipstream is found to be very useful for a deeper understanding of the ESD working principle as well as a source of new ideas regarding other possible energy saving measures.

The Pre-Swirl Stator (PSS) designed in the present work with the proposed method is the result of systematic evaluation of about 70 alternative solutions. According to CFD predictions, it provides about 4% of power savings through the speed range around the design point. Additional power savings can be achieved by designing a propeller wake-adapted for the operation behind the PSS, or by installation of a post-swirl device such as PBCF (Propeller Boss Cap Fins) on the propeller hub.

The verification of the PSS design by model tests confirms power savings of 2.2%–3.2%. These numbers agree well with CFD predictions on a model scale showing 2.5% savings. A better performance of PSS on a full scale is explained by the customization of fins design to full-scale wake conditions.

The propulsion factors of the ship without and with ESD found from model tests and CFD analyses are, in general, in good agreement. The thrust deduction factor and relative rotative efficiency do not change significantly with scale. Larger differences between EFD and CFD predictions are found in full-scale values of wake fraction WT. CFD results predict a larger scale effect and stronger influence of the PSS. Since determination and scaling of WT is the essential part of the scaling procedure for vessels equipped with ESDs, this problem requires further investigations. Acquiring reliable full-scale data regarding wake field with an operating propeller is, therefore, of paramount importance for both the CFD validation and improvement of scaling methods.

Author Contributions: All authors participated in the discussions and review of the article. L.S. participated in the discussions and review of the manuscript. V.K. and K.K. wrote the article. V.K. contributed a numerical investigation and analyses. K.K. and M.N. contributed with experimental methodology, analyses, and data. S.S. contributed with set-up for numerical optimisation. K.K. contributed with project management. All authors have read and agreed to the published version of the manuscript.

Funding: This research was funded by Norges Forsknigsråd, grant number 254797. In addition, industrial partners Odfjell Management and MAN Energy Solutions provided additional funding.

Acknowledgments: The work presented in this paper has been conducted within the frameworks of the R&D Project "NorSingProp" with funding from the Joint Program in Maritime Research between Singapore and Norway, Call 2015. The "NorSingProp" Consortium includes SINTEF Ocean (Norway), Odfjell Management (Norway), IHPC A*STAR (Singapore), SembCorp Marine (Singapore), and MAN Energy Solutions (Denmark). The support received from the Research Council of Norway for the Norwegian part of the project is greatly appreciated.

Conflicts of Interest: The authors declare no conflict of interest.

References

1. ABS. Ship Energy Efficiency Measures Advisory. Available online: https://ww2.eagle.org/content/dam/eagle/advisories-and-debriefs/ABS_Energy_Efficiency_Advisory.pdf (accessed on 12 December 2019).
2. Mewis, F. Six Years Mewis Duct®—Six years of hydrodynamic development. In *Jahrbuch der Schiffbautechnischen Gesellschaft*; Bohn, A., Ed.; Schiffahrts-Verlag Hansa GmbH & Co. KG.: Hamburg, Germany, 2014; Volume 108, pp. 157–164.
3. Guiard, T.; Leonard, S.; Mewis, F. The Becker Mewis Duct®—Challenges in full-scale design and new developments for fast ships. In Proceedings of the 3rd International Symposium on Marine Propulsors smp'13, Launceston, Australia, 5–7 May 2013.
4. Dang, J.; Dong, G.; Chen, H. An exploratory study on the working principles of energy saving devices (ESDs)—PIV, CFD investigations and design guidelines. In Proceedings of the 31st International Conference on Ocean, Offshore and Arctic Engineering OMAE 2012, Rio de Janeiro, Brazil, 1–6 July 2012.
5. Kim, J.-H.; Choi, J.-E.; Choi, B.-J.; Chung, S.-H.; Seo, H.-W. Development of energy-saving devices for a full slow-speed ship through improving propulsion performance. *Int. J. Nav. Archit. Ocean Eng.* **2015**, *7*, 390–398. [CrossRef]
6. Voermas, A. Development of the Wärtsilä EnergoFlow: An innovative energy saving device. In Proceedings of the 5th International Symposium on Marine Propulsors smp'17, Launceston, Finland, 12–15 June 2017.
7. Voermas, A. Experimental determination of hydrodynamic loads on the Wärtsilä pre-swirl stator EnergoFlow and validation of a prediction methodology for design loads. In Proceedings of the 6th International Symposium on Marine Propulsors smp'19, Rome, Italy, 26–30 May 2019.

8. Zhai, S.; Liu, D.; Hang, Y. Numerical study of hull pressure fluctuation with energy saving device PSV. In Proceedings of the 6th International Symposium on Marine Propulsors smp'19, Rome, Italy, 26–30 May 2019.
9. Nielsen, J.R.; Jin, W. Pre-swirl fins adapted to different operation conditions. In Proceedings of the 6th International Symposium on Marine Propulsors smp'19, Rome, Italy, 26–30 May 2019.
10. Ponkratov, D. 2016 Workshop on Ship Scale Hydrodynamic Computer Simulation. Lloyd's Register. 10 February 2017. Available online: http://info.lr.org/l/12702/2017-02-20/3m372v/12702/156863/Proceedings.zip (accessed on 12 December 2019).
11. Park, S.; Oh, G.; Rhee, S.H.; Koo, B.-Y.; Lee, H. Full scale wake prediction of an energy saving device by using computational fluid dynamics. *Ocean Eng.* **2015**, *101*, 254–263. [CrossRef]
12. ITTC. *The Specialist Committee on Unconventional Propulsors*; Final Report and Recommendations to the 22nd ITTC; ITTC: Zurich, Switzerland, 1999; Available online: https://ittc.info/media/1516/specialist-committee-on-unconventional-propulsion.pdf (accessed on 12 December 2019).

Journal of
Marine Science and Engineering

Article

Study on Propulsion Performance by Varying Rake Distribution at the Propeller Tip

Jin Gu Kang, Moon Chan Kim *, Hyeon Ung Kim and I. Rok Shin

Department of Naval Architecture & Ocean Engineering, Pusan National University, Busan 46287, Korea; wind0980@naver.com (J.G.K.); hwkim5491@naver.com (H.U.K.); junhoo7013@naver.com (I.R.S.)
* Correspondence: kmcprop@pusan.ac.kr; Tel.: +82-51-510-1424

Received: 26 September 2019; Accepted: 29 October 2019; Published: 30 October 2019

Abstract: This study provides a comparison of propulsion performance, with a particular focus on efficiency, by varying rake distribution at the tips of propellers. Owing to increased attention to environmental pollution, there is a significant interest in reducing the energy efficiency design index (EEDI) and SO_x emissions by improving the performance in the field of shipbuilding. The forward (Kappel) and backward tip rake propellers have been widely used to improve efficiency, as well as to reduce fluctuating pressure from the tip vortex cavitation. As there is almost no parametric and design research on tip rake propellers, this systematic parametric study was conducted to identify the optimal configuration by the potential code. For this performance comparison the KP505 (KCS propeller) was chosen as the reference propeller as the tips of that propeller have no rake. The model test and computational fluid dynamics (CFD) calculation confirmed the result by comparing the open water performances for the three optimally selected propellers (forward, backward, KP505). The differences of efficiency obtained from the potential analysis and the model test exhibit similar tendencies, but the result for the CFD is different. The difference would be investigated by changing the grid system around the tip as well as the turbulence model in the CFD analysis. An analysis of self-propulsion and pressure fluctuation is also expected to be conducted in the near future.

Keywords: tip rake propeller; EEDI; Energy saving device

---

1. Introduction

Interest in fossil energy depletion and global warming has increased in recent years. The International Maritime Organization (IMO) has been applying indicators for energy efficiency to ships constructed after 2013. In particular, the energy efficiency design index (EEDI) represents the amount of carbon dioxide emitted during the transportation of 1ton of cargo per mile. Emissions will need to be reduced 30% by 2025, beginning with a 10% reduction in January 2013. As a result, research is continuing to improve hull form and propulsion systems to reduce EEDI worldwide. Propulsion system performance has been greatly improved by developing a compound propulsion system. The Compound propulsion system can be classified as pre device, main device, and post device. The pre-swirl stator and Mewis ducts are known to be more effective in pre device, and high performance special propellers, contra rotating propellers, and duct propellers are known to be effective as main devices. Although the performance varies depending on the ship types, the approximate energy reduction effect is about 3%–4%. In addition, twisted rudders, rudder bulbs, and fins are effective post devices. The performance of the rudder has been improved by modifying the shape of the rudder or by installing an additive such as a fin or bulb.

This study addresses the Kappel [1] propeller (forward type) and backward tip rake propeller, which are recently developed propeller concepts. Figure 1 shows the shape of Kappel and CLT propeller. The backward tip rake propeller is popularly being used with Korean ship companies.

(Lee, et al., 2017) [2]. The present design concept of both forward and backward tip rake propellers came from the contracted loaded tip (CLT) propeller. The CLT idea originally came from winglets, which are widely used in aircraft. Winglets reduce the induced drag by weakening the vortex at the wing, leading to greater efficiency (Ha, et al., 2014) [3]. In other words, both propellers prevent the three-dimensional vortex effect by reducing fluctuating pressure changes at the wing tips. It should be noted that in the CLT propeller, contrary to the tip rake propeller, the end plates are unloaded and operate as barriers, preventing the cross flow of the pressure and suction side of the blades, with the finite load at the tip of the blade (G. Gennaro, et al., 2012) [4]. Although the CLT's effect of weakening the tip vortex is probably better than that of the tip rake propeller, the Kappel and tip rake propellers have been more widely used than the CLTs due to the risk of cavitation in the corners and also loss with the end plate drag at high speeds.

(a)

(b)

Figure 1. (a) Kappel propeller, (b) Contracted Loaded Tip (CLT) Propeller.

While the Kappel and tip rake propellers have been widely used for more than 15 years, their performance and design is not well understood (Yamasaki, et al., 2013) [5]. A parametric study for optimal rake shape has been conducted here. The performance of the optimized propellers was verified by computational fluid dynamics (CFD) and model tests.

Pressure fluctuation issues from the propeller should also be investigated and compared to verify the performance of the tip rake propeller because it is more effectively used to reduce the aft hull surface pressure fluctuation induced by propellers. These studies are expected to be conducted in the near future.

2. Study Methods of Tip Rake Propellers

2.1. Design Based on Potential Analysis

The KP505 propeller, which was designed for the Korea Research Institute of Ships & Ocean engineering (KRISO) Container ship (KCS), was selected as the reference propeller. This propeller is widely used for academic and comparative purposes as its performance and geometry are readily available. The KP505 is appropriate for this comparative study because no radial rake was applied. Rake was applied to the reference propeller by altering the starting radii, the maximum rake size, and the rake application (backward and forward).

The starting points of the rake distribution were set as 0.4 R, 0.5 R, 0.6 R, and continuously increased by 0.1 R to the blade tip. If the starting point is less than 0.4 R or greater than 0.7 R, it is difficult to introduce the rake effect. The maximum rake variation was 1%–10% of the propeller diameter. If the rake variation is greater than 10%, the propeller acquires excessive curvature and very difficult to be smooth along radii. Figure 2 represents the radial distribution of the rake applied to a forward and backward propeller, respectively. The rake shape along the radius was designed as a sine curve. The side view of tip rake propellers is shown in Figure 3.

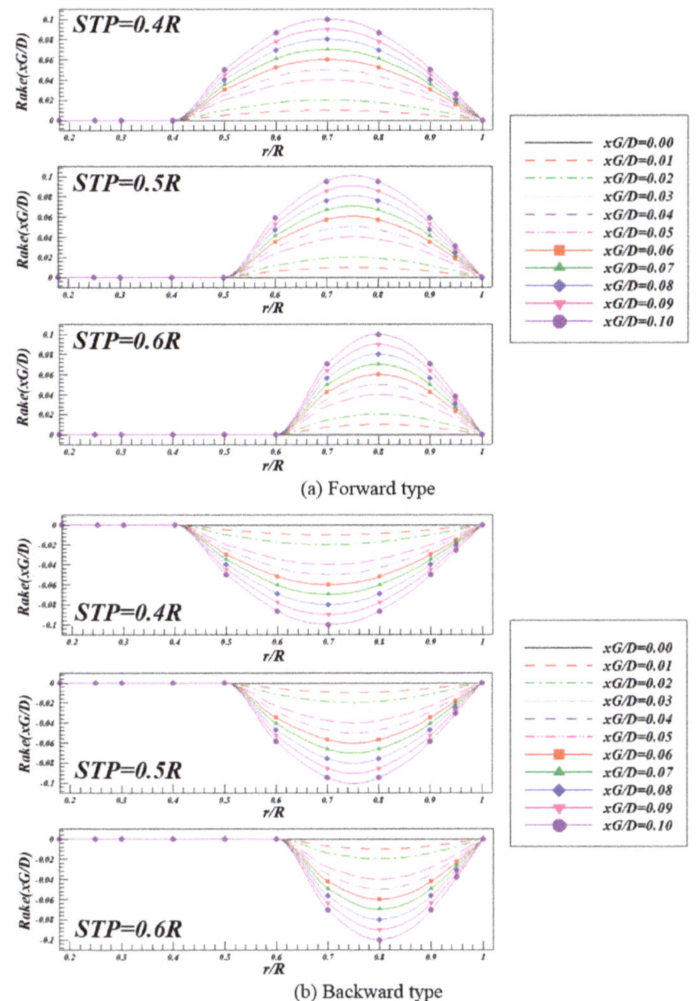

Figure 2. Rake distribution variability with radius: (**a**) Forward type, (**b**) Backward type.

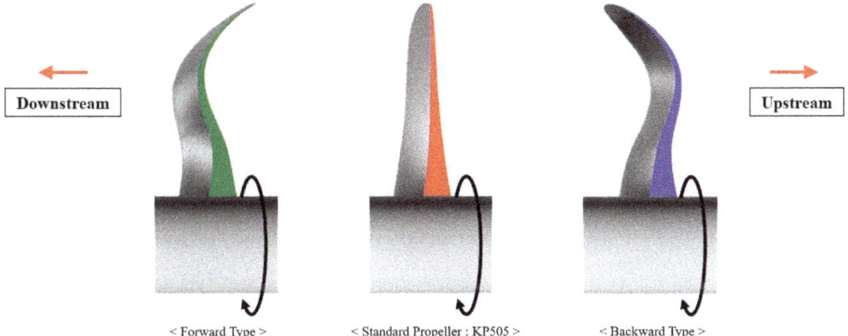

Figure 3. The side view of tip rake propellers.

Numerical analysis of the propeller's open water performance was performed using KPA4, a potential-based numerical analysis is commonly used in the study of open water propeller performance (Kim, et al., 1993; Kim and Lee, 2005) [6,7]. The KPA4 package was developed based on the vortex lattice method (VLM). The diameter of the model propeller for the propeller open water (POW) analysis was 250 mm and the propeller rotation speed was 16rps. The analysis was performed with $J = 0.05$ to $J = 1.00$ at an interval of 0.05. Finally, the geometry of the propellers is as shown in Table 1.

Table 1. Propeller geometry of forward- and backward-type propellers.

r/R	P/D	Rake (xG/D) Forward	Rake (xG/D) Backward	Skew(°)	C/D	f_0/C	T_0/D
0.18	0.8347	0.0000	0.0000	−4.72	0.2313	0.0284	0.0459
0.25	0.8912	0.0000	0.0000	−6.98	0.2618	0.0296	0.0407
0.30	0.9269	0.0004	0.0000	−7.82	0.2809	0.0295	0.0371
0.40	0.9783	0.0031	−0.0017	−7.74	0.3138	0.0268	0.0305
0.50	1.0079	0.0101	−0.0082	−5.56	0.3403	0.0220	0.0246
0.60	1.0130	0.0174	−0.0206	−1.50	0.3573	0.0173	0.0195
0.70	0.9967	0.0200	−0.0285	4.11	0.3590	0.0140	0.0149
0.80	0.9566	0.0180	−0.0285	10.48	0.3376	0.0120	0.0107
0.90	0.9006	0.0119	−0.0189	17.17	0.2797	0.0104	0.0069
0.95	0.8683	0.0070	−0.0113	20.63	0.2225	0.0101	0.0053
1.00	0.8331	0.0000	0.0000	24.18	0.0001	8.7000	0.0037

The computed POW efficiency was compared at $K_T/J^2 = 0.4725$ as shown in Figure 4. Based on these results, an optimal shape was selected for forward and backward propellers. The optimal rake for the backward propeller was STP = 0.4R and xG/D = 0.02 and was 2.15% greater than the reference propeller. The optimal rake for the forward propeller was STP = 0.5 R and xG/D = 0.03, and it was 3.35% greater than the reference propeller. After the optimal rake had been applied to the propeller, there were non-smooth surfaces during 3-D modeling. The observed discontinuity is shown in Figure 5. Therefore, the rake distribution was smoothed as shown in Figure 6. After the final modifications to the rake were applied, improvements of 1.9% and 2.2% by the forward and backward propellers, respectively, were obtained compared to the reference propeller, which is shown in Table 2.

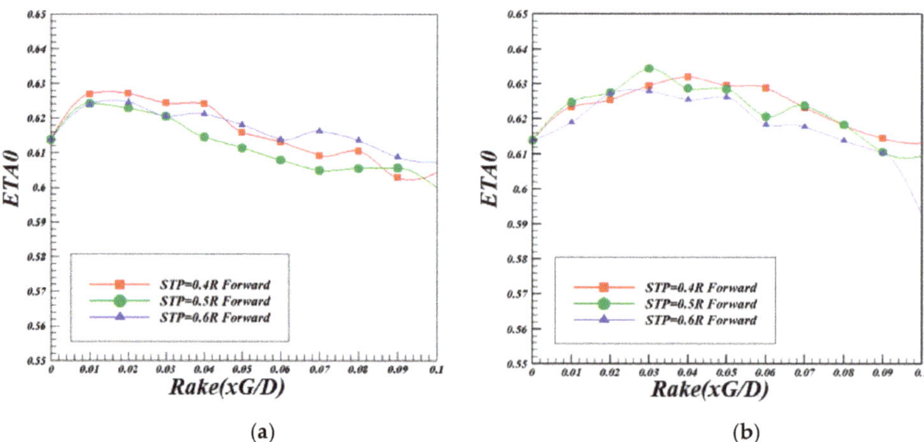

Figure 4. Comparison of open water propeller efficiency based on potential. (**a**) Forward propeller; (**b**) backward propeller.

Figure 5. Discontinuity of the 3-dimensional configuration.

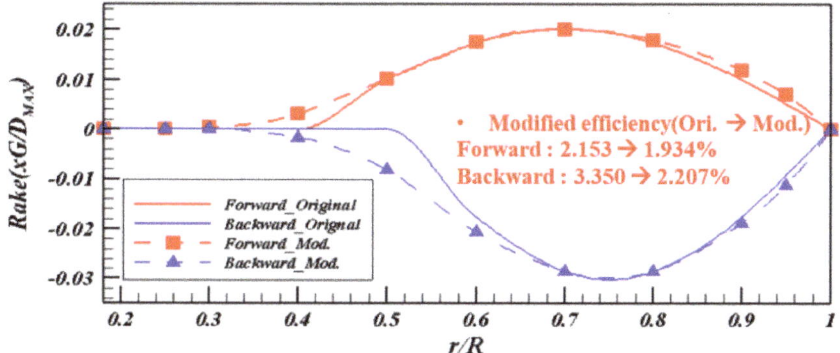

Figure 6. Modified rake distribution variability with radius.

Table 2. Comparison of open water propeller efficiency based on potential.

the Type of Propellers	PP016 (KP505)	PP033 (Forward)	PP034 (Backward)
J	0.659	0.662	0.661
η_0	0.614	0.626	0.627
Diff.(%)	-	1.934	2.207

2.2. Tip Rake Propeller Model Test

The aluminum model of the reference propeller (PP016, KP505) and the optimally designed forward (PP033), and backward propellers (PP034) were manufactured for the model test shown in Figure 7. The propeller's diameter is 0.2m. The propeller open water test was performed at intervals of 0.05 from J = 0.05–1.00. According to the International Towing Tank Conference (ITTC), the minimum Reynolds number for the POW test is 2×10^5, however some cases, that may not be large enough (Kim, et al., 1985) [8]. In this study, the minimum Reynolds number is more than 5×10^5 as shown in the Figure 8. The results of the self-propulsion test at the speed of 24 knots, which was conducted in the PNU towing tank, the target K_T/J^2=0.4725 was identified (Kwon, 2013) [9]. The open water efficiency of the three propellers is compared in Table 3 where the backward propeller is 1.28% greater and the forward propeller is 0.264% greater than the reference propeller (KP505) at the same point of K_T/J^2. The tendency of the efficiency gain is similar to the potential computation although there is little quantitative difference. Figure 9 shows a comparison with KP505 open water test results conducted by KRISO to verify reliability of PNU model test results. In the low J area, there was some difference in thrust and torque coefficients, but not much difference in open water efficiency with the KRISO results. The open water efficiency near target J (approximately 0.6) was almost identical to each other. Figure 10 shows the results of comparison of open water propeller efficiency based on experiments.

Figure 7. Model propellers.

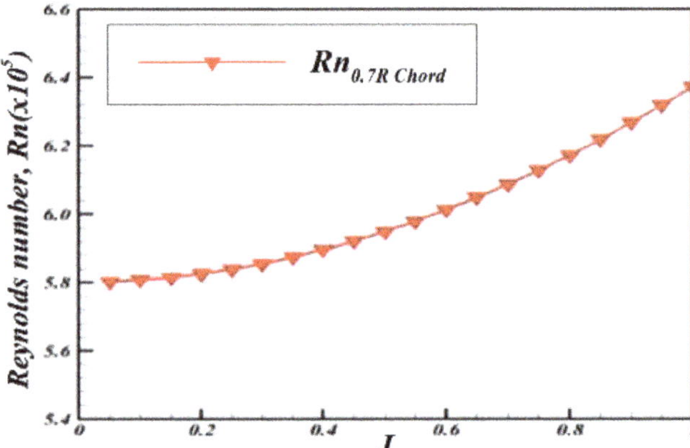

Figure 8. Reynolds number of J during experimental tests.

Table 3. Comparison of open water propeller efficiency based on experiments.

the Type of Propellers	PP016 (KP505)	PP033 (Forward)	PP034 (Backward)
J	0.646	0.651	0.651
η_0	0.607	0.608	0.614
Diff.(%)	-	0.264	1.280

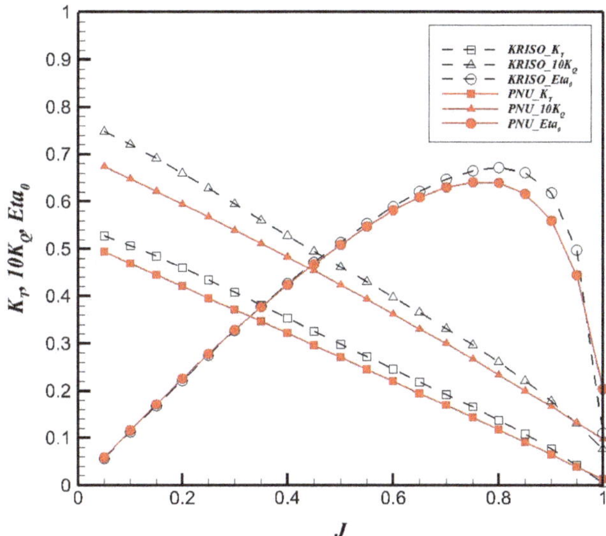

Figure 9. Comparison of model test results vs. KRISO.

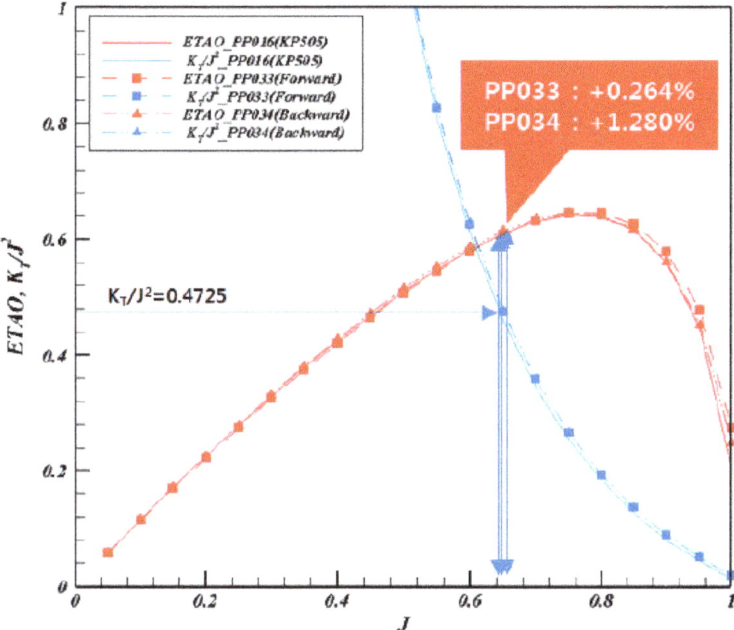

Figure 10. Comparison of open water propeller efficiency based on experiments.

2.3. Numerical Analysis of Propeller Open Water Performance by Computational Fluid Dynamics (CFD)

Numerical analysis has been performed for the reference propeller PP016 (KP505) and the optimized forward and backward propellers, PP033 and PP034, respectively. The configurations of the propellers are shown in Figure 11. Local flow analysis of wing tip vortices was also carried out to investigate the hydrodynamic phenomena manifested by the application of rake at the wing tip (Park, et al., 2011; Baek, et al., 2014) [10,11]. For all propellers, the revolution speed was 16 rps, the same as the potential analysis case. An interval of 0.2 from $J = 0.1$–0.9 was used, and local flow investigation was performed at $J = 0.7$. Table 4 shows the POW performance of the 3 propellers in CFD analysis.

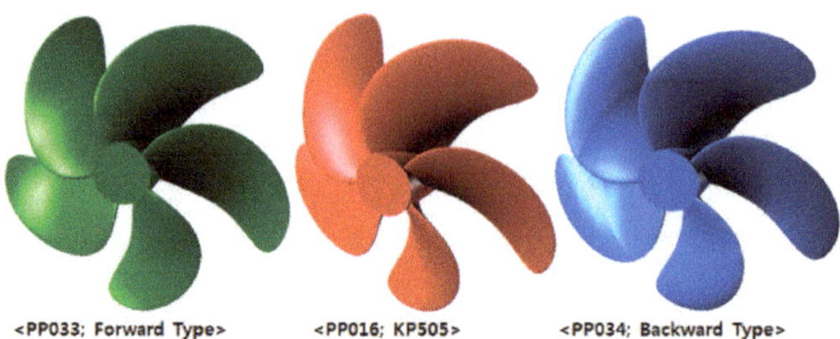

Figure 11. Three-dimensional modeling of the three selected propellers.

Table 4. The propeller open water (POW) performance of the 3 propellers.

J	KT	10KQ	ETAO
0.1	0.479	0.672	0.113
0.3	0.386	0.558	0.330
0.5	0.279	0.431	0.516
0.7	0.173	0.303	0.637
0.9	0.066	0.160	0.596
(a) Reference propeller (KP505)			
J	**KT**	**10KQ**	**ETAO**
0.1	0.472	0.668	0.112
0.3	0.384	0.561	0.327
0.5	0.281	0.436	0.512
0.7	0.175	0.308	0.635
0.9	0.068	0.164	0.597
(b) Forward propeller			
J	**KT**	**10KQ**	**ETAO**
0.1	0.480	0.672	0.114
0.3	0.385	0.557	0.330
0.5	0.278	0.429	0.515
0.7	0.172	0.302	0.633
0.9	0.065	0.158	0.586
(c) Backward propeller			

The analysis was performed using STAR CCM+, which is a commercial CFD program developed by CD-adapco. In the following equations,

$$\frac{\partial u_i}{\partial x_i} = 0 \tag{1}$$

$$\frac{\partial(\rho u_i)}{\partial t} + \frac{\partial(\rho u_i u_j)}{\partial x_i} = -\frac{\partial p}{\partial x_i} + \frac{\partial}{\partial x_i}\left(-\rho\overline{u_i' u_j'}\right) + \frac{\partial}{\partial x_i}\left[\mu\left(\frac{\partial u_i}{\partial x_j} + \frac{\partial u_j}{\partial x_i}\right)\right] \tag{2}$$

x_i, u_i, ρ, and μ denote the rectangular coordinate system, the velocity component, pressure, density, and viscosity, respectively. The Reynolds stress term in Equation (2) is analyzed using a κ-ε model. The Semi-Implicit Method for Pressure Linked Equations (SIMPLE) was used for velocity, pressure, and ductility, and a second order differential method was applied for calculating convection and diffusion. The computation domain and boundary conditions considered in this study are shown in Figure 12. The motion of the flow analysis area was analyzed by the sliding mesh method, which is the same method used for the local flow analysis. The analysis conditions used for CFD are shown in Table 5.

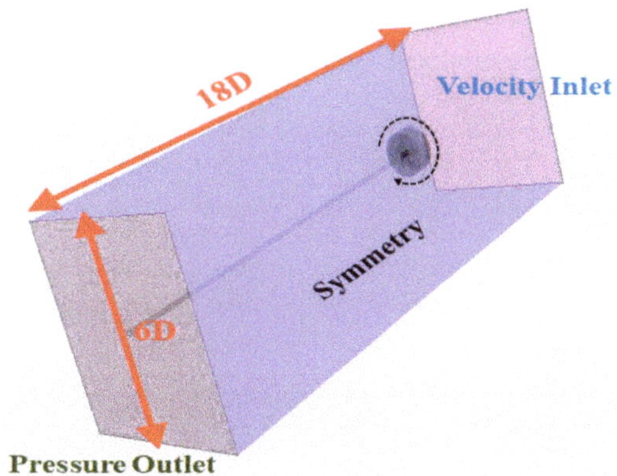

Figure 12. Computational domain and boundary conditions.

Table 5. Computational fluid dynamics (CFD) analysis conditions.

Program	Star CCM+ (Ver. 9.04)
Governing equation	Reynolds Averaged Navier Stokes (RANS) equation
Discretization	Cell centered FVM
Turbulence model	Realizable κ-ε model
Velocity-pressure coupling	SIMPLE algorithm
Rotation method	Sliding interface moving mesh
Cell number	1,600,000

The open water propeller efficiency of each propeller was compared at the same $K_T/J^2 = 0.4725$ as in the potential analysis study. The difference in open water propeller efficiency was 0.2% for the forward propeller (PP033) and 0.6% for the backward propeller (PP034) compared to the reference propeller (PP016, KP505). The results are summarized in Table 6 and Figure 13. The tendency of the computed results of the efficiency gain is different from the potential analysis and the experimental results. There is almost no quantitative difference among them. The results of the model test and potential analysis are thought to be more reliable, so it is necessary to investigate the reason for the different results. The denser grid around the tip might be necessary for capturing the tip vortex more accurately, and a parameter study of turbulence modeling might be also necessary. However, it might still be necessary to improve the present CFD analysis, the comparison of the tip vortices of the three propellers has been completed.

Table 6. Comparison of open water propeller efficiency based on CFD.

The Type of Propellers	PP016 (KP505)	PP033 (Forward)	PP034 (Backward)
J	0.669	0.670	0.668
η_0	0.618	0.617	0.614
Diff.(%)	-	−0.2	−0.6

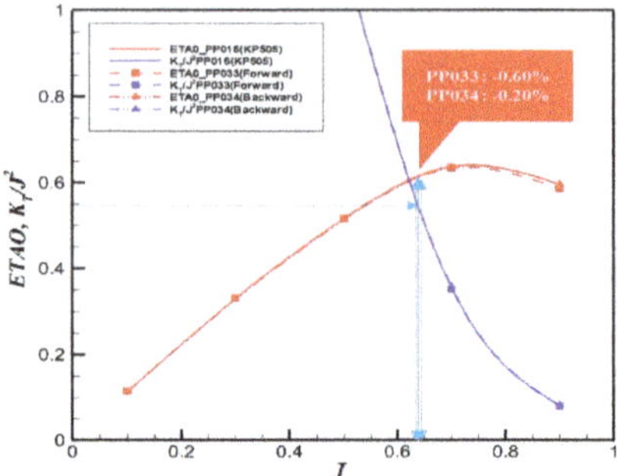

Figure 13. Comparison of open water propeller efficiency based on CFD.

The vorticity magnitude is shown in Figure 14 where the backward propeller is a little weaker than the other two propellers. Although the total efficiency of the tip rake propeller is slightly worse, the backward tip rake might locally work better than the others.

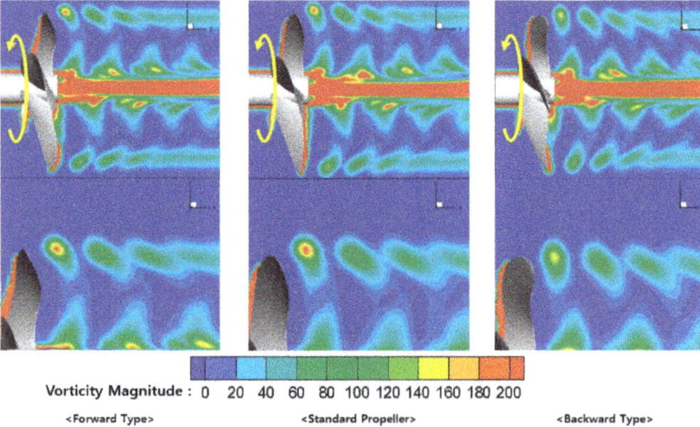

Figure 14. Vorticity magnitude ($J = 0.7$).

3. Conclusions and Discussion

A parametric study on tip rake propellers has been completed based on the KP505, the KCS propeller, by varying the maximum size and starting rake radius. A potential based analysis code was used for the present parametric study. The optimally designed forward and backward propellers had an efficiency gain of 1.9% and 2.2%, respectively, over the reference propeller (KP505) in POW conditions. The three propellers (reference, backward, and forward) were tested in POW conditions to confirm their performances experimentally. The results are similar to the potential analysis where the propellers performed 1.3% (backward) and 0.3% (forward) better than the reference propeller, although the efficiency gain is somewhat less than during the potential analysis.

Lastly, the CFD analysis was applied to validate the performance of the three propellers. The computed results are different from the previous two cases (potential analysis and experimental test). There was efficiency loss compared to the reference propeller of −0.2% (forward) and −0.6% (backward). While there is some discrepancy in efficiency by the CFD analysis, the tip vortex flow was also investigated CFD. The tip vortex strength of the backward propeller was slightly less than those of the other two propellers (Figure 15), which is a similar result to the potential based analysis and the experimental test. More investigation into CFD analysis is expected by varying the grid system and turbulence modelling. Table 7 shows the open water propeller efficiency difference according to analysis method.

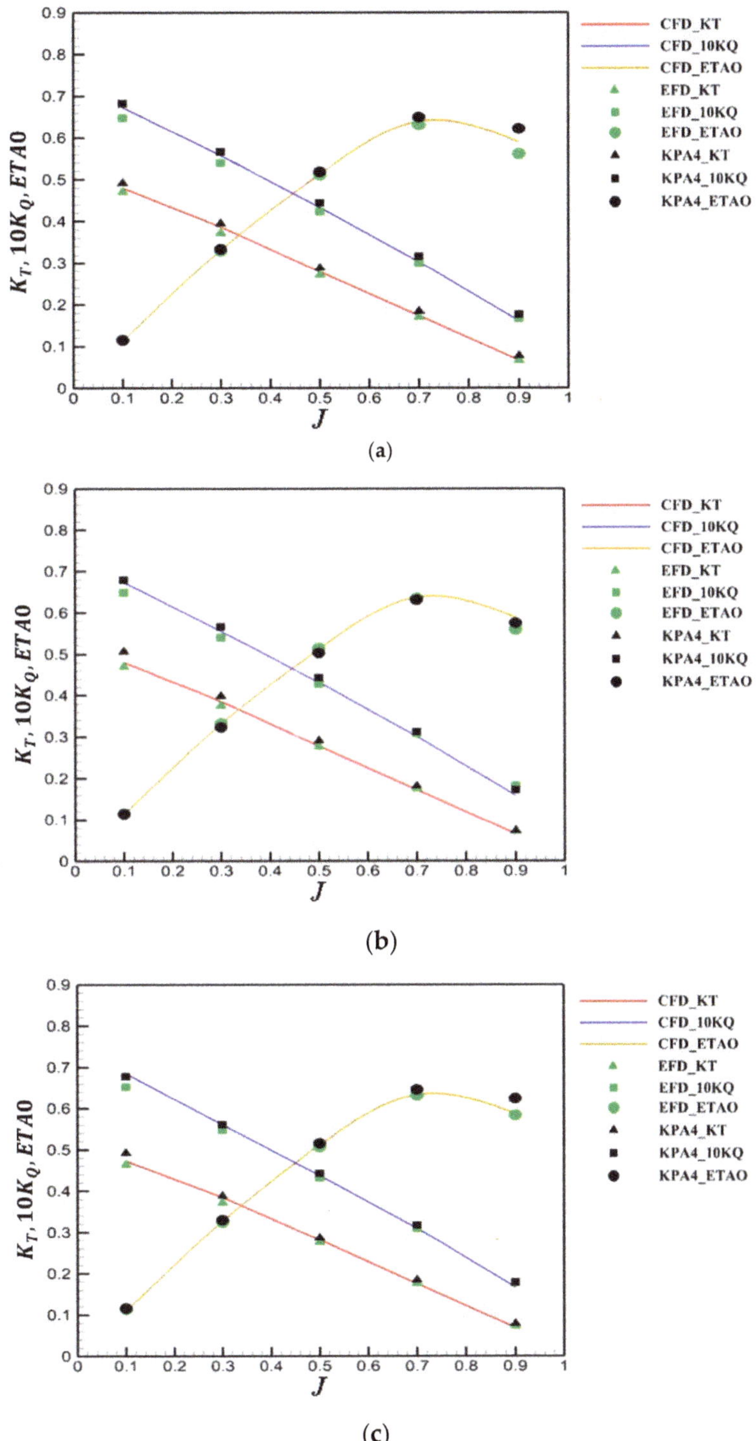

Figure 15. Comparison of results according to analysis method. (**a**) Reference propeller (KP505); (**b**) backward propeller; (**c**) forward propeller.

Table 7. Open water propeller efficiency difference according to analysis method compared to the results of the reference propeller.

	Potential Analysis	Model Test	CFD Analysis
Forward	+1.9%	+0.3%	−0.2%
Backward	+2.2%	+1.3%	−0.6%

Future work extending from this study is expected to include the study of pressure fluctuations and noise from the tip vortex by changing the configuration of the tip shape so that tip rake is more effectively applied to reduce hull pressure fluctuations, especially for container ships.

Author Contributions: This paper is the result of collaborative teamwork. Project administration, Validation, Writing–original draft, Writing–review & editing, J.G.K.; Conceptualization, Supervision, M.C.K.; Visualization, H.U.K.; Investigation, I.R.S.

Funding: This work was supported by the National Research Foundation of Korea (NRF) grant funded by the Korea government (MSIT) through GCRC-SOP (No. 2011-0030013 and NRF-2019R1F1A1058080).

Conflicts of Interest: The authors declare no conflict of interest.

References

1. Kappel, J.; Andersen, P. KAPPEL propeller development of a marine propeller with non-planar lifting surfaces. In Proceedings of the 24th Motor Ship Marine Propulsion Conference, Copenhagen, Denmark, 10–11 April 2002.
2. Lee, J.H.; Kim, M.C.; Shin, Y.J.; Kang, J.G.; Jang, H.G. A study on performance of tip rake propeller in propeller open water condition. *J. Soc. Nav. Archit. Korea* **2017**, *54*, 10–17. [CrossRef]
3. Ha, D.S.; Jeon, E.C.; Ahn, K.W.; Yeom, G.W. Aerodynamic analysis of various winglet shapes. In Proceedings of the Journal of the Korean Society of Mechanical Engineers 2014 Spring Conference, Busan, Korea, 15–16 May 2014; pp. 216–217.
4. Gennaro, G.; Adalid, J.G. *Improving the Propulsion Efficiency by Means of Contracted and Loaded Tip (CLT) Propellers*; The Society of Naval Architects & Marine Engineers: Athens, Greece, 13 September 2012.
5. Yamasaki, S.; Okazaki, A.; Mishima, T.; Kawanami, Y.; Yoshitaka, U. The effect of tip rakes on propeller induced pressure fluctuations first report: A practical formula to estimate the reduction rate of pressure fluctuations the application of backward tip rake. *J. Jpn. Soc. Nav. Archit. Ocean Eng.* **2013**, *17*, 9–17.
6. Kim, M.G.; Lee, J.T.; Lee, C.S.; Seo, J.C. Prediction of steady performance of a propeller by using a potential-based panel method. *J. Soc. Nav. Archit. Korea* **1993**, *30*, 73–86.
7. Kim, G.D.; Lee, C.S. Application of high order panel method for improvement of prediction of marine propeller performance. *J. Soc. Nav. Archit. Korea* **2005**, *42*, 113–123.
8. Kim, K.S.; Lee, C.S. *Influence of Reynolds Number on Propeller Open-Water Characteristics*; Korea Inst. Mach. Mater. Rep.; 1985.
9. Kwon, J.I. A Study on Biased Asymmetric Pre-Swirl Stator for Container Ship. Master's Thesis, Pusan National University, Busan, Korea, 2013.
10. Park, S.H.; Seo, J.H.; Kim, D.H.; Rhee, S.H.; Kim, K.S. Numerical analysis of a tip vortex flow for propeller tip shapes. *J. Soc. Nav. Archit. Korea* **2011**, *48*, 501–508. [CrossRef]
11. Baek, D.G.; Yoon, H.S.; Jung, J.H.; Kim, K.S.; Park, B.G. Effect of advance ration on the evolution of propeller wake. *J. Soc. Nav. Archit. Korea* **2014**, *51*, 1–7. [CrossRef]

Journal of
Marine Science and Engineering

Article

The Numerical Prediction and Analysis of Propeller Cavitation Benchmark Tests of YUPENG Ship Model

Chaosheng Zheng [1,2,*], Dengcheng Liu [1,2] and Hongbo Huang [1,2]

1 China Ship Scientific Research Center, Wuxi 214082, China
2 Jiangsu Key Laboratory of Green Ship Technology, Wuxi 214082, China
* Correspondence: zcszcs2005@163.com; Tel.: +86-0510-8555-5635

Received: 26 September 2019; Accepted: 19 October 2019; Published: 30 October 2019

Abstract: The numerical simulation of propeller cavitation benchmark tests of YUPENG ship model is studied based on OpenFOAM, an open-source CFD (Computational Fluid Dynamics) platform, and the benchmark tests are introduced as well. The propeller cavitation shape and the hull pressure fluctuation are measured and predicted, respectively. The uncertainty in hull pressure fluctuation measurement is also analyzed, and the analysis showed the uncertainty is below 10%. The cavitation shape and the hull pressure fluctuation predicted show quite good agreement with the observations and measured data, and the influences of the grid resolutions on the unsteady propeller cavitation and the hull pressure fluctuation are investigated as well. The monotonic convergence achieved and the quite small grid uncertainty illustrate the reliability of the numerical simulation methods.

Keywords: numerical simulation; propeller cavitation; benchmark tests; YUPENG

1. Introduction

Propeller cavitation is one important aspect of hydrodynamic performance for marine propellers, and recently it has been studied based on viscous numerical simulation methods, especially the viscous RANS (Reynolds Average Navier-Stokes) approach, and more and more research articles have been published. Salvatore [1] has studied the cavitation on the standard INSEAN E779A propeller and compared the cavitation predictions using several different solvers and cavitation models. The discrepancies in cavity extent are observed in the comparative analysis. Li [2] has also predicted the cavitating flows around the INSEAN E779A propeller in non-uniform wake, using a RANS approach and Zwart cavitation model based on commercial FLUENT. The numerical results show that the key features predicted, such as the attached cavity and the collapse of the main cavity, resemble well with the experiment observations. Paik [3] has conducted the simulations of cavitating flows for a marine propeller based on FLUENT and Schnerr-Sauer cavitation model. The simulation results show reliable performance on the prediction of cavitation shape and hull pressure fluctuation. Vaz [4] has compared more cavitation and pressure pulse predictions between different solvers and cavitation models, the results indicate that the pressure fluctuations predicted by half the participants are in reasonable agreement with measurements, but the remaining calculations predict pressure levels a factor of four or five too high. Fujiyama [5] has analyzed the unsteady cavitating flow of HSP-II and CP-II propeller at behind-hull condition both in model and full scale, using commercial software SC/Tetra v13. The results show that the unsteady propeller cavitation phenomena can be captured in the numerical calculation. Long [6] has researched the propeller cavitating flow behind the hull, analyzed the vorticity distribution and particle tracks as well, using commercial software CFX and Zwart cavitation model. The cavitation patterns predicted resemble well with the experimental observations, with some over-prediction of the cavitation area. Yilmaz [7] has investigated the hull–propeller–rudder interaction in the presence of tip vortex cavitation, using commercial software STAR-CCM+ and Schnerr-Sauer cavitation model with LES (Large eddy simulation) method. The investigation results demonstrate some capability of STAR-CCM+ for investigation in the presence of tip vortex cavitation.

In addition to commercial CFD software, an open-source CFD platform, OpenFOAM has been increasingly used to conduct the numerical simulation of propeller cavitating flows. Abolfazl [8] has performed the numerical prediction of the PPTC (Potsdam Propeller Test Case) propeller cavitation size in oblique wake based on ILES (Implicit Large Eddy Simulation) approach and Schnerr–Sauer cavitation model, using OpenFOAM. The simulation results show over-prediction of the propeller cavitation extent. Bensow [9] has carried out a prediction of a marine propeller behind a chemical tanker with ILES approach and Kunz cavitation model, based on OpenFOAM. The prediction of cavity area shows good agreement with the experiments. Zheng [10,11] has simulated the unsteady propeller cavitating flows in the stern of a ship using unsteady RANS approach, based on OpenFOAM. The predicted cavitation shape change shows quite good agreement with the experimental observation.

The present study aims to research the influence of the grid resolutions on the unsteady propeller cavitation and the induced hull pressure fluctuation behind the YUPENG benchmark ship model, which will illustrate the reliability of the numerical simulation methods.

2. Mathematic Base

2.1. Governing Equations

The unsteady RANS method is applied in the present study for the significantly lower computational resources needed compared with LES.

The time-averaged continuity equation and momentum equation are shown in Equations (1) and (2), respectively.

$$\frac{\partial \rho}{\partial t} + \nabla(\rho U) = 0 \tag{1}$$

$$\frac{\partial \rho U}{\partial t} + U \cdot \nabla(\rho U) = \nabla((\mu + \mu_t)\nabla U) - \nabla p - F_S \tag{2}$$

where U is the averaged velocity vector, p is the mean pressure. ρ is the fluid density, μ is the dynamic viscosity, and μ_t is the turbulent viscosity. F_S is the surface tension force which takes place only at the free surfaces. The turbulent viscosity μ_t is modeled by the SST kw [12] turbulence model combined with wall functions.

According to the VOF (Volume of Fluid) method, the physical properties of the cavitating flows are scaled by the volume fraction of water γ, with $\gamma = 1$ corresponding to pure water, $\gamma = 0$ corresponding to pure vapor, $0 < \gamma < 1$ corresponding to the interface between water and vapor.

Thus, the density and dynamic viscosity of the cavitating flows are scaled as follows. The subscripts w and v represent water and vapor, respectively.

$$\rho = \gamma \rho_w + (1-\gamma)\rho_v \tag{3}$$

$$\rho = \gamma \mu_w + (1-\gamma)\mu_v \tag{4}$$

The transport equation of the volume fraction of water γ can be written as follows.

$$\frac{D_\gamma}{D_t} = \frac{\partial \gamma}{\partial t} + \nabla(\gamma U) = \frac{\dot{m}}{\rho_w} \tag{5}$$

Here, the mass transfer rate, $\dot{m}$, between water and vapor is to be modeled by cavitation model.

Combining with Equations (3) and (5), the mass transport, Equation (1) can be re-written as follows.

$$\nabla \cdot U = \dot{m}\left(\frac{1}{\rho_w} - \frac{1}{\rho_v}\right) \tag{6}$$

It suggests that the mass transport rate, $\dot{m}$, between water and vapor participates in the coupled solving procedure of pressure and velocity. During the simulations of cavitating flows, $\dot{m}$ is calculated in the PISO (Pressure Implicit with Splitting of Operators) [13] loop at first, then the transport equation of the vapor volume fraction is solved and the standard PISO procedure is progressed.

2.2. Cavitation Model

Cavitation is a complex hydrodynamic phenomenon and is generally simplified as the transition of water into vapor in the region where the local pressure is lower than the water saturation pressure during the numerical simulation. The transition is caused by the small gas nuclei in the water. In order to mimic the phase change between water and vapor, the Schnerr–Sauer cavitation model is adopted.

$$\dot{m} = sign(\rho_v - \rho)\frac{n_0}{1 + n_0\frac{4}{3}\pi R^3}4\pi R^2\sqrt{\frac{2}{3}\frac{|\rho_v - \rho|}{\rho_w}} \tag{7}$$

Here, n_0 is the number of micro gas nuclei per water volume, and R is the initial radius of micro gas nuclei. The values of n_0 and R are default, e.g., $n_0 = 1.6 \times 10^{13}$, $R = 2.0 \times 10^{-6}$. Schnerr–Sauer cavitation model can be deduced based on Rayleigh–Plesset equation describing the bubble dynamics.

3. Calculation Modeling

3.1. Calculation Object

The investigated object is the YUPENG benchmark ship, which is a 30,000 DWT bulk carrier driven by a fixed-pitch four-bladed propeller. The benchmark propeller cavitation model test including cavitation observation and pressure fluctuation measurement is conducted in the Large Cavitation Channel (LCC) of the China Ship Scientific Research Center (CSSRC). The principal parameters of ship and propeller model are shown in Tables 1 and 2, respectively. The hull, rudder, and propeller are visualized in Figure 1.

Table 1. Principal parameters of YUPENG model.

L_{PP} (m)	7.0522
Design draught (m)	0.3843
Shaft height (m)	0.1343

Table 2. Principal parameters of YUPENG propeller model.

Diameter (m)	0.250	**Area ratio**	0.48
No. of blades	4	**P/D $_{(0.7R)}$**	0.80
Hub ratio	0.147	**Skew angle (°)**	24.5

The benchmark test setup inside the LCC is shown in Figure 2. The initial rotating angle, φ, is defined as 0° at twelve o'clock, and the propeller has been assembled and oriented at the initial position.

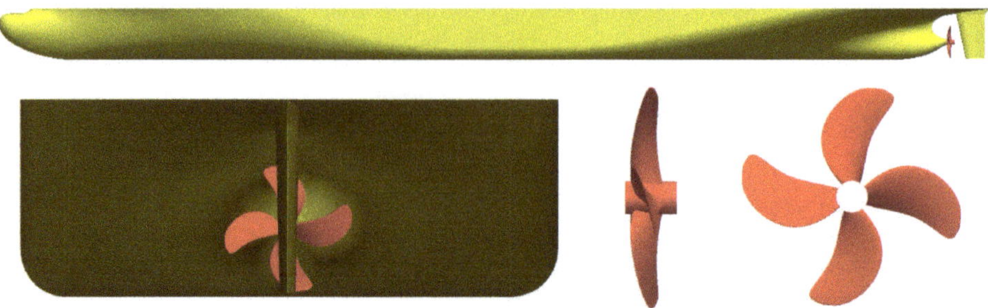

Figure 1. YUPENG ship and propeller geometry model.

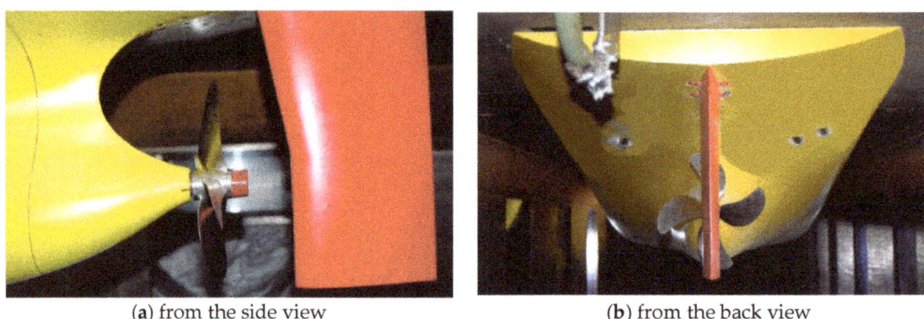

(**a**) from the side view (**b**) from the back view

Figure 2. The benchmark test set.

3.2. Computational Domain and Boundary Condition

The whole computational domain is a box of $7L_{pp} \times 4L_{pp} \times 2L_{pp}$. The boundary conditions on the front, bottom, and sides are velocity inlet, and the boundary condition on the top is symmetry, the boundary condition on the back is pressure outlet, as shown in Figure 3.

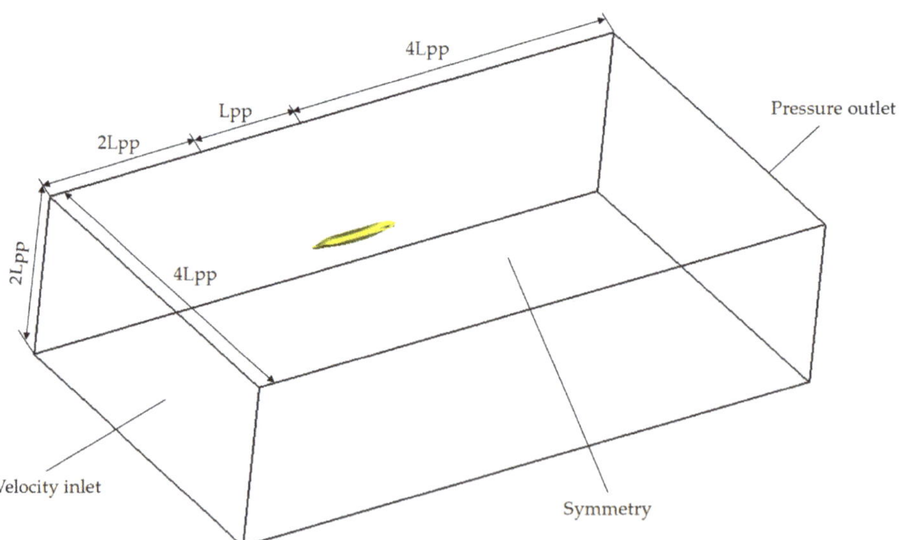

Figure 3. The computational domain and boundary conditions.

The whole computational domain consists of two sub-domains, the ship sub-domain and the propeller sub-domain. The ship sub-domain is a static region which includes the boundary of the inlet, outlet, hull, and rudder. The propeller sub-domain is a rotating region surrounding the propeller. All the grids of the two sub-domains consists of unstructured hexahedral cells of high quality. For more accurate prediction of the flow detail, several boundary layer cells are inserted near wall surfaces. The overview of the hull, rudder, and propeller surface is shown in Figure 4. The arbitrary mesh interface (AMI) method is applied to interpolate between the non-conforming interfaces of the two sub-domains.

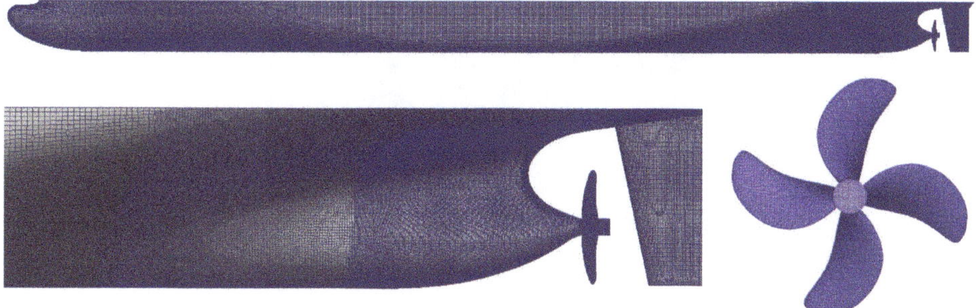

Figure 4. The overview of the hull, rudder, and propeller surface grid.

The velocity inlet condition is set to be a fixed value of velocity, U, at the inlet boundary, U = 5.481 m/s, and the turbulence intensity at the inflow is 0.1%. The pressure outlet condition is set to be a fixed value of pressure, which is constant based on the cavitation number, $\sigma_{n(0.8R)}$, at the outlet boundary. The no-slip wall condition is set at the hull, rudder, and propeller boundary.

4. Calculation Result Analysis

The benchmark test conditions numerically predicted in the present study are summarized in Table 3.

Table 3. The test conditions

Parameters	Design Draft
n (rps)	28
$\sigma_{n(0.8R)}$	0.3154
K_T	0.1371

n—rotating speed of the propeller, $\sigma_{n(0.8R)}$—cavitation number basing the rotating speed, K_T—the thrust coefficient of the propeller.

The formulae are as follows.

$$\sigma_{n(0.8R)} = \frac{p - p_v}{0.5\rho(0.8\pi nD)^2} \tag{8}$$

$$K_T = \frac{T}{\rho n^2 D^4} \tag{9}$$

The free surface is neglected, and the effect of the stern wave on pressure is not taken into account either in the experiment or in the numerical calculation, because the stern wave height of the low-speed vessel (Fr < 0.2) is generally less than 0.5 m at full scale, which will be negligible at model scale in the experiment.

For the computational convergence, the numerical simulation of the steady full wet flow is conducted to obtain a quasi-stable flow field using the multiple reference frame (MRF) method, then the unsteady calculation is started to simulate the propeller rotation using sliding mesh method, and the cavitation calculation is started to predict the propeller cavitating flows by activating the cavitation model, finally. Ten revolutions are simulated after the sliding mesh method is started, and 1° per time step is used both in unsteady and cavitation calculations.

4.1. Grid Independence

In order to investigate the grid independency, three grid sizes were chosen, as shown in Table 4 and Figure 5. All the ship and propeller regions consist of unstructured hexahedral cells with the same grid topology, generated by HEXPRESS, and the grid refinement ratio is $\sqrt{2}$.

(a) Grid 1

(b) Grid 2

(c) Grid 3

Figure 5. The surface mesh of ship stern and propeller.

Table 4. The three systematic refined grids.

Number of Cells	Grid 1	Grid 2	Grid 3
Ship region	16,136,913	6,797,553	2,804,256
Propeller region	3,019,932	1,096,318	523,144
Total	19,156,845	7,893,871	3,327,400

There are 120 CPU cores used in the numerical simulation, and the time cost for the unsteady cavitation prediction in the three refined grids is recorded in Table 5.

Table 5. The time cost for the unsteady cavitation simulation.

Time Cost	Grid 1	Grid 2	Grid 3
T (hours)	12.04	3.84	1.30

The variation of the thrust coefficient K_T over the last full propeller revolution is shown in Figure 6, and the results show that it is obviously periodic.

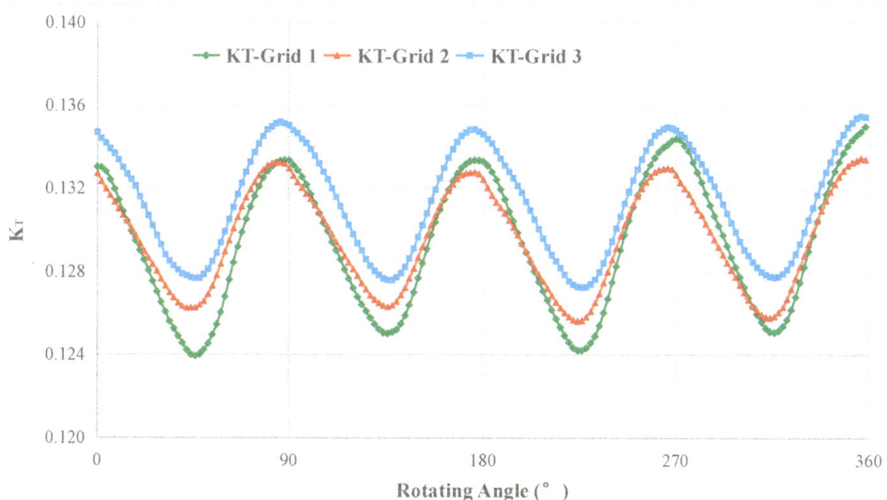

Figure 6. The variation of the thrust coefficient K_T over the last full propeller revolution.

The influence of grid resolution on K_T is assessed in Table 6. Based on the procedure recommended by ITTC [14]; the grid uncertainty, U_G; the corrected benchmark solution, S_C; and the convergence ratio, R_G, are also evaluated using the grid convergence index (GCI) approach proposed by Roache [15].

In Table 6, the K_T predicted using the three grids agrees well with the measured value, and the R_G obtained shows that monotonic convergence was achieved for $0 < R_G < 1$, the relative grid uncertainty, U_G/S_C, is also small, so all these predicted results illustrate the reliability of the numerical simulation methods adopted in the present study.

Table 6. The difference between prediction and measurement of K_T; grid uncertainty, U_G; and convergence ratio, R_G.

Parameters	Grid 1	Grid 2	Grid 3	Experiment
K_T	0.1293	0.1296	0.1314	0.1371
S_C		0.129		-
Difference	−5.65%	−5.47%	−4.19%	-
U_G		0.00024		-
U_G/S_C		0.186%		-
R_G		0.167		-

S_c—corrected benchmark solution.

4.2. Cavitation Shape

The predicted cavity is represented by vapor iso-surface, and it is true that different value of γ_v used for the iso-surface can affect the cavitation area, e.g., a larger area with the lower value and a smaller area with the higher value. The two values of γ_v, 0.5 and 0.1, are used to represent the cavitation pattern in Figure 7. The results show that the dependence is limited, and the cavitation retains the main shape.

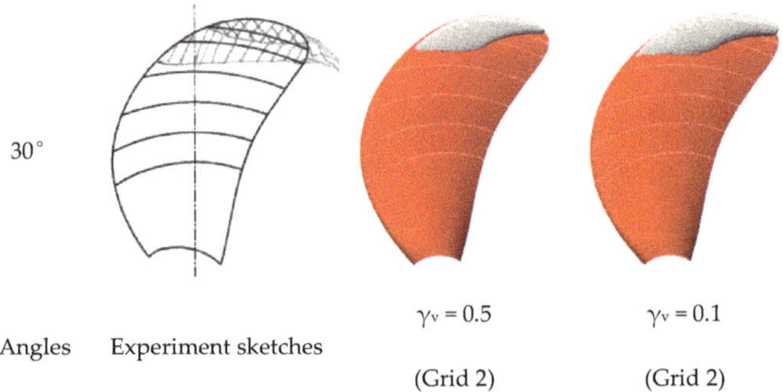

Figure 7. The predicted propeller cavitation represented by different values of vapor iso-surface.

The predicted propeller cavitation using three refined grids behind the stern is compared with benchmark test sketches at the corresponding rotating angles in Figure 8.

In Figure 8, the predicted cavity using three refined grids, represented by vapor iso-surface of 0.1, shows the same dynamic behavior (e.g., the change of the propeller cavitation shape) as the benchmark test observations. The main feature, the shape change of the attached cavity along with the rotating angles, correlates well with the test sketches, e.g. the cavitation appears at about the same angle, $\varphi \approx -10°$, and reaches the maximum size at $\varphi \approx 30°$. Nevertheless, the tip vortex cavity cannot be predicted well on account of the insufficient grid resolution near the blade tip. In addition, the difference of the cavity predicted among the three grids is not obvious, and it implies that Grid 2 can obtain satisfactory efficiency and precision of the unsteady propeller cavitation.

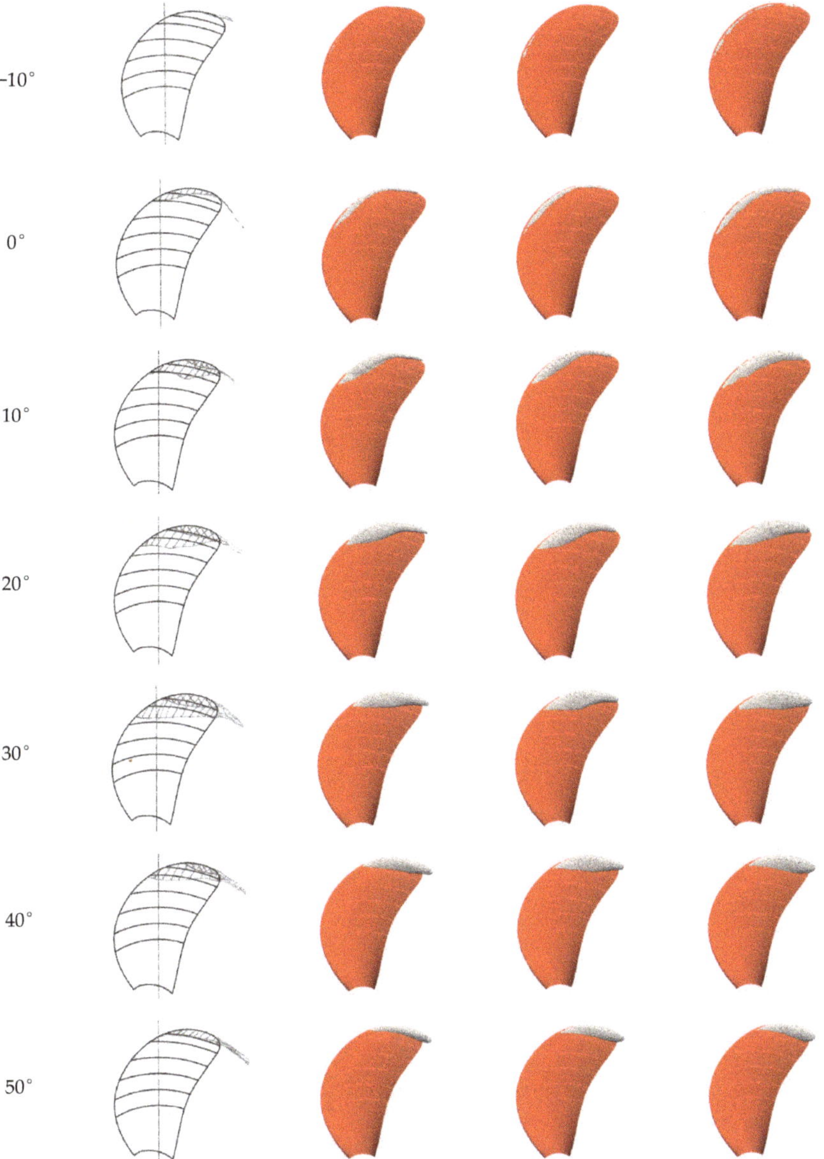

Figure 8. *Cont.*

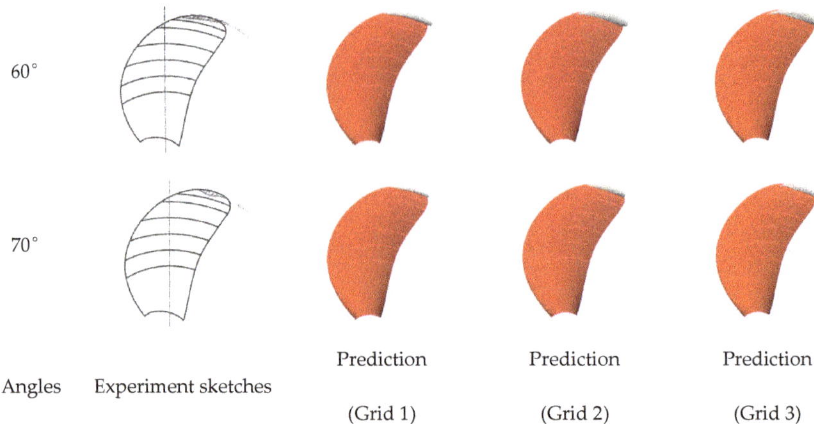

Figure 8. The predicted propeller cavitation shape compared with the benchmark test sketches.

4.3. Hull Pressure Fluctuation Induced by Cavitation

The hull pressure fluctuation induced by propeller cavitating flows is an important index that evaluates the propeller cavitation performance, so the monitoring points on the hull surface in the stern region are arranged as shown in Figure 9.

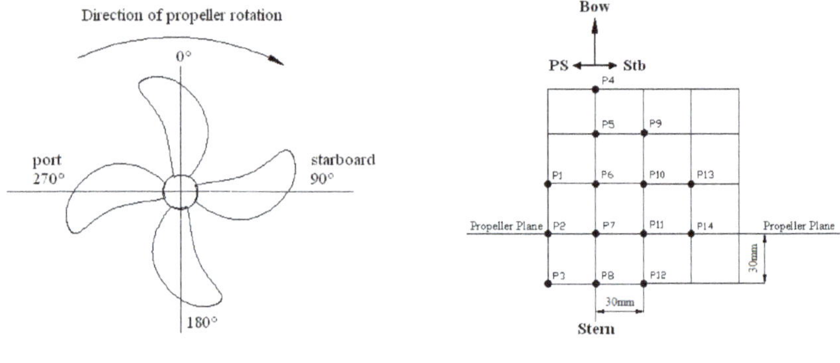

(**a**) the direction of propeller rotation (**b**) the location of monitoring points

Figure 9. The arrangement of monitoring points.

The variation of the pressure at P10 over the last full propeller revolution is shown in Figure 10, and the results show it is obviously periodic as well.

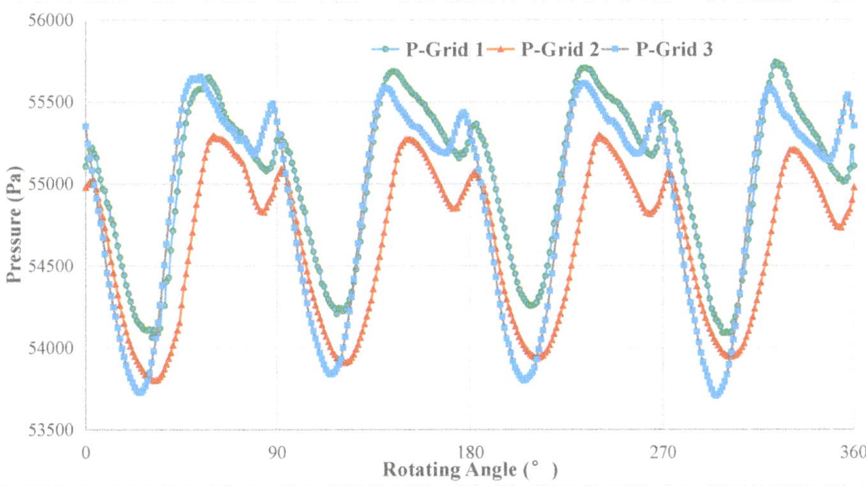

Figure 10. The variation of pressure at P10 over the last full propeller revolution.

Similar to experimental analysis, the pressure fluctuation calculated at model scale is firstly analyzed by the fast Fourier transformation (FFT) method, then converted to the one at full scale. The conversion formula is as follows.

$$K_{P_i} = \frac{\Delta P_{im}}{\rho_m n_m^2 D_m^2} \ (i = 1, 2, 3 \ldots .) \tag{10}$$

$$\Delta P_{is} = K_{P_i} \cdot \rho_{sea} \cdot n_s^2 \cdot D_s^2 \ (i = 1, 2, 3 \ldots .) \tag{11}$$

Firstly, the hull pressure fluctuation predicted in non-cavitating condition is compared with the experiment, as shown in Figure 11. The 1BF signifies the first blade frequency component of the pressure fluctuation.

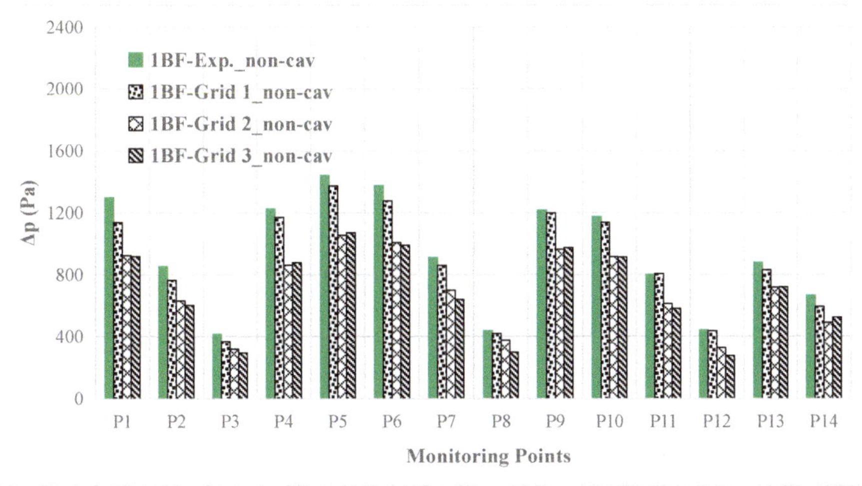

Figure 11. The 1BF (first blade frequency) component of hull pressure fluctuation predicted compared with the test data in non-cavitating condition.

It indicates that the maximum value of the 1BF component of the hull pressure fluctuation is at P5 in the non-cavitating condition, and the values decrease rapidly near the propeller in the vertical plane, such as from P1 to P3, from P5 to P8, and from P9 to P12. The results achieved with Grid 1 are much better than those of the other grids.

For the uncertainty analysis in cavitating condition, the hull pressure fluctuation measurement has been repeated three times, and the maximum value of the 1BF component of the hull pressure fluctuation is at P10 each time. The uncertainty contribution of the measurement, the combined uncertainty, and the expanded uncertainty [16] are counted in Table 7. The analysis shows that the expanded uncertainty is below 10%.

Table 7. The expanded uncertainty in measurement for the 1BF component of the hull pressure fluctuation in cavitating condition (at P10).

ΔP_S	Type	Uncertainty	Remark
Propeller diameter	B	0.04%	negligible
Operating pressure	B	0.11%	negligible
Saturated vapor pressure	B	0.29%	negligible
Water density	B	0.001%	negligible
Propeller rotate speed	B	0.02%	negligible
Dynamic pressure transducer	B	3.33%	dominant
Repeat test (N = 3)	A	0.70%	minor
Combined uncertainty		3.42%	-
Expanded uncertainty		6.83%	-

The amplitudes of the 1BF component of the hull pressure fluctuation converted at full scale are compared with the benchmark test data in Figure 12. The maximum value of the 1BF component of the hull pressure fluctuation is at P10 in the cavitating condition, while it is at P5 in the non-cavitating condition. This is mainly due to the rapid change of the cavitation volume at about $\varphi \approx 30°$, which causes the levels of the pressure pulse at P10 to be higher.

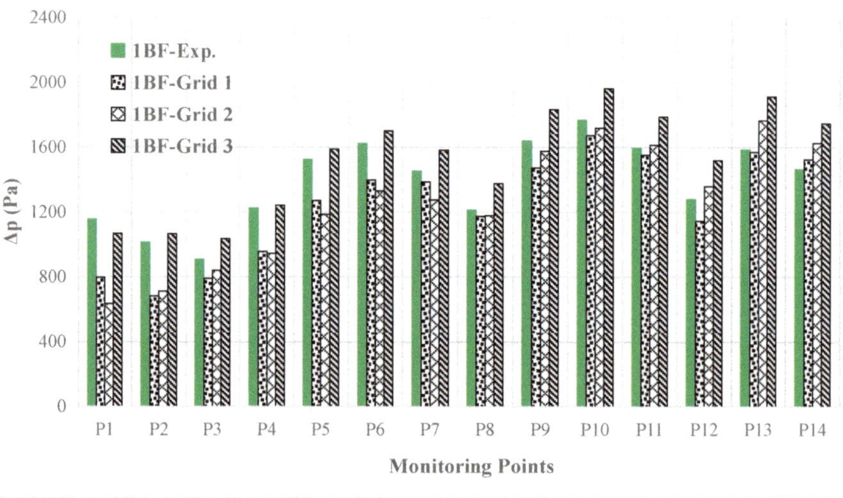

Figure 12. The 1BF component of hull pressure fluctuation predicted compared with the test data.

The predicted results obtained show quite good agreement with the benchmark test. It is found that not only the spatial laws but also the maximum value of 1BF component of the hull pressure fluctuation, which is mostly of great concern in engineering, can be predicted well. It seems that the calculated value of Grid 2 is a little closer to the test than that of Grid 1 at P10.

Moreover, the second and third blade frequency (2BF, 3BF) components are also compared with the benchmark test in Figures 13 and 14. The obvious differences can be found, and the cause may be the instability of the propeller cavitation at the same rotating angle during various periods, which can be observed in benchmark test, but cannot be simulated accurately. In order to capture the cavity shedding at high frequency, more work needs to be done in the future, such as the adoption of LES approach.

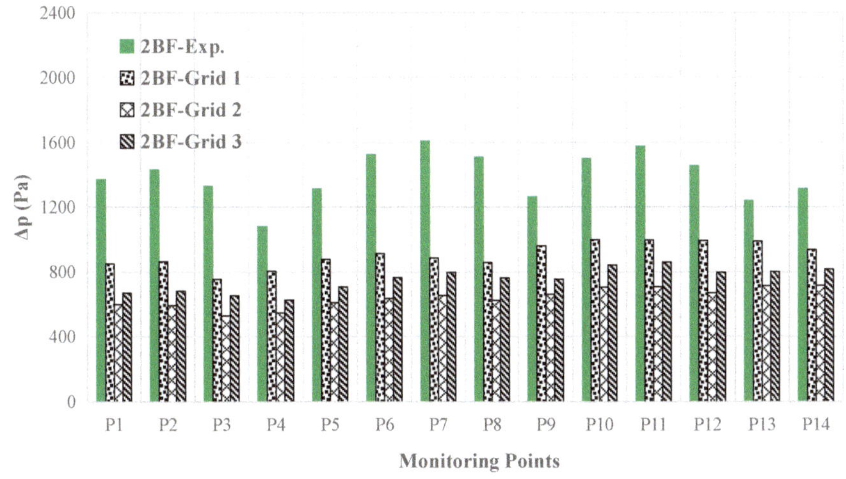

Figure 13. The 2BF component of hull pressure fluctuation predicted compared with the test data.

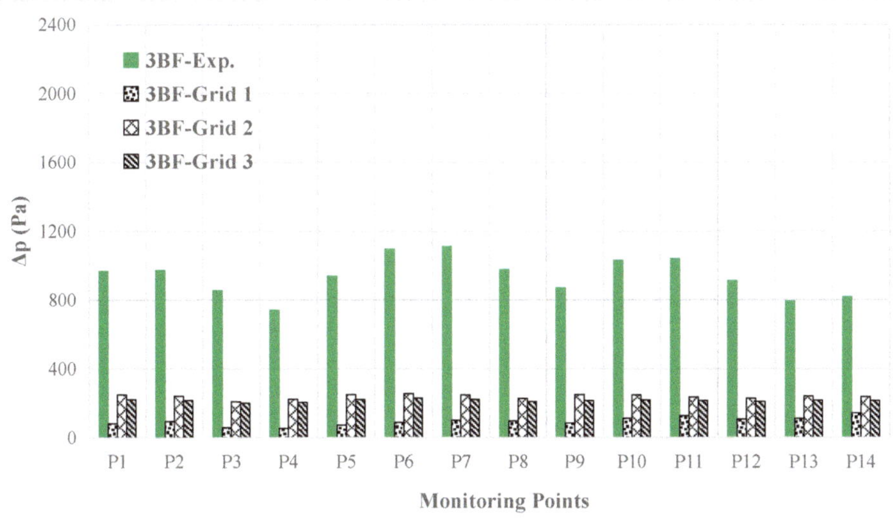

Figure 14. The 3BF component of hull pressure fluctuation predicted compared with the test data.

The influence of grid resolution on the hull pressure fluctuation at P10 is also assessed in Table 8. The predicted ΔP_S values using the three grids are all close to the experimental value, and the R_G

obtained shows that the monotonic convergence was achieved for $0 < R_G < 1$, the relative grid uncertainty U_G/S_C is also small, so all these predicted results illustrate the reliability of the numerical simulation methods again.

Table 8. The difference between prediction and measurement of 1BF component of hull pressure fluctuation ΔP_S, grid uncertainty U_G, and convergence ratio R_G.

Parameters	Grid 1	Grid 2	Grid 3	Experiment
ΔP_S (Pa)	1672.0	1719.34	1962.8	1767.2
S_C (Pa)		1624.6		-
Difference	−5.39%	−2.71%	11.07%	-
U_G		35.9		-
U_G/S_C		2.210%		-
R_G		0.194		-

5. Conclusions

The numerical simulation of propeller cavitation benchmark tests of the YUPENG ship model was conducted using the unsteady RANS approach based on OpenFOAM. The influence of grid resolutions on the unsteady propeller cavitation and hull pressure fluctuation was investigated in detail, and the conclusions obtained are as follows.

(1) The thrust coefficients predicted using the three grids are all close to the measured value, the monotonic convergence was achieved, and the grid uncertainty is also quite small.
(2) The difference of the cavity predicted among the three grids is not obvious, and Grid 2 can obtain satisfactory efficiency and precision of the unsteady propeller cavitation.
(3) The 1BF components of the hull pressure fluctuation predicted using the three grids are all still close to the benchmark test value, the monotonic convergence was also achieved, and the grid uncertainty is quite small as well.

There are some aspects that remain to be studied further, such as the more accurate prediction of tip vortex cavity and cavity shedding at high frequency, which is the point that our subsequent study will focus on.

Author Contributions: C.Z. conducted the numerical simulations and the result analysis; D.L. provided guidance throughout the research; H.H. provided the experimental data.

Funding: This research was funded by the National Natural Science Foundation of China, grant number 11772305.

Conflicts of Interest: The authors declare no conflict of interest.

References

1. Salvatore, F.; Streckwall, H.; Van Terwisga, T. Propeller Cavitation Modelling by CFD—Results from the VIRTUE 2008 Rome Workshop. In Proceedings of the First International Symposium on Marine Propulsors, Trondheim, Norway, 22–24 June 2009.
2. Li, D.; Grekula, M.; Lindell, P.; Hallander, J. Prediction of Cavitation for the INSEAN Propeller E779A Operating in Uniform Flow and Non-Uniform Wakes. In Proceedings of the 8th International Symposium on Cavitation, Singapore, 13–16 August 2012; pp. 368–373.
3. Paik, K.J.; Park, H.; Seo, J. URANS Simulations of Cavitation and Hull Pressure Fluctuation for Marine Propeller with Hull Interaction. In Proceedings of the 3nd International Symposium on Marine Propulsors, Launceston, Australia, 5–8 May 2013; pp. 389–396.
4. Vaz, G.; Hally, D.; Huuva, T.; Bulten, N.; Muller, P.; Becchi, P.; Korsström, A. Cavitating Flow Calculations for the E779A Propeller in Open Water and Behind Conditions: Code Comparison and Solution Validation. In Proceedings of the 4th International Symposium on Marine Propulsors, Austin, TX, USA, 31 May–4 June 2015; pp. 330–345.

5. Fujiyama, K.; Nakashima, Y. Numerical Prediction of Acoustic Noise Level Induced by Cavitation on Ship Propeller at Behind-Hull Condition. In Proceedings of the 5th Symposium on Marine Propulsors, SMP'17, Espoo, Finland, 12–15 June 2017; pp. 739–744.
6. Long, Y.; Long, X.; Ji, B.; Huang, H. Numerical simulations of cavitating turbulent flow around a marine propeller behind the hull with analyses of the vorticity distribution and particle tracks. *Ocean Eng.* **2019**, *189*, 106310. [CrossRef]
7. Yilmaz, N.; Aktas, B.; Sezen, S.; Atlar, M.; Fitzsimmons, P.A.; Felli, M. Numerical Investigations of Propeller-Rudder-Hull Interaction in the Presence of Tip Vortex Cavitation. In Proceedings of the 6th Symposium on Marine Propulsors, SMP'19, Rome, Italy, 26–30 May 2019; pp. 407–413.
8. Asnaghi, A.; Feymark, A.; Bensow, R.E. Computational Analysis of Cavitating Marine Propeller Performance using OpenFOAM. In Proceedings of the 4th International Symposium on Marine Propulsors, Austin, TX, USA, 31 May–4 June 2015; pp. 148–155.
9. Bensow, R.E. Large Eddy Simulation of a Cavitating Propeller Operating in Behind Conditions with and without Pre-Swirl Stators. In Proceedings of the 4th International Symposium on Marine Propulsors, Austin, TX, USA, 31 May–4 June 2015; pp. 458–477.
10. Chaosheng, Z. The unsteady numerical simulation of propeller cavitation behind a single-screw transport ship using OpenFOAM. In Proceedings of the 2nd Conference of Global Chinese Scholars on Hydrodynamics, Wuxi, China, 20–23 November 2016; pp. 117–123.
11. Chaosheng, Z. The numerical prediction of the propeller cavitation and hull pressure fluctuation in the ship stern using OpenFOAM. In Proceedings of the 5th Symposium on Marine Propulsors, SMP'17, Espoo, Finland, 12–15 June 2017; pp. 745–749.
12. Menter, F.R. Two-equation eddy-viscosity turbulence models for engineering applications. *AIAA J.* **1994**, *32*, 1598–1605. [CrossRef]
13. Issa, R.I. Solution of the implicitly discretised fluid flow equations by operator-splitting. *J. Comput. Phys.* **1986**, *62*, 40–65. [CrossRef]
14. ITTC. CFD General Uncertainty Analysis in CFD Verification and Validation Methodology and Procedures. In ITTC Procedure 2002 7.5-03-01-01, Revision 01. Available online: https://www.marin.nl/storage/uploads/25027/files/7.5-03-01-01_ITTC_Uncertainty_proc_for_CFD.pdf (accessed on 10 August 2019).
15. Roache, P.J. *Verification and Validation in Computational Science and Engineering*; Hermosa: Albuquerque, NM, USA, 1998.
16. JCGM. Evaluation of measurement data—Guide to the expression of uncertainty in measurement, JCGM 100:2008 GUM 1995 with minor corrections. In *Joint Committee for Guides in Metrology*; Bureau International des Poids Mesures (BIPM): Sevres, France, 2008.

Journal of
Marine Science and Engineering

Article

Morphing Hydrofoil Model Driven by Compliant Composite Structure and Internal Pressure

Mohammed Arab Fatiha [1,†], Benoît Augier [2,†], François Deniset [1,†], Pascal Casari [3,†] and Jacques André Astolfi [1,*,†,‡]

1 French Naval Academy Research Institute—IRENav EA3634, 29200 Brest, France; fatiha.mohammed_arab@ecole-navale.fr (M.A.F.); francois.deniset@ecole-navale.fr (F.D.)
2 The French Research Institute for Exploitation of the Sea, 29200 Brest, France; benoit.augier@ifremer.fr
3 Research Institute in Civil Engineering and Mechanics—GeM, 44606 Saint-Nazaire, France; pascal.casari@univ-nantes.fr
* Correspondence: jacques-andre.astolfi@ecole-navale.fr; Tel.: +33-298-234-017
† These authors contributed equally to this work.
‡ This paper is an extended version of our paper published in SMP'19.

Received: 27 October 2019; Accepted: 18 November 2019; Published: 21 November 2019

Abstract: In this work, a collaborative experimental study has been conducted to assess the effect an imposed internal pressure has on the controlling the hydrodynamic performance of a compliant composite hydrofoil. It was expected that the internal pressure together with composite structures be suitable to control the hydrodynamic forces as well as cavitation inception and development. A new concept of morphing hydrofoil was developed and tested in the cavitation tunnel at the French Naval Academy Research Institute. The experiments were based on the measurements of hydrodynamic forces and hydrofoil deformations under various conditions of internal pressure. The effect on cavitation inception was studied too. In parallel to this experiment, a 2D numerical tool was developed in order to assist the design of the compliant hydrofoil shape. Numerically, the fluid-structure coupling is based on an iterative method under a small perturbation hypothesis. The flow model is based on a panel method and a boundary layer formulation and was coupled with a finite-element method for the structure. It is shown that pressure driven compliant composite structure is suitable to some extent to control the hydrodynamic forces, allowing the operational domain of the compliant hydrofoil to be extended according to the angle of attack and the internal pressure. In addition, the effect on the cavitation inception is pointed out.

Keywords: smart-structure; hydrofoil; morphing; compliant; composite; cavitation

---

1. Introduction

In naval applications, it is crucial to make strategic decision to reduce the fuel oil consumption of ships and therefore to decrease their CO_2 emissions. The demand for the reduction of fuel oil consumption and CO_2 emissions is greater than ever before [1]. Underlying the need for improved performance, better comfort, and stability, the use of new concepts of innovative hydrofoils or propeller blades can be an option to enhance the hydrodynamic performance and reduce the consumption of ships.

Using this new concept should allow to control the forces (lift and drag) for various operating conditions to be controlled. However, this can lead to cavitation onset at high speed and moderate angles of incidence but also at low speed and high angles. Improving the hydrodynamic performances and delaying the cavitation inception requires the modification of shape, hence the idea of using morphing hydrofoils.

The aim objective of the current research focuses on aerodynamic and hydrodynamic performance 31 enhancement. In the energy field, we find the works of Aramendia, which analyzed the effect of Gurney flap (GF) length on the lift/drag ratio C_L/C_D of blades. They have proved numerically the capability of the GF to improve lift/drag ratio of passive and active flow control [2,3].

Currently, hydrofoils use mechanical systems as a flap to modify their shape and to control their performance. Morphing structures could be an interesting route to change the hydrofoil performance [4].

In aerodynamic applications, the use of morphing structures has proved its effect in flying performance [5]. Brailovski et al. [6] have studied the effect on the aerodynamic performance and foil mechanical properties of a flexible suction side powered by two actuators numerically. In another study [7], the gap present at the spanwise ends of the control surfaces is replaced by a smooth, three-dimensional morphing transition section. The passive control of this compliant morphing flap transition has the advantage of increasing the lift and reducing the drag. We can not talk about the benefits of morphing structures on aerodynamic performance, not to mention the effect of various variable camber continuous trailing edge flap (VCCTEF) on the lift and drag forces [8]. It was noted that the best stall performance (L/D) was demonstrated by the circular and parabolic arc camber flaps.

Even if, the main objectives of the hydrodynamic applications are similar to those of the aerodynamics, many of the techniques employed in aerodynamic applications cannot be transferred to naval ones. So, it is necessary to take into account the differences between the fluid properties and the cavitation phenomena in naval applications.

To meet hydrodynamic requirements, adaptive composites are used in many marine technologies, including propulsive devices, underwater vehicles, and propellers. In [9], the authors summarized the progress on the numerical modeling, the experimental studies, design, and optimization of adaptive composite marine propulsors and turbines. Firstly, they have presented the differences between adaptive aerodynamic and hydrodynamic lifting surfaces. Therefore in the hydrodynamic applications, the local pressure fluctuations led to the formation of cavitation [10], which can lead to load fluctuations, vibrations and performance decay. Furthermore, they discussed the current challenges in the numerical modeling, experimental studies, design, and optimization of the adaptive marine propulsors and turbines. The major challenge in the numerical and experimental modeling is the three-dimensional viscous fluid-structure interaction [11] and the cavitation [12].

The cavitation inception is influenced by several factors. Amromin [13] indicate that the flow-induced vibration of hydrofoils affects pressure pulsations on their surfaces which depends on the hydrofoil material and influences the cavitation inception and desinence.

Many of the recent developments have focused on the use of composite materials over traditional metallic materials. The composite materials have many advantages, including higher strength-to-weight ratios, better fatigue characteristics, higher durability. They provide resistance to salt water and improve the resistance to corrosion [14].

The effect of material and Reynolds number on the hydrodynamic performance of hydrofoils was investigated experimentally by Zarruk et al. [15]. They studied the performances of flexible hydrofoils of similar geometry made of stainless steel, aluminum and a composite of carbon-fiber reinforced plastic with layup orientations at 0° and 30°. They concluded that the composite hydrofoils have the best hydrodynamic performance, showing the potential of a tailored hydroelastic composite hydrofoil. The hydro-elastic behavior of flexible propellers has also been analyzed by Maljaars et al. [16]. They compared the results of a boundary element method (BEM) and a Reynolds-averaged-Navier-Stokes (RANS) simulations with the calculation of measured open water diagram and the open water curves. They found that the two results of these analyses concurred.

In [17], the cantilevered rigid and compliant three-dimensional hydrofoils were studied in a cavitation tunnel in order to analyse the cloud cavitation behavior. The rigid hydrofoil was made of stainless steel and the compliant one of a carbon and glass fiber-reinforced epoxy resin with the structural fibers aligned along the spanwise direction to avoid material bend-twist coupling. The tests were carried out at a Reynolds number of 0.7×10^6, an incidence of 6° and cavitation number of 0.8. The compliant hydrofoil was seen to dampen the higher frequency force fluctuations while showing a strong correlation between normal force and tip deflection. Furthermore, the 3D nature of the flow field causes complex cavitation behavior with two shedding modes on both models. Another type of cavitation has been studied by Zhu et al. [18]. They evaluated the hydrodynamic performances of a propeller with winglets numerically and they compared them to those of the benchmark propeller (MAU5-80). They concluded that the presence of the winglets reduces the vapor volume and alleviates the tip vortex cavitation (TVC).

The bend-twist coupling effects on the hydroelastic response of composite hydrofoils have been experimentally studied [19]. The authors concluded that bend-twist coupling affects the deformation of the hydrofoils which modify the hydrodynamic performance. The effect of material bend-twist coupling on the cavitating response of adaptive composite hydrofoils has also been analyzed experimentally [20] for three identical unloaded hydrofoils. Two hydrofoils of composite material and another rigid one of stainless steel (SS) were tested in the same cantilevered configuration. They concluded that material bend-twist coupling has an effect on the hydrodynamic load coefficients, cavitation inception and the maximum cavity length compared to a SS hydrofoil.

In order to assess the effect of the cavitation on the structural response, Ducoin et al. [21] have studied the displacement of a flexible hydrofoil in a cavitating flow. They found that the hydrodynamic loading unsteadiness increases the vibrations experienced by the hydrofoil. Numerically, Garg et al [22,23] have developed a shape optimization tool to predict the hydrodynamic performance including the cavitation inception conditions.

In order to control the lift generated by hydrofoils on boats, Giovannetti et al. [24] have numerically and experimentally analysed hydrofoil geometry designed to reduce the lift coefficient passively by increasing the flow velocity. This study was conducted with the use of wind tunnel experiments including deformation measurements, which concurred with the numerical results. They found that the twist deformations resulted in a reduction in the effective angle of attack by 30% at higher flow velocities, which significantly reduced the foil's lift and drag.

Numerical predictions of the hydrodynamic forces, deformations and cavitation performance for a NACA 0009 hydrofoil and an optimized hydrofoil which have been studied by Garg et al. [22,25] are compared to the experimental ones. The predicted hydrodynamic coefficients (C_L, C_D, and C_M) and the tip bending deflections are concur with measured values for both the baseline and the optimized hydrofoils across a wide range of lift conditions. The mean difference between the numerical predictions and the experimental measurements for mean C_L, C_D, and C_M for the optimized hydrofoil is noted. This indicates that values are 2.96%, 5.10%, and 3.0%, respectively. The mean difference in the tip bending deflections is 3.45%.

The French Naval Academy Research Institute (IRENav) is interested in the study of the deformed hydrofoils and their performances. Fluid-Structure Interaction has been investigated experimentally by studying the structural response of a flexible lightweight hydrofoil undergoing various flow conditions including cavitating flow by Lelong et al. [26,27]. An optimization of the design of the shape and the elastic characteristics of a hydrofoil equipped with deformable elements giving flexibility to the trailing edge was developed by Sacher et al. [28].

IRENav, the Research Institute in Civil Engineering and Mechanics (GeM) and IFREMER have initiated a research program related to compliant hydrofoils for naval applications. The objective is to characterize a compliant composite hydrofoil driven by an internal low pressure regarding the lift and drag forces as well as cavitation inception. This paper presents the experimental study performed in the hydrodynamic tunnel at IRENav. The hydrofoil manufactured at GeM was firstly tested in the open air to assess the effect of the internal pressure on hydrofoil deformations by the use of the digital image correlation (DIC-3D) system. Then, the hydrofoil was tested in the cavitation tunnel at IRENav where the lift and drag coefficients and the hydrofoil deformations were measured. In accordance with the experiments, a numerical approach based on a fluid-structure coupling algorithm has been developed. The paper describes the experimental setup, the numerical fluid-structure interaction (FSI) algorithm and presents the main results.

2. Experimental Setup

The experiments are carried out in the cavitation tunnel at IRENav (Figure 1). The tunnel test section is 1 m long with a square section of 0.192 m side (Figure 2). The inflow velocity ranges between 0.5 and 12 m/s. The pressure in the tunnel test section ranges from 100 mbar and 3 bar to control the cavitation which is given by a cavitation number defined by Equation (1) and the measured turbulence intensity in the test section is 2% at 5 m/s. This cavitation number can therefore be compared with the opposite of pressure coefficient $-C_{pmin}$ defined as the minimum of the pressure coefficient (Equation (2)).

Figure 1. Hydrodynamic tunnel test section at IRENav with the compliant composite hydrofoil.

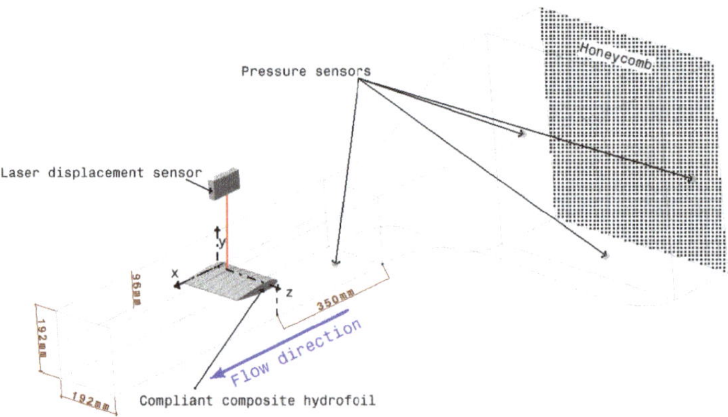

Figure 2. Tunnel test section characteristics with the compliant composite hydrofoil and Laser displacement measurement system.

$$\sigma = \frac{P_{ref} - P_v}{\frac{1}{2}\rho V^2} \tag{1}$$

$$C_p = \frac{P - P_{ref}}{\frac{1}{2}\rho V^2} \tag{2}$$

where P_{ref} is the pressure in the test section, P_v is the vapor pressure at the water temperature, P is the local pressure, V is the inflow velocity, and ρ is the water density. Thus, when $\sigma < -C_{pmin}$, that is to say when $P < P_v$, cavitation is expected to appear in the flow at the point where the pressure coefficient is the lowest.

The compliant composite hydrofoil was manufactured at the Research Institute in Civil Engineering and Mechanics (GeM). At rest, the compliant composite hydrofoil has a NACA 0012 section and a rectangular plan-form of 0.191 m span and 0.15 m chord length. It is cantilevered and clamped on a cylindrical aluminum beam fitted to the hydrodynamic balance. The axis of rotation is at $X/c = 0.25$.

The hydrofoil is composed of two walls, one rigid and one compliant, providing a cavity in which vacuum can be applied to vary the shape. In the following part, pressure variation will always correspond to a suction inside the hydrofoil compared to the atmospheric reference in the test section. It is defined as ΔP (called internal pressure in the paper) and has a positive value as it decreases. The pressure system, the compliant composite hydrofoil and the pneumatic actuator are presented in Figure 3. The pressure inside the cavity is measured using a manometer.

The compliant wall is laminated with three carbon/epoxy plies oriented at $0°/90°$ in the middle, and thinner at the leading and trailing edges. These two edges are composed of two plies: one of carbon/epoxy $[0°/90°]$ and the second of glass fiber with an orientation at 45°. The rigid side is composed of five plies of carbon/epoxy with the layups orientations at 0° and 90° (Figure 4).

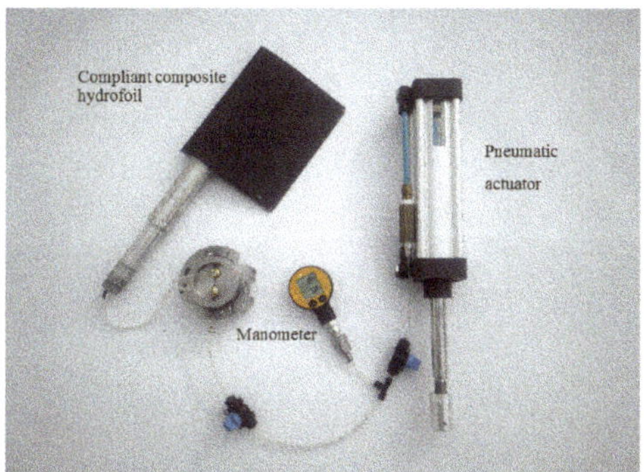

Figure 3. Compliant composite hydrofoil equipped with the control internal pressure system.

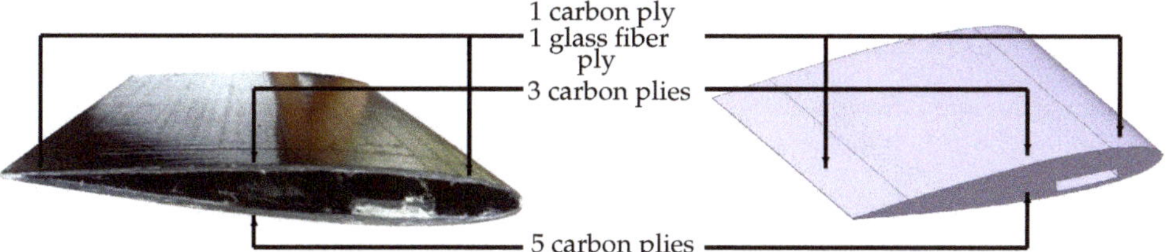

Figure 4. Laminate structure of the compliant composite hydrofoil.

To assess the effect of the internal pressure on hydrofoil deformations in open air, the chordwise displacement of skin is investigated using the digital image correlation (DIC-3D) system at GeM laboratory (Figure 5). The digital image correlation is well known as an effective method of obtaining field surface displacements. This method is based on optimal strain measurements. It is a non-contact technique and it is particularly suitable for flexible materials. It is based on the comparison of two digital images features of the composite surface before and after loading, and total displacement and strain fields can be obtained. For this purpose, a two digital camera system has been used to monitor the strain pre-gressing on the surface and a computer with DIC software. In order to produce fine and exploitable details, a random pattern of paint is usually applied to the surface of the hydrofoil (Figure 6). The software selects points in the reference image and follows them in the following images thanks to a window defined around the points. The window consists of some pixels which grant a unique greyscale intensity distribution to the window. To detect the area with the most similar intensity distribution, which contains the required points, a cross-correlation method is used to scan the next image. Thus, the movements of two separate points are followed by DIC and the last computes the change in the distance between these points [29].

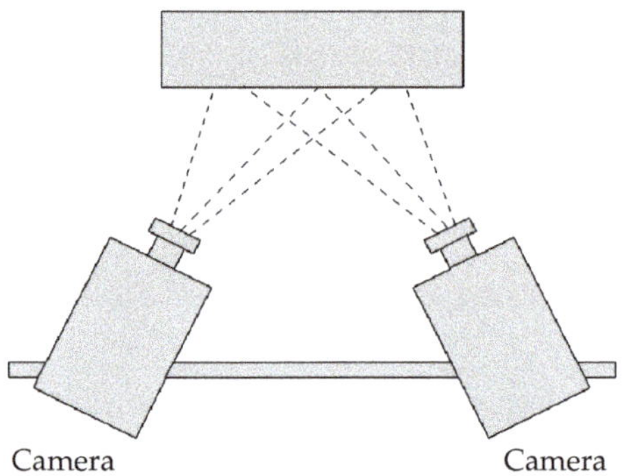

Figure 5. Digital image correlation (DIC) schematic at GeM (Saint-Nazaire, France).

Figure 6. Specimen prepared for DIC with random pattern paint on the surface.

The hydrofoil displacement is also investigated in open air by using a micrometric touch probe at a discrete position ($Z/c = 0.33$, $X/c = 0.63$) for various imposed internal pressures of $\Delta P = 0.15$ bar, 0.4 bar and 0.51 bar.

In the hydrodynamic tunnel, the static deformation is measured using a Laser distance measurement system mounted on a 2D translation system on the upper side of the test section. The system measures the vertical position of the hydrofoil suction side along sections scanned through the span from the root to the tip. It continuously monitors displacement according to a sampling frequency chosen by the user (set to 50 Hz). The system allows us to scan the hydrofoil surface for a given flow condition along various sections selected along the span. In this case, nine sections from the root to the tip are selected. At first and for the $\Delta P^* = 0$, the distance measurements are carried out without a flow, in order to determine the rest position of the hydrofoil. Measurements are repeated under a flow at 5 m/s. The results of the laser distance measurements are presented in Figure 7. It is shown that the velocity has no effect on the hydrofoil displacement. After, for the same inflow velocity, the distance measurements are carried out for various internal pressures. The deformation is obtained by comparing the scans between referenced and tested internal pressures.

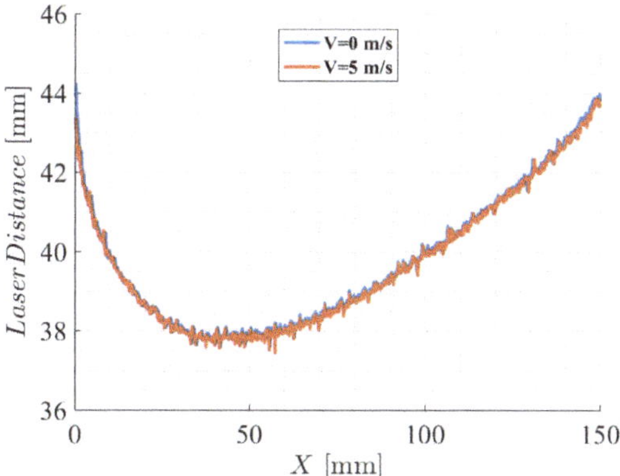

Figure 7. Results of the laser displacement measurement system without flow and $Re = 0.75 \times 10^6$.

In the hydrodynamic tunnel, measurement of hydrodynamic forces is conducted using a hydrodynamic balance at various conditions of internal pressure and angle of attack and different flow velocities. Firstly, the hydrodynamic forces are measured for a velocity of 5 m/s corresponding to a Reynolds number of 0.75×10^6. Secondly, they are measured for a velocity of 6.67 m/s and 9 m/s which correspond to 10^6 and 1.35×10^6 Reynolds numbers respectively. This increase in velocity is in order not to increase the incidence too much and in order to analyze the cavitation inception for low cavitation numbers.

The 5-component hydrodynamic balance has a range of up to 1700 N for the lift force, 180 N for the drag and a 43 N m for the pitching moment. It is fixed into a supporting frame, mounted on bearings, and driven in rotation by a Baldor motor. The stepper motor allows for 600,000 impulsions per rotation, meaning a resolution of $6 \times 10^{-4\circ}$. The foil is fastened into the balance, secured by a tightly fitted key/nut system [30]. As the test section is horizontal, the geometric 0° angle of attack of the hydrofoil is visually controlled using the water surface at mid height of the test section when filling the tunnel. Furthermore, as the hydrofoil is symmetric, the zero-lift angle is used for the positioning of the final angle of attack.

3. Uncertainties

The experimental uncertainties consist the precision of the hydrodynamic balance, pressure sensors, Laser distance measurement system, digital image correlation (DIC-3D) and micrometric touch probe.

In the cavitation tunnel, the uncertainties of velocity and pressure measurements are based on the accuracy of the pressure sensors. The latter is about 0.04 bar. For the measurements of hydrodynamic forces and from the document provided by the manufacturer of the hydrodynamic balance, the uncertainties are about ±1.02 N for the lift, ±0.324 N for the drag and ±0.26 N.m for the pitching moment. The uncertainty on the displacement measurements in the test section is about ±0.046 mm.

For open air measurements, the uncertainty of the digital image correlation (DIC-3D) system is about ±0.002 mm for the in-plane displacements and ±0.004 mm for the displacements out of the plane. The micrometric touch probe precision is about ±0.01 mm.

4. Numerical Approach

The numerical study consists of 2D simulation to investigate the effect of a static internal pressure on the structural response of the compliant hydrofoil as well as the impact on hydrodynamic performances.

The flow model of the XFOIL solver and the finite-element method of the ANSYS-Mechanical solver are used for the FSI analysis. The FSI algorithm is based on an iterative method between the two solvers with a small perturbation hypothesis. The flow model is based on the coupling between a panel method with a boundary layer model. More details concerning Xfoil are given in [31]. The panel method accelerates the flow calculations as compared to finite volume methods.

The numerical model of the hydrofoil structure is calculated by ANSYS Mechanical using ANSYS APDL (Ansys Parametric Design Language) with solid elements according to the preliminary hydrofoil design. The original APDL script was modified to handle distinct geometry changes and meshing. Solid elements PLANE183 (quadratic) are used to mesh the hydrofoil geometry.

The coupling algorithm is developed using Python-scripts. The fluid-structure coupling algorithm is described in Figure 8. The FSI algorithm is initialized by a structural computation as the cantilevered hydrofoil is submitted to an internal pressure only. It leads to displacements which produce to a new hydrofoil shape. Then, the viscous flow around the new foil is solved. The computation returns a C_p distribution and the forces of the hydrofoil. The external hydrodynamic pressure resulting from the C_p distribution is applied during the structural resolution. The problem is solved using an iterative method until the convergence on the maximum structural displacement and the lift coefficient C_L is reached.

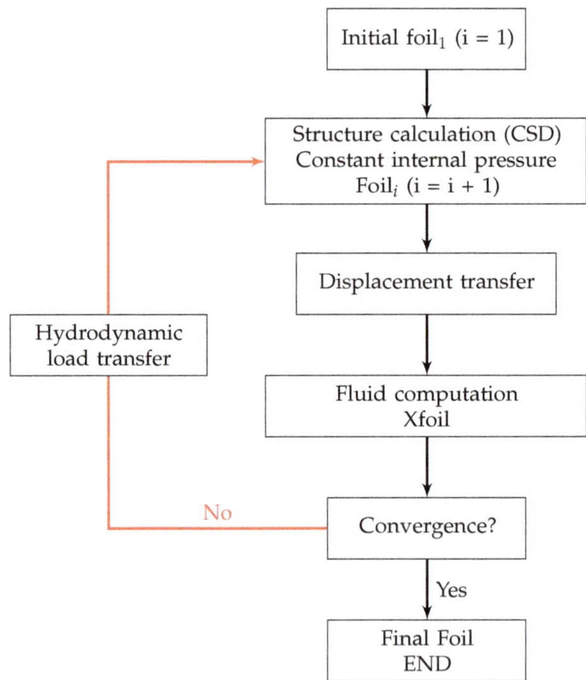

Figure 8. Schematic algorithm of fluid-structure coupling with imposed internal pressure.

The convergence to the equilibrium of the hydrofoil is obtained after a small number of iterations showing that the method developed in this work has an advantage when compared to advanced CFD-CSD solvers that require very significant CPU (Central Processing Unit) times.

In a first approach and to simplify the calculation, the hydrofoil material is considered as an homogeneous elastic equivalent. The Young's modulus input for the 2D section of the compliant hydrofoil was adjusted until the maximum displacement of the hydrofoil in the simulations coincided with the maximum displacement of the hydrofoil during the experiments for various internal pressures. The equivalent Young modulus used in the computation was set to $E = 70,000$ MPa and the equivalent Poisson coefficient was set to 0.34.

The initial and deformed shapes of the hydrofoil at a 3° angle of attack, $Re = 0.75 \times 10^6$ and an imposed internal pressure of $\Delta P^* = 1.92$ are presented if Figure 9. The calculated maximum chordwise displacement of the hydrofoil was found to be 3.3 percent of the chord length.

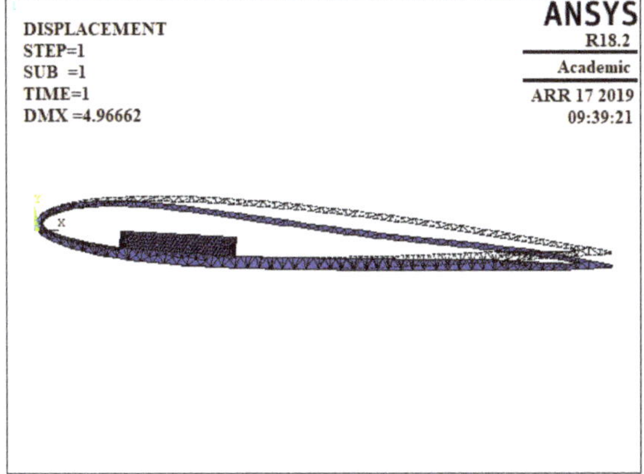

Figure 9. Initial shape and deformed shape of the hydrofoil at the first iteration, $\Delta P^* = 1.92$, $\alpha = 3°$ and $Re = 0.75 \times 10^6$.

5. Results and Discussion

5.1. Hydrofoil Deformation and Hydrodynamic Forces

The effect of internal pressure on the hydrofoil deformation is investigated by using digital image correlation (DIC-3D). For $\Delta P = 0.415$ bar, the displacement field plotted against the foil coordinates is presented in Figure 10. It is observed that the maximum displacement is 8.06 mm (5.3%c) located at the center of the hydrofoil. The deformation is not uniform along the spanwise direction due to the structural boundary conditions at the root and the tip.

The displacement of the hydrofoil as a function of the chord is taken at $Z/c = 0.55$ for four internal pressures. It is plotted in Figure 11. It confirms the results of the 3D displacement field presented in Figure 10. The hydrofoil displacement is not uniform along the chord which explains the modification of the camber.

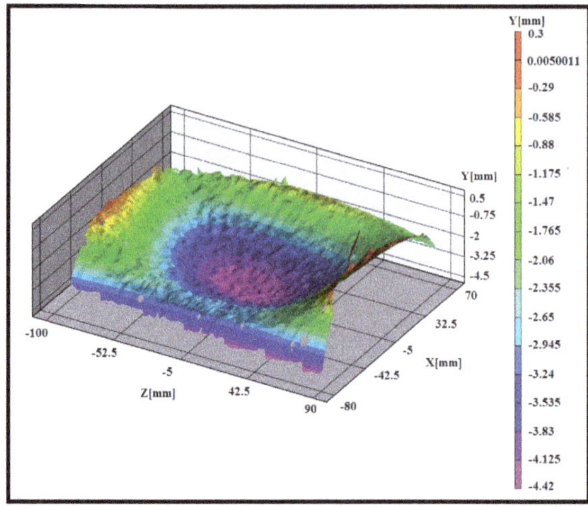

Figure 10. DIC-3D displacement, $\Delta P = 0.415$ bar.

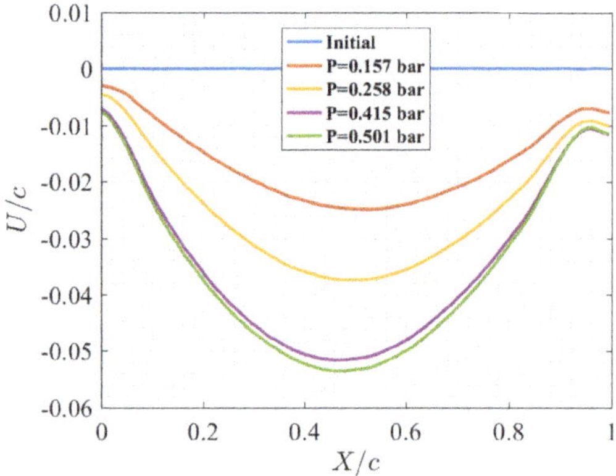

Figure 11. Experimental hydrofoil displacement in open air plotted against the chord and the internal pressure at $Z/c = 0.55$.

The hydrofoil displacements are also measured with a micrometric touch probe at $X/c = 0.33$ and $Z/c = 0.63$. The results are compared to the numerical simulation and to the measurements of the DIC-3D system at the same point (Figure 12). As shown in Figure 10, the displacement is linear for an internal pressure of up to $\Delta P = 0.4$ barand reaches a limit value of about $0.045c$. The numerical results correspond to experimental ones for an internal pressure of up to $\Delta P = 0.4$ bar showing that the equivalent homogeneous model is fairly consistent with the experiment. Beyond $\Delta P = 0.4$ bar, the predicted deformation increases linearly exhibiting no saturation.

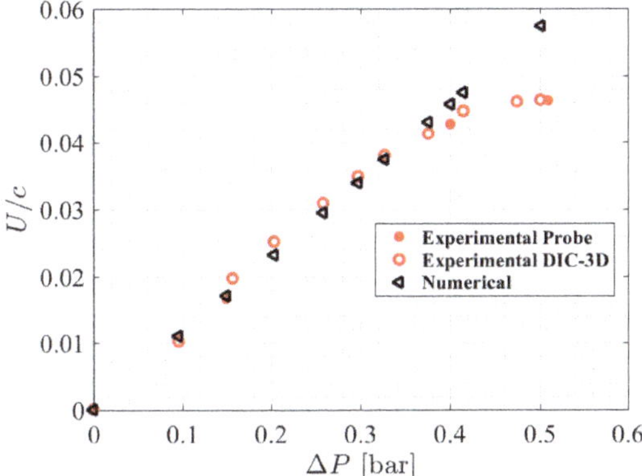

Figure 12. Experimental and numerical displacement in open air at $Z/c = 0.33$ and $X/c = 0.63$.

The foil deformation is then analyzed using the scanning measurement system in the hydrodynamic tunnel for a velocity of 5 m/s. The experimental deformed section is presented in Figure 13 for an internal pressure $\Delta P^* = 1.92$ and compared to the numerical one. As depicted in Figure 13, the experimental and numerical displacements have the same trend excepted at the trailing edge where a significant difference is observed. This difference is due to the twist of the hydrofoil in the experiment that is not observed with a 2D computation in the numerical study. The connection between the core beam (Figure 9) and the low pressure side skin is indeed flexible and allows the whole section to rotate around the cantilevered one.

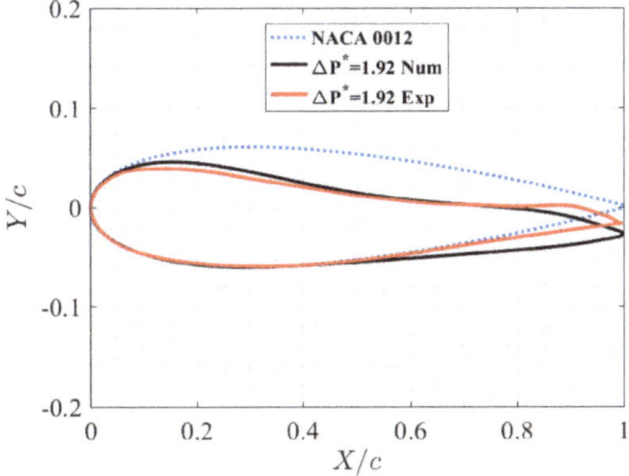

Figure 13. Experimental and numerical foil sections, $\Delta P^* = 1.92$ and $V = 5$ m/s.

The hydrodynamic forces for different internal pressures are presented in Figure 14a–c. It is reminded that when the pressure inside the cavity decreases, ΔP^* increases. It is shown that the lift coefficient and lift/drag ratio shift upwards as ΔP^* increases with a slight change in the slope. For the ranges of angles of attack and internal pressure of the present experiments, it is found that the lift coefficient and lift/drag ratio increase linearly for both parameters. The results of Figure 14a can be explained by the response surface of the compliant hydrofoil in terms of lift coefficient versus angle of attack and the non-dimensional internal pressure. The operating domain of the compliant hydrofoil is plotted as a function of the two independent variables in Figure 15.

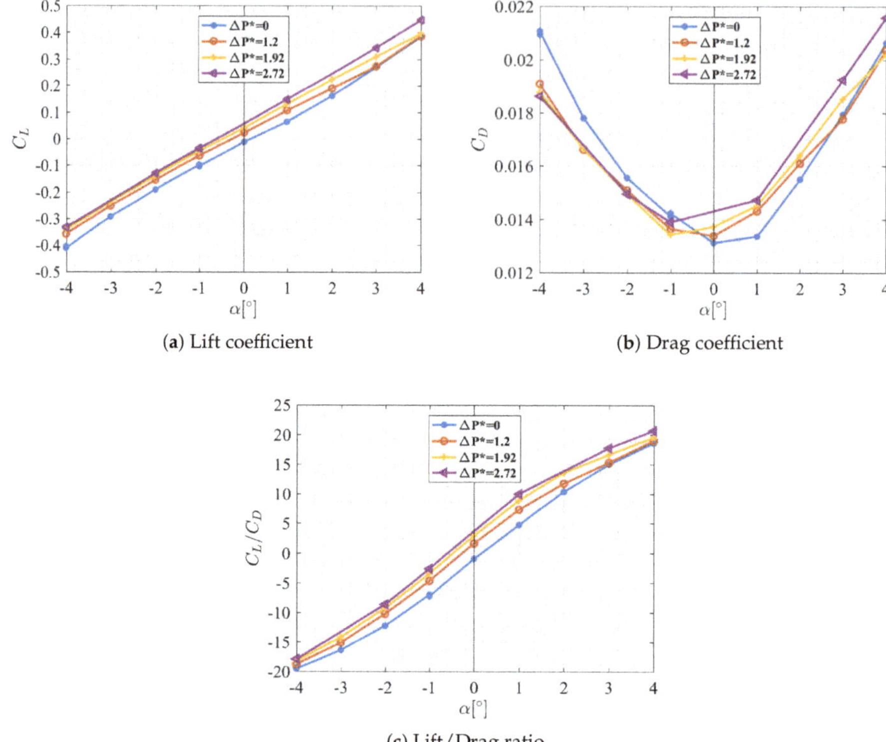

(**a**) Lift coefficient (**b**) Drag coefficient

(**c**) Lift/Drag ratio

Figure 14. Experimental lift and drag coefficients as a function of angle of attack and an internal pressure at $Re = 0.75 \times 10^6$.

According to theory, the lift coefficient depends on the maximum thickness and angle of attack (Equation (3)). In this study, it depends on maximum thickness, angle of attack and internal pressure (Equation (4)).

$$C_L = 0.109(1 - k\tau)\alpha \tag{3}$$

$$C_L(\alpha, \Delta P^*) = C_L(\alpha, \Delta P^* = 0) + \Delta C_L(\alpha, \Delta P^*) \tag{4}$$

If it is considered that the internal pressure changes the thickness (see Figure 12) and the camber linearly, the maximum thickness and ΔC_L depend only on internal pressure variation. A linear approximation of the response surface (Figure 15) is shown in Equation (5).

$$C_L(\Delta P^*, \alpha) = 0.109(1 - k(\tau + a\Delta P^*))\alpha + b\Delta P^* \tag{5}$$

where $\Delta P^* = \frac{\Delta P}{q}$ is the non dimensional internal pressure and $a = 0.013$, $b = 0.025$ and $k = 0.95$ are constants, which depend on material and hydrofoil design, determined from linear regressions on experimental data.

This simple expression highlights the effect of the internal pressure on the lift slope and lift coefficient described experimentally. The latter is equivalent to the shift of the zero lift angle as a result of camber modification. The expression of the surface response depending on flow conditions and design parameters can be very useful in the context of foil optimization methods. The same approach could be used regarding the drag coefficient and the lift to drag ratio.

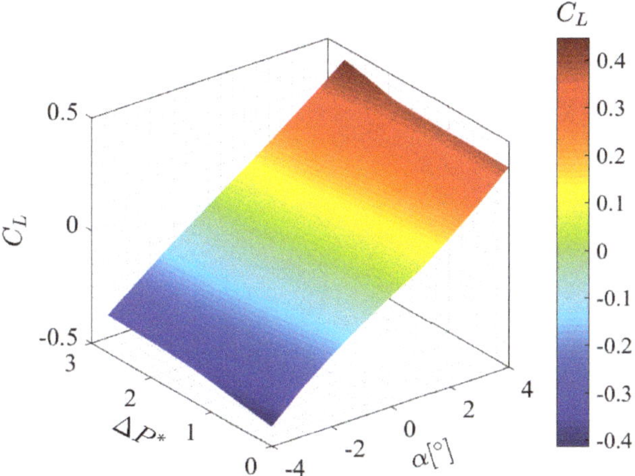

Figure 15. Experimental surface response of the lift coefficient plotted against the angle of attack and internal pressure at $Re = 0.75 \times 10^6$.

The experimental and numerical lift coefficients are presented in Figure 16. The numerical lift coefficient obtained from the FSI algorithm concurs with the experimental one but departs progressively from the experiment as the angle of attack progressively increases. The reason fro the discrepancies can be found in the analysis of the structure. Indeed, the experimental section shapes and displacements of the compliant wall are extracted at mid-span for α ranging from $-4°$ to $4°$ and they are shown in Figure 17a,b for the positive and negative angles of attack respectively.

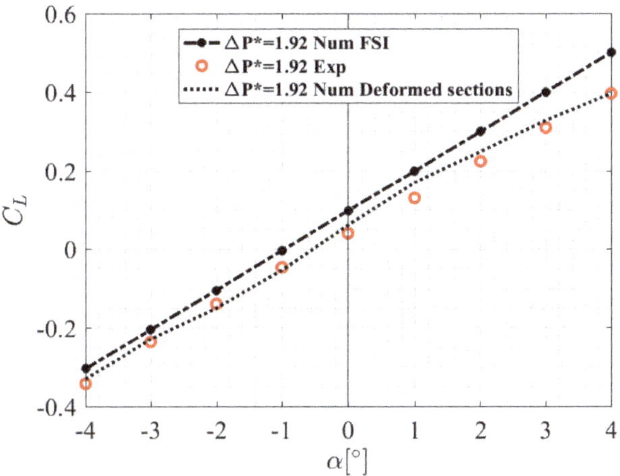

Figure 16. Experimental and numerical lift coefficients as a function of the angle of attack for an internal pressure $\Delta P^* = 1.92$ and $Re = 0.75 \times 10^6$. FSI computation and flow computation over the experimentally deformed sections at mid-span.

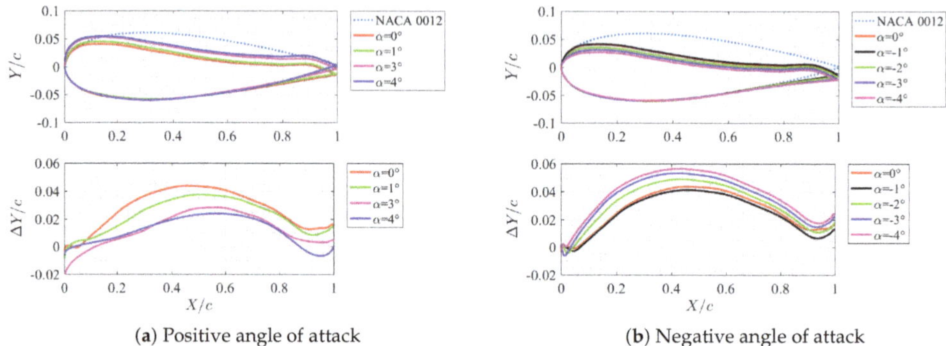

(a) Positive angle of attack (b) Negative angle of attack

Figure 17. Experimental mid-span shapes and associated displacements for $\Delta P^* = 1.92$, $Re = 0.75 \times 10^6$ and positive and negative angles of attack.

A close examination shows that a small flap effect is observed experimentally at the trailing edge affecting the lift coefficient. This effect is not shown by the 2D structural solver where the connection between the skin and the structural beam is considered as rigid. Moreover, experimentally the whole section rotates around 0.25 X/c due to a slight twisting of the structural internal beam as previously described. This is particularly observed for a positive angle of attack. The foil twisting tends to reduce the angle of attack, therefore to decrease the lift coefficient. It is observed that the FSI computation concurs very well the experiments for negative angles of attack when the twist can be neglected. Furthermore, for positive angles of attack, the twist observed experimentally is not seen into the FSI simulation. This is clearly shown in Figure 16 where the lift coefficient is computed directly on the experimental deformed sections at mid-span. In this case, the numerical lift concurs well the experiments all over the angle of attack range. The following parameters should be explored as they can impact the numerical-experimental comparison as a structural model: geometry, material properties (isotropic or orthotropic), structural boundary conditions, flow 3D effects and confinement in the test section.

5.2. Cavitation Control

In addition, the effect of the internal pressure on cavitation is analyzed. The effect of the internal pressure on the theoretical cavitation inception is numerically predicted using the FSI algorithm. The lift coefficient versus the opposite of the minimum pressure coefficient ($-C_{pmin}$) for the compliant hydrofoil under various internal pressures is shown in Figure 18. The internal pressure has a direct influence on the theoretical cavitation inception, particularly for the lift coefficients above -0.1.

Experimentally, the cavitation inception for an internal pressure equal to $\Delta P^* = 3.1$, an angle of attack $\alpha = 7.4°$ and a cavitation parameter $\sigma = 3.8$ at a Reynolds number of 10^6 is shown in Figure 19a. For identical conditions, it can be shown that cavitation disappears (Figure 19b) by only decreasing the internal pressure to $\Delta P^* = 1.71$. The corresponding hydrofoil shapes at cavitation inception and desinence is shown in Figure 20. This shape modification of the hydrofoil leads to a slight decrease in the lift coefficient from 0.805 to 0.754.

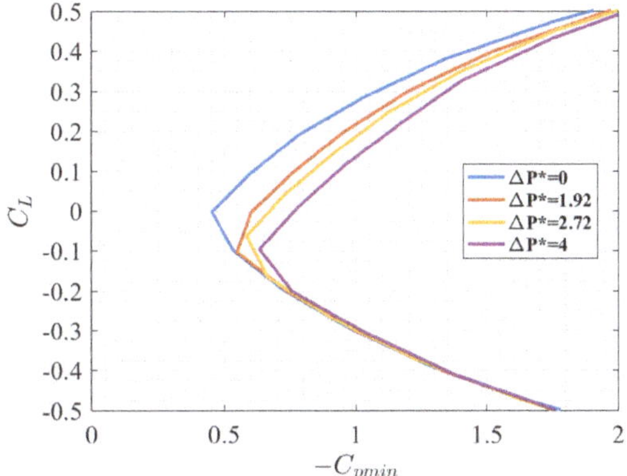

Figure 18. Numerical cavitation bucket, $Re = 0.75 \times 10^6$, $\Delta P^* = 0$, $\Delta P^* = 1.92$, $\Delta P^* = 2.72$ and $\Delta P^* = 4$.

(**a**). Cavitation inception
$C_L = 0.805$
$\Delta P^* = 3.1$

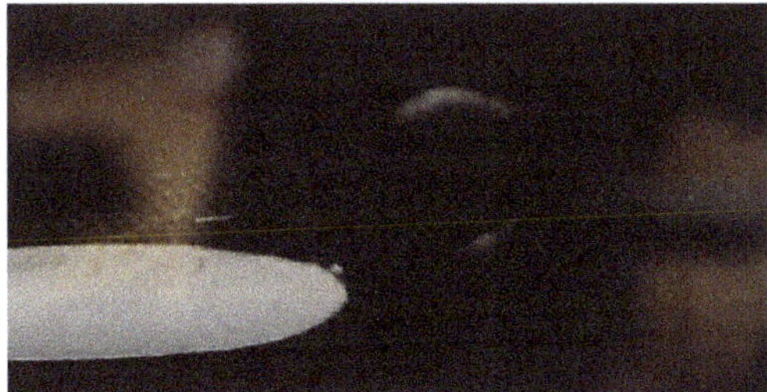

(**b**). Cavitation desinence
$C_L = 0.754$
$\Delta P^* = 1.71$

Figure 19. Cavitation inception and desinence on a compliant composite hydrofoil, $Re = 10^6$, $\alpha = 7.4°$ and $\sigma = 3.8$.

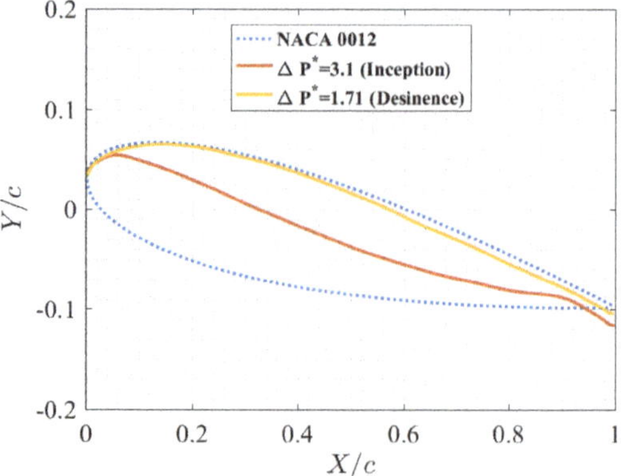

Figure 20. Experimental hydrofoil shapes at cavitation inception and desinence at $Re = 10^6$, $\alpha = 7.4°$ and $\sigma = 3.8$.

From the theoretical cavitation bucket (Figure 18), it is noted that the internal pressure has an effect on the cavitation inception for a positive lift coefficient. For this reason, the numerical and the experimental cavitation buckets are compared for a Reynolds number $Re = 1.35 \times 10^6$, positive lift coefficient and two internal pressures. The numerical and experimental cavitation buckets of a the symmetrical hydrofoil and a deformed one under internal pressure $\Delta P^* = 0.59$ are presented in Figure 21. It is shown that the numerical results concur with the experimental ones for the range of lift coefficient of the present analysis. Furthermore, it can be noted that the internal pressure has an effect on the non-cavitation domain. The differences between the numerical and the experimental results could be explained by some pressure fluctuations in the test section.

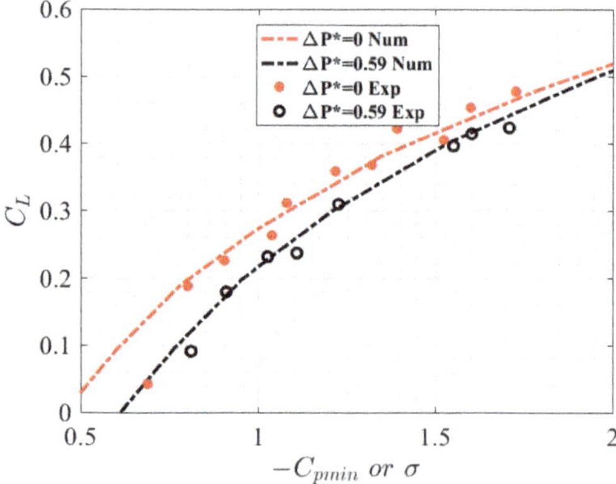

Figure 21. Experimental and numerical cavitation buckets of a symmetrical hydrofoil and a deformed one under internal pressure $\Delta P^* = 0.59$, $Re = 1.35 \times 10^6$.

6. Conclusions

In this paper, an experimental and numerical study has been presented in order to assess the effect of the internal pressure on the hydrodynamic performance of a compliant composite hydrofoil. The compliant hydrofoil was controlled by a cavity driven by pressure (suction). It was tested in a cavitation tunnel at $Re = 0.75 \times 10^6$ at different angles of attack. Firstly, the hydrofoil deformations were measured in open air using digital image correlation (DIC-3D) and a micrometric touch probe.

Secondly, experiments were performed in a hydrodynamic tunnel where lift and drag coefficients were measured using a hydrodynamic balance together with the compliant skin deformation using a scanning laser measurement system.

A 2D numerical fluid-structure coupling is developed in accordance with the experiments. It was based on an iterative method coupling the flow solver Xfoil and the structural solver ANSYS-Mechanical. Experiments and simulations were carried out at different angles of attack and various imposed internal pressures.

The experiments show that internal low pressure variation has a significant effect on the hydrofoil shape and thus on the hydrodynamic forces. Experimentally, the internal low pressure variation leads to section variations (thickness, camber) together with a slight overall twisting and a small flap effect at the trailing edge. The increase of the suction in the cavity tends to increase the lift coefficient and the lift/drag ratio but at the expense of increased minimum pressure and hence the incipient cavitation number. It is also shown that cavitation can be controlled to some extent by changing only the internal pressure for a given angle of attack and a given flow velocity. The increase of the suction in the cavity tends also to decrease the slopes of the lift coefficient and the lift/drag ratio. A response surface depending on the the internal pressure and the angle of attack can be simply derived from a linear approximation.

The numerical cavitation bucket predicted from the FSI algorithm was compared to the experimental one and the results correlated. The morphing hydrofoil model developed in this work based on a compliant composite structure driven by an internal pressure has provided encouraging results. It allows the hydrodynamics forces and the cavitation to be controlled, paving the way for optimization methods to enhance hydrodynamic performances based on Fluid-Structure Interactions.

Author Contributions: M.A.F. carried out the experiments, developed the numerical tool with a help from F.D. for the whole software part, and wrote the manuscript under the supervision of all authors. P.C. manufactured the hydrofoil and characterized it with digital image correlation. J.A.A. set up the different measurement systems at IRENav (cavitation tunnel with the laser distance measurement system and the hydrodynamic scale with a help from B.A.). All authors discussed the results and commented on the manuscript.

Funding: This research received no external funding.

Acknowledgments: The authors would like to thank the technical staff of the French Naval Academy Research Institute, as well as Research Institute in Civil Engineering and Mechanics (Saint-Nazaire) for their support to this study.

Conflicts of Interest: The authors declare no conflict of interest.

Nomenclature

CFD	Computational Fluid Dynamics.
CSD	Computational Structural Dynamics.
FSI	Fluid-Structure Interaction.
α	angle of attack [°].
ϵ	convergence criteria.
ρ	fluid density $[kg/m^3]$.
σ	cavitation number: $\sigma = \frac{P-P_v}{\frac{1}{2}\rho V^2}$.
c	hydrofoil chord [m].
C_D	drag coefficient: $C_D = \frac{D}{\frac{1}{2}\rho V^2 s}$.
C_L	lift coefficient: $C_L = \frac{L}{\frac{1}{2}\rho V^2 s}$.
D	drag force [N].
$\Delta P = P_{atm} - P_{internal}$	difference between the atmospheric reference and internal pressure [bar].
$\Delta P^* = \Delta P/q$	non dimensional internal pressure.

e	hydrofoil span [m].
L	lift force [N].
P	pressure [bar].
q	dynamic pressure: $q = \frac{1}{2}\rho V^2$.
Re	Reynolds number: $Re = Vc/\nu$.
s	hydrofoil planform.
U	total displacement [m].
V	inflow velocity [m/s].
X, Y, Z	foil coordinates [m].
ν	kinematic viscosity [m^2/s].
τ	maximum thickness [m].

References

1. Coraddu, A.; Oneto, L.; Baldi, F.; Anguita, D. Vessels fuel consumption: A data analytics perspective to sustainability. In *Soft Computing for Sustainability Science*; Springer: Cham, Switzerland, 2018; pp. 11–48.
2. Aramendia, I.; Saenz-Aguirre, A.; Fernandez-Gamiz, U.; Zulueta, E.; Lopez-Guede, J.M.; Boyano, A.; Sancho, J. Gurney Flap Implementation on a DU91W250 Airfoil. *Multidiscip. Dig. Publ. Inst. Proc.* 2018, *2*, 1448. [CrossRef]
3. Aramendia, I.; Fernandez-Gamiz, U.; Zulueta, E.; Saenz-Aguirre, A.; Teso-Fz-Betoño, D. Parametric study of a gurney flap implementation in a du91w (2) 250 airfoil. *Energies* **2019**, *12*, 294. [CrossRef]
4. Weisshaar, T.A. Morphing aircraft systems: historical perspectives and future challenges. *J. Aircr.* **2013**, *50*, 337–353. [CrossRef]
5. Dimino, I.; Lecce, L.; Pecora, R. *Morphing Wing Technologies: Large Commercial Aircraft and Civil Helicopters*; ELSEVIER: Amsterdam, The Netherlands, 2017.
6. Brailovski, V.; Terriault, P.; Coutu, D.; Georges, T.; Morellon, E.; Fischer, C.; Bérubé, S. Morphing laminar wing with flexible extrados powered by shape memory alloy actuators. In Proceedings of the ASME 2008 Conference on Smart Materials, Adaptive Structures and Intelligent Systems, Ellicott City, MD, USA, 28–30 October 2008; pp. 615–623.
7. Woods, B.K.; Parsons, L.; Coles, A.B.; Fincham, J.H.; Friswell, M.I. Morphing elastically lofted transition for active camber control surfaces. *Aerosp. Sci. Technol.* **2016**, *55*, 439–448. [CrossRef]
8. Kaul, U.K.; Nguyen, N.T. Drag characterization study of variable camber continuous trailing edge flap. *J. Fluids Eng.* **2018**, *140*, 101108. [CrossRef]
9. Young, Y.L.; Motley, M.R.; Barber, R.; Chae, E.J.; Garg, N. Adaptive composite marine propulsors and turbines: progress and challenges. *Appl. Mech. Rev.* **2016**, *68*, 060803. [CrossRef]
10. Brennen, C.E. *Cavitation and Bubble Dynamics*; Oxford University Press: New York, NY, USA, 1995.
11. Young, Y.L.; Chae, E.J.; Akcabay, D.T.; Deniz, T. Hybrid algorithm for modeling of fluid-structure interaction in incompressible, viscous flows. *Acta Mech. Sin.* **2012**, *28*, 1030–1041. [CrossRef]
12. Akcabay, D.T.; Chae, E.J.; Young, Y.L.; Ducoin, A.; Astolfi, J.A. Cavity induced vibration of flexible hydrofoils. *J. Fluids Struct.* **2014**, *49*, 463–484. [CrossRef]
13. Amromin, E. Impact of hydrofoil material on cavitation inception and desinence. *J. Fluids Eng.* **2017**, *139*, 061304. [CrossRef]
14. Mouritz, A.P.; Gellert, E.; Burchill, P.; Challis, K. Review of advanced composite structures for naval ships and submarines. *Compos. Struct.* **2001**, *53*, 21–42. [CrossRef]
15. Zarruk, G.A.; Brandner, P.A.; Pearce, B.W.; Phillips, A.W. Experimental study of the steady fluid–structure interaction of flexible hydrofoils. *J. Fluids Struct.* **2014**, *51*, 326–343. [CrossRef]
16. Maljaars, P.; Bronswijk, L.; Windt, J.; Grasso, N.; Kaminski, M. Experimental Validation of Fluid–Structure Interaction Computations of Flexible Composite Propellers in Open Water Conditions Using BEM-FEM and RANS-FEM Methods. *J. Mar. Sci. Eng.* **2018**, *6*, 51. [CrossRef]
17. Smith, S.M.; Venning, J.A.; Giosio, D.R.; Brandner, P.A.; Pearce, B.W.; Young, Y.L. Cloud cavitation behavior on a hydrofoil due to fluid-structure interaction. *J. Fluids Eng.* **2019**, *141*, 041105. [CrossRef]
18. Zhu, W.; Gao, H. A Numerical Investigation of a Winglet-Propeller using an LES Model. *J. Mar. Sci. Eng.* **2019**, *10*, 333. [CrossRef]

19. Young, Y.L.; Garg, N.; Brandner, P.A.; Pearce, B.; Butler, D.; Clarke, D.; Phillips, A.W. Load-dependent bend-twist coupling effects on the steady-state hydroelastic response of composite hydrofoils. *Compos. Struct.* **2018**, *189*, 398–418. [CrossRef]
20. Young, Y.L.; Garg, N.; Brandner, P.; Pearce, B.; Butler, D.; Clarke, D.; Phillips, A. Material Bend-Twist Coupling Effects on Cavitating Response of Composite Hydrofoils. In Proceedings of the Tenth International Cavitation Symposium (CAV2018), Baltimore, MD, USA, 14–16 May 2018; pp. 14–16.
21. Ducoin, A.; Astolfi, J.A.; Sigrist, J.F. An experimental analysis of fluid structure interaction on a flexible hydrofoil in various flow regimes including cavitating flow. *Eur. J. Mech.* **2012**, *36*, 63–74. [CrossRef]
22. Garg, N.; Lyu, Z.; Dhert, T.; Martins, J.; Young, Y.L. High-fidelity hydrodynamic shape optimization of a 3-d morphing hydrofoil. In Proceedings of the Fourth International Symposium on Marine Propulsors, Austin, TX, USA, 31 May–4 June 2015.
23. Garg, N.; Kenway, G.K.; Martins, J.R.; Young, Y.L. High-fidelity multipoint hydrostructural optimization of a 3-D hydrofoil. *J. Fluids Struct.* **2017**, *71*, 15–39. [CrossRef]
24. Giovannetti, L.M.; Banks, J.; Ledri, M.; Turnock, S.R.; Boyd, S.W. Toward the development of a hydrofoil tailored to passively reduce its lift response to fluid load. *Ocean. Eng.* **2018**, *167*, 1–10. [CrossRef]
25. Garg, N.; Pearce, B.W.; Brandner, P.A.; Phillips, A.W.; Martins, J.R.; Young, Y.L. Experimental investigation of a hydrofoil designed via hydrostructural optimization. *J. Fluids Struct.* **2019**, *84*, 243–262. [CrossRef]
26. Lelong, A.; Guiffant, P.; Astolfi, J.A. An Experimental Analysis of the Structural Response of Flexible Lightweight Hydrofoils in Cavitating Flow. *J. Fluids Eng.* **2018**, *140*, 021116. [CrossRef]
27. Lelong, A.; Guiffant, P.; Astolfi, J.A. An Experimental Analysis of the Structural Response of Flexible Lightweight Hydrofoils in Various Flow Conditions. In Proceedings of the 16th International Symposium on Transport Phenomena and Dynamics of Rotating Machinery, Honolulu, HI, USA, 10–15 April 2016.
28. Sacher, M.; Durand, M.; Berrini, E.; Hauville, F.; Duvigneau, R.; Le Maitre, O.; Astolfi, J.A. Flexible hydrofoil optimization for the 35th America's Cup with constrained EGO method. *Ocean. Eng.* **2018**, *157*, 62–72. [CrossRef]
29. De Barros, S.; Fadhil, B.M.; Alila, F.; Diop, J.; Reis, J.M.L.; Casari, P.; Jacquemin, F. Using blister test to predict the failure pressure in bonded composite repaired pipes. *Compos. Struct.* **2019**, *211*, 125–133. [CrossRef]
30. Marchand, J.B.; Astolfi, J.A.; Bot, P. Discontinuity of lift on a hydrofoil in reversed flow for tidal turbine application. *Eur. J. Mech.* **2017**, *63*, 90–99. [CrossRef]
31. Drela, M. XFOIL: An analysis and design system for low Reynolds number airfoils. In *Low Reynolds Number Aerodynamics*; Springer: Berlin/Heidelberg, Germany, 1989; pp. 1–12.

Journal of
Marine Science and Engineering

Article

Transitional Flow on Model Propellers and Their Influence on Relative Rotative Efficiency

Da-Qing Li *, Per Lindell and Sofia Werner

SSPA Sweden AB, Box 24001, 40022 Göteborg, Sweden; per.lindell@sspa.se (P.L.); sofia.werner@sspa.se (S.W.)
* Correspondence: da-qing.li@sspa.se

Received: 31 October 2019; Accepted: 19 November 2019; Published: 25 November 2019

Abstract: Unexpected low value of the relative rotative efficiency η_R is sometimes noted when scaling the towing tank model-test result with the ITTC-78 method to obtain the propulsive efficiency factors of propellers. The paper explains the causes of this phenomenon. The boundary layer state of three propellers was studied by a paint test and a RANS method. The paint tests showed that the propellers in behind conditions at low Reynolds number (Rn) are covered mainly with laminar flow, which is different from open water tests conducted at a high Rn. Apart from that a moderate difference in Rn between the open water and the self-propulsion test may lead to a low η_R value, the paper points out that flow separation in behind conditions could be another significant reason for the drop of η_R for some propellers. Therefore, two factors will lead to an unexpected decrease of η_R: (1) A slightly lower open water torque interpolated from an open water test carried out at a high Rn and (2) a slightly higher torque in a self-propulsion test due to laminar flow separation near the trailing edge. The phenomenon is caused by the Rn scaled effect and closely associated with design philosophy like the blade section profile, the chord length, and chordwise load distribution.

Keywords: Reynolds number; propeller; laminar/transitional/turbulent flow; scale effects; transition model; separation; paint test method

---

1. Introduction

In 1978 the International Towing Tank Conference (ITTC) published its recommended power performance prediction method, known as ITTC-78 method, to scale the towing tank model-test results to full-scale ship performance. This method has been used for four decades. The method works well in most cases. However, when applied for propellers designed with non-conventional design philosophies, e.g., propeller with extremely low blade area ratios, unusual radial and chord wise load distributions, unusual sectional shapes, and tip fins of various types, the scaling method encounters new challenges on the Reynolds number (Rn) scaling procedure. Throughout the paper, the Rn number is based on the chord length and the resultant velocity of the blade section at 0.75R radius, with R being the propeller radius. The paper pays special attention to the low blade area propellers, where a drop of the relative rotative efficiency η_R is sometimes encountered when using the ITTC-78 method. If the drop of η_R is not physical but due to a too simple Rn-scaling scheme, it will lead to an unfavourable full-scale performance prediction. A number of investigators, for example, Tamura and Sasajima [1], Tsuda et al. [2], and Hasuike et al. [3–5] attributed this phenomenon to an Rn number difference (i.e., Reynolds scale effects) between the propeller open water test (POT) and the self-propulsion test (SPT) that causes a difference in the boundary layer flow between the two tests. Due to the low operational speed of the model ship and the propeller during an SPT test (as required by Froude's scaling law), the Rn number in an SPT is lower than that in a POT. When determining the relative rotative efficiency η_R, an open water torque coefficient K_{Q0} needs to be extracted from the open water (POW) characteristic curves obtained from a POT. If the POT is carried out at a high Rn as proposed in the ITTC-78 method, the extracted K_{Q0} may not be fully representative of the propeller characteristics at the low Rn as used in the SPT. For this reason, an alternative scaling method, called "2POT method," was proposed and used by a few ITTC member organizations. It consists of performing two sets of POTs, one at a low Rn equal to the Rn in an SPT test and the other POT at a sufficiently high Rn. The low Rn POW data is used to analyse the SPT results (i.e., determining wake fraction w_{Tm} and relative rotative efficiency η_R etc.) whereas the high Rn POW is used to scale the POW to full scale. The rest of the scaling procedure for powering prediction in the 2POT method is the same as the ITTC-78 method.

Apart from the 2POT method, there are other scaling methods that have been developed and are being tested to account for scale effects of propellers especially those of unconventional design. For example, the "Strip Method" by Streckwall et al. [6], in which a local skin friction coefficient is used to compute the sectional drag by integrating the laminar and the turbulent part of skin friction contribution. For this purpose, two friction lines were developed for each blade section, with one line applicable in the POT condition and the other applicable in the SPT condition. Helma [7] presented an extrapolation method suitable for scaling the POW of various propellers. Lücke and Streckwall [8] applied the paint test on three model propellers in behind and open water conditions to visualise the boundary layer flow structure on blades. The scaling issue and propulsion performance method were discussed accordingly. They found a high amount of laminar flow on all propellers in POT and SPT conditions at a Rn number corresponding to the SPT condition. They also noted the streamlined orientation in the SPT condition has a more gradual change in contrast to a sharp separation zone in the POT condition. Heinke et al. [9] applied an interpolation scaling procedure based on a set of POT tests carried out at multiple Reynolds numbers. Analyses of Rn scale effects and scaling methods for ducted propellers and non-conventional propellers have been addressed by Bhattacharyya1 et al. [10–12], Sánchez-Caja et al. [13], Shin and Andersen [14], and Moran-Guerrero et al. [15] recently.

With regard to CFD RANS methods for analysis of transitional flow around propellers, the most relevant work is due to Hasuike et al. [3–5], where the authors applied a laminar-kinetic-energy based k-kl-ω model to determine the laminar flow on many low blade area propellers and compared it with available experimental data. The results demonstrated the presence of a large amount of laminar flow on blades in open water and behind conditions when the propeller is operated at low Rn numbers. Baltazar et al. [16] employed a γ-Re_θ transition model to predict POW performance of a conventional and a high-skew propeller in model scale and compared the results with the model test and an SST k-ω model. Improvement on transition model in itself is still an ongoing research, as discussed, e.g., in papers by Gaggero and Villa [17], Colonia et al. [18], Lopes et al. [19] and Moran-Guerrero et al. [20].

In the standard ITTC-78 method, the POT shall be carried out at a sufficiently high Rn and the Rn should, in any case, be greater than the critical Reynolds number Rnc (normally considered to be Rnc $\approx 3 \times 10^5$). During a self-propulsion test where Froude scaling must be applied, the ship and the propeller model are operated at rather low speeds. Typically, the Rn based on sectional chord length at nominal radius 0.75R (R is the propeller radius) is limited to a range of 1~3 $x10^5$. Clearly, there is a difference in Rn between the POT and the SPT, yet the difference is moderate, and the two Rn numbers are within the same order of magnitude (10^5). This difference is much smaller than the Rn difference between a model scale (10^5) and a full-scale propeller (10^7). One question is how much the Rn scale effect will be introduced with this moderate difference for model scale propellers. Would different propellers react to this Rn difference differently? Operating a propeller at a Rn number lower than Rnc would imply the occurrence of laminar flow and possibly flow separation. On the other hand, for a propeller in the behind condition, the elevated turbulence intensity in the wake of the ship model would make the propeller more susceptible to turbulence transition. One would wonder how large a blade area is covered by laminar flow in an SPT. What is the difference in the boundary layer characteristics for a propeller in the behind condition and in the open water condition at the same Rn? What is the reason for the unexpected low η_R value and what type of propellers have such an issue?

The aim of the present work is to address the above five questions by means of model testing, combined with a RANS method coupled with a transition model. It investigated the boundary layer flow over the blades of three model propellers. The chosen propellers represent a low blade-area high-efficiency propeller (of NAKASHIMA design), and two conventional designs with a medium and a high blade area, respectively. The POT test and paint test were conducted for these propellers at a low and a high Rn number. The paint test was also performed in the behind condition to look at the difference in the near wall boundary layer flow. For the RANS method, the Menter's SST k-ω turbulence model [21] coupled with the turbulence intermittency transition model by Menter et al. [22] was used. The model is called the γ-model. The bypass transition and separated-flow transition are the most dominant transition phenomena occurring on model scale propellers. They are triggered by a variety of possible flow disturbances such as free-stream turbulence, pressure gradient, wall roughness, flow instabilities in the form of elongated streamwise streaks, and intermittent turbulent spots. For rotating bodies like the propeller, cross flow instability plays an important role in the transition onset. Since the γ-model [22] includes a term to account for the cross flow effect, it is considered appropriate for the present study.

2. Propeller Model and Model Tests

The main particulars of propellers are presented in Table 1, in which D is the propeller diameter and R is the radius. The $t_{max}/C_{0.75R}$ is the maximum thickness-to-chord ratio at the 0.75R blade section. For confidentiality reason, some data for Prop A is not shown in the table.

In a POT, the propeller model was mounted on a horizontal shaft and towed through the water at an immersion depth of the shaft center equal to one propeller diameter. During the test, the rate of revolutions was kept constant while the advanced speed was varied to achieve a series of advance ratios (J). In SPT tests, the same hull model was used for the propellers. Prop A was designed for this hull form. Therefore, the SPT for Prop A was performed at design speed whereas the SPT for the other two propellers was performed at an off-design point.

The paint test is used to detect the boundary layer state of the flow over blades. Before the test, the propellers were painted with a black paint along the leading edge (LE). During the test, the flow direction of paint is governed by the skin frictional force and affected by the inertia. For a blade area covered with laminar flow where the skin frictional force is not strong enough to make the paint following the circular movement on the blade, the inertia (apparent centrifugal force) of the paint makes it move in the outward radial direction. Therefore, the traces of the paint will mainly follow the radial direction. For a blade area covered with fully turbulent flow where the shear stress force is large enough to hold the paint along the circumferential direction, the paint traces will look more like circular lines. In the paint test, the propeller was mounted in the same way as in POT and SPT, but only one run at one speed was made each time. The test was performed at a high and a low Rn number for each propeller according to the loading conditions defined in Table 2. Photographs were taken after each test. The chosen low Rn numbers corresponded to a typical Rn used in an SPT.

Table 1. Main particulars of the propellers.

Propeller	$A_E/A_0/Z$	$P/D_{0.75R}$	$t_{max}/C_{0.75R}$	$C_{0.75R}/D$	Number of Blades
Prop A	0.10	-	-	-	-
Prop B	0.13	0.83	0.0350	0.27	4
Prop C	0.16	1.00	0.0438	0.33	5

Table 2. Loading conditions for POT.

	Prop A		Prop B		Prop C	
Condition	**Low Rn**	**High Rn**	**Low Rn**	**High Rn**	**Low Rn**	**High Rn**
N [rps]	8.1	20	6	18	5	15
V_A [m/s]	1.214	3	1	3	1	3
J [-]	0.646	0.646	0.694	0.694	0.824	0.824
Rn [x10^5]	2.06	5.08	2.15	6.46	2.24	6.71

3. Numerical Methods

3.1. Numerical Models

The viscous flow is solved by an incompressible Reynolds Averaged Navier-Stokes (RANS) method using ANSYS FLUENT 18.2. The Menter's Shear Stress Transport (SST in short) k-ω turbulence model [21] is employed for simulation of the propeller in a fully turbulent flow. The result serves as a reference for comparison with the transition model.

To simulate the laminar flow and transition to turbulence flow, Menter et al.'s intermittency transition model [22] is employed. The fraction of the time during which the flow over any point on the surface is turbulent is defined as "turbulence intermittency." This transition model, denoted as the γ-model, is a further development of the γ-Re_θ model based on the coupling of the SST model with a transport equation for the intermittency γ. It inherits the same underlying modelling concept as the γ-Re_θ model, e.g., local-correlation based transition modelling, and the correlation between transition phenomena and free stream turbulence intensity and pressure gradient. The detail is referred to in Reference [22]. The model has an option important for cross flow transition, which is used in the present work.

3.2. Numerical Schemes

The following schemes are used:

- Incompressible pressure-based solver.
- Pressure and velocity solved in a coupled manner.
- Second order discretisation for pressure gradient at face.
- QUICK scheme for all transport equations.
- Single Reference Frame for POT calculation in a steady mode.
- Sliding mesh grid interface for calculation in the behind condition in an unsteady mode.

3.3. Computational Domain and Mesh

For simulation in the POT condition, the CFD domain consists of a blade passage by utilizing the circumferential periodic boundary condition (BC) and single rotational reference frame (SRF). The distance of the domain boundaries to the center of the propeller are defined as a multiple of propeller diameter D. Namely, Inlet = 2D, Outlet = 3D, and Outer radial = 3D. The meshes generated on the two periodic surfaces are fully conformal. All meshes are of hexahedral type and generated by ICEMCFD Hexa. The non-dimensional wall-normal distance (y^+) of the first cell layer to the blade surface is kept below 1. Figure 1 shows the CFD domain for the open water calculations.

For simulation in the behind condition (only with Prop A), the domain is a rectangular box. The distance of the domain boundaries is defined as a multiple of ship length between perpendiculars Lpp as follows: Inlet = 0.7Lpp, Outlet = 2Lpp, Sides = 1Lpp from the central plane, and Bottom = 1Lpp from the free surface plane. The mesh for the computation in behind condition is generated by HexPress. The mesh on blades is refined in the wall-normal direction to achieve a $y^+ < 1$ whereas the near-wall distance to the hull surface is set to a $y^+ = 2$. The mesh count for the simulations is given in Table 3. Figure 2 shows the surface mesh on the suction side (SS) of the blade for each propeller.

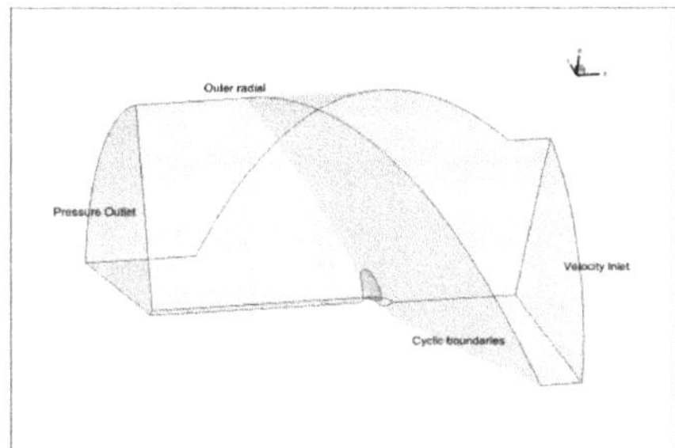

Figure 1. Computational domain for the POT calculations.

Table 3. The mesh count for simulations.

Propeller	A	B	C	A (in behind)
Number of cells [million]	5.028	3.840	4.091	8.193

3.4. Boundary Conditions

Constant velocity, turbulence intensity (*Tu*), and turbulent viscosity ratio (*TVR*) were specified at the inlet boundary, whereas a zero pressure was set at the outlet boundary. No-slip wall was set on all body surfaces. It is known that transition models are sensitive to the turbulence quantities prescribed at the inlet. A number of turbulence quantities (*Tu* and *TVR*) were tested and compared with the model test result during the initial study. The adopted *Tu* and *TVR* values are a compromise that gives reasonable agreement with the paint test results and the measured POW data.

For simulation in the POT condition, the turbulence quantities at velocity inlet were set as $Tu = 2\%$, $TVR = 6$, and an intermittency $\gamma = 1$ for the γ-model. For simulation in the behind conditions, the free surface, side, and bottom boundaries are treated as symmetry boundary conditions. The turbulence quantities at the velocity inlet were prescribed as $Tu = 5\%$, $TVR = 10$, and $\gamma = 1$.

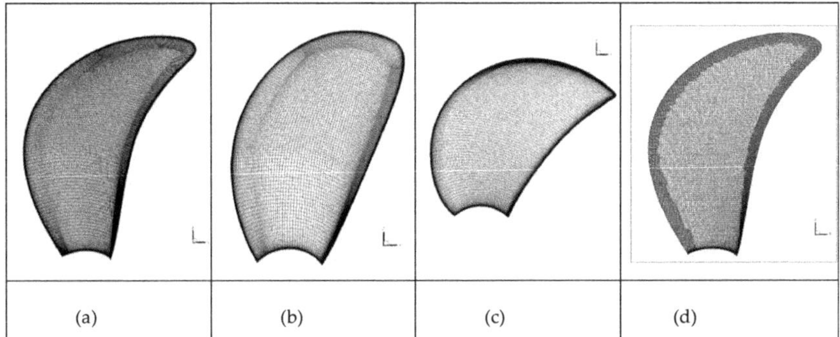

Figure 2. Surface mesh on the suction side of the propellers. (**a**) Prop A; (**b**) Prop B; (**c**) Prop C; (**d**) Prop A in SPT.

4. Results and Discussions

Computational results are compared with the model test data with regard to the near-wall boundary layer flow and the POW curves in the following. In addition, the computed pressure and skin friction coefficients are compared along the blade sections of different propellers.

4.1. Convergence of the Solution

The flow solver has stable convergence behaviour. The steady computations were run generally for about 8000 iterations. An example convergence history of the residuals from the solved equations for a POT calculation with the γ-model and the SST model is given in Figure 3. As seen in the figure, the residuals for all the equations have dropped to a level below 10^{-4} as compared with their initial values. Therefore, the iterative solution is considered well converged after 8000 iterations.

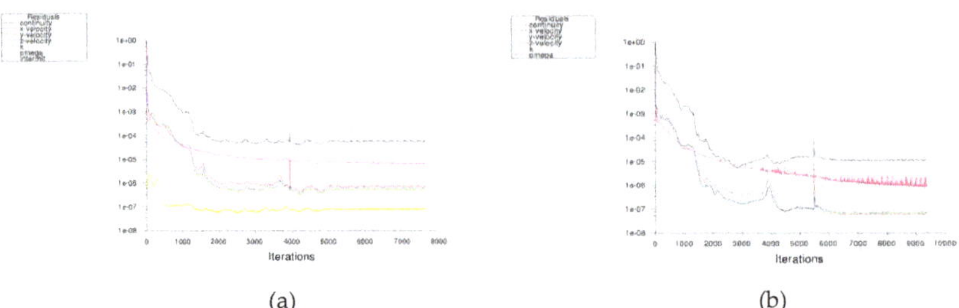

(a) (b)

Figure 3. Convergence history of residuals for the γ-model (**a**) and the SST model (**b**).

4.2. In Open Water Conditions

4.2.1. Paint Test vs. Predicted Near-Wall Flow

In each of the following figures (Figures 4–9), paint test results are presented in the first column while a contour plot of the skin friction coefficient C_f overlaid by limiting streamlines is presented in the second column. The contour plot of turbulence intermittency γ in the range of 0.02 to 0.08 is displayed in the third column. The limiting streamlines by the SST model is shown in the fourth column. The turbulence intermittency of a value equal to or less than 0.02 is used for laminar regimes according to ANSYS Fluent, so any dark blue areas with $\gamma \leq 0.02$ implies a laminar zone in the contour plot for γ. The value in the range $\gamma = [0.02, 0.08]$ corresponds roughly to the transitional zone from laminar to turbulent flow. Values above 0.08 is regarded as turbulent flow.

Prop A

For Prop A, the RANS results are compared with the photos of the paint test in Figure 4 for the low Rn and Figure 5 for the high Rn condition. In the low Rn condition (Figure 4), the paint test reveals a strong radially-going flow on both sides of blades, which indicates a status of laminar flow. Near the trailing edge (TE) on the suction side (SS), the radially-going streamlines are concentrated at the ca 0.8C chordwise location, which shows the presence of a typical cross flow separation. A very similar pattern is reproduced by the γ-model. A transitional zone is visible on the SS near the TE and near the root on the pressure side (PS) for Prop A. The streamlines predicted by the SST model show a very good agreement with the paint test on SS but are somewhat different from the experiment on PS near the root and tip region. They exhibit a more tangentially going flow.

In the high Rn case (Figure 5), the paint test shows that the flow is directed more in the circumferential direction, especially at the outer radii SS, which indicates the expected turbulent flow structure. A large discrepancy is seen for the γ-model, which computed more radially-oriented streamlines on the PS. This means that the laminar flow region is overpredicted. On the SS close to the TE region, a strong radially-going flow and a flow separation along the TE are predicted by the γ-model, but no separation can be verified from the paint test result. In the outer blade area (0.75R< r < 1.0R), the paint test indicates a fully turbulent flow whereas the γ-model exhibits an overpredicted laminar flow on the SS. Compared with the γ-model, the SST model seems to have a closer agreement with the experiment for the high Rn condition.

Prop B

Limiting streamlines on Prop B are compared with the paint test results in Figure 6 for the low Rn case and compared with the paint results in Figure 7 for the high Rn condition.

In the low Rn POT case in Figure 6, the paint test reveals that the flow is going in the radial and tangential direction on both sides. Separation occurs on the SS at a distance upstream the TE and the separation line is parallel to the TE. This concentration of streamlines is well captured by the γ-model and SST model. The latter model gave a somewhat different pattern between the concentration line and TE. The streamlines show a slightly stronger radial orientation than the paint test traces. The models also predicted a reattachment along the LE on the PS. The intermittency quantity γ shows a transitional flow on a large part of PS and a laminar flow on SS.

In the high Rn condition (Figure 7), the streamlines on the inner blade area (with r < 0.65R) still have a radially orientated velocity component. The flow on the outer blade area, after a short transition distance from the LE, turns quickly to tangential (turbulent) flow. The intermittency contour plot shows a high amount of transitional/turbulent flow on the PS. There is about 40% transitional and 60% laminar flow on the SS. Compared with the SST model, the γ-model gives better agreement with the paint test on the SS.

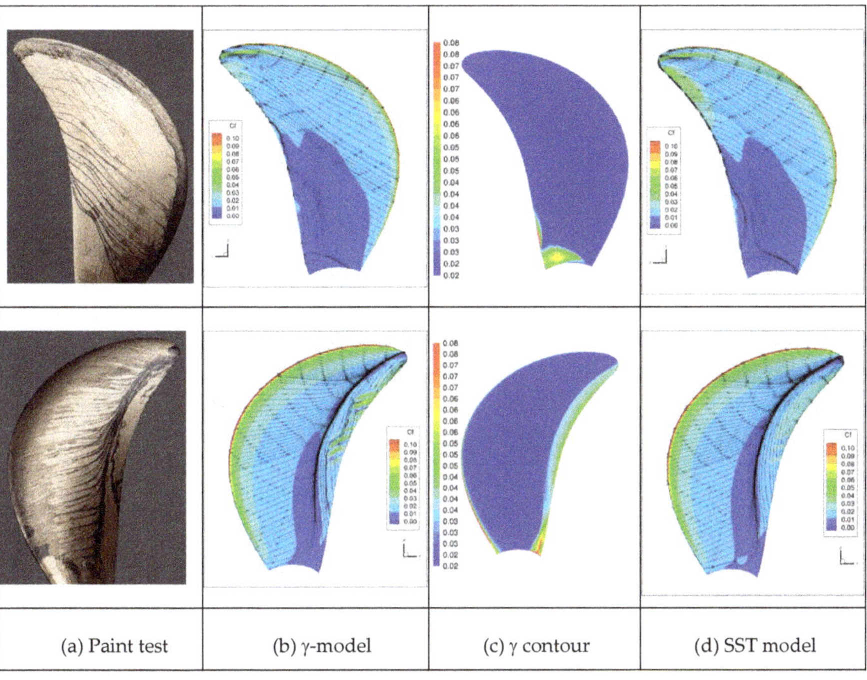

Figure 4. Prop A at low Rn. Paint test (**a**) versus limiting streamlines by the γ-model (**b**), turbulence intermittency, (**c**) and limiting streamlines by the SST model (**d**).

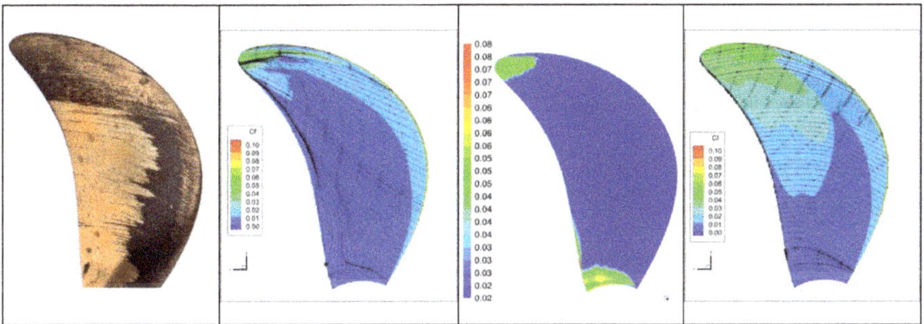

Figure 5. *Cont.*

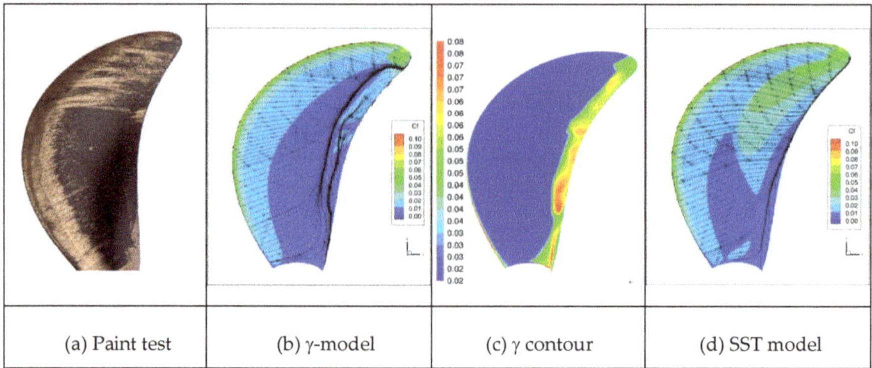

Figure 5. Prop A at high Rn. Paint test (**a**) versus limiting streamlines by the γ-model (**b**), turbulence intermittency, (**c**) and limiting streamlines by the SST model (**d**).

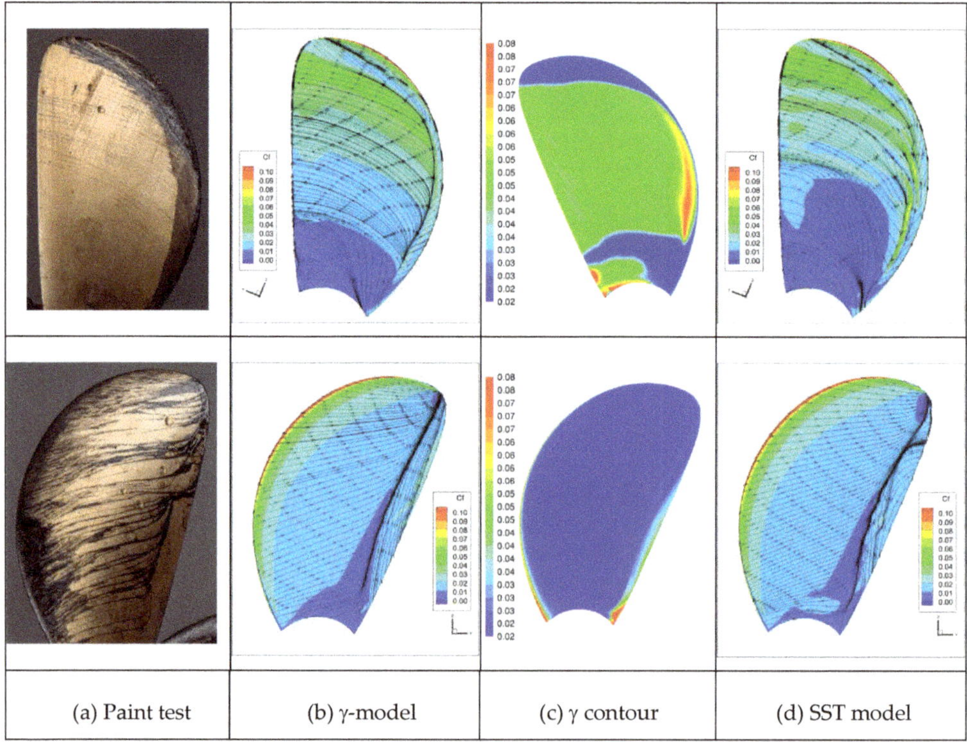

Figure 6. Prop B at low Rn. Paint test (**a**) versus limiting streamlines by the γ-model (**b**), turbulence intermittency, (**c**) and limiting streamlines by the SST model (**d**).

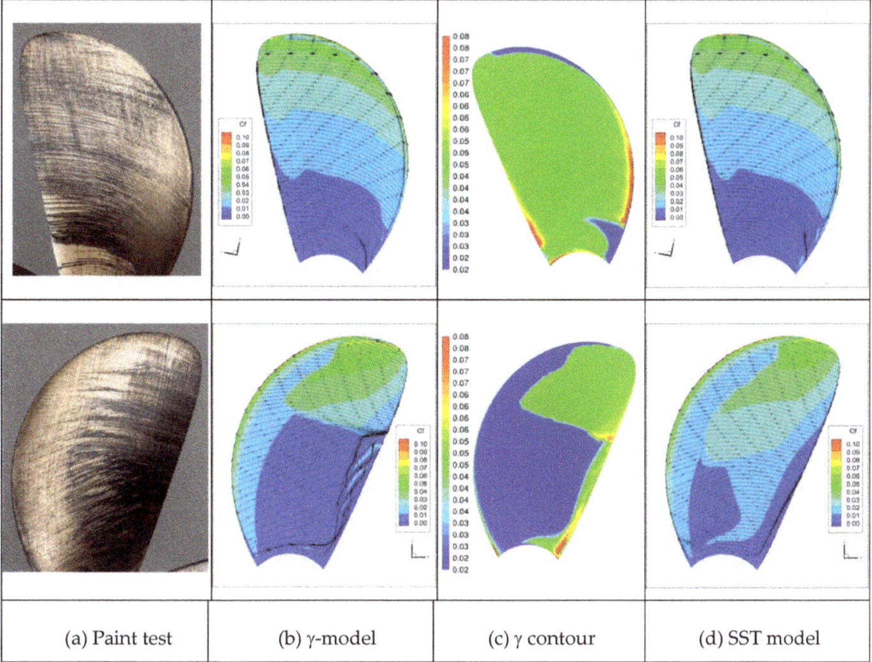

Figure 7. Prop B at high Rn. Paint test (**a**) versus limiting streamlines by the γ-model (**b**), turbulence intermittency (**c**), and limiting streamlines by the SST model (**d**).

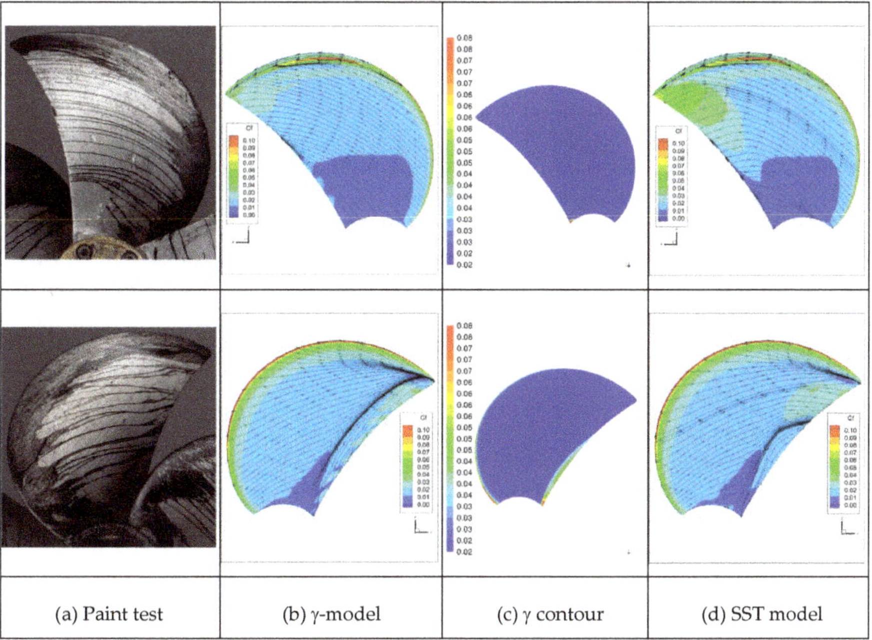

Figure 8. Prop C at low Rn. Paint test (**a**) versus limiting streamlines by the γ-model (**b**), turbulence intermittency (**c**), and limiting streamlines by the SST model (**d**).

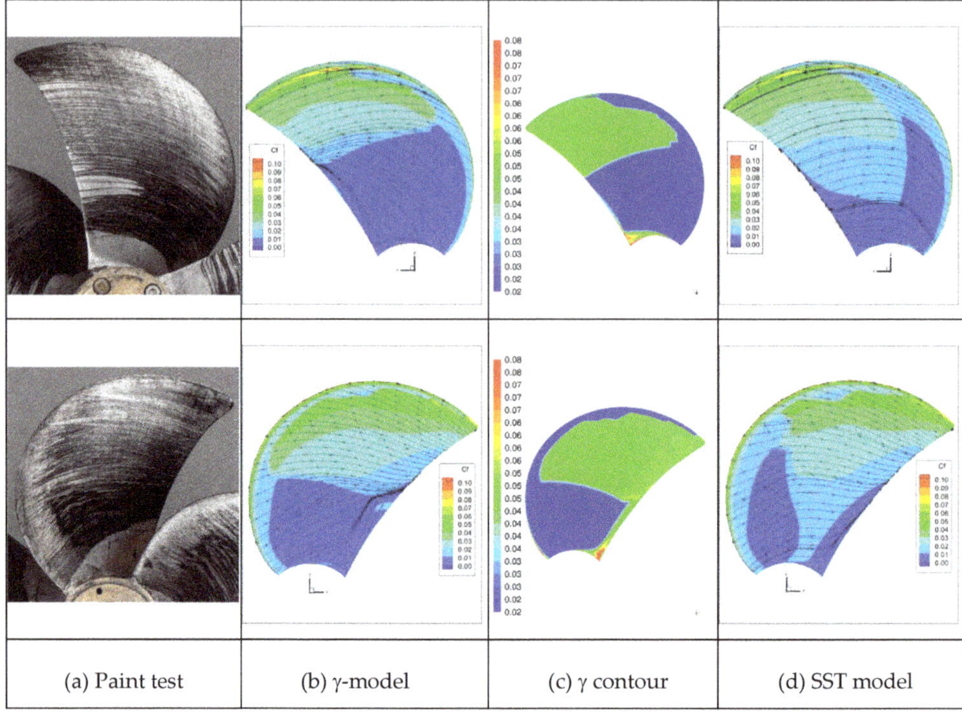

Figure 9. Prop C at high Rn. Paint test (**a**) versus limiting streamlines by the γ-model (**b**), turbulence intermittency (**c**), and limiting streamlines by the SST model (**d**).

Prop C

Limiting streamlines on Prop C are compared with the paint test results in Figure 8 for the low Rn case and in Figure 9 for the high Rn case. At low Rn, the paint traces imply the presence of both laminar and transitional flow on either side. On the SS, there are a few traces of flow separation before the flow reaches the TE. Compared with the paint test, the limiting streamlines by the γ-model demonstrate a stronger radial orientation, which means that the laminar part is over-predicted. This can also be seen from the intermittency contour plot (being entirely blue) in Figure 8c. The streamlines by the SST model exhibit some more transitional flow content near the TE on the outer blade. Both models predicted flow separation on the SS near the TE, but the separation line predicted by the SST model is shorter than that by the γ-model.

At the high Rn, more tangentially going streamlines are observed from the paint test, particularly on the PS. At the inner radii on the SS starting from the LE, some streamlines show a radial velocity component, which indicates a region of laminar flow. The limiting streamlines predicted by the γ-model are oriented more to the radial direction at the inner radii, which shows a dominance of laminar flow on the inner blade surface, whereas those predicted by the SST model appear to have closer agreement with the experiment. On the outer blade area, both models and the paint test suggest a fully developed turbulent boundary layer.

To summarize the prediction by the γ-model: at low Rn, the agreement with the paint test result is fairly good for Prop A and B, but is over-predicted in the laminar zone for Prop C. At high Rn, the agreement with the paint test is good for Prop B, but less satisfactory for Prop A and C where an overestimation of laminar flow is noted. Flow separation appears to be slightly over-predicted for all propellers at high Rn. The flow predicted by the SST model are similar with the γ-model for the low Rn cases. For the high Rn cases, the streamlines in the transition to the turbulence region correspond fairly well with the paint traces. The separation region predicted by the SST model is less than the prediction by the γ-model.

4.2.2. Flow Feature along Blade Sections

The pressure coefficients -C_p at blade section r = 0.5R and 0.7R, predicted by the γ-model in the low Rn case, are compared for Prop A and Prop C in Figure 10. The skin coefficients C_f at these sections are compared in Figure 11.

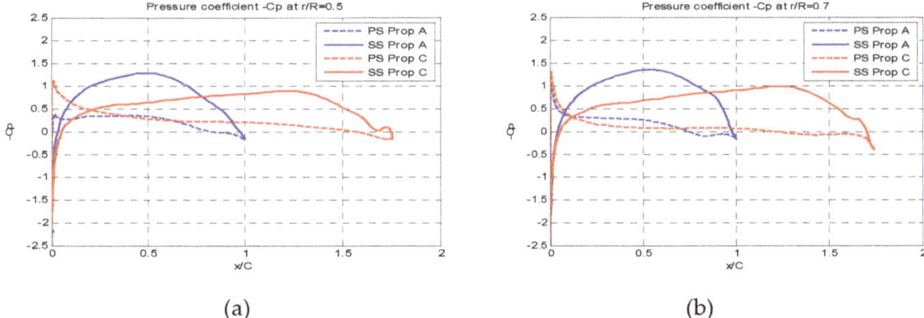

Figure 10. Comparison of pressure coefficient -C_p for Prop A and C at radius r = 0.5R (**a**) and r = 0.7R (**b**).

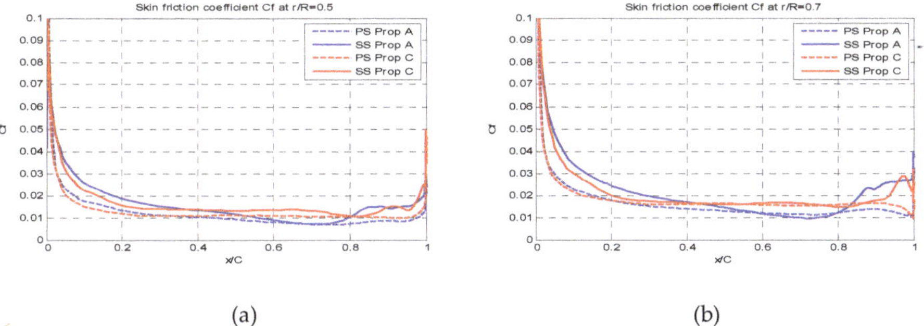

Figure 11. Comparison of skin friction coefficient C_f for Prop A and C at r = 0.5R (**a**) and r = 0.7R (**b**).

In Figure 10a, the non-dimensional chord length (x/C) of Prop C is rescaled along the abscissa with regard to that of Prop A, to show a relative length relationship with Prop A. The K_T for Prop C is a bit higher than Prop A, which means that the sectional loading on Prop C is higher than Prop A. However, looking at the downstream part of C_p curves for the two propellers in Figure 10a,b, we see that the adverse pressure gradient on the suction side (SS) of Prop A is relatively higher than Prop C. This is a rather significant difference between Prop A and C. It implies that Prop A is more prone to flow separation, particularly when the blade is working in a laminar flow that is highly sensitive to free-stream disturbance. Severe separation can be confirmed from the streamlines for Prop A in the paint test but not for Prop C (Figures 4 and 8). From the C_f curves in Figure 11, we can see that the transition onset starts at a chordwise location at x/C≈ 0.70 for Prop A and x/C≈ 0.82 for Prop C.

4.3. In Behind Condition

Paint tests of the propeller fitted to a ship model were conducted for all propellers but CFD computation was performed only for Prop A. It should be mentioned that the paint traces from the paint test in an SPT represent the averaged boundary layer flow characteristics resulting from many blade rotations and an inhomogeneous wake.

4.3.1. Paint Test Results and Predicted Streamlines

The paint test results for Prop A are compared with the limiting streamline predicted by the γ-model and the SST model in Figure 12. First, we noted from the paint test result that the near wall flow on the PS inner part of the blade goes more along the tangential direction than in the low Rn POT case (Figure 4a). Second, the near wall flow on the SS is similar yet different from that in Figure 4a. It has more radially-going content on the inner part of the blade, and more tangentially-orientated flow structure at the outer radii. The strong separation in parallel to the TE observed in the low Rn POT is not seen here. It is, however, not clear in the photo whether there is a separation near TE in the region 0.6R < r < 0.8R in Figure 12a. Similar characters were also reported by Hasuike et al. [5] and Lücke et al. [8]. The near wall flow pattern in the behind condition is much more complicated, due to a variable angle of attack to blade sections in the radial and circumferential directions, and a high turbulence level in the ship wake. The high turbulence intensity must have stabilized the flow to some extent but is not enough to change the boundary layer state from laminar to fully turbulent, so the blade is primarily covered with laminar and transitional flow.

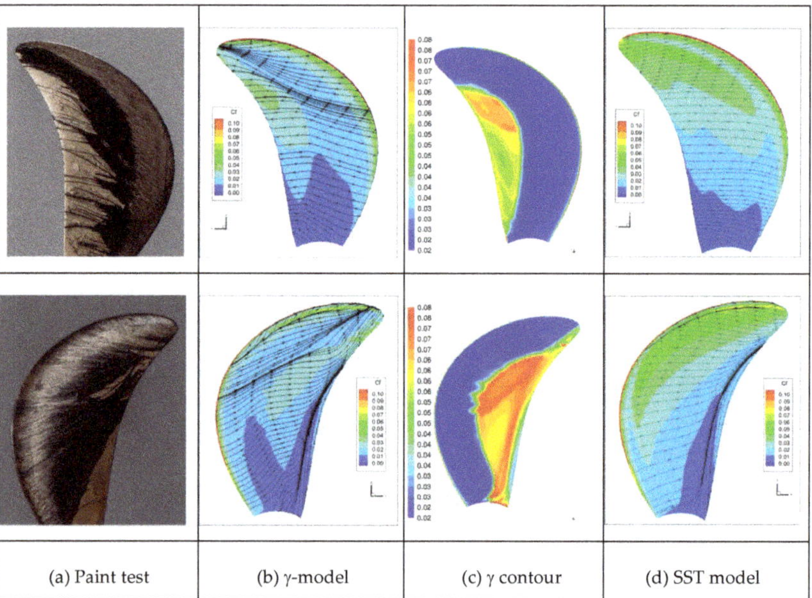

Figure 12. Prop A in the behind condition. Paint test (**a**) versus limiting streamlines by the γ-model (**b**), turbulence intermittency (**c**), and limiting streamlines by the SST model (**d**).

The γ-model predicted a laminar flow in the outer blade area on the SS whereas the paint test revealed a turbulent flow in the region r > 0.9R. In the inner blade area, the trend is the opposite: the streamlines predicted by the γ-model exhibit a lesser extent of radial orientation compared with the paint test, which indicates an under-estimation of laminar flow and an earlier transition onset (Figure 12b). This feature can also be seen in the contour plot of intermittency in Figure 12c: the red area (high γ value) means a state of transitional/turbulent flow. The paint traces, however, indicated that the area is dominated by laminar flow. Nevertheless, compared with the SST model in Figure 12d, the γ-model has slightly closer agreement with the paint test.

Both the γ-model and the SST model predicted a flow separation near TE in the region with r < 0.75R, as implied by the converging streamlines in Figure 12b,d. In addition, the γ-model predicted an LE vortex separation and its subsequent reattachment along the LE at the outer radii with r > 0.75R.

The paint test results for Prop B and C are presented in Figure 13, and they are compared with the respective open water cases in Figures 6 and 8. For Prop B, starting from LE, a laminar flow develops over about half of the blade area on the SS. Then the flow gradually changes direction toward circumferential, which indicates a transitional state of flow. However, unlike the open water case, no separation is observed in the behind condition. For Prop C, the paint traces have some similarity with those observed in the low Rn open water condition (Figure 8), yet have two discrepancies. They start with a clear and strong laminar flow pattern from the LE, spread about 1/4 chordwise over the blade. Then the flow direction is changed more rapidly to a circumferential region due to the disturbance of a high turbulence intensity in the wake. As a result, the flow over the rest of 3/4 blade area exhibits a transitional flow character. No separation is found on the SS.

(a) PS of Prop B | (b) SS of Prop B | (c) PS of Prop C | (d) SS of Prop C

Figure 13. Paint test results for Prop B (left two) and Prop C (right two) in the behind condition.

To summarize the paint test results for propellers in low-Rn open water and in the behind conditions, we noted that the near wall flow is still dominated by laminar and transitional content in the behind condition, but the amount of laminar flow has decreased and the amount of turbulent flow increased. For the two high blade area propellers (B and C), there is no separation on SS. Although separation cannot be verified for Prop A from the present paint test result, both the oil flow visualization and computational results from Hasuike et al. [3–5] confirmed that flow separation is constantly present in SPT for low blade area propellers. Examples of oil flow visualization from [5] are given in Figure 14 for two narrow blade propellers tested in POT and SPT conditions. The paint tests carried out at HSVA [8] also revealed such a difference between low and high blade area propellers. We believe that flow separation in SPT has made a low blade area propeller different from a high blade area propeller. The large blade-thickness-to-chord ratio and the high adverse pressure gradient near the TE of the low blade area propellers have made the blade more vulnerable to laminar flow separation. It is likely that the flow separation has resulted in an increased pressure drag, which leads to a higher K_{Qb} in the behind condition. This may have played a role for the unexpected drop of η_R when the standard ITTC-78 method is used. (to be explained further in §4.5)

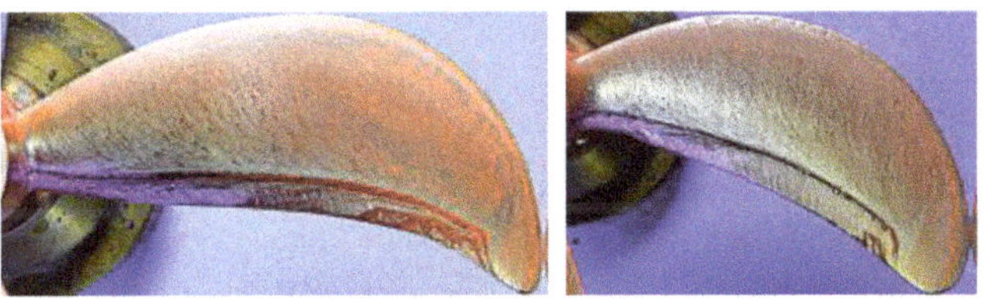

(a) Rn = $3x10^5$ in POT condition

Figure 14. *Cont.*

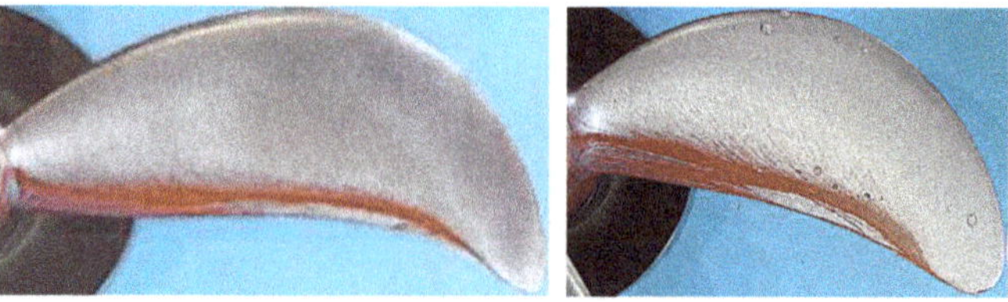

(b) Rn = 3 × 10^5 in SPT condition

Figure 14. Oil flow visualization on the suction side of two propellers in (**a**) POT conditions, and (**b**) SPT conditions (Reproduced from [5] with permission from the publisher of International Symposium on Marine Propulsors, 2017).

4.3.2. Flow Feature along Blade Sections

The pressure (C_p) and skin friction coefficient (C_f) at blade section r = 0.7R of Prop A is shown in Figure 15, and they are compared with the values in the open water case at the same J-value. Due to a higher angle of attack to blade sections when the propeller works behind the ship wake, the C_p envelop in the open water and the behind condition (Figure 15a) looks somewhat different but the integration over the entire blade should lead to the same K_T for the two cases. Compared with the C_f in an open water case (blue lines) in Figure 15b, the transition location is moved upstream to x/C≈0.42 on the SS and x/C≈0.64 on the PS in the behind condition (red lines), due to the increased turbulence intensity level in the ship wake as well as a difference in the angle of attack. It means that the near wall flow feature in POT and SPT is different despite the fact that the propeller is operating at the same Rn.

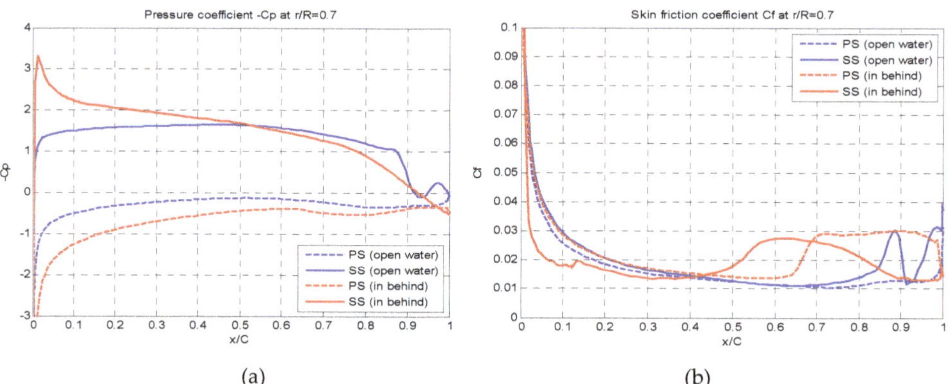

Figure 15. C_p (**a**) and C_f (**b**) at blade section 0.7R of Prop A compared in between the open water and in behind condition.

The velocity vector at a cylindrical surface of radius r = 0.7R is plotted in Figure 16, which shows the flow field near the TE of that blade section. On the SS near wall region, the velocity vectors look shorter than those on the PS because the flow has a radially-going component (i.e., the cross-flow component) not visible on the plotting plane. Separation took place near the TE on the SS, as marked by the red oval in the figure. The vector plot reveals the same character as shown in Figure 12b from another perspective.

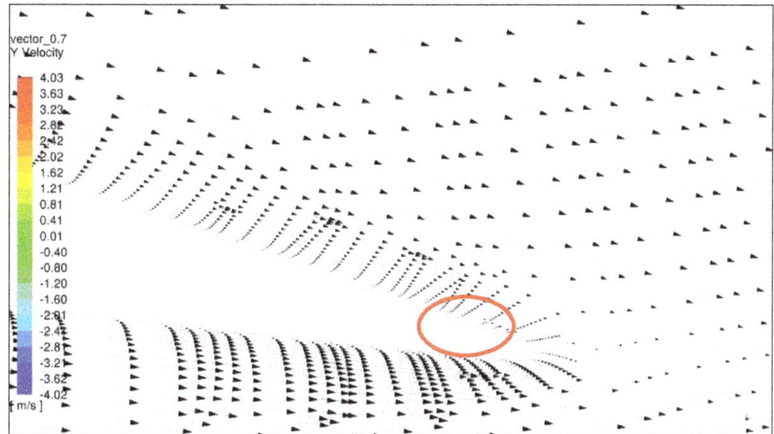

Figure 16. Velocity vector at blade section r = 0.7R for Prop A in the behind condition.

4.4. POW Characteristics at Two Rn Numbers

To show the difference in POW between two POTs tested at a low Rn and a high Rn (see Table 2 for the Rn difference), the measured POW data for Prop A, B, and C are presented in Figure 17, Figure 18, and Figure 19, respectively.

For Prop A, the K_Q at the low Rn is slightly higher than in the high Rn case for the lower half of the advance-ratio (J) range, ca J < 0.50. The K_Q becomes lower in the high J-range. Difference in K_T occurs only in the high J-range, where K_T at the low Rn is lower. For Prop B, the scale effect occurs mainly in the J-range of J > 0.45 where both K_T and K_Q at the low Rn are a bit lower than those at the high Rn, which is a typical sign that the propeller tested in the low Rn case is working below the critical Reynolds number. For Prop C, K_Q at the low Rn is higher than that at the high Rn in the range of J < 0.6. K_T and K_Q at the low Rn are lower than those at the high Rn for the range with J > 0.6. The response of POW to the Rn variation is clearly dependent on the propeller design. Different propellers exhibit different extent of scale effects due to a moderate change of Rn. Assuming a propeller is normally operated in an open water efficiency range of $0.45 < \eta_0 < 0.65$ in model tests, the difference in K_T (and K_Q) between the two sets of POW is not greater than 2.5%. The difference in η_0 can be larger, e.g., for Prop A, and η_0 is always higher at the high Rn.

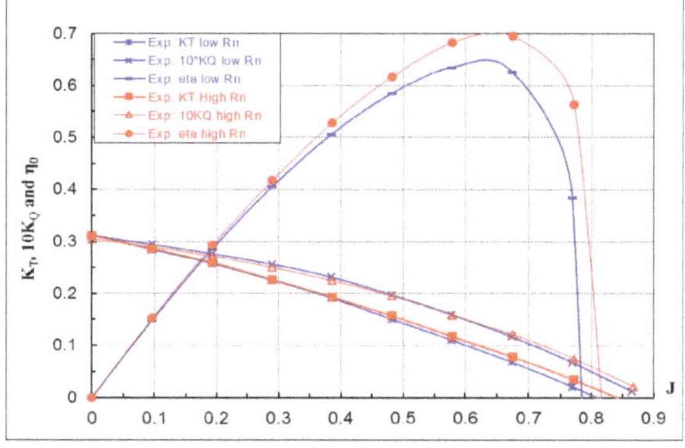

Figure 17. POW of Prop A at low and high Rn.

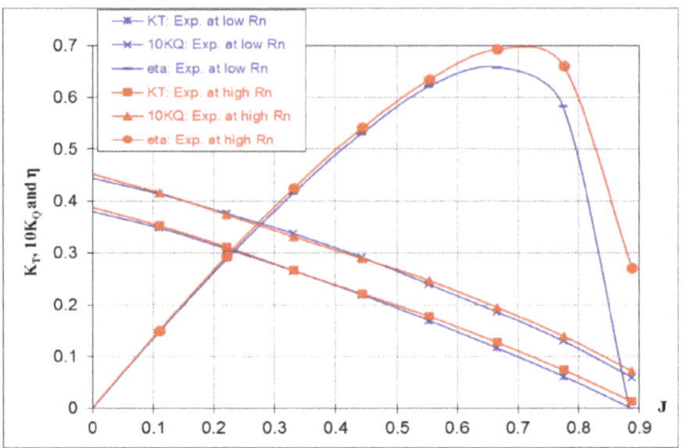

Figure 18. POW of Prop B at low and high Rn.

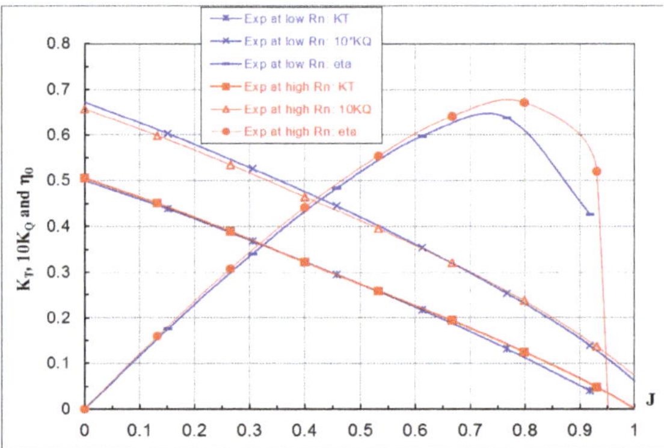

Figure 19. POW of Prop C at low and high Rn.

The K_T and K_Q predicted by the RANS method for the same J values as in the paint tests are presented for the γ-model in Table 4 and for the SST model in Table 5. The errors (ΔKT, ΔKQ, and $\Delta\eta_o$) relative to experimental data, defined as e.g., ΔKT = (K_{T_cfd}/K_{T_exp} -1), are also shown in the tables. It is seen that both models underpredict the K_T and K_Q, but the POW predicted by the γ-model has slightly closer agreement with the measured data than the SST model. On the other hand, considering that the J values in these computations are off-design values located in the high-J range. A large deviation of computed POW is, to some extent, expected. For Prop A and Prop B, the prediction errors of the γ-model are larger in the low Rn case than the high Rn case. Detailed analysis of computational results is beyond the scope of the paper and will be discussed in another paper.

Table 4. POW predicted by the γ-model.

Prop	R_n [x10^5]	J	K_T [-]	$10K_Q$ [-]	η_o [-]	ΔK_T [%]	ΔK_Q [%]	$\Delta\eta_o$ [%]
A	2.06	0.646	0.077	0.122	0.650	−4.3	−6.3	2.2
	5.08	0.646	0.088	0.126	0.715	−2.7	−4.8	2.2
B	2.15	0.694	0.098	0.169	0.642	−4.5	−2.4	−2.2
	6.46	0.694	0.109	0.179	0.671	−3.6	−1.4	−2.2
C	2.24	0.824	0.099	0.203	0.638	1.4	−3.5	5.1
	6.71	0.824	0.101	0.212	0.627	−9.2	−3.5	5.9

Table 5. POW predicted by the SST k-ω model.

Prop	R_n [x10^5]	J	K_T [-]	$10K_Q$ [-]	η_o [-]	ΔK_T [%]	ΔK_Q [%]	$\Delta\eta_o$ [%]
A	2.06	0.646	0.072	0.115	0.641	−10.4	−11.3	1.0
	5.08	0.646	0.081	0.122	0.682	−9.8	−7.5	−2.5
B	2.15	0.694	0.103	0.174	0.654	0.4	0.8	−0.4
	6.46	0.694	0.106	0.176	0.665	−5.9	−2.9	−3.1
C	2.24	0.824	0.097	0.203	0.629	−0.3	−3.6	3.5
	6.71	0.824	0.098	0.206	0.626	−12.0	−6.3	−6.0

4.5. Influence on Performance Prediction

An unusual low value of η_R is sometimes obtained when the ITTC-78 method is used in the performance prediction for low blade area propellers, as reported by Tsuda et al. [2] and Hasuike et al. [3–5]. The authors attributed this phenomenon to a Reynolds number (Rn) difference between the POT and the SPT. This is certainly one reason. However, there might be other reasons.

In Table 6, the propulsive efficiency factors predicted by the ITTC-78 method and the 2POT method (for Prop A in one project) are presented for a few cases where these propellers are used in the SPT in a number of projects. The Rn level used in the POT is marked with low Rn or high Rn. The n_m in the table is the rate of revolution during SPT. Note that, except for Prop A, each case for Prop B and Prop C was extracted from a different project with a different ship model. It represents the result at the design speed of the vessel in that project. There is no connection between cases, except for Case 1a and Case 1b of Prop A, which are from the same project and the same vessel.

Table 6. ITTC-78 vs. 2POT predicted propulsive factors.

Prop	Case	Method	V_s [kn]	POT at	n_m [1/s]	η_{om} [-]	η_R [-]	η_H [-]	Ship η_o [-]	Ship η_D [-]
A	1a	ITTC-78	14.5	high Rn	8.1	0.540	0.991	1.170	0.614	0.711
	1b	2POT	14.5	low Rn	8.1	0.508	1.029	1.180	0.610	0.741
B	2	ITTC-78	14	high Rn	6.6	0.575	1.022	1.224	0.631	0.790
	3	ITTC-78	14.5	high Rn	7.4	0.592	1.019	1.149	0.634	0.742
	4	ITTC-78	15	high Rn	7.5	0.609	1.023	1.145	0.647	0.758
C	5	ITTC-78	24	high Rn	8.9	0.657	1.020	1.080	0.689	0.759
	6	ITTC-78	22	high Rn	9.6	0.614	1.021	1.119	0.648	0.740

For Prop A, we see in Table 6 (Case 1a) an unusual low η_R (η_R = 0.991) is obtained when the ITTC-78 method is used. Using the 2POT method with a POT tested at a low Rn, an expected level of magnitude is obtained, i.e., η_R =1.029 (Case 1b). This is the situation with the low blade area propeller. If we now look at the two conventional propellers, B and C, they are both run at about the same low Rn as Prop A during SPTs for all the cases (Case 2–6). POT is run at a high Rn and the standard ITTC-78 method is applied. In all these cases, the predicted η_R has an expected order of magnitude. From the POW curves in Figures 18 and 19, we see that there is also some Rn scale effect on Prop B and C, particularly for Prop C, but why is it that Prop B and Prop C do not suffer from the drop of η_R? The answer is related to the flow separation phenomenon in SPT for the low blade area propellers, as discussed in Section 4.3.2. We observed a significant difference between the low and the high blade area propeller from the paint test result in SPT. Flow separation is present on the SS of the low blade area propeller but no separation on the high blade area propellers is present. For airfoils, laminar flow separation on the suction side leads to reduced lift and increased pressure drag. The same applies for propeller blades. The separation reduces the thrust and increases the torque. To generate the required thrust, a propeller suffering flow separation must rotate at a bit higher rpm, which results in a slightly lower J value and a lower efficiency of the propeller. The low open water efficiency value of η_{om}

in Table 6 for Prop A indicates that the Prop A seems to work less efficiently in model scale, when compared with Prop B and Prop C.

There are two factors that can lead to a decrease of η_R: (a) A difference in POW curve due to an Rn difference in a POT and an SPT, and (b) flow separation on blades during an SPT. As per definition, the relative rotative efficiency η_R is defined as:

$$\eta_R = \frac{K_{Q0}}{K_{Qb}} \tag{1}$$

where K_{Q0} is the torque coefficient interpolated from the POW curve using the K_T-identity method and K_{Qb} is the measured torque coefficient during the SPT. If a POW curve from a POT carried out at a high-Rn is used for interpolation in the ITTC-78 method, the minor Rn difference between POT and SPT can give rise to a lower K_{Q0}, which leads to a decrease of η_R. On the other hand, a higher K_{Qb} can also lead to a drop of η_R. Occurrence of flow separation near the TE implies a higher pressure drag on blades and, hence, an increased K_{Qb}, which leads to a lower η_R. In reality, both a higher K_{Qb} in SPT and a lower K_{Q0} interpolated from a POT tested at a high Rn might have collaboratively resulted in a decrease of η_R. The two conventional propellers do not suffer from this issue despite working at the same Rn as Prop A in SPT because their near-wall flow is not separated in the SPT due to a longer chord length and a gradual change of the adverse pressure gradient on the SS. Both factors are closely related to Rn scale effects, but what makes the difference in response to a moderate Rn variation in POT and SPT among different propellers is the design philosophy. If separation is the decisive factor on the change of η_R, in principle, any propeller that suffers from laminar separation in an SPT will likely get a low η_R with the ITTC-78 method. Therefore, this phenomenon does not necessarily occur only on low blade area propellers, and it can happen to other propellers. So far, it seems that low blade area propellers suffer more from the Rn scale effects than conventional propellers.

Using a POW curve from a low-Rn POT will influence the intersected J_{Tm} and K_{Q0}. It generally leads to a slightly higher η_R and a higher hull efficiency η_H, a slightly lower J_{Tm} and η_{mo}, and a few percentage increase of η_D. The consequence of applying the 2POT method can be seen in Table 6. The 2POT method may be able to bring the η_R back to a normal value for the low blade area propellers. However, the previous work by others [2,8] and SSPA's paint test results find that the near wall flows in the POT and SPT, performed at the same Rn, show a similarity of laminar flow character, but also reveal some differences in the flow orientation. The high turbulence intensity in ship wake has made the flow more transitional and turbulent. The shift of the transition onset location towards LE in the SPT discussed in Section 4.3.2 is clear evidence of such a difference. This means that, if a 2POT method is to be adopted, a somewhat higher Rn than that used in SPT should be chosen for use in the low Rn POT test, in order to achieve some consistency for the near-wall flow between the POT and SPT. This calls for a thorough calibration work like the one mentioned in Reference [8].

On the other hand, we have to be aware the disadvantages in model testing with the 2POT method. For example, it can be troublesome to carry out a model test in laminar and transitional flow regimes, as this type of flow is very sensitive to a change of model speed (hence, Rn number), which sets a very high requirement on the accuracy of the model speed. Laminar flow is vulnerable to the free-stream turbulence disturbance and is prone to separation. Testing in a laminar/transitional flow regime may result in flow instability and oscillating forces, and it requires special equipment to measure the forces reliably. Therefore, it is preferable not to perform a model test in the low-Rn laminar flow regime (unavoidable for SPT). This was one of the reasons to require that a POT be performed at an Rn higher than the critical Rn_c when the ITTC-78 method was developed. The challenge that ITTC members are facing is whether a better scaling method or a modified ITTC-78 method can be developed for non-conventional propellers in the future. Some initial work has just been started at SSPA.

5. Conclusions

The aim of this work was to explain the causes of the unexpected low value of η_R resulted from the scaling of model test result with the ITTC-78 method for the low blade area propellers. Model testing and a RANS method with the intermittency transition γ-model were used to study the near-wall flow characteristics in open water and behind conditions for three propellers, which represent a modern low blade area design (Prop A) and two conventional designs (Prop B and Prop C).

For the open water tests at the low Rn, the paint test results show a laminar flow dominance on blades for all the propellers. Separation is observed on the SS of Prop A and Prop B. For the POT at the high Rn, the near wall flow is a combination of laminar, transitional, and turbulent flow. There is no separation on the SS for all the propellers during the POT at the high Rn.

In the behind condition, the paint tests show that laminar and transitional flow coexist, and the amount of laminar flow has decreased. No separation occurs on Prop B and Prop C. It is unclear from the paint test if there is a separation on the SS of Prop A, but the CFD result predicts a flow separation for this propeller. SSPA's paint test results showed a similar tendency as observed at HSVA regarding the different response to the Rn difference in POT and SPT tests for the low and high blade area propellers.

We observe that two factors can lead to a decrease of η_R for low blade area propellers. (a) A moderate difference in Rn between the POT (following standard ITTC-78) and SPT that causes a small difference in POW curves. (b) The flow separation on the SS of the blade near TE. Both phenomena are caused by the Reynolds scale effects. If the experiment had been done at full scale, the phenomena would not have happened. Furthermore, the occurrence of the unexpected low η_R is closely associated with design philosophy like the blade section profile, length, thickness-to-chord-ratio, and chordwise load distribution.

The 2POT method offers a solution to bring back a too-low η_R for low blade area propellers, as it attempts to make the boundary layer state (Rn) in a POT similar with that in an SPT. If an unusual drop of η_R is observed in a prediction with the ITTC-78 method, the 2POT method may be considered as an alternative method to predict propulsive efficiency factors for propellers that suffer from the η_R problem. However, if the 2POT method is used for power performance prediction, then the correlation factors like C_N and C_P need to be re-established. With the awareness of the potential problems with the 2POT method (discussed in Section 4.5), we encourage the continuing effort in the ITTC committees and member society to develop new scaling methods for unconventional propellers. Some initial work has been started at SSPA.

The limiting streamlines predicted by the γ-model do not always agree well with the paint test results in the computed cases. An overpredicted laminar zone and delayed transition onset are observed for Prop A in the high Rn case, for Prop B in the low Rn case, and for Prop C at both Rn numbers. The inconsistency in streamline prediction is partially associated with the sensitivity of the transition model to the free-stream turbulence quantities at the inlet, and partly caused by numerical errors. The latter issue will be discussed in a separate paper. Overall, the streamlines predicted by the γ-model have slightly better agreement with the paint tests compared with the SST model in the low Rn cases. Using the turbulence intermittency threshold value as an indicator, the amount of laminar flow area on blade surfaces can be quantified approximately with the γ-model. This information confirms the laminar flow dominance on blades for all propellers at low Rn. Transition models are known to be sensitive to *Tu* and *TVR* prescribed at the velocity inlet boundary. Without the correct info of *Tu* and *TVR* at some distance upstream of the propeller LE, it is difficult for a transition model in its present form to accurately predict the onset of transition and skin friction at a low Rn. Nevertheless, a transition model provides a qualitatively better insight in the near-wall flow on blades than a fully turbulence model.

The paint test is still the most reliable and efficient method to detect the presence of laminar and turbulent flow, the onset of transition, and the occurrence of flow separation. One future work is to study the near wall flow with the present method for a propeller with other section profiles such as an Eppler-like section.

Author Contributions: All the authors contributed to the discussion of results. D.-Q.L. carried out the CFD simulations, analyses, and wrote the manuscript. P.L. and S.W. contributed to the paper with valuable comments and suggestions for revisions.

Funding: SSPA fund "Hugo Hammars fond för sjöfartsteknisk forskning" (grant number HHS 258) and "Fru Martina Lundgrens fond för sjöfartsteknisk forskning" (grant number ML 105) financed this work.

Acknowledgments: The permission by Nakashima Propeller Co. Ltd. to use one of their propellers in the study is gratefully acknowledged. The financial support from the funding organizations are gratefully acknowledged.

Conflicts of Interest: The authors declare no conflict of interest. The funders had no role in the design of the study, in the collection, analyses, or interpretation of data, in the writing of the manuscript, or in the decision to publish the results.

Abbreviations

The following abbreviations are used in this manuscript.

ITTC	International Towing Tank Conference
POT	Propeller Open water Test
2POT	Two Sets of Propeller Open Water Tests
POW	Propeller Open Water Characteristics
SPT	Self-Propulsion Test
LE	Leading Edge
TE	Trailing Edge
PS	Pressure Side
SS	Suction Side

References

1. Tamura, K.; Sasajima, T. *Some Investigations on Propeller Open-Water Characteristics for Analysis of Self-Propulsion Factors*; Mitsubishi Heavy Industries, Ltd.: Tokyo, Japan, 1977.
2. Tsuda, T.; Konishi, S.; Asano, S.; Ogawa, K.; Hayasaki, K. Effect of propeller Reynolds number on self-propulsion performance. *Jpn. Soc. Nav. Archit. Ocean Eng.* **1978**, *169*, 127–136.
3. Hasuike, N.; Okazaki, A.; Yamasaki, S.; Ando, J. Reynolds effect on Propulsive Performance of Marine Propeller Operating in wake flow. In Proceedings of the 16th NuTTS 2013, Mulheim, Germany, 2–4 September 2013.
4. Hasuike, N.; Okazaki, M.; Okazaki, A. Flow characteristics around marine propellers in self-propulsion test condition. In Proceedings of the 19th NuTTS, St. Pierre d'Oleron, France, 3–4 October 2016.
5. Hasuike, N.; Okazaki, M.; Okazaki, A.; Fujiyama, K. Scale effects of marine propellers in POT and self-propulsion test conditions. In Proceedings of the 5th International Symposium on Marine Propulsors, SMP'17, Espoo, Finland, 12–15 June 2017.
6. Streckwall, H.; Lucke, T.; Bugalski, T.; Felicjancik, J.; Goedicke, T.; Greitsch, L.; Talay, A.; Alvar, M. Numerical Studies on Propellers in Open Water and behind Hulls aiming to support the Evaluation of Propulsion Tests. In Proceedings of the 19th NuTTS, St. Pierre d'Oleron, France, 3–4 October 2016.
7. Helma, S. An Extrapolation Method Suitable for Scaling of Propellers of any Design. In Proceedings of the 4th International Symposium on Marine Propulsors, SMP'15, Austin, TX, USA, 1–4 June 2015.
8. Lücke, T.; Streckwall, H. Experience with Small Blade Area Propeller Performance. In Proceedings of the 5th International Symposium on Marine Propulsors, SMP'17, Espoo, Finland, 12–15 June 2017.
9. Heinke, H.J.; Hellwig-Rieck, K.; Lübke, L. Influence of the Reynolds Number on the Open Water Characteristics of Propellers with Short Chord Lengths. In Proceedings of the 6th International Symposium on Marine Propulsors, SMP'19, Rome, Italy, 17–21 May 2019.
10. Bhattacharyya, A.; Neitzel, J.C.; Steen, S.; Abdel-Maksoud, M.; Krasilnikov, V. Influence of Flow Transition on Open and Ducted Propeller Characteristics. In Proceedings of the 4th International Symposium on Marine Propulsors, SMP'15, Austin, TX, USA, 1–4 June 2015.

11. Bhattacharyya, A.; Krasilnikov, V.; Steen, S. A CFD-based scaling approach for ducted propellers. *Ocean Eng.* **2016**, *123*, 116–130. [CrossRef]
12. Bhattacharyya, A.; Krasilnikov, V.; Steen, S. Scale effects on open water characteristics of a controllable pitch propeller working within different duct designs. *Ocean Eng.* **2016**, *112*, 226–242. [CrossRef]
13. Sánchez-Caja, A.; González-Adalid, J.; Pérez-Sobrino, M.; Sipilä, T. Scale Effects on Tip Loaded Propeller Performance Using a RANSE Solver. *Ocean Eng.* **2014**, *88*, 607–617. [CrossRef]
14. Shin, K.W.; Andersen, P. CFD Analysis of Scale Effects on Conventional and Tip-Modified Propellers. In Proceedings of the 5th International Symposium on Marine Propulsors, SMP'17, Espoo, Finland, 12–15 June 2017.
15. Moran-Guerrero, A.; Gonzalez-Adalid, J.; Perez-Sobrino, M.; Gonzalez-Gutierrez, L. Open Water results comparison for three propellers with transition model, applying crossflow effect, and its comparison with experimental results. In Proceedings of the 5th International Symposium on Marine Propulsors, SMP'17, Espoo, Finland, 12–15 June 2017.
16. Baltazar, J.; Rijpkema, D.; De Campos, J.F. On the use of the γ-Re_θ transition model for the prediction of propeller performance at model-scale. In Proceedings of the 5th International Symposium on Marine Propulsors, SMP'17, Espoo, Finland, 12–15 June 2017.
17. Gaggero, S.; Villa, D. Improving model scale propeller performance prediction using the k-kL-w transition model in OpenFOAM. *Intl. Shipbuild. Prog.* **2018**, *65*, 187–226. [CrossRef]
18. Colonia, S.; Leble, V.; Steijl, R.; Barakos, G. Assessment and Calibration of the g-Equation Transition Model at Low Mach. *AIAA J.* **2017**, *55*, 1126–1139. [CrossRef]
19. Lopes, R.; Eca, L.; Vaz, G. On the Decay of Free-stream Turbulence Predicted by Two-equation Eddy-viscosity Models. In Proceedings of the 20th NuTTS, Wageningen, The Netherlands, 1–3 October 2017.
20. Moran-Guerrero, A.; Gonzalez-Gutierrez, L.M.; Oliva-Remola, A.; Diaz-Ojeda, H.R. On the influence of transition modeling and crossflow effects on open water propeller simulations. *Ocean Eng.* **2018**, *156*, 101–119. [CrossRef]
21. Menter, F.R. Two-Equation Eddy-Viscosity Turbulence Models for Engineering Applications. *AIAA J.* **1994**, *32*, 1598–1605. [CrossRef]
22. Menter, F.R.; Smirnov, P.E.; Liu, T.; Avancha, R. A one-equation local correlation-based transition model. *Flow Turbul. Combust.* **2015**, *95*, 583–619. [CrossRef]

Journal of
Marine Science and Engineering

Article

The Effect of Rudder Existence on Propeller Eccentric Force

Gisu Song, Hyounggil Park and Taegoo Lee *

Samsung Heavy Industries Co., Ltd., 217 Munji-ro, Yuseong-gu, Daejeon 34051, Korea; chisu.song@samsung.com (G.S.); h-gil.park@samsung.com (H.P.)
* Correspondence: tag.lee@samsung.com; Tel.: +82-42-865-4377

Received: 3 November 2019; Accepted: 4 December 2019; Published: 12 December 2019

Abstract: In order to design a safe shafting system in a ship, it is vital to precisely predict load on stern tube bearing. It is well known that load on stern tube bearing is directly influenced by the eccentric force of a propeller. In this paper, the effect of rudder existence on propeller eccentric force was studied based on numerical analysis with a 10,000 TEU class container vessel. To obtain propeller eccentric force, numerical simulations including propeller rotation motion using a sliding mesh technique were carried out. When a ship is turning, propeller eccentric force significantly changes compared to those of straight run. For starboard turning especially, the propeller vertical moment was decreased by about 50% due to the existence of a rudder compared to that without a rudder. In contrast, as for port turning, the results of simulations with and without a rudder were similar to each other. This difference is fundamentally due to the interaction between the direction of propeller rotation and the inflow direction to a propeller. Based on this study, it is inferred that the influence of appendages around a propeller need to be considered to ensure the reliable prediction of propeller eccentric force.

Keywords: propeller eccentric force; shaft force; maneuvering; rudder

---

1. Introduction

The safe design of a shafting system on a ship has been a critical issue for a long time. If unexpected problems on a shafting system occur on the voyage under harsh sea conditions, the safety of people on the ship would be at risk. With this reason in mind, a shafting system has been designed conservatively according to the strict guidance of classification societies. A shafting system consists of a propeller, a shaft, bearings and so on. The load on stern tube bearing located in front of a propeller is directly related to the eccentric force of a propeller [1]. Since a propeller rotates in a non-uniform wake field, eccentric force on a propeller is continuously provided and they are changed dramatically during ship turning. Although a number of processes to calculate or check for the design of a shafting system exist, unintended problems, such as stern tube bearing damage, are rarely but still addressed. Moreover, classification societies such as ABS [2] or DNV-GL [3] have published a revised guideline for the safe design of a shafting system. The predicted eccentric force of a propeller from numerical simulations is used to calculate the stern tube bearing load to ultimately secure the safety of a shafting system. For this purpose, detail conditions, including draft and ship speed for reliable numerical simulations, are strictly clarified in the guidelines of most of the classification societies. Besides, some researchers have investigated a safe shafting system for a long time in line with academic interests or engineering needs. Kuroiwa et al. [4] proposed a quasi-steady method to estimate the wake into the propeller during ship turning and showed that the propeller shaft force evaluated by the proposed method was similar to the measured value from the model test. Vartdal et al. [5] analyzed the lateral force on the propeller during ship turning motion by full-scale measurement in various ship types. They calculated the contact pressure in the tail shaft bearing using the FEM simulation. Moreover, they asserted that the CFD simulation can be an effective method of predicting these forces with reasonable accuracy. Ui [1] investigated the influence of propeller force on shaft alignment based on the data from 35 vessels of 10 ship-yards participating in the JIME research committee. He examined two representative trouble cases; (1) A hard turn to starboard side, (2) An extremely shallow draught. Coraddu et al. [6] showed asymmetric propeller behavior obtained by thrust and torque measurement on a twin-screw vessel during a free running test. Different wake distribution into two propellers (leeward and windward) during the ship-turning motion was simulated by numerical method. Based on this wake, the propeller loads were evaluated using BEMT model. Shin [7] studied the effect of propeller force on propeller shaft bearing during straight run and turning. He simulated the nominal wake using commercial RANS solver and evaluated the eccentric force of a propeller using a potential code, MPUF3A. In particular, he considered the hull deflection in the shaft alignment. Ortolani et al. [8] investigated the radial bearing force exerted by the propeller during model ship operation. In particular, a free running test was carried out and the realistic radial bearing forces were measured. Dubbioso et al. [9] analyzed the turning capability of a navy supply vessel with different rudder arrangements; twin rudders with a centerline skeg and single rudder. Due to the centerline skeg in the case of the twin rudders, the inflow to the propeller was fairly different to that of the single rudder during ship turning. Lee et al. [10] investigated the propeller eccentric force and the load of the stern tube bearing on a CRP system, compared to those of conventional single propellers. They implemented the sliding mesh technique to directly obtain the eccentric force on the propeller. They showed the shaft behavior inside the Aft-stern bush bearing in the CRP system. Dubbioso et al. [11] investigated the propeller bearing load of a twin-screw frigate vessel in a straight ahead and steady turning motion. They evaluated the wake distributions using a URANS simulation with an overlapping grid approach. The bearing loads evaluated by the BEMT model and the measurements were compared, and they tended to be similar to each other. Muscari et al. [12] examined the effect of the centerline skeg and the propeller rotation direction on the propeller bearing load in a twin-screw vessel. From their study, it can be inferred that the wake is significantly affected by the appendages around the propeller during ship turning. Ortolani and Dubbioso [13,14] measured the single blade loads on propellers during a free-running operation; a straight ahead and a steady turning motion. Their experimental data provided the insight to understand the interaction between the wake and the propeller.

From many previous studies, it can be inferred that several factors are influencing propeller eccentric force. Among them, this paper focused on the effect of a rudder. To analyze the effect of rudder existence, numerical simulations involving two cases with and without a rudder were conducted under the same conditions for straight run or turning motion, and then the propeller eccentric force in these simulations was compared to each other.

2. Numerical Details

In this study, the target vessel was a 10,000 TEU class container ship, constructed in SHI and delivered in 2014. The main particulars of the ship and propeller are briefly summarized in Table 1, and Figure 1 shows the target vessel.

Table 1. Main particulars of target vessel and propeller.

Items	Values
L	286.0
B	48.2
T_d	12.5
D	9.4
Z	5

Figure 1. The target vessel.

2.1. Numerical Methodology

2.1.1. Numerical Setup

This study was mainly performed by a numerical simulation with the commercial code, STAR-CCM+. The fundamental information used in the numerical simulations is presented in Table 2.

Table 2. Numerical Setup.

Items	Description
Code	STAR-CCM+ Ver. 10.06
Governing equation	RANS
Temporal discretization	1st order
Convection term	2nd order upwind
Gradient method	Green-Gauss
Pressure–Velocity coupling	SIMPLE
Turbulence model	Reynolds stress model
Wall treatment	Wall function

2.1.2. Case I: Resistance Test

At first, the difference of resistance for the target ship between the model test and the numerical simulation was compared in the model scale. As shown in Table 3, the ship speed was defined as from 19 to 23 Kn. For this simulation, the computational domain, boundary condition and grid system are presented in Figures 2–4, respectively. To reduce the computational time and cost, half of the hull was modeled and the symmetry condition was defined in the centerline plane. The hexahedral unstructured grid system, called trimmer mesh in STAR-CCM+, was applied and the grids were clustered for the realistic elevation and propagation of waves around a ship. The total number of grids was about 3.2M. The free surface was captured by the VOF method with HRIC scheme. The time step was defined as 0.001 s and total computational time was 80 s. The number of inner iterations per time step was defined as 5. The value of y+ on thew hull surface was defined as 40, which was located in the recommended range to apply the wall function for the high y+ wall treatment model [15,16].

Table 3. Ship speed and the Froude number for the resistance test in the numerical simulation.

V_s	Fr
19	0.185
21	0.205
23	0.224
24	0.234
25	0.243

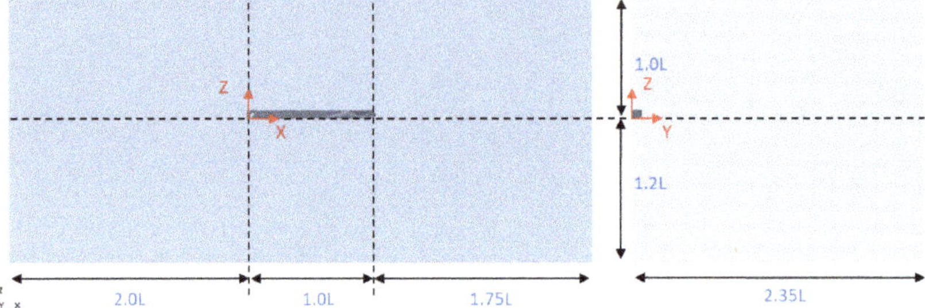

Figure 2. Computational domain (side view and front view).

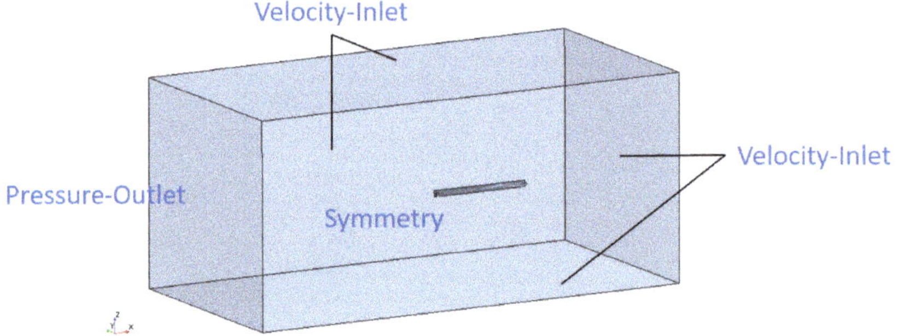

Figure 3. Boundary conditions.

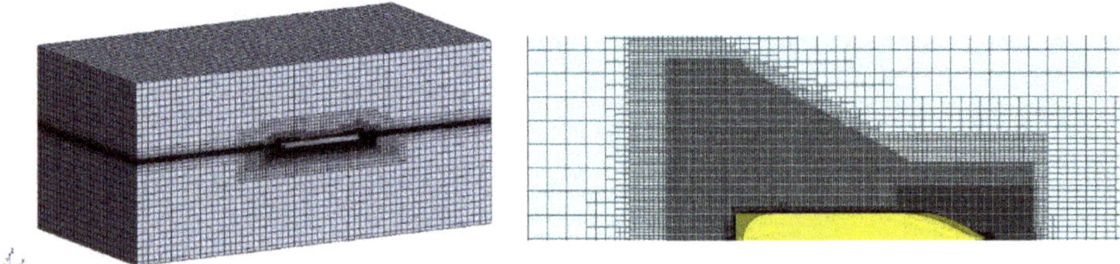

Figure 4. Grid system for the resistance simulation.

Figure 5 shows the time history of the resistance in the model scale (Rtm) for the 23Kn simulation and we judged that the value was converged at 80 s. As presented in Figure 6, the difference in the effective horsepower (EHP) calculated from the Rtm value between the model test and the numerical simulation in all ship speed range was within 2%. Figure 7 shows that the wave elevation and propagation around the Aft-hull was similar to that observed in the model test. Figure 8 presents the y+ range simulated at 23Kn ship speed; most of y+ value on the hull was distributed in the range of 30 to 50.

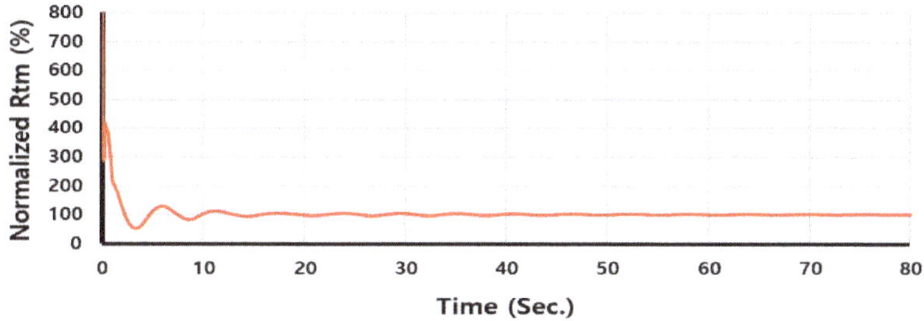

Figure 5. Time history of the resistance at the model scale (Rtm) at 23Kn.

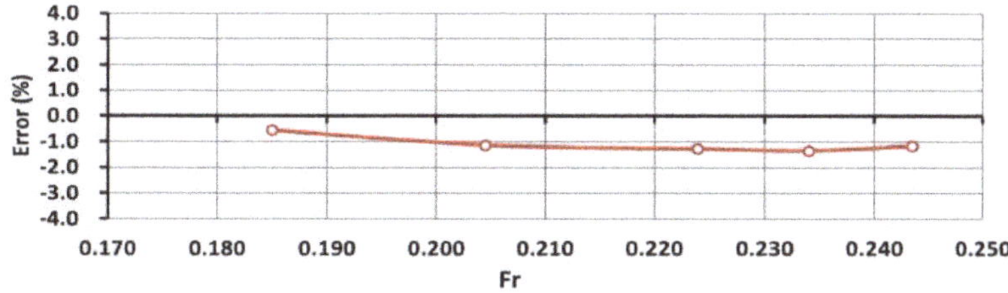

Figure 6. The difference of effective horsepower (EHP) between the model test and the CFD simulation at various ship speeds.

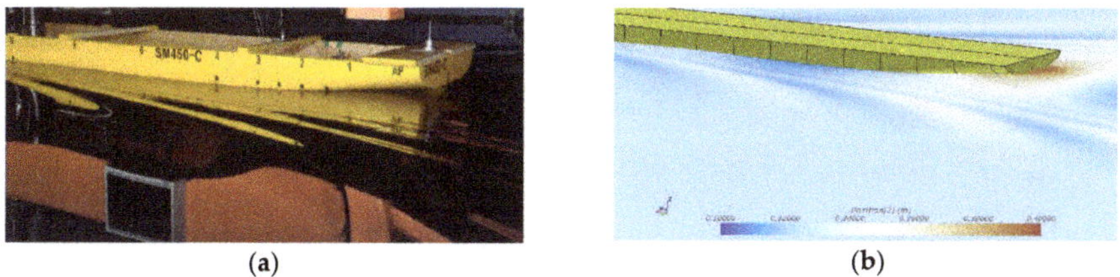

Figure 7. Wave pattern around Aft-hull at 24Kn in the (**a**) model test; (**b**) CFD simulation.

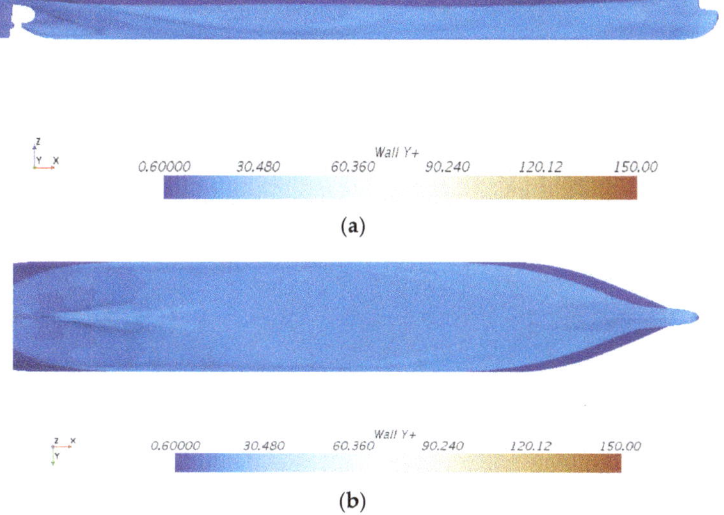

Figure 8. Ranges of wall y+ values at 23Kn ship speed in resistance test (**a**) side view; (**b**) bottom view.

2.1.3. Case II: Propeller Open Water (POW) Test

Since the propeller's eccentric force is the main topic in this study, the accurate prediction of the propeller's thrust or torque is essential. Thus, the POW test was carried out to check the reliability of the numerical simulation for the target propeller. The computational domain, boundary condition and grid system for the POW simulation are presented in Figures 9–11, respectively. In this simulation, the hybrid grid system was implemented. As shown in Figure 9, two regions in the computational domain were defined; the inner part and the outer part. In the inner part, the propeller was modeled and the diameter and width were 1.1 D and 0.25 D, respectively, and the polyhedral unstructured grid system was applied for better presenting the propeller geometry [17]. On the other hand, the trimmer mesh was used likewise for the resistance simulation in the outer part. The total number of grids was about 2.2M and the grids were clustered to describe the leading curvature. The value of y+ on the propeller surface was defined as 40, which was same value on the hull in resistance test in this study. Since the computation domain was modeled as an axis-symmetric shape, likewise a cylinder, the MRF method is a reasonable choice for the POW simulation.

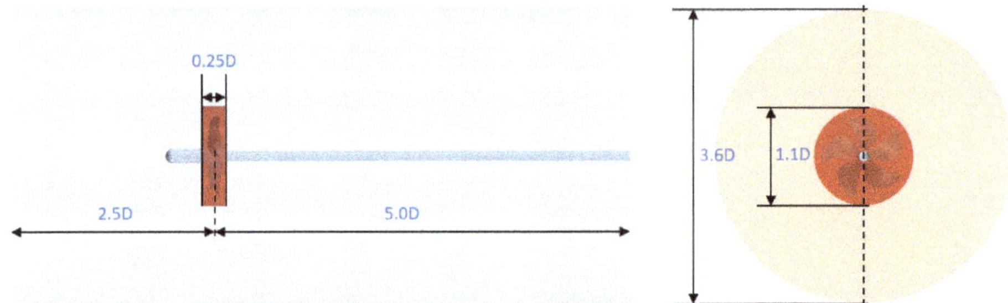

Figure 9. Computational domain (side view and front view).

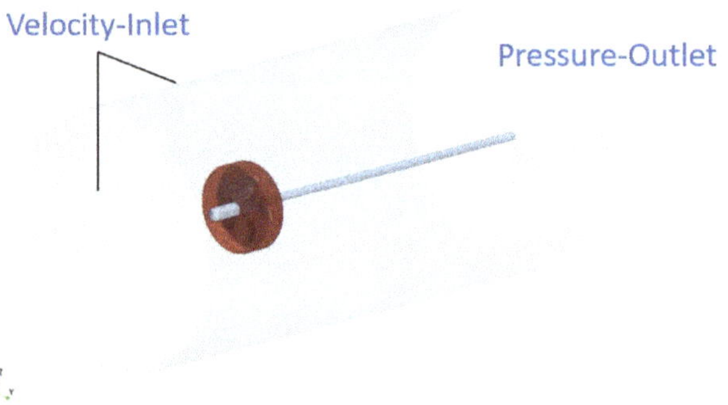

Figure 10. Boundary conditions for POW simulation.

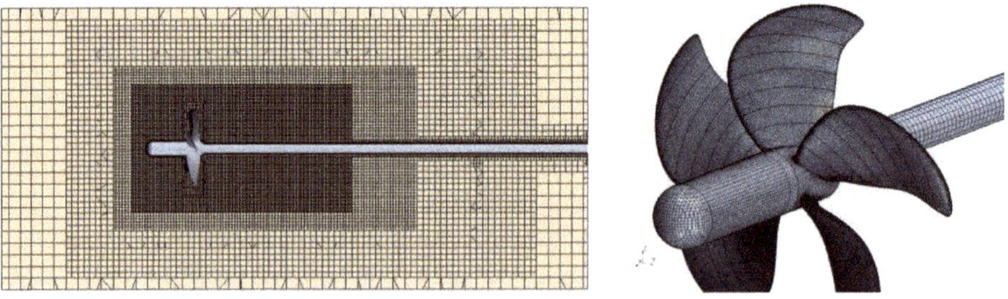

Figure 11. Grid system for POW simulation.

The advance ratio, generally denoted by J, was defined from 0.1 to 1.0. The K_T, K_Q and EtaO from the numerical simulation were compared to those from the model test in Figure 12 and Table 4. The results from the numerical simulation agreed well with experimental data. The differences of K_T, K_Q and EtaO were less than 2% in most of the J range (0.1–0.8), respectively.

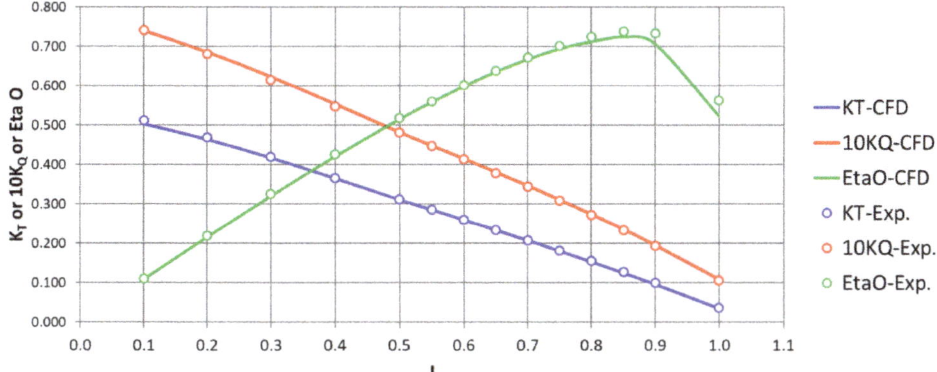

Figure 12. Comparison of K_T, 10 K_Q and EtaO from the model test and the CFD simulation at various advance ratios (J).

Table 4. The accuracy of the POW test results from the numerical simulation and model test.

J	K_T (CFD/EXP.)	$10K_Q$ (CFD/EXP.)	EtaO (CFD/EXP.)
0.10	98.4%	99.8%	98.6%
0.20	99.0%	100.8%	98.2%
0.30	99.7%	101.4%	98.3%
0.40	99.8%	101.0%	98.8%
0.50	99.8%	100.4%	99.4%
0.55	100.0%	100.4%	99.6%
0.60	100.2%	100.5%	99.7%
0.65	100.3%	100.7%	99.6%
0.70	100.2%	100.8%	99.4%
0.75	99.9%	100.9%	99.0%
0.80	99.2%	100.9%	98.3%
0.85	98.3%	100.8%	97.6%
0.90	97.4%	100.9%	96.5%
1.00	96.3%	103.5%	93.1%

Based on the previous two simulations; CASE I: resistance test and CASE II: POW test, it can be concluded that the numerical setup applied in this study is reliable to predict the eccentric force on the propeller.

2.2. Definition for Maneuvering Condition

As mentioned above, a propeller during ship turning experiences continuous change in inflow. To simply describe this circumstance, the quasi-steady approach suggested by Kuroiwa et al. [4] was applied. It is a method to define a flow angle against a propeller plane at a specific instant during ship turning by calculating an actual drift angle. The actual drift angle is calculated by Equation (1). The definition of variables in Equation (1) is schematically presented in Figure 13.

$$\alpha = \beta + tan^{-1}\left(\frac{X_p}{V_s}\cdot r\right) \approx \beta + \frac{X_p}{V_s}\cdot r \tag{1}$$

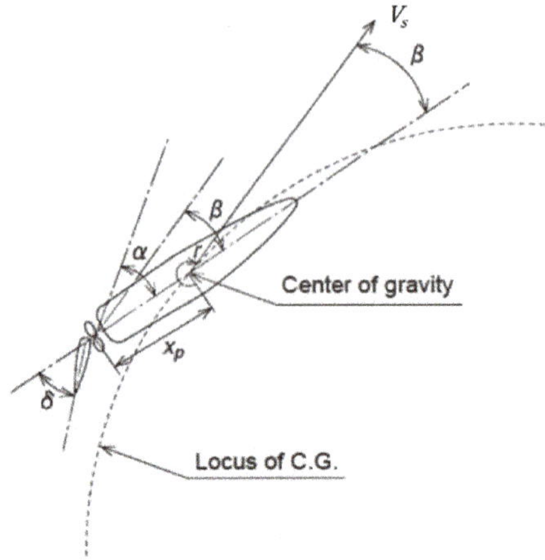

Figure 13. Reproduced from [4], with permission from Mitsubishi Heavy Industries, Ltd., 2007.

Figures 14 and 15 show measured values, including ship speed, yaw rate, and propeller rpm of a full-scale ship during the turning circle test in an official sea trial, respectively. The rotation angle of the rudder was +35° or −35°. From these figures, it can be recognized that a ship in turning motion reaches to the point of maximum yaw rate status, called "yaw rate max.", in a short period of time just after a turning test starts. About 300 s later, the yaw rate converges, and ship speed also becomes stable. This status is called "steady". It is known that inflow conditions to a propeller of these two situations are quite different from that of a straight run, also making eccentric force induced by a propeller be different. Table 5 represents the inflow conditions.

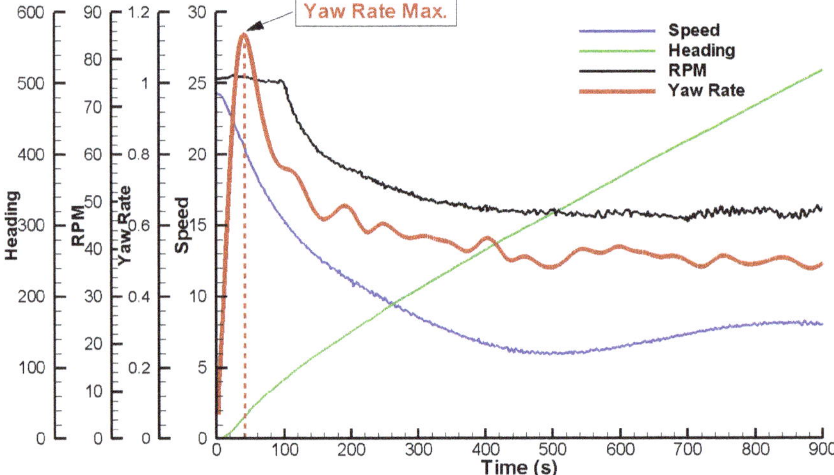

Figure 14. Ship motion during the turning circle test (starboard turning) in the sea trial.

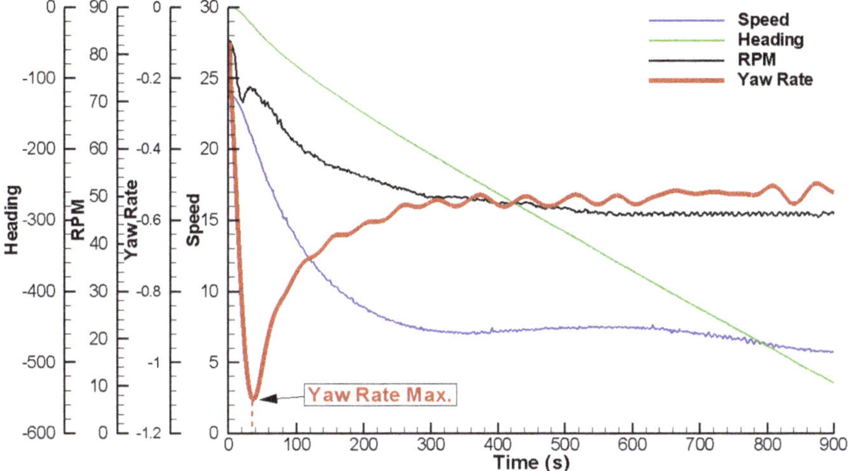

Figure 15. Ship motion during the turning circle test (port turning) in the sea trial.

Table 5. Information of the turning rate at three representative inflow conditions.

Turning Status	Turning Direction	r	α
Straight run	-	0.00	0.00
Yaw rate max.	Port turn	−1.01	22.45
	Starboard turn	0.99	−22.15
Steady	Port turn	−0.80	27.04
	Starboard turn	0.79	−26.20

Numerical simulations for the prediction of propeller eccentric force in these conditions were carried out based on methodologies presented in the previous Section 2.1. Approach ship speed and the propeller's rotation speed were determined by referring to official sea trial data in Figures 14 and 15. Although the wave around the ship is naturally generated in the straight run and turning of a ship, the free surface effect on the propeller eccentric force didn't consider several previous studies [7,10] on reducing the computational time. In this regard, the boundary condition on the plane representing a design draught was defined as the symmetry condition. The computational domain was expanded by reflecting the symmetric plane used in the previous resistance test to consider a propeller behind the hull, therefore the total number of grids became about 7.2M. The rotation of a propeller is realized by implementing a sliding technique, and the propeller eccentric force could be directly calculated during the simulation. Unlike previous studies with high-speed navy vessels [6,8,9,11–14], the target vessel had a relatively low speed and was operated conservatively meaning that, ship motion such as a heeling during the turning was not considered. The rudder angle for each turning direction in the numerical simulation was defined as being the same as the value of official turning circle test.

2.3. Rudder Configuration

The schematic diagram of an asymmetric spade rudder installed on a target vessel is presented in Figure 16. The section shape differs on the rudder's upper part and lower part in order to prevent cavitation at the leading edge of the rudder. Energy-saving devices such as rudder bulb or post-swirl stator were not considered.

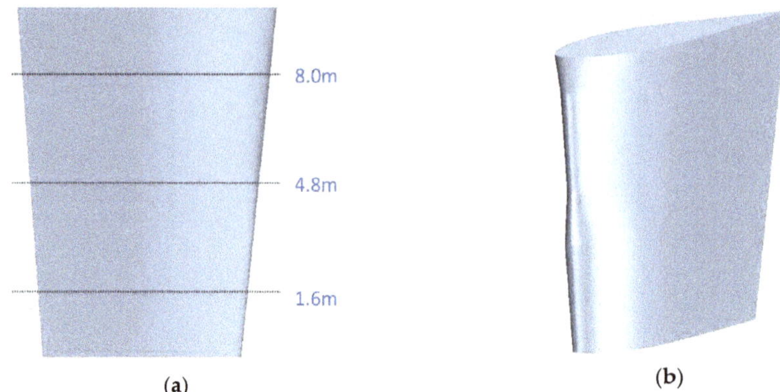

Figure 16. Asymmetric rudder configurations (**a**) Side view; (**b**) three-dimensional (3D) view.

2.4. Coordinate System

A coordinate system used for the analysis of propeller eccentric force was defined as shown in Figure 17.

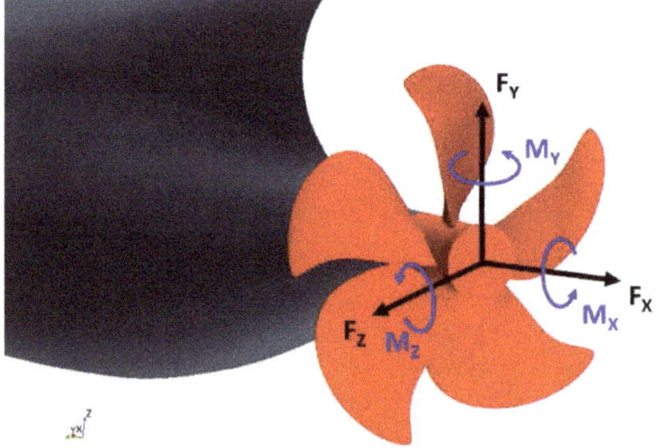

Figure 17. The coordinate system for propeller lateral forces and moments.

3. Results

To investigate rudder effect on propeller eccentric force, simulation cases were categorized into two conditions: straight run and turning motion, including starboard and port turning. In addition, propeller lateral force and moment were evaluated by averaging for the last one revolution of a propeller. The propeller's rotation angle per time step was defined as 1° for realistic simulation.

3.1. Straight Run

Nominal wake distributions evaluated by numerical simulations and measured by model tests on the straight run are compared in Figure 18. In model tests, a rudder does not exist due to measuring equipment consisting of a pitot tube and traverse system. To compare the rudder effect, two numerical simulations were carried out with and without a rudder.

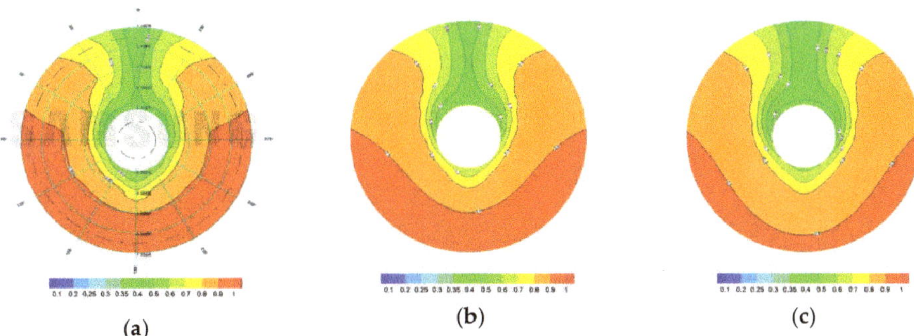

Figure 18. Nominal wake distribution. (**a**) Measurement; (**b**) simulation without rudder; (**c**) simulation with rudder.

As can be seen in Figure 18, the wake distribution in the upper part of a propeller (from 9 o'clock to 3 o'clock) is quite similar to the numerical simulation without a rudder and model tests. If a rudder exists, axial velocity is decelerated on the vertical center line. In Table 6, the thrust and torque of a propeller with and without a rudder were presented, respectively.

Table 6. Change of thrust and torque by rudder existence on the straight run.

Item	With Rudder	Without Rudder
Fx	112%	100%
Mx	107%	100%

The thrust and torque of a propeller with a rudder were larger than those of a propeller without a rudder. Due to rudder existence, inflow velocity to the propeller was reduced, consequently making the angle of attack of a propeller relatively larger. Since the increment of thrust was bigger than that of the torque with a rudder, the propulsive efficiency of a propeller would be increased. Figure 19 shows instantaneous pressure distribution normalized by P_d; $\frac{1}{2}\rho {V_s}^2$ on a propeller face side as well as the difference. Moreover, the pressure on the propeller cap-end surface is also increased. This is a well-known phenomenon, known as the "wake effect".

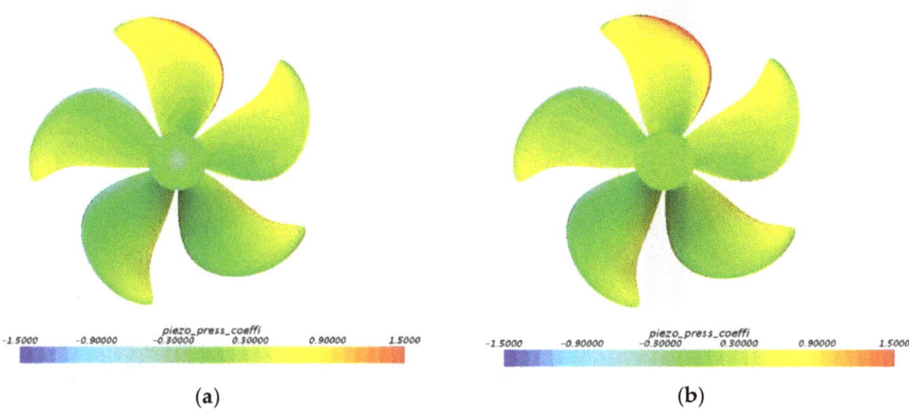

Figure 19. Instantaneous normalized pressure distribution on the propeller face side (looking upstream) during the straight run (**a**) without a rudder; (**b**) with a rudder.

3.2. Turning Condition

The ratio of propeller lateral force and moment in the turning condition for thrust and torque in the straight run is presented in Figures 20–23, including the sets with and without a rudder.

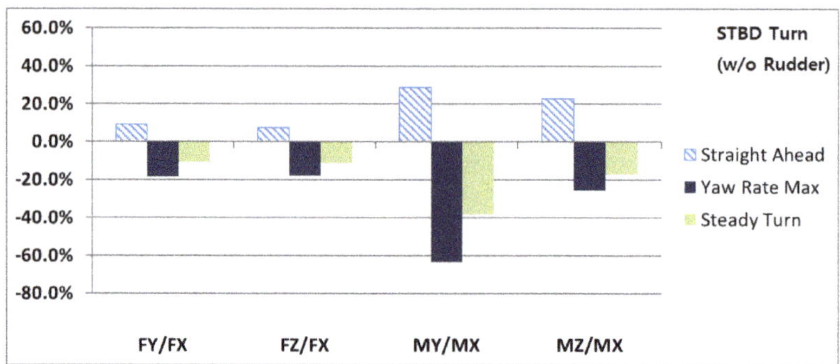

Figure 20. Propeller lateral force and moment during the starboard turn (without rudder).

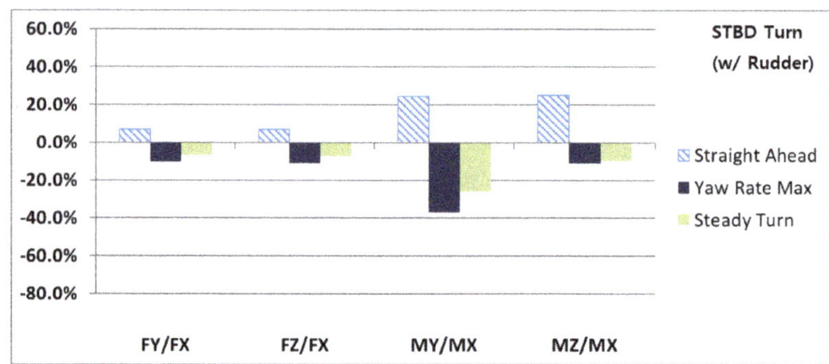

Figure 21. Propeller lateral force and moment during the starboard turn (with rudder).

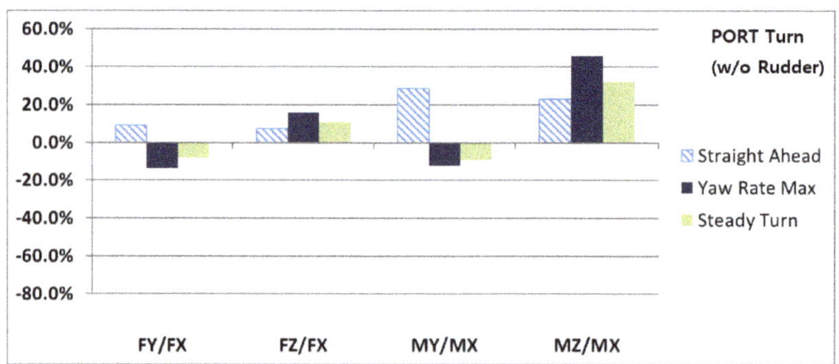

Figure 22. Propeller lateral force and moment during the port turn (without rudder).

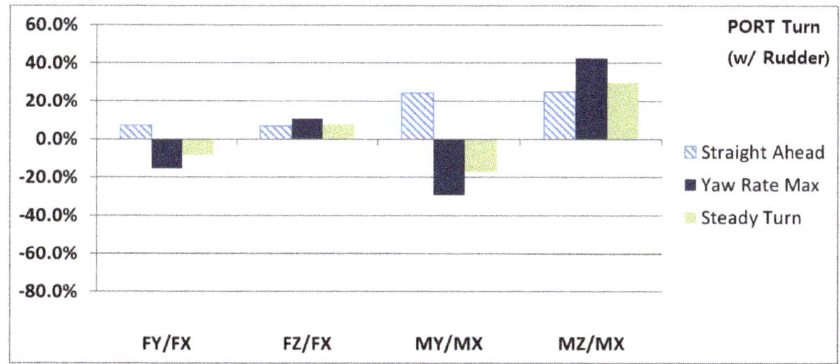

Figure 23. Propeller lateral force and moment during the port turn (with rudder).

During starboard turning with a rudder, the change of the Mz directly related to load of stern tube bearing is remarkable compared to those without a rudder. Due to the existence of a rudder, the Mz was reduced by about 50%, and this is effective for alleviation of load on stern tube bearing. This tendency is similar regardless of turning status (yaw rate max. or steady). In contrast, during port turning, the Mz with a rudder is almost similar to that without a rudder. As indicated in the results above, it can be summarized that the effect of rudder existence on the Mz is asymmetric in line with turning direction. This is schematically presented in Figure 24 to show the asymmetric effect of rudder on the Mz in the case of the ship changing direction.

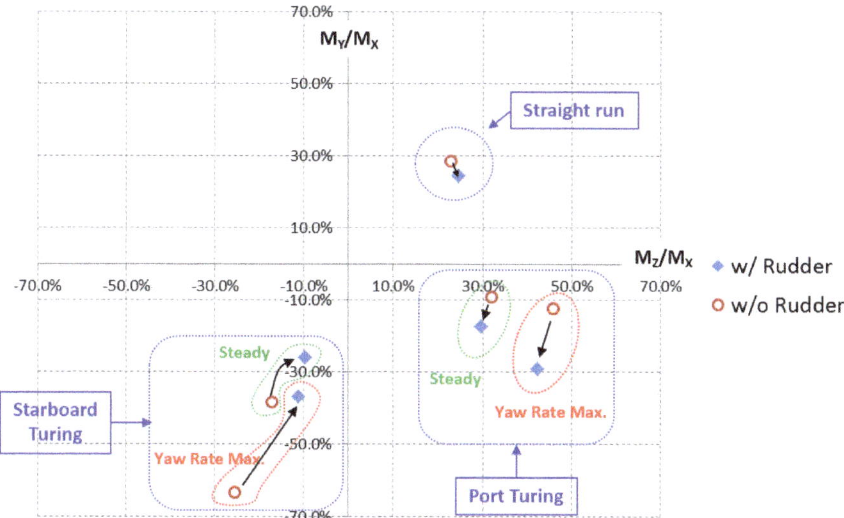

Figure 24. The rudder effect on the propeller's lateral moment during the straight run or the turning motion.

This is because of the interaction between the direction of the propeller's rotation and inflow direction. During the starboard turning motion, the direction of flow into a propeller is from the port side to the starboard side. Since a propeller rotates clockwise, it continuously experiences a count-swirl flow in the lower part of a propeller (from 3 o'clock to 9 o'clock) during ship turning motion. Figure 25 presents the instantaneous pressure distribution on the face side of a propeller without a rudder in both turning status, and it is known that the thrust center of a propeller is naturally located in the lower part of a propeller.

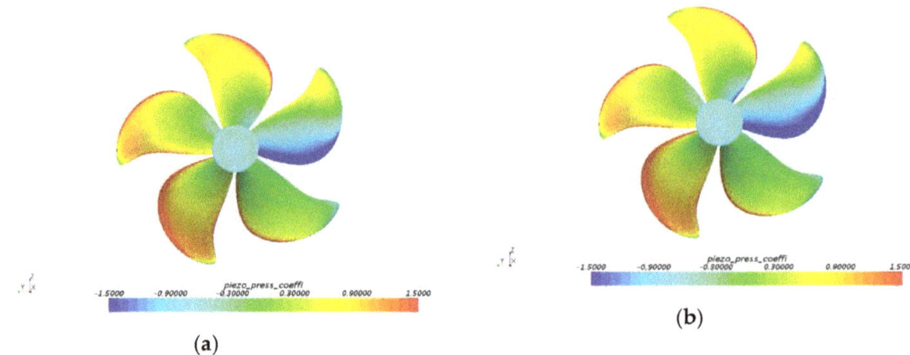

Figure 25. Instantaneous normalized pressure distribution on the propeller face side without rudder during starboard turning (**a**) At yaw rate max.; (**b**) at steady.

If a rudder exists, in contrast, the pressure on the upper part of a propeller was much increased, as shown in Figure 26. The pressure on the lower part of a propeller was somewhat increased, but the effect was limited, as the upper region of the rudder is relatively close to the propeller due to the sweep angle of the rudder. This tendency is similar regardless of turning status, namely, yaw rate max. or steady. Figure 27 shows the instantaneous pressure distribution on a plane defined at several heights above a baseline in yaw rate max. condition. Notably, the difference of results simulated with and without a rudder is clearly contrasted at the upper region of a rudder.

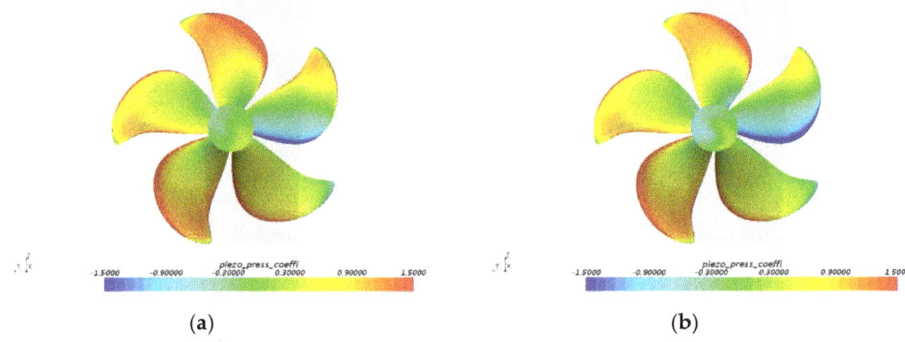

Figure 26. Instantaneous normalized pressure distribution on the propeller face side with rudder during starboard turning (**a**) At yaw rate max.; (**b**) at steady.

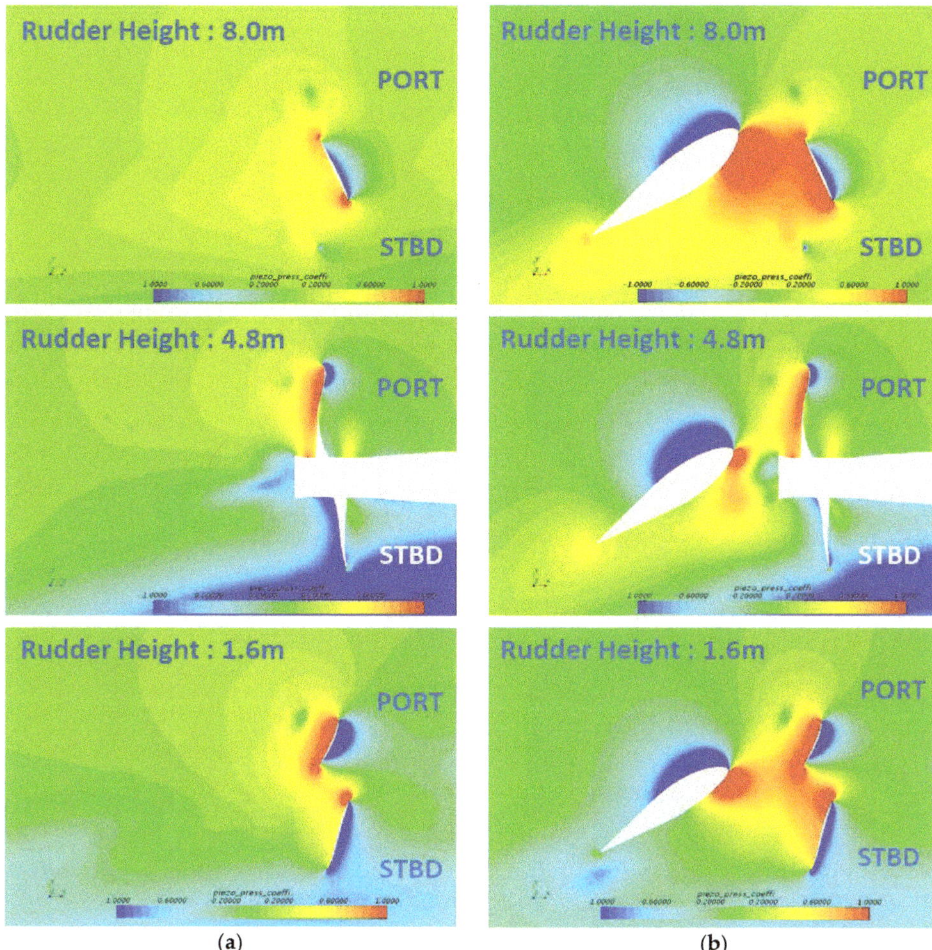

Figure 27. Instantaneous normalized pressure distributions on a plane at three different heights during starboard turning at yaw rate max. status (**a**) without rudder; (**b**) with rudder.

During port turning motion, the direction of the flow into a propeller is from the starboard side to the port side. Due to propeller rotation direction, a propeller experiences a count-swirl flow in the upper part of a propeller during ship turning motion. Consequently, as shown in Figure 28, it is inferred that the center of thrust is located in the upper part of the propeller without a rudder, and the Mz essentially works on reducing the bearing load of the stern tube in this condition.

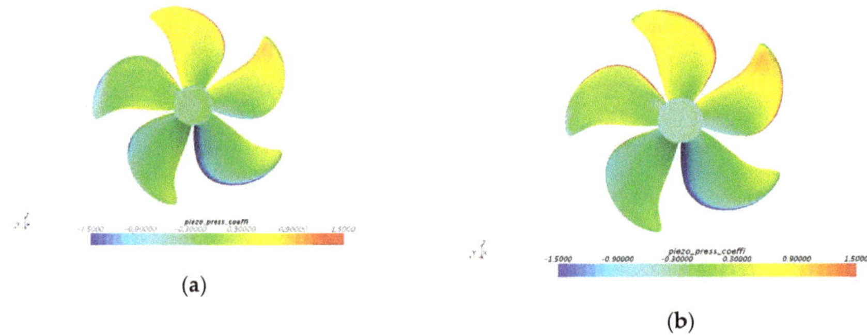

Figure 28. Instantaneous normalized pressure distribution on the propeller face side without rudder during port turning (**a**) At yaw rate max.; (**b**) at steady.

As shown in Figure 29, the pressure on the upper part of a propeller was also increased by rudder, similar to in the starboard turning cases. Although the pressure on the lower part of a propeller was also increased, the effect by a rudder was relatively less than that of the starboard turning cases. This can be explained by Figure 30. During port turning, the pressure on the lower section of a rudder is not significantly increased, unlike Figure 27, at yaw rate max. status.

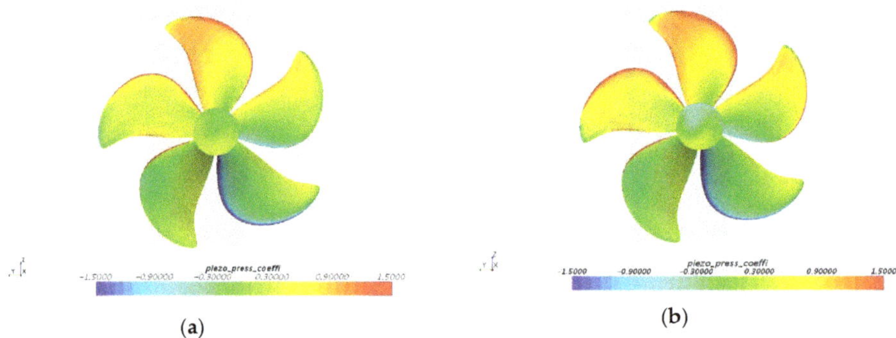

Figure 29. Instantaneous normalized pressure distribution on the propeller face side with rudder during port turning (**a**) at yaw rate max.; (**b**) at steady.

Moreover, the My of a propeller is more sensitive to rudder effect rather than Mz for ship turning motion. If a rudder is operating for the starboard or port turning motion of a ship, a rudder area facing a propeller is larger, and the rudder is respectively weighted to a propeller, as shown in Figures 27 and 29. In this context, the pressure distribution on the part of a propeller blade facing the biased rudder is mainly changed, and the My is consequently affected. During starboard turning, this value is reduced by 20% by the rudder; however, it is increased about two times during port turning, as represented from Figures 20–24.

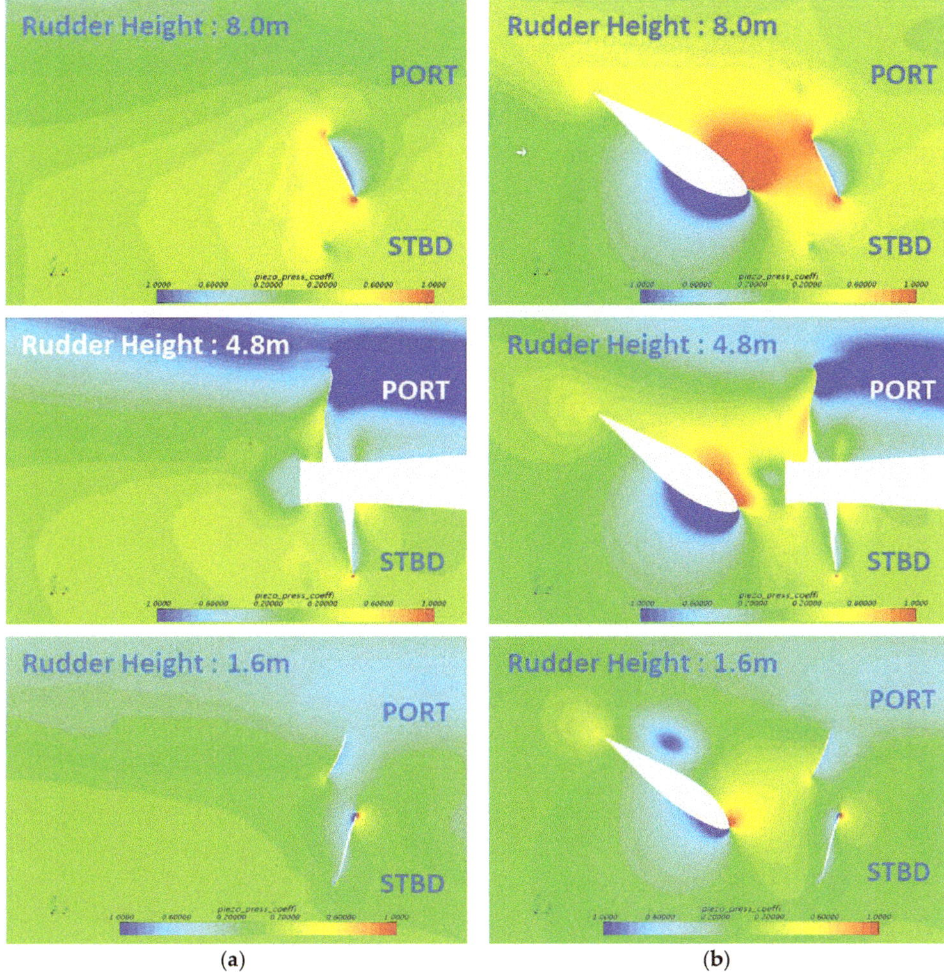

Figure 30. Instantaneous normalized pressure distributions on a plane at three different heights during port turning at yaw rate max. status (**a**) without a rudder; (**b**) with a rudder.

As mentioned in Section 2.3, the rudder used in this study was composed of different sections along the rudder height to prevent rudder cavitation. This characteristic is also related to the asymmetrical effect of a rudder on propeller eccentric force at each turning direction. The comparison between Figures 31 and 32 would explain why the rudder's effect on propeller eccentric force is asymmetrical as per the turning direction of a ship. The instantaneous pressure distributions around the leading edge of the rudder were clearly different to the turning direction (port and starboard). In contrast, the turning status, represented by yaw rate max. or steady, did not significantly affect the pressure distribution on the rudder in each turning direction.

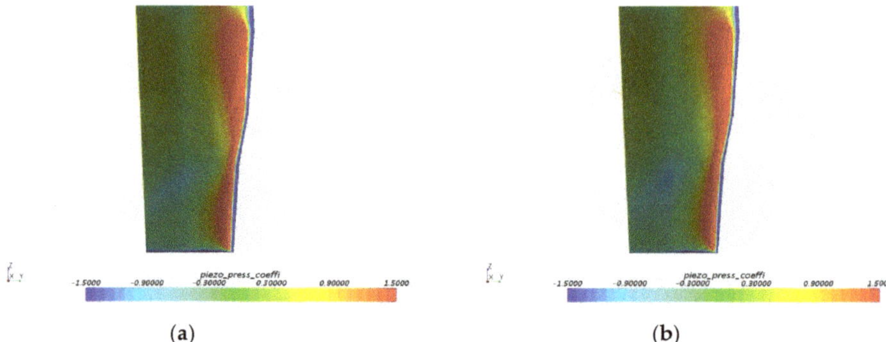

Figure 31. Instantaneous normalized pressure distributions on a rudder during starboard turning (**a**) At yaw rate max. (**b**) at steady (looking downstream).

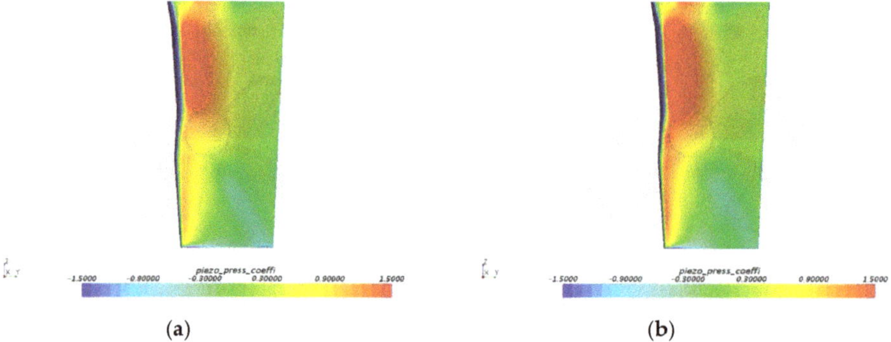

Figure 32. Instantaneous normalized pressure distributions on a rudder during port turning. (**a**) At yaw rate max., (**b**) at steady turning (looking downstream).

4. Conclusions

In this study, numerical simulations were performed to investigate the effect of rudder existence on propeller lateral forces and moments. A target vessel, about a 10,000TEU class container ship, was chosen. For the straight run, the thrust and torque of a propeller were increased due to rudder existence and the propulsive efficiency on a propeller was improved by the wake effect of a rudder. The effect of a rudder on propeller eccentric force was different in the turning direction. The Mz directly related to the bearing load of a stern tube was diminished by about 50% owing to the existence of a rudder for starboard turning, while the values were similar to each other regardless of rudder existence for port turning. This difference basically results from the interaction between the direction of a propeller's rotation and the inflow direction to a propeller. In particular, the effect of a rudder on propeller lateral force and moment depends on where flow straightness behind a propeller is relatively strong or weak. Although the center of thrust on a propeller was located in the lower part of a propeller for starboard turning, the magnitude of the Mz was compensated by rudder effect in the upper part of a propeller. In contrast, the center of thrust on a propeller is already located in the upper part of a propeller for port turning so the effect of a rudder was relatively limited on propeller eccentric force. Additionally, a rudder affects the My of a propeller, consequently, that value with rudder showed similar level regardless of the ship's turning direction. A rudder's main configurations, including the leading-edge sweep angle, aspect ratio, section shape or distance between a propeller and a rudder, might have an impact on propeller eccentric force. Furthermore, appendages located around a propeller, including a rudder and various energy saving devices, need to be considered in order to estimate reliable propeller eccentric force.

Author Contributions: Conceptualization, G.S.; validation, G.S., and T.L.; analysis, G.S., and T.L.; writing—original draft preparation, G.S.; writing—review and editing, T.L. and H.P.; supervision, H.P.; project administration, H.P.

Funding: This research received no external funding.

Acknowledgments: This research was carried out as part of an internal research project.

Conflicts of Interest: The authors declare no conflict of interest.

Nomenclature

B	Breath [m]	BEMT	Blade Element and Momentum Theory
D	Propeller diameter [m]	CFD	Computational Fluid Dynamics
Eta O	Propeller efficiency	CRP	Contra-Rotating Propeller
Fr	Froude number	EHP	Effective Horse Power
K_T	Thrust coefficient	FEM	Finite Element Method
K_Q	Torque coefficient	Fx	Thrust of the propeller
Kn	Knots	Fy	Force to the vertical axis
L	Length Between Perpendiculars [m]	Fz	Force to the horizontal axis
P_d	Dynamic pressure [Kg/(m·s^2)]	HRIC	High Resolution Interface Capturing
r	Rate of turn [°/s]	JIME	Japan Institute of Marine Engineering
s	Second	MRF	Moving Reference Frame
T_d	Design draught [m]	Mx	Moment based on the propeller-shaft axis
V_s	Ship speed [Kn]	My	Moment based on the vertical axis
x_p	Distance from the propeller to the gravitational center of the hull [m]	Mz	Moment based on the horizontal axis
Z	Number of blades	POW	Propeller Open Water
α	Water flow angle against propeller plane [°]	Rtm	Total resistance in model scale
β	Drift angle [°]	SIMPLE	Semi-Implicit Method for Pressure Linked Equations
δ	Rudder angle [°]	TEU	Twenty-foot Equivalent Unit
ρ	Density of fluid [Kg/m^3]	URANS	Unsteady Reynolds Averaged Navier-Stokes
		VOF	Volume of Fluid

References

1. Ui, T. Influence of propeller force to shaft alignment. In Proceedings of the International Symposium on Marine Engineering (ISME), Kobe, Japan, 17–21 October 2011.
2. American Bureau of Shipping (ABS). *Guide for Enhanced Shaft Alignment*; American Bureau of Shipping: Houston, TX, USA, 2018.
3. Det Norske Veritas-Germanischer Lloyd (DNV-GL). *Shaft Alignment, DNVGL-CG-0283*; Det Norske Veritas-Germanischer Lloyd: Høvik, Norway, 2018.
4. Kuroiwa, R.; Oshima, A.; Nishioka, T.; Tateishi, T.; Ohyama, T.; Ishijima, T. Reliability improvement of stern tube bearing considering propeller shaft forces during ship turning. *Mitshbishi Heavy Ind. Ltd. Tech. Rev.* **2007**, *44*, 1–5.
5. Vartdal, B.J.; Gjestland, T.; Arvidsen, T.I. Lateral propeller forces and their effects on shaft bearings. In Proceedings of the First International Symposium on Marine Propulsors, Trondheim, Norway, 22–24 June 2009; pp. 475–481.
6. Corraddu, A.; Dubbioso, G.; Mauro, S.; Viviani, M. Analysis of twin screw ship's asymmetric propeller behavior by means of free running model tests. *Ocean Eng.* **2013**, *68*, 47–64. [CrossRef]
7. Shin, S.H. Effects of Propeller Forces on the Propeller Shaft Bearing during Going Straight and Turning of Ship (in Korean). *J. Soc. Nav. Archit. Korea* **2015**, *52*, 61–69. [CrossRef]

8. Ortolani, F.; Mauro, S.; Dubbioso, G. Investigation of the radial bearing force developed during actual ship operation. Part 1: Straight ahead sailing and turning maneuvers. *Ocean Eng.* **2015**, *94*, 67–87. [CrossRef]
9. Dubbioso, G.; Durante, D.; Di Mascio, A.; Broglia, R. Turning ability analysis of a fully appended twin screw vessel by CFD. Part II: Single vs. twin rudder configuration. *Ocean Eng.* **2016**, *117*, 259–271. [CrossRef]
10. Lee, T.; Song, G.; Park, H. Effect of Propeller Eccentric Forces on the Bearing Loads of the Complicated Shafting System for Large Container Ships. In Proceedings of the Fifth International Symposium on Marine Propulsors, Helsinki, Finland, 12–15 June 2017.
11. Dubbioso, G.; Muscari, R.; Ortolani, F.; Di Mascio, A. Analysis of propeller bearing loads by CFD. Part I: Straight ahead and steady turning maneuvers. *Ocean Eng.* **2017**, *130*, 241–259. [CrossRef]
12. Muscari, R.; Dubbioso, G.; Ortolani, F.; Di Mascio, A. CFD analysis of the sensitivity of propeller bearing loads to stern appendages and propulsive configurations. *Appl. Ocean Res.* **2017**, *69*, 205–219. [CrossRef]
13. Ortolani, F.; Dubbioso, G. Experimental investigation of single blade and propeller loads by free running model test. Straight ahead sailing. *Appl. Ocean Res.* **2019**, *87*, 111–129. [CrossRef]
14. Ortolani, F.; Dubbioso, G. Experimental investigation of single blade and propeller loads: Steady turning motion. *Appl. Ocean Res.* **2019**, *91*, 101874. [CrossRef]
15. De Luca, F.; Mancini, S.; Miranda, S.; Pensa, C. An extended verification and validation study of CFD simulations for planning hulls. *J. Ship Res.* **2016**, *60*, 101–118. [CrossRef]
16. CD-Adapco. *STAR-CCM+ User's Guide Version 10.06*, CD-Adapco: Plano, TX, USA, 2015.
17. Krasilnikov, V. CFD modeling of hydroacoustic performance of marine propellers: Predicting propeller cavitation. In Proceedings of the 22nd Numerical Towing Tank Symposium, Tomar, Portugal, 29 September–1 October 2019.

Journal of
Marine Science and Engineering

Article

Theoretical Performance Evaluation of a Marine Solid Propellant Water-Breathing Ramjet Propulsor

Nachum E. Eisen * and Alon Gany

Faculty of Aerospace Engineering, Technion—Israel institute of Technology, Haifa 3200003, Israel; gany@tx.technion.ac.il
* Correspondence: nachum.eisen@gmail.com

Received: 24 November 2019; Accepted: 18 December 2019; Published: 20 December 2019

Abstract: This work analyzes and presents theoretical performance of a marine water-breathing ramjet propulsor. A conceptual scheme of the motor is shown, the equation of thrust is presented, and the dependence on cruise velocity and depth are discussed. Different propellant compositions, representing a wide variety of formulations suitable for propelling a water-breathing ramjet, are investigated. The theoretical results reveal that the specific impulse of a water-breathing ramjet can increase by as much as 30% compared to a standard rocket, when using a conventional hydroxyl terminated polybutadiene (HTPB)-ammonium perchlorate (AP) propellant, which does not react chemically with the water. When employing a water-reactive propellant containing metal particles such as magnesium or aluminum, the specific impulse may be more than doubled. The thrust coefficient of the propulsor was computed at different cruise velocities and depths and was found to be greater than the predictable drag even at significant depth.

Keywords: underwater propulsion; water-breathing ramjet; marine ramjet; water ducted rocket

---

1. Introduction

The search for underwater high-speed cruising reveals certain high-thrust marine propulsion concepts which are parallel in principle to their aeronautical counterparts [1]. Rocket engines can propel high-speed underwater vehicles just like aeronautical vehicles. Independent of speed and ambience medium, rocket engines provide very high thrust and do not need air for their operation, hence, they can suite underwater vehicles. However, their specific fuel (propellant) consumption is very high and, thus, they enable only a limited range. Therefore, propulsion principles utilizing the surrounding water as a working fluid as well as a chemical reactant (usually oxidizer), have been considered [2–8]. One of the interesting concepts for propelling high-speed underwater vehicles is the solid propellant water-breathing ramjet [4–8]. It is a powerful propulsion means, independent of atmospheric air, hence, it can operate fully and deeply underwater.

The basic principle of the marine water-breathing ramjet is to utilize the high dynamic (ram) pressure, resulting from the high-speed motion, in order to ingest water into the combustion chamber of the motor, thus increasing the thrust and energetic performance of the propulsor. This work focuses on theoretical prediction of the performance of a water-breathing underwater solid propellant ramjet motor.

The design of an underwater solid propellant water-breathing ramjet is similar to the design of an aeronautical ducted rocket. Water enters the motor through an inlet located in the front of the vehicle and is sprayed through a porous wall into the combustion chamber. Due to the high temperature in the combustion chamber, the incoming water boils and then mixes with the combustion products, forming a high-speed exhaust jet that provides the thrust necessary to propel the vehicle. A conceptual illustration of a solid propellant water-breathing ramjet is presented in Figure 1.

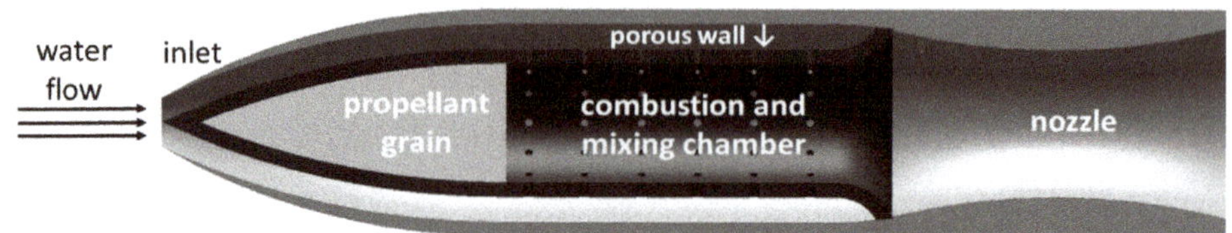

Figure 1. Conceptual illustration of a marine solid propellant water-breathing ramjet.

Similarly to solid rocket motors (SRMs), the solid propellant grain can provide high thrust without the addition of water, as well. Nevertheless, the mere addition of water, entering through an inlet due to the vehicle motion, and its further conversion into steam because of the high heat release of the burning solid propellant, increases the thrust and the specific impulse, even with no chemical interaction. Greater performance improvement can be obtained when employing water-reactive propellant compositions, which utilize the incoming water as an oxidizer to increase heat release in addition to using it as a source for steam which increases the mass flow rate of the exhaust jet. Certain metals are known to react very exothermically with water and may be considered for inclusion as water reactive ingredients in propellants. Typically, the reaction of metals with water generates hydrogen besides metal oxides. Table 1 presents theoretical physical and thermochemical properties of selected metals. While the gravimetric heat of reaction is related to the mass of fuel (propellant) carried on board, the volumetric heat of reaction reflects the energy per unit volume and is mainly significant in volume-limited systems.

Table 1. Maximum theoretical gravimetric and volumetric heat of reaction and hydrogen generation for selected metals with liquid water, listed in the order of their volumetric heat of reaction (data adapted from [9]).

Fuel	Density $[g/cm^3]$	Gravimetric Heat of Reaction $[kJ/g]$	Volumetric Heat of Reaction $[kJ/cm^3]$	Specific H_2 Mole Production $[mol/g]$	Specific Volumetric H_2 Mole Production $[mol/cm^3]$
Be	1.85	36.1	66.7	0.111	0.21
B	2.35	18.4	43.2	0.139	0.34
Al	2.70	15.2	40.9	0.055	0.15
Zr	6.49	5.7	37.2	0.022	0.14
Mg	1.74	13.0	22.6	0.041	0.072
Li	0.54	25.5	13.8	0.072	0.039
Na	0.97	2.8	2.7	0.022	0.021

It is apparent from Table 1 that beryllium reacts extremely exothermically with water both in terms of weight and volume, yet, beryllium and its products are notoriously toxic and therefore are ruled out. Second to beryllium is boron, which demonstrates a promising theoretical heat and gas (i.e., hydrogen) release. In practice, it is hard to obtain high combustion efficiency of boron (though Rosenband et al. [10] have demonstrated combustion of boron in steam), and in addition boron is a relatively expensive ingredient. Hence, for practical applications, aluminum is commonly considered. Aluminum reacts with water to form aluminum oxide (alumina) or aluminum hydroxide, and hydrogen, while releasing a significant amount of heat, as presented in Equation (2). However, aluminum does not combust readily as a thin layer of aluminum oxide naturally forms on the exposed surfaces of the aluminum particles and passivates the metal substrate to inhibit oxidation progression. Different approaches for overcoming the alumina coating problem are discussed extensively in the literature [11–14]. Zirconium has high volumetric heat of reaction, yet lower than that of aluminum. In addition, its other energetic parameter, gravimetric heat of reaction, is substantially inferior to that of more regularly used metals such as aluminum and magnesium. Relatively to aluminum, magnesium has a lower energy density, but the magnesium-water reaction is easier to initiate. Therefore, some researchers suggested using magnesium [4,5,7,8] or using an aluminum-magnesium mixture [6]. The high-temperature stoichiometric reactions of boron, aluminum, and magnesium with water are presented here:

$$2B + 3H_2O \rightarrow B_2O_3 + 3H_2 + 397\frac{kJ}{mol}, \tag{1}$$

$$2Al + 3H_2O \rightarrow Al_2O_3 + 3H_2 + 818\frac{kJ}{mol}, \tag{2}$$

$$Mg + H_2O \rightarrow MgO + H_2 + 312\frac{kJ}{mol}. \tag{3}$$

One may note that lithium has a very high gravimetric heat of reaction in addition of its high reactivity with water. However, besides the difficulty in handling lithium, it has been considered for application only as a liquid fuel in its molten phase and not as solid fuel [15].

The chamber pressure in the marine water-breathing ramjet is dictated by the cruise speed, and its maximum value at a given speed is equal to the stagnation pressure of the incoming water flow:

$$P_0 = P_a + \frac{1}{2}\rho_w u^2, \tag{4}$$

where, P_0 = stagnation pressure; P_a = ambient pressure; ρ_w = water density; u = cruise velocity.

The stagnation pressure increases with the cruise speed (as shown in Figure 2); the increased chamber pressure at higher velocities enable enhanced conversion of thermal energy to kinetic energy in the exhaust nozzle, increasing the energetic performance of the motor.

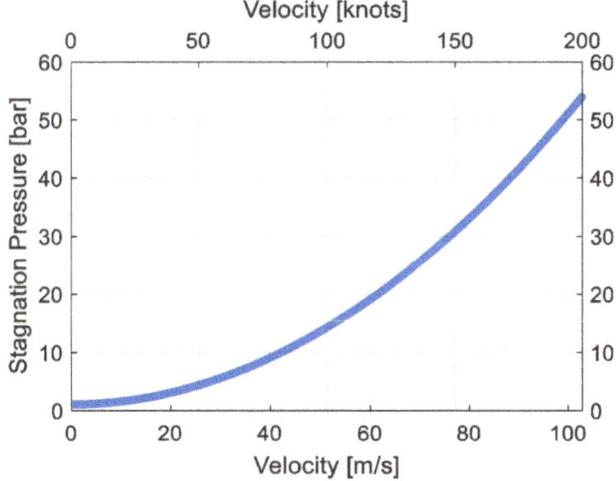

Figure 2. Stagnation pressure vs. cruise speed in water (assuming: $\rho_w = 1000\ kg/m^3$).

2. Theoretical Model

2.1. Governing Equations and Evaluating Performance

In order to evaluate the performance of a marine solid propellant water-breathing ramjet, the governing equations are presented. The general thrust equation of a jet engine is:

$$F = \dot{m}_e u_e - \dot{m}_w u + (P_e - P_a) \tag{5}$$

where, F = thrust; $\dot{m}_e$ = exhaust mass flow rate; u_e = exhaust jet velocity; $\dot{m}_w$ = incoming water mass flow rate; P_e = exhaust jet static pressure.

The specific impulse (I_{sp}) of a propulsion system is an important index defined as the thrust divided by the weight flow rate of propellant consumed in order to produce the thrust, (as shown in Equation (6)). The density specific impulse (ρI_{sp}) of a propulsion system is defined as the specific impulse multiplied by the density of the propellant. For volume-limited vehicles (such as many underwater marine vehicles), the density specific impulse may be of more significant since it represents the thrust produced per volume flow rate of propellant. Thus, the specific impulse and density specific impulse can be used to compare and grade the performance of different propulsion methods. Assuming the nozzle is adapted so that the exhaust pressure is equal to ambient pressure (i.e., optimal expansion ratio for maximal thrust), the specific impulse of a marine ramjet can be calculated as following:

$$I_{sp} = \frac{F}{\dot{m}_p g_0} = \frac{\dot{m}_e u_e - \dot{m}_w u}{\dot{m}_p g_0}, \tag{6}$$

where, I_{sp} = specific impulse; $\dot{m}_p$ = propellant mass flow rate; and g_0 = universal gravity constant.

The exit (jet) flow rate is equal to the sum of the propellant combustion products and water flow rates, thus:

$$\dot{m}_e = \dot{m}_w + \dot{m}_p. \tag{7}$$

Defining:

$$\frac{w}{p} = \frac{\dot{m}_w}{\dot{m}_p}, \tag{8}$$

one obtains:

$$I_{sp} = \frac{\left[1 + \left(\frac{w}{p}\right)\right]u_e - \left(\frac{w}{p}\right)u}{g_0}. \tag{9}$$

Assuming that the combustion chamber pressure is equal to its maximal theoretical value (i.e., the stagnation pressure) and that the ambient pressure is 1 bar (i.e., cruising at sea level), the specific impulse of a marine solid propellant water-breathing ramjet was calculated. The calculations were done according to Equation (9) with the aid of the Computer Program for Calculation of Complex Chemical Equilibrium Compositions and Applications (CEA) [16], which yields the equilibrium flow conditions and jet performance (adapted nozzle). The specific impulse and combustion chamber temperature were calculated for different propellant compositions and with different water/propellant mass ratios.

In this work, a number of propellant compositions were examined. The first composition was a common solid propellant formulation of 85% ammonium perchlorate (AP) and 15% hydroxyl terminated polybutadiene (HTPB), for which the water ingested into the motor undergoes only a phase change into steam with no chemical interaction with the burning propellant. In addition to the non-hydro-reactive AP-HTPB composition, this work deals with certain water-reactive fuels and compositions, representing the potential range of propellants and hydro-reactive propellant ingredients suitable for propelling a marine water-breathing ramjet. One hydro-reactive fuel is pure aluminum, presenting the maximum theoretical performance of aluminized propellants, since it reacts solely with the incoming water with no oxidizer stored onboard. Additional elements examined as potential fuels or propellant ingredients, were boron because of its high heat of reaction (although its practical application may be questionable), and magnesium due to its better reactivity relatively to aluminum and boron. At last, a more practical solid propellant composition, i.e., 70% Mg and 30% polytetrafluoroethylene (PTFE, Teflon), was examined. The small amount of PTFE oxidizer enables initial combustion to take place without the addition of water, but being a fuel-rich propellant, the excessive fuel can react with incoming water.

2.2. Thrust Coefficient and Influence of Cruise Depth and Velocity on Performance

In addition to examining the specific impulse obtained by different propellant compositions at different water to propellant mass ratios, the thrust coefficient and the influence of cruise depth and velocity on performance have been studied. It should be mentioned, that in contrast to the specific impulse which highly depends on the propellant composition and water/propellant ratio, the thrust, and therefore the thrust coefficient, weakly depends on these parameters. Thus, it was chosen in this work to present only the thrust coefficient of the more practical fuel-rich hydro-reactive propellant composition containing 70% Mg and 30% PTFE, with a constant water/propellant ratio of 3.5 (where maximal specific impulse is achieved). Yet, similar results and trends of the thrust coefficient have been observed with other compositions and at different water to propellant ratios.

Assuming the marine water-breathing ramjet is used to propel a cylindrical torpedo-shaped vehicle, the cross-section area of the body may be used as the reference area in computing the thrust and drag coefficients (Equations (10) and (11)). In addition, it was assumed that the nozzle throat diameter is 50% of the body's external diameter.

$$C_T = \frac{F}{\frac{1}{2}A\rho_w u^2}, \quad (10)$$

where, C_T = thrust coefficient; A = reference area.

It should be mentioned that for a cylindrical torpedo-shaped vehicle with a length to diameter ratio of 15, the drag coefficient is presumed to be smaller than 0.25 at all cruise velocities [17] and may be significantly smaller in a supercavitating vehicle cruising at a high velocity [18]. In order for the vehicle to overcome the drag force, the thrust coefficient has to be equal or larger than the drag coefficient, which is related to the same reference area:

$$C_D = \frac{D}{\frac{1}{2}A\rho_w u^2}, \quad (11)$$

where, C_D = drag coefficient; D = drag force.

Assuming isentropic supersonic flow in the divergent section of the nozzle, the data obtained from the thermochemical code, mentioned above, enable to calculate the exit velocity and pressure as presented in Sutton and Biblatz [19], and thus the thrust (Equation (5)).

The expansion ratio is defined as the ratio between the nozzle's exit and throat cross-section areas. The results presented in this study refer to a ramjet with an expansion ratio of two cruising at a velocity in the range of 50 to 150 m/s at a depth of 0 to 100 m below sea level. Atmospheric pressure (1 bar) is assumed at sea level and a pressure increase by 1 bar every 10 m below sea level.

3. Results and Discussion

3.1. Performance of Different Propellant Compositions

3.1.1. Non-Hydro-Reactive Composition

Figure 3a presents the calculated specific impulse and density specific impulse for a solid propellant water-breathing ramjet motor with a propellant grain consisting of 85% AP and 15% HTPB versus the water to propellant (w/p) mass ratio for two cruise speeds, 25 m/s (implying chamber pressure of 4 bar) and 100 m/s (implying chamber pressure of 50 bar). The values at a zero w/p ratio represent the specific impulse of a pure rocket (no water addition). In general, the w/p ratio is limited by the decrease of the overall chamber temperature to the water equilibrium boiling (condensation) temperature (Figure 3b). It is obvious that the addition of water increases the specific impulse (and thus the density specific impulse) substantially. However, at the low cruise speed (low chamber pressure), the specific impulse is inferior to that of a rocket motor operating at standard conditions (chamber pressure of about 69 bar).

Nevertheless, the performance of the water-breathing ramjet at the higher chamber pressure (50 bar, corresponding to a cruise speed of 100 m/s) is noticeably higher than that of the lower pressure due to the contribution of the greater expansion in the nozzle. When cruising at 100 m/s, the specific impulse exceeds that of a standard rocket motor by as much as 30% when adding water at a w/p ratio of 2.5.

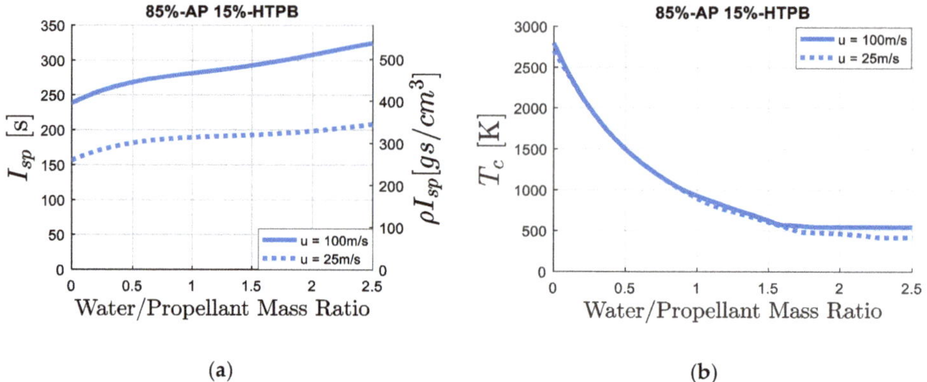

Figure 3. (**a**) Calculated specific impulse—I_{sp}, density specific impulse—ρI_{sp}, and (**b**) combustion chamber temperature—T_c, of an underwater water-breathing ramjet using a non-hydro-reactive solid propellant (85% ammonium perchlorate (AP) + 15% hydroxyl terminated polybutadiene (HTPB)) vs. water/propellant mass ratio, for two different cruise velocities.

3.1.2. Hydro-Reactive Fuels

Figure 4a presents the theoretical specific impulse and density specific impulse for a water-breathing motor with aluminum as a fuel versus the water to propellant (fuel) mass ratio, for two cruise speeds, 25 and 100 m/s, corresponding to low (4 bar) and high (50 bar) chamber pressures, respectively. In this case, even at the low speed cruise conditions, the theoretical specific impulse of the water-breathing ramjet exceeds that of a standard rocket, the density specific impulse exceeds that of a standard rocket even more so due to the relatively high density of aluminum (compared to HTPB-AP based propellants). Figure 4b presents the calculated combustion chamber temperature versus water/propellant mass ratio. It can be observed that at low w/p ratios (bellow the stoichiometric ratio) the addition of water increases the chamber temperature due to the increase in available oxidizer that enables more aluminum to react and release heat. On the other hand, at large w/p ratios, the addition of water reduces the temperature, similarly to the trend shown above in the case of a non-hydro- reactive propellant.

To be more realistic, one can take into account the fact that the reaction products of aluminum with water contain large amounts of condensed material (Al_2O_3 or $Al(OH)_3$) that may be retained within the motor. Hence, calculations were repeated for the case where the condensed material remains in the reaction chamber and does not exit the nozzle. Even in such case, the motor performance may be doubled and more, compared to standard solid rockets.

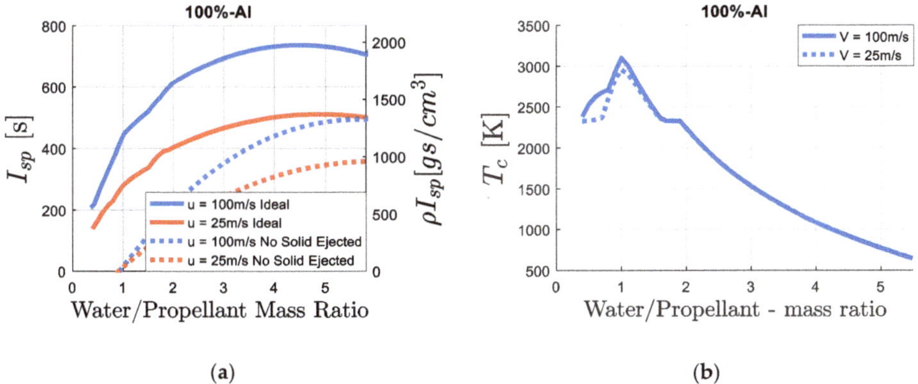

Figure 4. (**a**) Calculated specific impulse—I_{sp}, density specific impulse—ρI_{sp}, and (**b**) combustion chamber temperature—T_c, of an underwater water-breathing ramjet using aluminum as a hydro-reactive fuel vs. water/propellant mass ratio, for two different cruise velocities.

As mentioned above, theoretically, boron as a fuel may provide the uppermost performance of a water-breathing ramjet propulsor (assuming beryllium is ruled out due to toxicity), though there is uncertainty regarding practical application. Additionally, magnesium is widely considered as a fuel for a marine ramjet due to its reactivity. Figures 5a and 6a present the theoretical specific impulse and density specific impulse of a ramjet using boron and magnesium as fuels at two cruise speeds (100 m/s and 25 m/s). It is apparent from Figure 5a that at a cruise speed of 100 m/s boron enables to achieve superior theoretical performance of up to four times the specific impulse achieved by an SRM, yet, even at 25 m/s a significant specific impulse may be achievable. It should be noted, that due to the immense amount of heat released from the boron-water reaction, maximal specific impulse is achieved at a very large w/p ratio. Figure 6a reveals that the specific impulse of magnesium is inferior to that of aluminum by roughly 15% at both cruise speeds. However, due to magnesium's reactivity, which possibly implies increased combustion efficiency, magnesium is a promising fuel and can be used as an energetic propellant ingredient to improve performance of solid propellant marine ramjets. Figures 5b and 6b present the combustion chamber temperature of a ramjet using boron and magnesium as fuels, respectively. The trends discussed above in the case of aluminum, of an increase in temperature at a low w/p ratio and a decrease in temperature at a large w/p ratio, can be observed also in the cases of boron and magnesium. It should be commented, that while aluminum (Figure 4b) and magnesium (Figure 6b) present roughly an equal peak temperature of 3100 K, the peak temperature obtained with boron (Figure 5b) is significantly lower due to the large latent heat and the relatively low temperature of evaporation of the water-boron reaction product, boron oxide. In all cases, where phase change occurs (observed as "plateaus" in the temperature figures), the phase change temperature is greater for the higher velocity due to the increased chamber pressure, which results in an elevated equilibrium vapor pressure.

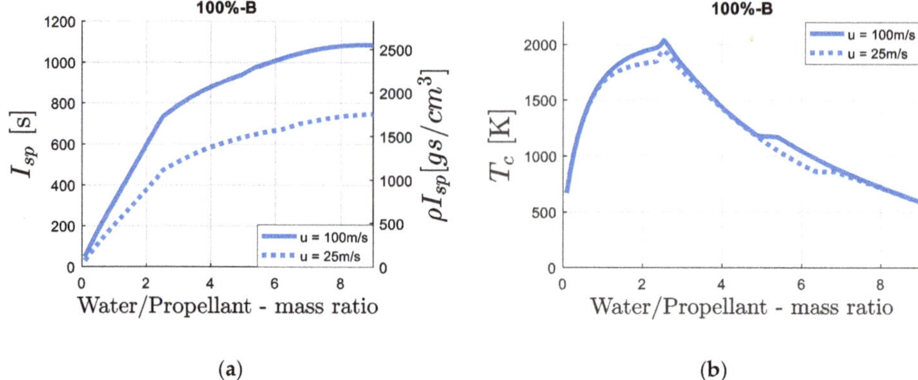

Figure 5. (**a**) Calculated specific impulse—I_{sp}, density specific impulse—ρI_{sp}, and (**b**) combustion chamber temperature—T_c, of an underwater water-breathing ramjet using boron as a hydro-reactive fuel vs. water/propellant mass, for two different cruise velocities.

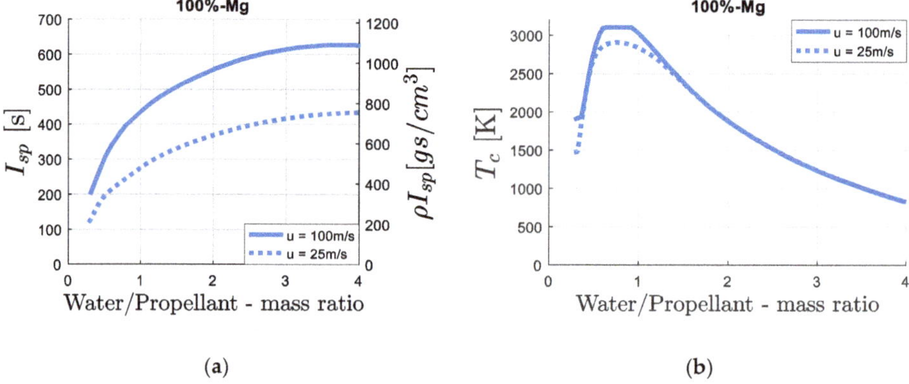

Figure 6. (**a**) Calculated specific impulse—I_{sp}, density specific impulse—ρI_{sp}, and (**b**) combustion chamber temperature—T_c, of an underwater water-breathing ramjet using magnesium as a hydro-reactive fuel vs. water/propellant mass ratio, for two different cruise velocities.

3.1.3. Practical Hydro-Reactive Composition

In practice, water-reactive ingredients may be embedded within a fuel-rich propellant matrix containing a small fraction of oxidizer. Figure 7a presents the calculated specific impulse and density specific impulse for a water-breathing ramjet motor with a propellant grain consisting of 70% Mg and 30% PTFE versus the water to propellant mass ratio for two cruise speeds. This propellant composition is fuel-rich, thus also in this case maximum temperature is achieved when water is introduced in the combustion chamber as an additional oxidizer reacting with the excess magnesium, as can be seen in Figure 7b. It is apparent from Figure 7a, that the addition of a hydro-reactive fuel ingredient such as magnesium, increases significantly the performance relatively to the case of a non-hydro-reactive composition presented at the beginning. At a cruise speed of 100 m/s it enables one to achieve a specific impulse double that achieved with a standard solid propellant rocket.

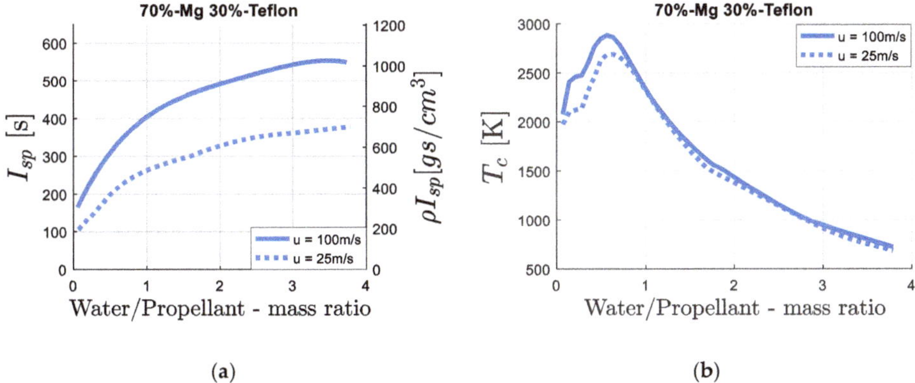

Figure 7. (**a**) Calculated specific impulse—*Isp*, density specific impulse—ρI_{sp}, and (**b**) combustion chamber temperature—T_c, of an underwater water-breathing ramjet using a hydro-reactive fuel-rich solid propellant (70% Mg + 30% PTFE) vs. water/propellant mass ratio, for two different cruise velocities.

3.2. Thrust Coefficient and Influence of Cruise Depth and Velocity on Performance

Figure 8 presents the thrust coefficient and the specific impulse of a marine water-breathing ramjet propelled by a solid propellant grain containing 70% magnesium and 30% PTFE with a w/p ratio of 4 versus the cruise speed at different depths. It is apparent from Figure 8a,b that with the increase in depth, the thrust coefficient and the specific impulse decrease. The reason is that the increase in ambient pressure reduces the exit nozzle performance (see Equation (5)). Yet, even at a depth of 100 m below sea level, the thrust coefficient is greater than the drag coefficient predicted on a torpedo-like vehicle discussed above, thus the motor can suite a broad range of velocities and depths. If necessary, the selection of a different throat diameter will enable to achieve a lower or higher thrust level (that will dictate a different propellant consumption rate).

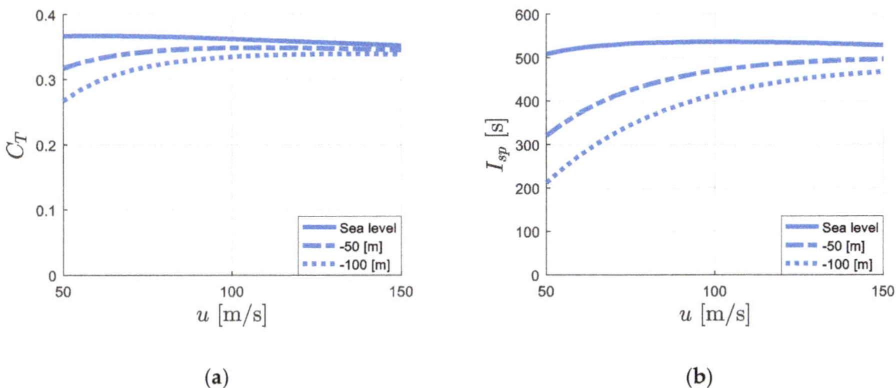

Figure 8. Calculated (**a**) thrust coefficient and (**b**) specific impulse of an underwater water-breathing ramjet with a nozzle expansion ratio of two and a throat diameter 50% of the vehicle's external diameter, using a hydro-reactive fuel-rich solid propellant (70% Mg + 30% PTFE) with a water/propellant mass ratio of 3.5 vs. cruise velocity, cruising at a depths of 0, 50, and 100 m below sea level.

4. Conclusions

The concept of a water-breathing ramjet for underwater propulsion has been evaluated theoretically based on thermochemical calculations and available practical propellants and energetic components. It is revealed that theoretically, even without chemical interaction, a solid propellant ramjet motor augmented by water entering from the surroundings through an inlet, can deliver specific impulse higher by 30% than a standard solid rocket. By introducing a fuel-rich propellant containing large amounts of a water-reactive metal such as magnesium, aluminum, or boron, the specific impulse may be doubled or more. At cruise speeds as low as 25 m/s, the performance of a water-breathing ramjet with a non-water-reactive propellant is inferior to that of a standard rocket, however, when the propellant is enriched with water-reactive ingredients, improvement in performance can be achieved even at relatively low velocities. In all cases, the specific impulse increases significantly with speed because of the increase in chamber pressure. At sea level, the thrust coefficient is almost constant at velocities between 50–150 m/s, whereas at greater depths there is a decrease in both the thrust coefficient and the specific impulse. Yet, at high velocities the thrust coefficient decreases mildly with depth, thus, the marine solid propellant water-breathing ramjet is a powerful propulsion method suitable for propelling high-speed underwater vehicles at a wide range of depths.

Author Contributions: Conceptualization, N.E.E. and A.G.; Data curation, N.E.E.; Formal analysis, N.E.E. and A.G.; Funding acquisition, A.G.; Investigation, N.E.E.; Methodology, N.E.E. and A.G.; Project administration, A.G.; Resources, A.G.; Software, N.E.E.; Supervision, A.G.; Validation, A.G.; Visualization, N.E.E.; Writing—original draft, N.E.E.; Writing—review & editing, A.G. All authors have read and agreed to the published version of the manuscript.

Funding: This research was funded by PMRI—Peter Munk Research Institute—Technion.

Acknowledgments: The authors wish to thank David Albagli for his advice with the thermochemical calculations.

Conflicts of Interest: The authors declare no conflict of interest.

References

1. Muench, R.K.; Garrett, J.H. A Review of Two-Phase Marine Propulsion. In Proceedings of the 1972 AIAA/SNAME/USN Advanced Marine Vehicles Meeting, Annapolis, MD, USA, 17–19 July 1972.
2. Greiner, L.; Hansen, F.A. Sea-Water-Aluminum Torpedo Propulaion System. In *Underwater Missile Propulsion*; Greiner, L., Ed.; Compass Publications: Arlington, VA, USA, 1967; pp. 289–300.
3. Miller, T.F.; Walter, J.L.; Kiely, D.H. A Next-Generation AUV Energy System Based on Aluminum-Seawater Combustion. In Proceedings of the 2002 Workshop on Autonomous Underwater Vehicles, San Antonio, TX, USA, 20–21 June 2002; IEEE: Piscataway, NJ, USA, 2002; pp. 111–120.
4. Yang, Y.J.; He, M.G. A Theoretical Investigation of Thermodynamic Performance for a Ramjet Based on a Magnesium—Water Reaction. *J. Eng. Marit. Environ.* **2010**, *224*, 61–72. [CrossRef]
5. Huang, L.; Xia, Z.; Hu, J.; Zhu, Q. Performance Study of a Water Ramjet Engine. *Sci. China Technol. Sci.* **2011**, *54*, 877–882. [CrossRef]
6. Huang, H.T.; Zou, M.S.; Guo, X.Y.; Yang, R.J.; Li, Y.K. Analysis of the Aluminum Reaction Efficiency in a Hydro-Reactive Fuel Propellant Used for a Water Ramjet. *Combust. Explos. Shock. Waves* **2013**, *49*, 541–547. [CrossRef]
7. Hu, J.; Han, C.; Xia, Z.; Huang, L.; Huang, X. Experimental Investigation on Combustion of High-Metal Magnesium-Based Hydroreactive Fuels. *J. Propuls. Power* **2013**, *29*, 692–698. [CrossRef]
8. Ghassemi, H.; Farshi, F.H. Investigation of Interior Ballistics and Performance Analysis of Hydro-Reactive Motors. *Aerosp. Sci. Technol.* **2015**, *41*, 99–105. [CrossRef]
9. Gany, A. Innovative Concepts for High-Speed Underwater Propulsion. *Int. J. Energ. Mater. Chem. Propuls.* **2018**, *17*, 83–109. [CrossRef]
10. Rosenband, V.; Hafied, A.; Gany, A.; Timnat, Y.M. Magnesium and Boron Combustion in Hot Steam Atmosphere. *Def. Sci. J.* **1998**, *48*, 309–315. [CrossRef]

11. Risha, G.A.; Huang, Y.; Yetter, R.A.; Yang, V.; Son, S.F.; Tappan, B.C. Combustion of Aluminum Particles with Steam and Liquid Water. In Proceedings of the 44th AIAA Aerospace Sciences Meeting and Exhibit, Reno, NV, USA, 9–12 January 2006; AIAA: Reston, VA, USA, 2006; pp. 14007–14014. [CrossRef]
12. Franzoni, F.; Milani, M.; Montorsi, L.; Golovitchev, V. Combined Hydrogen Production and Power Generation from Aluminum Combustion with Water: Analysis of the Concept. *Int. J. Hydrogen Energy* **2010**, *35*, 1548–1559. [CrossRef]
13. Rosenband, V.; Gany, A. Application of activated aluminum powder for generation of hydrogen from water. *Int. J. Hydrogen Energy* **2010**, *35*, 10898–10904. [CrossRef]
14. Elitzur, S.; Rosenband, V.; Gany, A. Study of Hydrogen Production and Storage Based on Aluminum–Water Reaction. *Int. J. Hydrogen Energy* **2014**, *39*, 6328–6334. [CrossRef]
15. Chan, S.H.; Tan, C.C.; Zhao, Y.G.; Janke, P.J. Li-SF_6 Combustion in Stored Chemical Energy Propulsion Systems. In *Symposium (International) on Combustion, Proceedings of the Twenty-Third Symposium (International) on Combustion, Orleans, France, 22–27 July 1990*; Elsevier: Amsterdam, The Netherlands, 1991; pp. 1139–1146.
16. Gordon, S.; McBride, B.J. *Computer Program for Calculation of Complex Chemical Equilibrium Compositions and Applications*; Rept. RP-1311; NASA Lewis Research Center: Cleveland, OH, USA, 1994.
17. Hoerner, S.F. *Fluid-Dynamic Drag*, 2nd ed.; Hoerner Fluid Dynamics: Bakersfield, CA, USA, 1965; p. 177.
18. Wolfe, W.P.; Hailey, C.E.; Oberkampf, W.L. Drag of Bodies of Revolution in Supercavitating Flow. *J. Fluids Eng.* **1989**, *111*, 300–305. [CrossRef]
19. Sutton, G.P.; Biblarz, O. *Rocket Propulsion Elements*, 8th ed.; John Wiley & Sons, Inc.: Hoboken, NJ, USA, 2010; pp. 53–74.

Journal of
Marine Science and Engineering

Article

A Comparison of Physical and Numerical Modeling of Homogenous Isotropic Propeller Blades

Luca Savio [1,2,*,†], Lucia Sileo [1,†] and Sigmund Kyrre Ås [1,2]

1 SINTEF Ocean, NO-7465 Trondheim, Norway; Lucia.Sileo@sintef.no (L.S.); sigmund.k.aas@sintef.no (S.K.Å.)
2 Norwegian University of Science and Technology (NTNU), Faculty of Engineering, Department of Marine Technology , NO-7491 Trondheim, Norway
* Correspondence: luca.savio@sintef.no
† These authors contributed equally to this work.

Received: 10 December 2019; Accepted: 26 December 2019; Published: 3 January 2020

Abstract: Results of the fluid-structure co-simulations that were carried out as part of the FleksProp project are presented. The FleksProp project aims to establish better design procedures that take into account the hydroelastic behavior of marine propellers and thrusters. Part of the project is devoted to establishing good validation cases for fluid-structure interaction (FSI) simulations. More specifically, this paper describes the comparison of the numerical computations carried out on three propeller designs that were produced in both a metal and resin variant. The metal version could practically be considered rigid in model scale, while the resin variant would show measurable deformations. Both variants were then tested in open water condition at SINTEF Ocean's towing tank. The tests were carried out at different propeller rotational speeds, advance coefficients, and pitch settings. The computations were carried out using the commercial software STAR-CCM+ and Abaqus. This paper describes briefly the experimental setup and focuses on the numerical setup and the discussion of the results. The simulations agreed well with the experiments; hence, the computational approach has been validated.

Keywords: fluid structure interaction; propulsion; CFD

---

1. Introduction

The topic of hydroelastic behavior of marine propellers is nowadays often associated with the behavior of composite propellers that aim at using some sort of anisotropy to achieve some set goal. The goal may be reducing vibrations or cavitation (Young [1,2]). Already in 2001, Mouritz et al. [3] published a review article citing a quite extensive list of applications of composite materials in naval applications, including propellers, where the benefits of composite materials could offset the increased costs that are associated with their design and production. However, this paper covers isotropic propellers as a first step towards anisotropic propellers. In fact, as it will be clear further in the paper, the simulation strategy adopted here should in principle work independently whether the material is isotropic or anisotropic. The choice of starting from isotropic materials was mainly made because the production methods needed for making the model scale blades were easier than for anisotropic materials.

The elastic behavior of isotropic blades has received less attention although it may in same cases be noticeable also for full-scale metallic propellers. In the past, Atkinson and Glover [4] have shown numerically how full-scale metallic propellers could bend under the action of hydrodynamic forces to an extent that some particular propeller geometries, namely blades with large skew angles, could have some significance in terms of propeller performance. The numerical method adopted was a combination of lifting line and finite element methods (FEM) in what we would call a weak co-simulation in modern terms. Nowadays, co-simulations are increasingly carried out coupling CFD

(Computational Fluid Dynamics) with FEM (Finite Element Method) code. The coupling between the two codes vary from weak to strong, according to how the two simulations communicate in terms of time increments and variables that are transferred. While this has been a research area for quite some time, modern software implementations have reached a level of maturity that makes these types of analysis attractive also at the level of industrial design engineers. The increased ease of setting up co-simulations should not however be taken as a guarantee for obtaining accurate results. In fact, the accuracy of the computations should be verified and compared with results obtained in controlled scenarios, as, for example, during open water tests in a towing tank or cavitation tunnel. Despite some recent attempts of fill in the gap [5,6], there is still a scarcity of experimental data. The validation of numerical results by means of experiments presents a series of challenges that require careful preparation and interpretation of the results. It is important to limit the scope of what can be validated by means of experiments; it is practically impossible to design experiments that allow for correctly scaling to laboratories size displacements and forces of some full-scale hydroelastic phenomenon. A thorough review of what are the limitations of mode scale experiments can be found in Young [7]. What is possible is to design experiments that allow for their correct representation in the simulation codes with the aim of validating the latter. In this paper, we present the results from experiments carried out on three propeller designs that have been produced with blades both in aluminum and in a casting resin resembling an epoxy resin , hence presenting a rigid and flexible behavior, respectively. The three propellers share the same design parameters apart from the skew distribution. The tests have been performed in a classical open-water configuration at different propeller speeds and at different pitch settings. The open-water configuration was chosen so that the hydroelastic behavior was limited to static deflection, i.e., vibrations of the blades were avoided by performing the tests in a homogenous inflow.

2. Model Tests

2.1. Propeller Geometries

Three propeller geometries were generated starting from a master geometry varying the skew distribution. The master geometry is labeled P1374 and was designed using SINTEF Ocean proprietary propeller design code AKPD by Achkinadze et al. as presented in [8,9]. Propeller P1374 has a balanced skew distribution that amounts to 23 degrees. The first variation, P1565, was obtained by removing entirely any skew distribution. In the second variation, P1566, the same total skew as propeller P1374 was kept but in an unbalanced form. The geometries of the blades are available, since the FleksProp consortium decided to allow disclosure of the geometries of the three propellers. They are reported in Appendix A, while the geometrical definition of the propellers are described in Appendix B. A visual impression of how the three propellers look like is given in Figure 1.

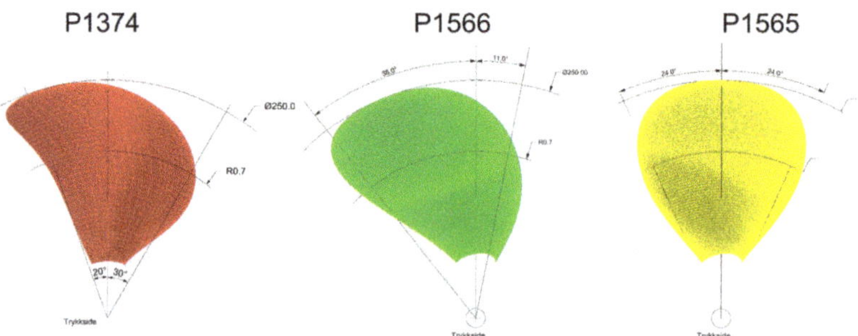

Figure 1. Outline of the three geometries showing the different skew distributions.

2.2. Production of Flexible Blades

The flexible blades were manufactured using resin casting under vacuum, which is a commonly used technique in rapid prototyping. There are several variants of the resin casting technique. In the variant that was adopted in this case, the first step is to produce a positive object that is then used to form a silicon mold where the resin is cast under vacuum and left to cure at constant temperature. Three complete controllable pitch propellers, including the hubs, were produced in aluminum according to the production standard used in SINTEF Ocean. One blade per kind was then sent to a company specializing in rapid prototyping that produced 5 copies of the blade.

According to the producer of the resin used for casting, Young's modulus is equal to 2.2 GPa. Poisson's ratio is not reported, but given the nature of the material, it was assumed to be equal to 0.33. The material properties reported by the producer have not been checked by mechanical testing; however, the good agreement between computations and experiments indicates that the reported values can be considered accurate.

2.3. Test Executions

The model tests have been carried in the SINTEF Ocean large towing tank between the months of February and June 2018. The setup adopted was the classical open-water configuration where the propeller is placed in front of the dynamometer. The dynamometer used in the test was a Kempf & Remmers H29 model. Figure 2 shows that only the surfaces in contact with water located on the left side of the 1-mm gap contribute to the forces measured by the dynamometer.

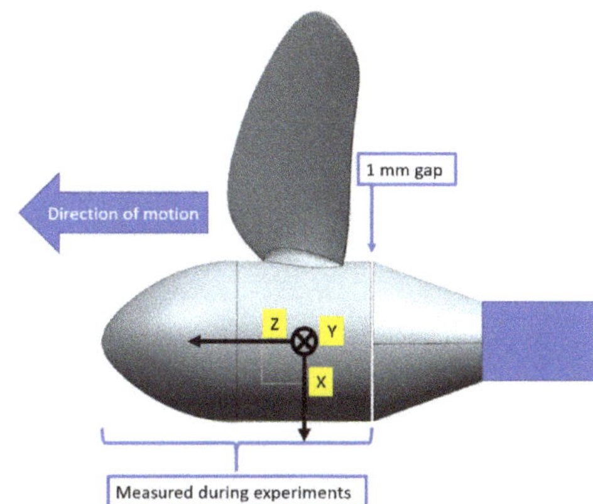

Figure 2. Experimental setup showing which surfaces contribute to the forces measured during the experiments.

According to standard procedures for open-water tests, a dummy hub without blades was tested to obtain forces on the cap and hub at different velocities. The forces measured during the test with the dummy are then subtracted from the forces measured with the actual propeller to obtain what is traditionally referred to as open water curves. The traditional way of presenting the data tries to isolate the blades from the propeller. When the experimental data is to be compared with CFD results, it is more consistent to use the uncorrected (raw) force measured during the experiments and to compute the forces on the abovementioned surface on the measuring side of the gap.

The open-water tests were performed at 3 different propeller rotational speeds: 7, 9, and 11 rps in the range of advance coefficients from 0 to 1.2 in 0.2 steps for the pitch setting $P/D = 1.1$ and from 0 to 1.05 in 0.15 steps for the pitch setting 0.9. Tests were also performed at $P/D = 1.2$, but the results from this pitch setting have not been simulated yet.

The experimental relative uncertainity at 95% confidence for the thrust and torque coefficients has been evaluated using the methodology described in Reference [10] and found to be weakly dependent on the advance coefficient J as Figure 3 shows for P1374 at n = 7 rps. Slightly smaller values of the uncertainty were found for the other rps. The uncertainty reported here is relative only to propeller P1374, but similar values were found also for the other propellers.

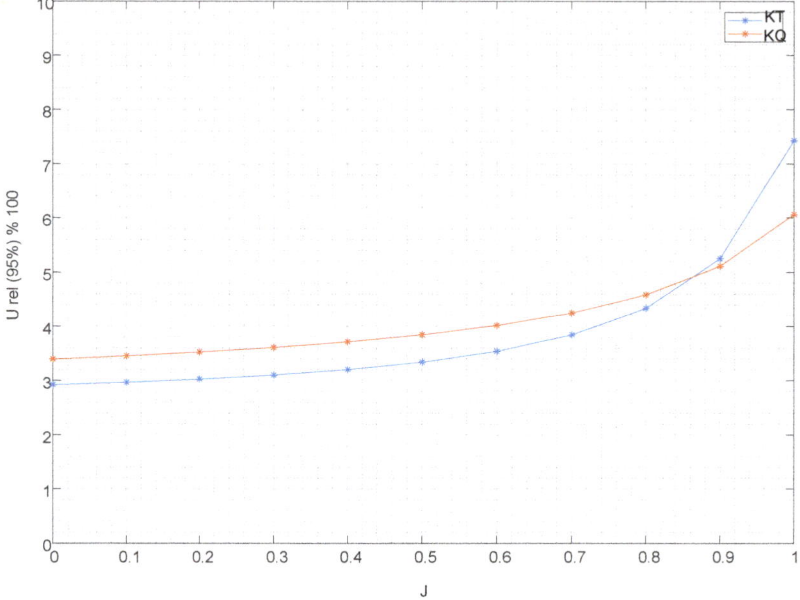

Figure 3. Example of relative uncertainity for propeller P1374 at P/D 1.1 and $n = 7$.

3. FSI Simulations

Fluid-structure interaction (FSI), in the widest sense, is the thermomechanical interaction of one or more solid structures with an internal or surrounding fluid flow. FSI problems play prominent roles in many scientific and engineering fields, such as mechanical, aerospace, and biomedical engineering. Thanks to the recent advances of commercial software, numerical simulations have become a feasible and efficient way to investigate the fundamental physics involved in the complex interaction between fluids and solids. Nevertheless, there is still the need for validation when these tools are used in new applications.

When using a numerical procedure, an FSI problem is usually described in terms of a fluid domain and a structural domain, communicating through an in-between fluid-structure interface. This multi-physics problem with adjacent domains can be simulated in a monolithic or in a partitioned way [11]: the former simultaneously solves the governing equations for fluid and structure within a unified algorithm, and it requires a code developed for this particular combination of physical problems. In the partitioned approach, on the other hand, the fluid and the structure are treated as two computational fields which can be solved separately with their respective mesh discretizations and numerical algorithms, and the boundary conditions at the interface are used explicitly to transfer information between the fluid and structure solutions. The partitioned approach enables the use of existing well-established fluid and structural solvers. This represents a significant motivation for adopting this approach in addition to the fact that, for many problems, the staggered approach works well, is very efficient, and preserves software modularity. Difficulties may arise, however, in terms of robustness and accuracy for specific problems like flutters or parachute problems [12,13].

In the present case, we consider the problem of viscous incompressible fluid flow interacting with an elastic body (the propeller blade) immersed in the flow and being deformed by the fluid action. A staggered approach is used, coupling two different solvers that use different methods and codes: a CFD/FVM (Computational Fluid Dynamics/Finite Volume Method) solver for fluids and FEA/FEM (Finite Element Analysis/Method) for structure mechanics.

In "two way" FSI, the results of the CFD solver, in terms of hydrodynamic forces on the interface, are mapped from the fluid model to the structural model and are provided to the structural equation solver; the results obtained by the structural solver are mapped back to the first model in terms of displacements of the interfacial surface, involving the modification/morphing of the mesh of the fluid model. The procedure is repeated until convergence. How often this mapping needs to be performed depends on the degrees of coupling, and it is related to the response times of the structure compared to the fluid: in case of strong coupling, the mapping may be needed in the internal time-step iterations inside each time step, following an implicit method.

In the present case, the fluid and structure are "weakly" coupled because the response of the structure to a disturbance in the fluid is slow compared to the fluid. As seen from the experiments, a "static" solution is expected, which means that one can search for a steady-state condition, corresponding to the shape the structure takes in the stationary fluid when it reaches the hydroelastic equilibrium, corresponding to the steady-state flow around the deformed blade. All these considerations have important consequences on the flexible-blade simulations:

- explicit algorithm can be used, which means that the fluid and the structure solvers are not necessarily in the processor memory at the same time and that the mapping between the two codes may happen at each time-step (and not at each internal iteration)
- the analysis does not need to be time accurate, as an equilibrium state with a static solution is desired
- material damping can be increased
- first-order integration in time may be used

Numerical Setup

In the present work, STAR-CCM+ v12.04 by Siemens PLM Software [14] is used as CFD code and Abaqus 2016 by Dassault Systèmes is used as FEM code. An open-water simulation was used as a starting point, with the following characteristics: the computational domain is corresponding to one-blade passage only, the MRF (Multi-Reference Frame) approach is used. Using the MRF approach means that the propeller rotation is not simulated with the actual rotation of the mesh but that it is modeled using a reference system rotating with the blades. The equations of motion are solved in this reference system, once the boundary conditions have been correctly set up on the propeller surfaces with the proper value of velocity.

The computational domain and boundary conditions are illustrated in Figures 4 and 5. It is a 90° sector of a cylinder, of which the axis corresponds to the propeller axis of rotation and radius equal to 10 propeller diameters. The propeller is located in the region highlighted in red in Figure 5, where a polyhedral mesh is generated and then extruded further upstream and downstream toward the inlet and outlet boundary respectively.

Figure 4. Computational domain used in the simulation: The picture shows the size of the computational domain compared to the blade size. Note the fact that only one of the four blades was simulated.

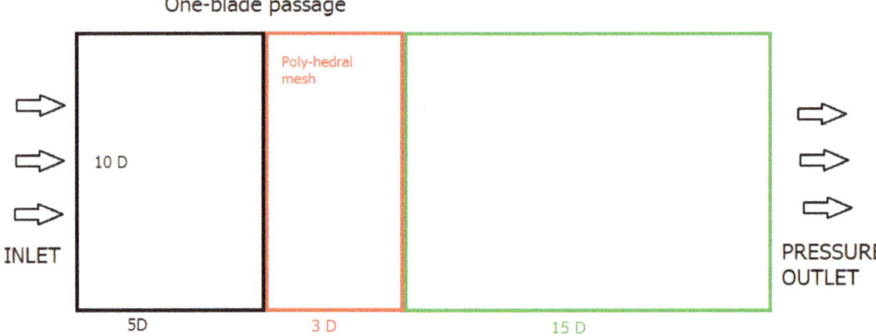

Figure 5. Domain size and boundary conditions: The figure depicts the domain size in terms of propeller diameters and the function of the different blocks used in the meshing procedure.

The geometry of the numerical propeller model is kept as close as possible to the one used in the experiments, including the gap of 1 mm between the hub and the cone; the housing is modeled with infinite length, extended up to the outlet.

A longitudinal section of the volume mesh is depicted in Figure 6. The mesh for all the considered cases consists of about 1.6 M cells, including 10 layers of prismatic layers at the wall. The thickness of the first prismatic layer is set to approximately have Y^+ of the order of 1 on the blades. The Gamma–ReTheta transition model is used for turbulence, which performed better than the Shear Stress Turbulent (SST)-K-ω and the Reynolds Stress Turbulent (RST) models; the reason is because, at the model scale, there are likely areas of the blades subjected to laminar flow with transition to turbulent flow.

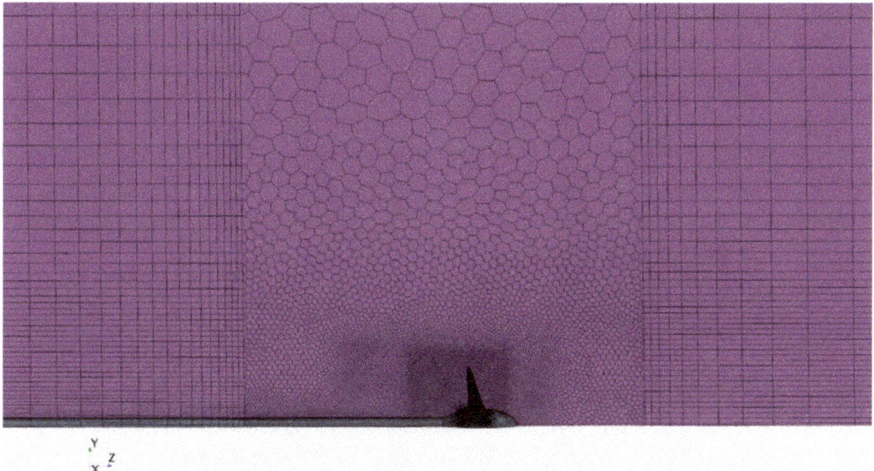

Figure 6. Longitudinal section of the volume mesh of the fluid dynamic solver.

The structural mesh consisted of 2.1×10^5 tetrahedral elements and was generated by the Abaqus embedded mesher. To determine the number of cells that was necessary for a correct structural simulation, a mesh refinement study was performed where the mesh was refined until the displacements at the blade tip were no longer changing with the mesh size. Mesh refinements were applied at the leading and trailing edges of the blade. While the mesh target size was set to 0.5 mm, the mesh refinement at the edges dominated the mesh size as it can be seen in Figure 7, where the structural mesh of propeller P1565 is shown as an example.

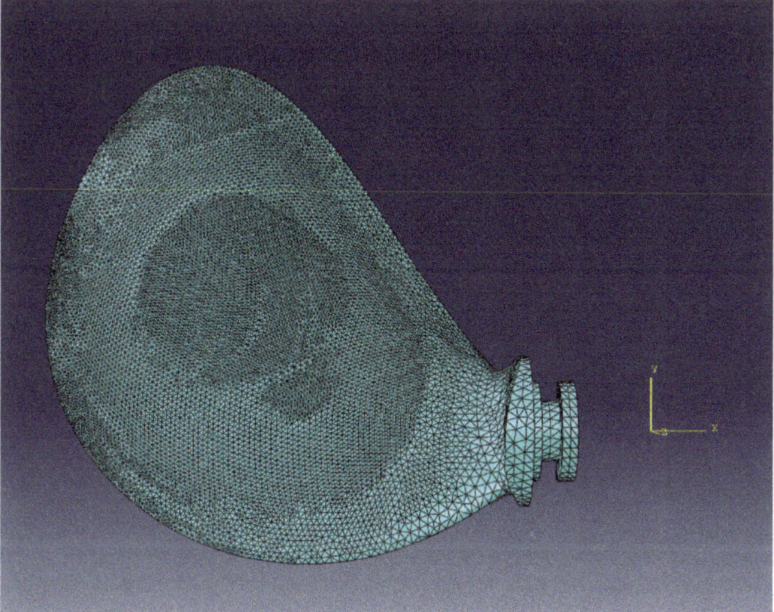

Figure 7. As an example of the mesh used by the strucural solver, the mesh relative to propeller P1565 is presented.

For the structural part of the computations, it is important to model the root of the blade and the fastening system. The blade root is built in the same resin as the rest of the blade and has several contact surfaces that constrain it to the propeller hub. The different contact surfaces determine different boundary conditions. In Figure 8, the different boundary conditions are identified with different colors; the red surfaces are an encastre-type boundary condition, while the green surface identifies a surface that allows only rotation along the direction perpendicular to the surface and translations in the plane identified by the surface.

Figure 8. Boundary condition at the blade root: The color codes can be found in the text.

The three propellers designs, P1374, P1566, and P1565, are analyzed. Initially the open-water calculations are carried out for the rigid case, with reference to the model tests with the aluminium propellers. The resulting flow field is then used as initial flow state for the CFD-FEM co-simulations, activating the Abaqus and the morpher solvers.

The mapping is managed by StarCCM+, where the co-simulation setup is accomplished by identifying the coupled model parts and the exported/imported fields (pressure/displacements) and by specifying the external code execution details (commands, input files, etc.).

4. Results

A converged solution of the co-simulation is reached well before the simulation time, initially fixed to 10 s, with time step of 0.005 s, as shown in Figure 9, where the axial displacements at different locations of the blade are monitored during the simulation for the propeller P1374, $P/D = 1.1$, and $n = 11$ rps. The results obtained for the three propellers at $P/D = 1.1$ and $P/D = 0.9$ are reported in Figures 10 and 11, respectively: k_T, k_Q, and η obtained by the co-simulation are plotted as a function of the advance ratio J both for the aluminum propeller and for the (flexible) resin propeller and compared to the experimental values. The agreement is very satisfactory, also in terms of the relative increase of the coefficients in the flexible case with respect to the rigid one. Except for the bollard condition and for the $J = 1$ case, where the blades can experience flow with "negative" angle of attack and subsequent separation and unsteadiness, the difference between the numerical results and the experimental data is less than 2% in terms of thrust coefficients, as shown for example in Figure 12, where the differences for k_T, k_Q, and η_0 are shown for the propeller P1374, $P/D = 1.1$. Note that, with reference to Figure 3, the deviation seen between CFD and experiments is, in most cases, within the uncertainty bands of the experiments. The only case where the deviations seem to be large is for the propeller efficiency at $P/D = 0.9$ and $J = 1.0$; however, it should be noted that the KT coefficient is slightly negative, leading to a negative efficiency, which is strictly speaking not consistent with the definition of propeller efficiency, and hence, the propeller efficiency for that specific condition should be disregarded; further, the thrust and torque coefficients are correctly captured by the simulations.

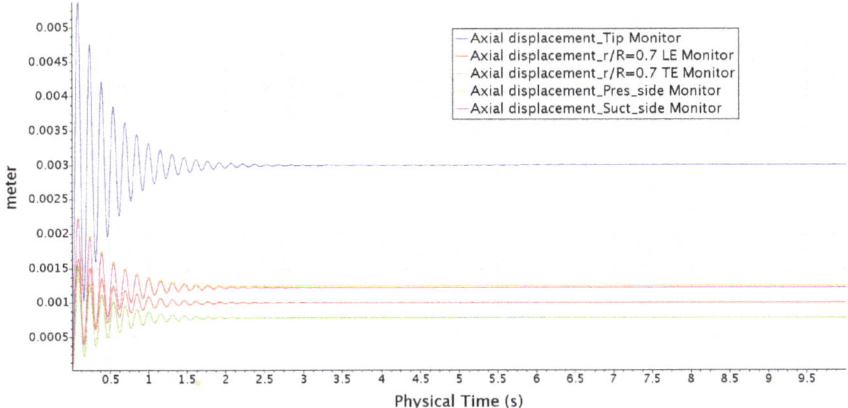

Figure 9. Propeller P1374, $P/D = 1.1$, and $n = 11$ rps: axial displacements at different locations of the blade monitored during the simulation.

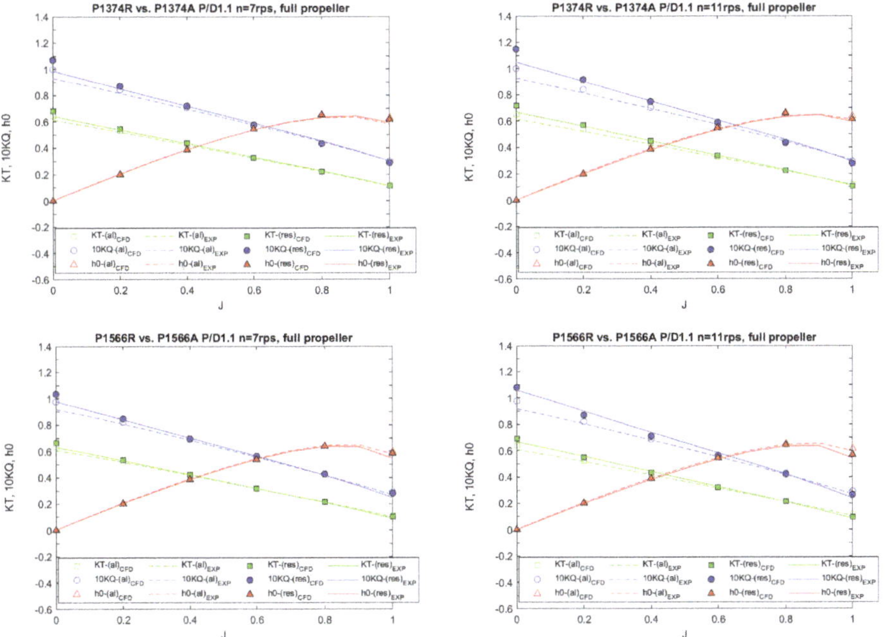

Figure 10. *Cont.*

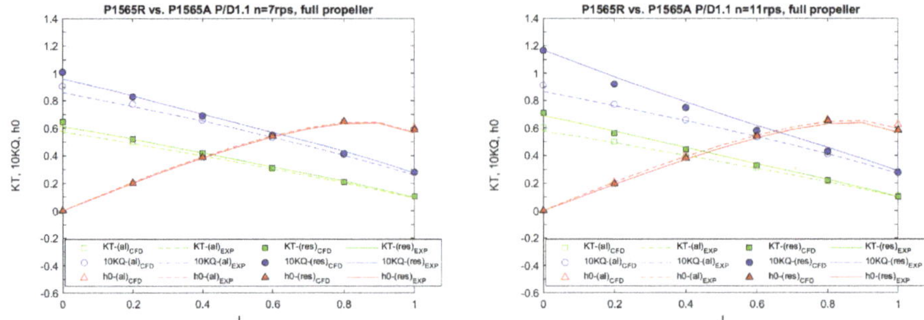

Figure 10. Results for $P/D = 1.1$: The results are presented in a matrix form where the different geometries are arranged in rows and the rps is in columns. The continuous lines are relative to the experimental results, while the dots are relative to the simulations. Further, the dashed lines refer to the aluminum models and the solid lines refer to the resin models; the hollow dots are relative to the computation with the rigid propeller, whereas the the solid dots are relative to the flexible propeller.

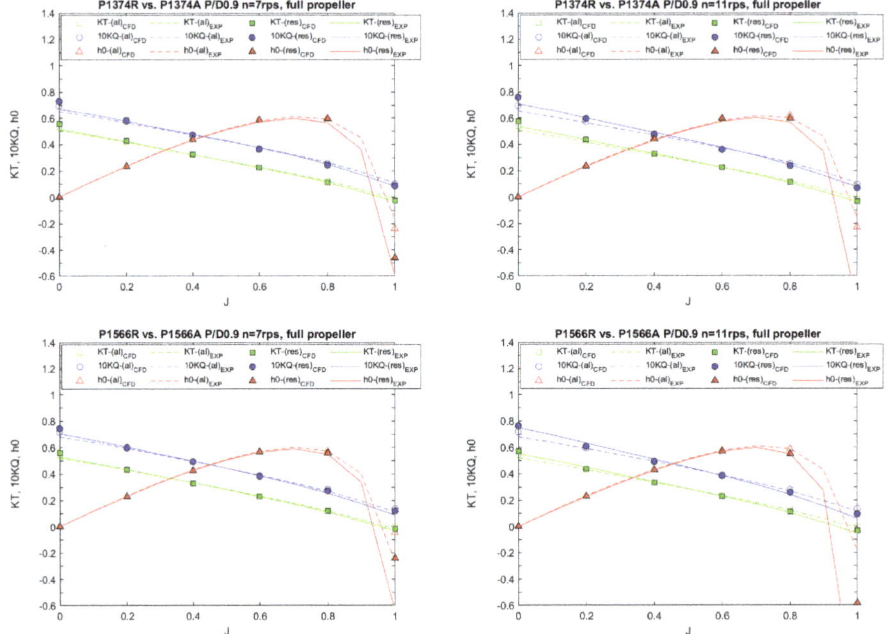

Figure 11. *Cont.*

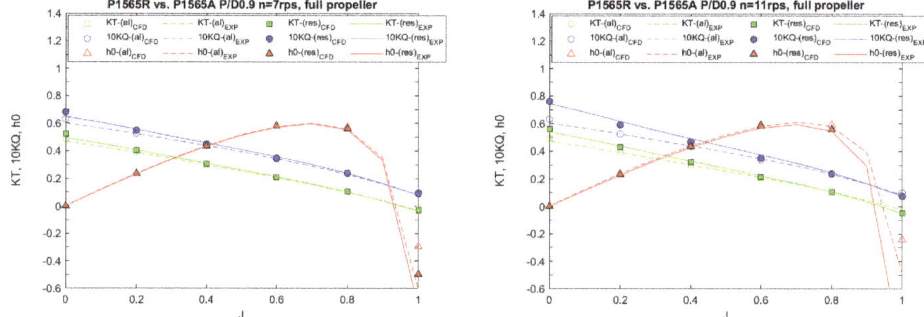

Figure 11. Results of $P/D = 0.9$: The results are presented in a matrix form where the different geometries are arranged in rows and the rps is in columns. The continuous lines are relative to the experimental results, while the dots are relative to the simulations. Further, the dashed lines refer to the aluminum models and the solid lines refer to the resin models; the hollow dots are relative to the computation with the rigid propeller, whereas the the solid dots are relative to the flexible propeller.

Figure 12. *Cont.*

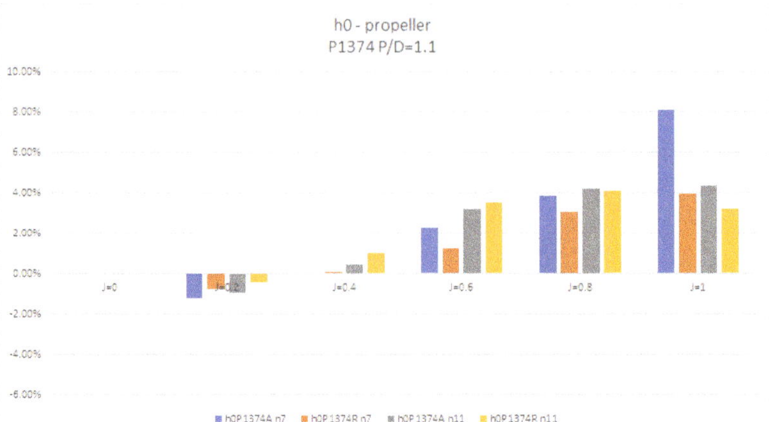

Figure 12. Propeller P1374, $P/D = 1.1$, $n = 7$, and $n = 11$ rps: relative differences between numerical and experimental data in terms of thrust and torque coefficients and propeller efficiency.

Blade Displacements

It is rather straightforward to extract the blade cylindrical section displacements from the numerical computations and to compare the effect of the different skew distributions of the three blades. One way that seems quite natural is to decompose the displacements of the sections in terms of bending and twisting. In this approximation, the bend is defined as the average displacement of the blade sections in the YZ plane, where Z is the propeller advance direction, X is the radial direction, and Y is the direction perpendicular to both X and Z. Twist is defined as the rotation of the blade section in the same YZ plane. Some results for this twist definition are shown in Figures 13 and 14. Close to the tip, the twist component is correctly identified, while erroneous results appear closer to the blade root. Still, this presentation will be used here since it offers a compact way of presenting the data. Note also that the reservations on the bend-twist presentation of the data apply only to twist.

It was expected that the elastic behavior of the low aspect ratio propeller blades could only partially be represented by pure bend and twist of the blade sections. It is important to keep in mind this finding when trying to implement design codes that are based on lower order theories, as for example lifting line theory coupled with beam theory. It may be beneficial to consider also the warping in addition to the bending and twisting of the blade sections, as shown in Figure 14.

Even with the reservation that has just been described, it is possible to use the bend-twist approximation to explain the effect of the propeller skew on propeller blades that are otherwise identical. In Figures 15 and 16, the bend and twist of the sections from the radial location 0.4 to 0.975 are plotted.

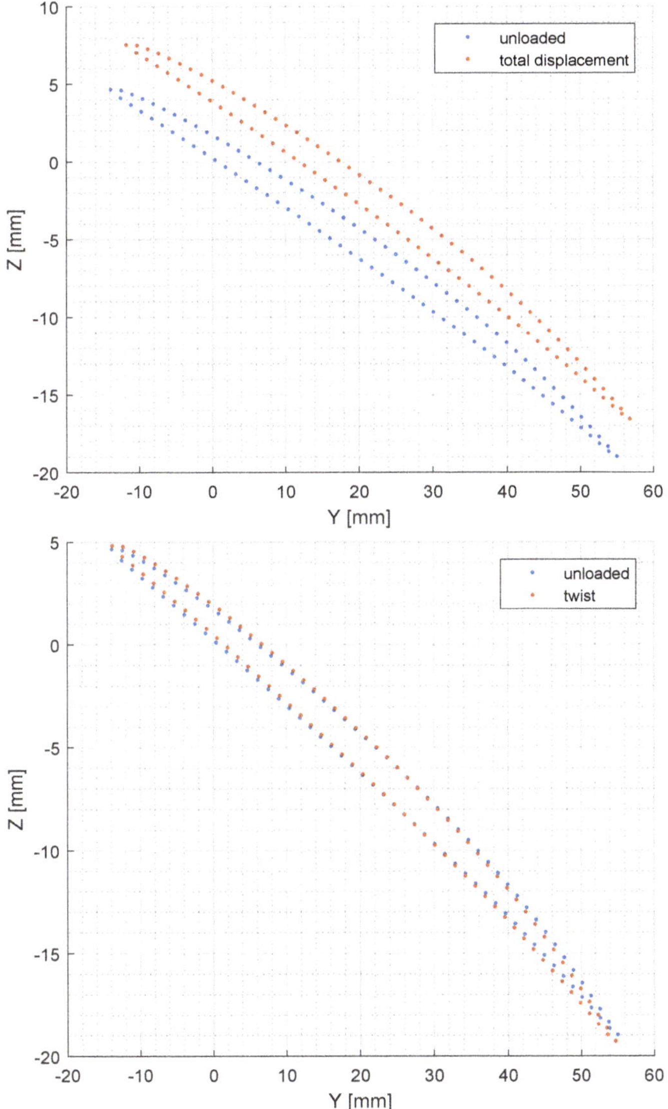

Figure 13. Bend and twist of a typical blade section towards the tip: The top plot shows the same cylindrical section unloaded and loaded. The bottom picture shows the same sections once the bending is removed from the loaded blade so that the blade twist can be more easily seen.

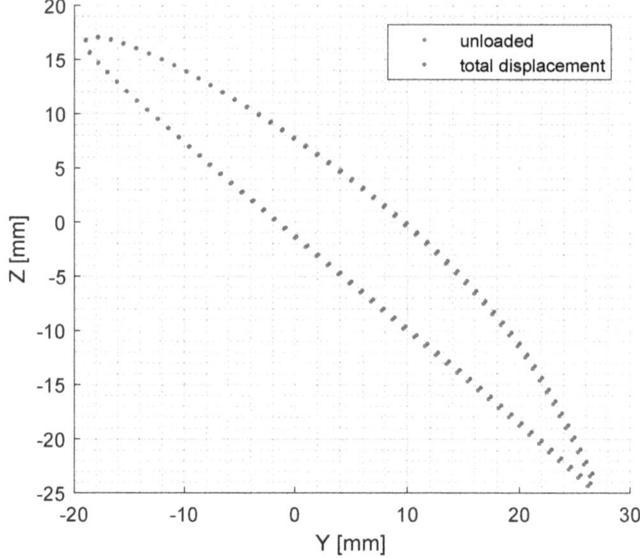

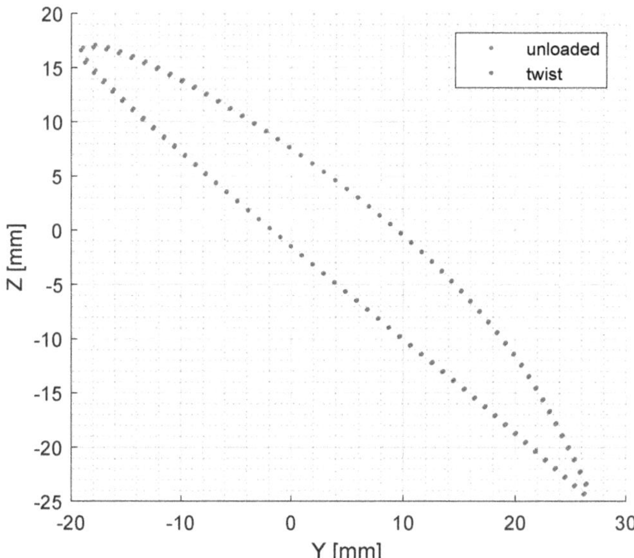

Figure 14. Total displacement and twist of a section close to the blade root: The top plot shows the same cylindrical section unloaded and loaded. In the bottom picture, the procedure of removing the section bending is applied showing that, close to the blade root, the sections are warping rather than twisting.

Although it may appear to an extent fictitious, since the skew distribution is often used to control cavitation and little room is left for other uses of it, in a hypothetical case where cavitation is not a concern, it may be in principle possible to think using the skew distribution to achieve an hydroelastic response of some sort. The mechanism by which the bend and twist are coupled in isotropic materials is the so-called geometrical bend twist coupling. In the geometrical bend twist coupling, the skew distribution influences the relative position of the pressure center and the elastic center leading to different elastic behaviors even for the same, for the sake of simplicity, uniform pressure distribution. In reality, the skew distribution influences also the pressure distribution on the blade, but in a first approximation ,we assume the geometrical bend-twist coupling as the main driving factor for the different behavior of the three geometries. In Figures 15 and 16, the bend and twist distribution along the blade radius for the three propellers at two operating conditions are shown. The operating conditions are chosen to be close to the design point and an off-design condition where the propeller is heavy-loaded. Not surprisingly, the different skew distributions of the three geometries have limited effects on the propeller bending, while the effect is clearly seen on the twist of blade sections. The blade section twist is defined as positive if, as a result of the deformation, the blade section ends up having a larger pitch than in the unloaded condition. At a design point, the balanced skew (P1374) and the zero skew (P1565) propellers are subjected to almost no twist while the unbalanced skew propeller (P1566) has a slight negative twist. In the off-design condition, the skew distribution affects to a larger extent the elastic behavior of the propellers; in this condition, all blades show a twist distribution that is increasing from the root towards the tip of the blade. The zero skew blade shows an almost linear, comparatively large twist of the blade sections. The two blade designs with skew show a somewhat similar behavior with the section towards the tip, bending considerably more than those at the root. Even though, as it has already been pointed out, the bend-twist decomposition of the blade deformation may not be accurate especially for the sections close to the blade root, a few conclusion can be drawn. The first conclusion is that it is indeed possible to use the skew distribution to obtain an hydroelastic bend-twist coupling for isotropic materials. The second conclusion, at least for the geometries that have been tested, it is that the bend-twist coupling that is obtained is hardly useful for achieving the goals that are typically expected from self-adaptive blades; in fact, all the skew distributions tend to show a tendency to increasing the twist angle and, hence, the pitch, with increasing loading, a result that is the opposite of what is generally desirable. These result are far from being surprising; however, it may be useful in some special applications to consider geometric bend-twist coupling as well when designing isotropic blades that are expected to work in off-design conditions.

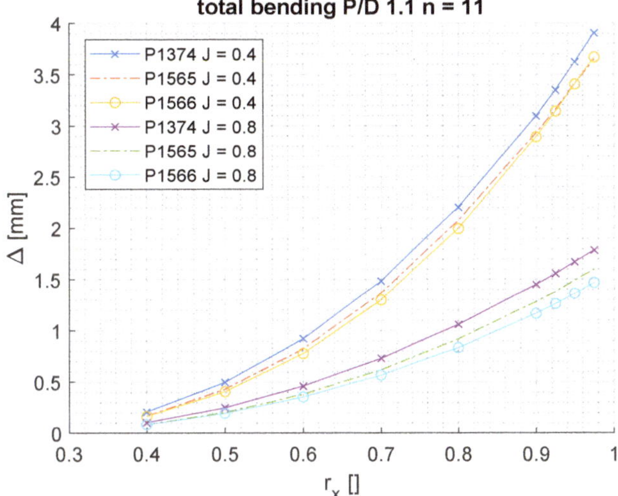

Figure 15. Bending of the three different blade geometries at $P/D = 1.1$ and 11 rps.

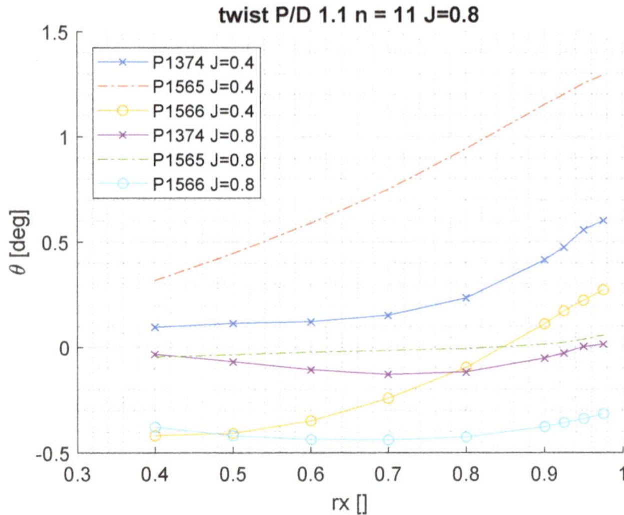

Figure 16. Twist of the three different blade geometries at P/D = 1.1 and 11 rps.

5. Discussion

In this paper, the validation of fluid-structure interaction computations for flexible propeller blades carried out using the co-simulation approach has been described. The validation data come from a series of ad hoc tests performed with the specific aim of serving as a reference for numerical computations. For the experiments, three propellers were manufactured both in a metallic and plastic material in order to have rigid and flexible models. The propeller geometries and the experimental data are public domain and can be obtained upon request. The simulated values of thrust and torque coefficients lie mostly within the bands of experimental uncertainty, showing how CFD-FEM co-simulations can be used to accurately represent the behavior of homogeneous isotropic blades tested in open water. Furthermore, the thrust coefficients seem to compare better than the torque coefficients. It is often reported that, in model-scale CFD simulation, the torque coefficient suffers from deviation that are larger than those observed on the thrust coefficient, a phenomenon that is often attributed to the laminar-to-turbulent transition of the flow on the model-scale blades. Since the Reynolds numbers of full-scale propellers are higher, it is to be expected that CFD can better predict the full-scale torque coefficient, since the flow is mainly turbulent, a fact that makes the approach described here even more valuable. Finally, in the paper, it is shown that warping of the blade sections towards the root of the blade may be a better representation of the actual deformation of the blade than twist.

Author Contributions: Conceptualization, L.S. (Luca Savio) and L.S. (Lucia Sileo); methodology, L.S. (Lucia Sileo); software, L.S. (Lucia Sileo) and S.K.Å.; validation, L.S. (Luca Savio); resources, L.S. (Luca Savio); writing—review and editing, L.S. (Lucia Sileo), L.S. (Luca Savio), and S.K.Å.; project administration, L.S. (Luca Savio); funding acquisition, L.S. (Luca Savio). All authors have read and agreed to the published version of the manuscript.

Funding: This research was funded by Research Council of Norway and Kongsberg Maritime grant number 267495/O80.

Acknowledgments: The present work has been fully supported by the FleksProp project. The FleksProp project is a cooperation between SINTEF Ocean, the Norwegian University of Science and Technology NTNU, and Kongsberg Maritime with the economic support of the Research Council of Norway (RCN) and Kongsberg Maritime. The economic support of the Research Council of Norway and Kongsberg Maritime is greatly appreciated.

Conflicts of Interest: The authors declare no conflict of interest.

Abbreviations

The following abbreviations are used in this manuscript:

FSI	Fluid-Structure Interaction
CFD	Computational Fluid Dynamics
FVM	Finite Volume Method
FEA	Finite Element Analysis
FEM	Finite Element Method
MRF	Multi-Reference Frame
SST	Shear Stress Turbulent model
RST	Reynolds Stress Turbulent model
rps	revolutions per second

Appendix A. Propeller Geometry

Appendix A.1. P1374

Main propeller elements

Propeller diameter................................ : 250.000 (mm)
Hub diameter.. : 60.000 (mm)
Number of blades..................................: 4
Expanded area ratio...............................: 0.602
Expanded "total" skew angle..............: 23.067 (deg.)

Table A1. Main geometrical parameters of propeller P1374.

-	mm	mm	mm	mm	mm	mm	mm	mm	mm	deg
r/R	r	Lfor	Laft	b	cs	xr	eo	fo	P	Fi
0.24	30	16.67	−16.67	33.35	0	0	9.51	0.28	269.5	55.03
0.25	31.25	18.12	−17.22	35.34	0.45	0	9.38	0.85	269.74	53.95
0.3	37.5	25.21	−19.78	44.99	2.71	0	8.68	1.69	270.84	48.98
0.35	43.75	31.96	−22.13	54.09	4.91	0	8	2.19	271.81	44.68
0.4	50	38.26	−24.34	62.6	6.96	0	7.35	2.56	272.66	40.96
0.5	62.5	48.98	−28.62	77.6	10.18	0	6.12	3.08	273.96	34.9
0.6	75	55.93	−33.46	89.39	11.24	0	5	3.35	274.74	30.24
0.7	87.5	57.33	−39.65	96.98	8.84	0	3.98	3.36	275	26.57
0.8	100	50.68	−47.81	98.49	1.44	0	3.05	3	272.05	23.41
0.9	112.5	31.88	−57.23	89.1	−12.68	0	2.23	2.08	259.41	20.15
0.95	118.75	14.71	−60.34	75.05	−22.81	0	1.85	1.35	248.38	18.41
0.975	121.88	1.87	−59.42	61.29	−28.77	0	1.68	0.9	241.52	17.51
0.99	123.75	−9.69	−55.61	45.92	−32.65	0	1.56	0.54	236.97	16.95
0.995	124.38	−15.66	−52.31	36.65	−33.99	0	1.54	0.38	235.39	16.76
1	125	−25.99	−44.74	18.75	−35.36	0	1.5	0	233.75	16.57

Appendix A.2. P1565

Main propeller elements

Propeller diameter................................ : 250.000 (mm)
Hub diameter.. : 60.000 (mm)
Number of blades..................................: 4
Expanded area ratio...............................: 0.602
Expanded "total" skew angle..............: 0.0 (deg.)

Table A2. Main geometrical parameters of propeller P1565.

-	mm	mm	mm	mm	mm	mm	mm	mm	mm	deg
r/R	**r**	**Lfor**	**Laft**	**b**	**cs P1566**	**xr**	**eo**	**fo**	**P**	**Fi**
0.24	30	16.67	−16.67	33.35	0	0	9.51	0.28	269.5	55.03
0.25	31.25	17.67	−17.67	35.34	0	0	9.38	0.85	269.74	53.95
0.3	37.5	22.49	−22.49	44.99	0	0	8.68	1.69	270.84	48.98
0.35	43.75	27.04	−27.04	54.09	0	0	8	2.19	271.81	44.68
0.4	50	31.3	−31.3	62.6	0	0	7.35	2.56	272.66	40.96
0.5	62.5	38.8	−38.8	77.6	0	0	6.12	3.08	273.96	34.9
0.6	75	44.69	−44.69	89.39	0	0	5	3.35	274.74	30.24
0.7	87.5	48.49	−48.49	96.98	0	0	3.98	3.36	275	26.57
0.8	100	49.24	−49.24	98.49	0	0	3.05	3	272.05	23.41
0.9	112.5	44.55	−44.55	89.1	0	0	2.23	2.08	259.41	20.15
0.95	118.75	37.53	−37.53	75.05	0	0	1.85	1.35	248.38	18.41
0.975	121.88	30.64	−30.64	61.29	0	0	1.68	0.9	241.52	17.51
0.99	123.75	22.96	−22.96	45.92	0	0	1.56	0.54	236.97	16.95
0.995	124.38	18.32	−18.32	36.65	0	0	1.54	0.38	235.39	16.76
1	125	9.38	−9.38	18.75	0	0	1.5	0	233.75	16.57

Appendix A.3. P1566

Main propeller elements

Propeller diameter................................ : 250.000 (mm)
Hub diameter.. : 60.000 (mm)
Number of blades...................................: 4
Expanded area ratio..............................: 0.602
Expanded "total" skew angle..............: 23.000 (deg.)

Table A3. Main geometrical parameters of propeller P1566.

-	mm	mm	mm	mm	mm	mm	mm	mm	mm	deg
r/R	**r**	**Lfor**	**Laft**	**b**	**cs P1565**	**xr**	**eo**	**fo**	**P**	**Fi**
0.24	30	16.67	−16.67	33.35	0	0	9.51	0.28	269.5	55.03
0.25	31.25	17.39	−17.95	35.34	−0.28	0	9.38	0.85	269.74	53.95
0.3	37.5	20.68	−24.3	44.99	−1.81	0	8.68	1.69	270.84	48.98
0.35	43.75	23.47	−30.62	54.09	−3.58	0	8	2.19	271.81	44.68
0.4	50	25.71	−36.9	62.6	−5.59	0	7.35	2.56	272.66	40.96
0.5	62.5	28.34	−49.27	77.6	−10.46	0	6.12	3.08	273.96	34.9
0.6	75	28.19	−61.2	89.39	−16.51	0	5	3.35	274.74	30.24
0.7	87.5	24.72	−72.26	96.98	−23.77	0	3.98	3.36	275	26.57
0.8	100	17.01	−81.48	98.49	−32.23	0	3.05	3	272.05	23.41
0.9	112.5	2.77	−86.33	89.1	−41.78	0	2.23	2.08	259.41	20.15
0.95	118.75	−9.41	−84.46	75.05	−46.94	0	1.85	1.35	248.38	18.41
0.975	121.88	−18.97	−80.26	61.29	−49.61	0	1.68	0.9	241.52	17.51
0.99	123.75	−28.29	−74.21	45.92	−51.25	0	1.56	0.54	236.97	16.95
0.995	124.38	−33.47	−70.12	36.65	−51.8	0	1.54	0.38	235.39	16.76
1	125	−42.98	−61.73	18.75	−52.35	0	1.5	0	233.75	16.57

Appendix A.4. Blade Section Profiles for P1374/P1565/P1566

Blade section profiles

x/b—chordwise coordinate, [0;1]; Yu,Yl—ordinates of the upper/lower sides of the profile, Yc—ordinates of the profile mean line, given in mm

Table A4. Geometry of the blade sections—P1374/P1565/P1566.

r/R	x/b	0	0.005	0.0075	0.0125	0.025	0.05	0.1	0.2	0.3	0.4	0.5	0.6	0.7	0.8	0.9	0.95	1
	Yu	0	0.65	0.81	1.05	1.47	2.06	2.86	3.89	4.53	4.91	5.03	4.9	4.42	3.52	2.1	1.24	0.3
0.24	Yl	0	−0.62	−0.77	−1	−1.39	−1.91	−2.62	−3.5	−4.06	−4.38	−4.48	−4.36	−3.93	−3.14	−1.9	−1.14	−0.3
	Yc	0	0.01	0.02	0.03	0.04	0.07	0.12	0.19	0.24	0.26	0.28	0.27	0.25	0.19	0.1	0.05	0
	Yu	0	0.66	0.83	1.09	1.55	2.19	3.08	4.24	4.97	5.39	5.54	5.39	4.87	3.88	2.29	1.34	0.3
0.25	Yl	0	−0.59	−0.73	−0.93	−1.28	−1.73	−2.32	−3.05	−3.5	−3.75	−3.83	−3.72	−3.36	−2.68	−1.66	−1.02	−0.3
	Yc	0	0.04	0.05	0.08	0.14	0.23	0.38	0.6	0.74	0.82	0.85	0.83	0.76	0.6	0.31	0.16	0
	Yu	0	0.65	0.82	1.09	1.57	2.27	3.26	4.56	5.38	5.86	6.03	5.88	5.32	4.23	2.46	1.42	0.3
0.3	Yl	0	−0.51	−0.62	−0.78	−1.04	−1.35	−1.74	−2.19	−2.45	−2.6	−2.64	−2.56	−2.3	−1.85	−1.21	−0.8	−0.3
	Yc	0	0.07	0.1	0.15	0.27	0.46	0.76	1.18	1.46	1.63	1.69	1.66	1.51	1.19	0.62	0.31	0
	Yu	0	0.63	0.79	1.06	1.55	2.27	3.28	4.64	5.5	6.01	6.19	6.03	5.46	4.35	2.51	1.44	0.3
0.35	Yl	0	−0.44	−0.53	−0.66	−0.86	−1.08	−1.32	−1.58	−1.72	−1.8	−1.81	−1.75	−1.57	−1.27	−0.9	−0.64	−0.3
	Yc	0	0.09	0.13	0.2	0.35	0.59	0.98	1.53	1.89	2.1	2.19	2.14	1.94	1.54	0.81	0.4	0
	Yu	0	0.6	0.76	1.02	1.51	2.23	3.26	4.65	5.53	6.05	6.23	6.08	5.5	4.38	2.52	1.44	0.3
0.4	Yl	0	−0.38	−0.46	−0.56	−0.7	−0.84	−0.97	−1.07	−1.11	−1.13	−1.12	−1.07	−0.95	−0.79	−0.64	−0.5	−0.3
	Yc	0	0.11	0.15	0.23	0.41	0.69	1.15	1.79	2.21	2.46	2.56	2.5	2.27	1.8	0.94	0.47	0
	Yu	0	0.54	0.69	0.94	1.41	2.11	3.14	4.53	5.42	5.95	6.14	5.99	5.42	4.32	2.48	1.41	0.3
0.5	Yl	0	−0.28	−0.33	−0.38	−0.43	−0.45	−0.39	−0.23	−0.11	−0.03	0.01	0.03	0.05	0	−0.21	−0.28	−0.3
	Yc	0	0.13	0.18	0.28	0.49	0.83	1.38	2.15	2.66	2.96	3.08	3.01	2.74	2.16	1.13	0.57	0
	Yu	0	0.47	0.61	0.84	1.28	1.95	2.94	4.28	5.15	5.66	5.85	5.71	5.17	4.13	2.36	1.35	0.3
0.6	Yl	0	−0.19	−0.22	−0.24	−0.22	−0.14	0.06	0.4	0.63	0.78	0.85	0.85	0.78	0.58	0.11	−0.11	−0.3
	Yc	0	0.14	0.2	0.3	0.53	0.91	1.5	2.34	2.89	3.22	3.35	3.28	2.98	2.35	1.23	0.62	0

Table A4. *Cont.*

r/R	x/b	0	0.005	0.0075	0.0125	0.025	0.05	0.1	0.2	0.3	0.4	0.5	0.6	0.7	0.8	0.9	0.95	1
	Yu	0	0.41	0.53	0.73	1.13	1.74	2.65	3.89	4.69	5.17	5.34	5.22	4.73	3.78	2.16	1.25	0.3
0.7	Yl	0	−0.12	−0.13	−0.12	−0.07	0.08	0.36	0.8	1.1	1.29	1.37	1.35	1.24	0.94	0.31	−0.01	−0.3
	Yc	0	0.14	0.2	0.3	0.53	0.91	1.5	2.35	2.9	3.23	3.36	3.28	2.98	2.36	1.24	0.62	0
	Yu	0	0.33	0.43	0.6	0.93	1.45	2.22	3.28	3.96	4.37	4.52	4.42	4	3.21	1.85	1.09	0.3
0.8	Yl	0	−0.08	−0.08	−0.06	0.02	0.17	0.46	0.91	1.21	1.39	1.47	1.45	1.33	1	0.36	0.02	−0.3
	Yc	0	0.13	0.18	0.27	0.48	0.81	1.34	2.1	2.59	2.88	3	2.93	2.66	2.11	1.1	0.55	0
	Yu	0	0.24	0.31	0.43	0.66	1.03	1.57	2.32	2.8	3.08	3.19	3.11	2.82	2.28	1.35	0.83	0.3
0.9	Yl	0	−0.06	−0.06	−0.05	−0.01	0.1	0.29	0.59	0.79	0.91	0.96	0.95	0.87	0.64	0.18	−0.07	−0.3
	Yc	0	0.09	0.12	0.19	0.33	0.56	0.93	1.45	1.79	2	2.08	2.03	1.85	1.46	0.77	0.38	0
	Yu	0	0.18	0.23	0.32	0.49	0.75	1.14	1.66	2	2.2	2.28	2.22	2.01	1.64	1.02	0.66	0.3
0.95	Yl	0	−0.07	−0.07	−0.08	−0.06	−0.02	0.07	0.23	0.33	0.4	0.43	0.42	0.39	0.26	−0.02	−0.16	−0.3
	Yc	0	0.06	0.08	0.12	0.21	0.37	0.61	0.95	1.17	1.3	1.35	1.32	1.2	0.95	0.5	0.25	0
	Yu	0	0.15	0.19	0.26	0.39	0.59	0.89	1.28	1.53	1.68	1.74	1.69	1.53	1.26	0.82	0.56	0.3
0.975	Yl	0	−0.07	−0.09	−0.1	−0.11	−0.11	−0.08	−0.02	0.02	0.05	0.06	0.06	0.06	0	−0.15	−0.23	−0.3
	Yc	0	0.04	0.05	0.08	0.14	0.24	0.4	0.63	0.78	0.86	0.9	0.88	0.8	0.63	0.33	0.17	0
	Yu	0	0.13	0.16	0.22	0.32	0.47	0.69	0.98	1.17	1.28	1.32	1.29	1.17	0.97	0.66	0.48	0.3
0.99	Yl	0	−0.08	−0.1	−0.12	−0.15	−0.18	−0.21	−0.23	−0.24	−0.24	−0.24	−0.23	−0.21	−0.21	−0.26	−0.28	−0.3
	Yc	0	0.02	0.03	0.05	0.09	0.15	0.24	0.38	0.47	0.52	0.54	0.53	0.48	0.38	0.2	0.1	0
	Yu	0	0.12	0.15	0.2	0.29	0.42	0.61	0.86	1.02	1.11	1.15	1.12	1.01	0.85	0.6	0.45	0.3
0.995	Yl	0	−0.09	−0.11	−0.13	−0.17	−0.22	−0.27	−0.33	−0.37	−0.39	−0.39	−0.38	−0.34	−0.32	−0.32	−0.31	−0.3
	Yc	0	0.02	0.02	0.03	0.06	0.1	0.17	0.26	0.33	0.36	0.38	0.37	0.33	0.26	0.14	0.07	0
	Yu	0	0.1	0.12	0.16	0.23	0.31	0.43	0.58	0.68	0.73	0.75	0.73	0.66	0.56	0.42	0.34	0.25
1	Yl	0	−0.1	−0.12	−0.16	−0.23	−0.31	−0.43	−0.58	−0.68	−0.73	−0.75	−0.73	−0.66	−0.56	−0.42	−0.34	−0.25
	Yc	0	0	0	0	0	0	0	0	0	0	0	0	0	0	0	0	0

Appendix B. AKPD/AKPA Geometrical Definitions and Sign Conventions

Elements of cylindrical sections describing propeller blade

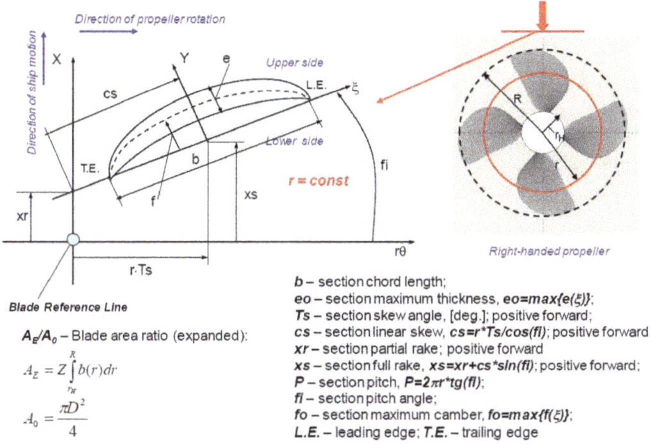

Figure A1. AKPD-AKPA geometrical definitions.

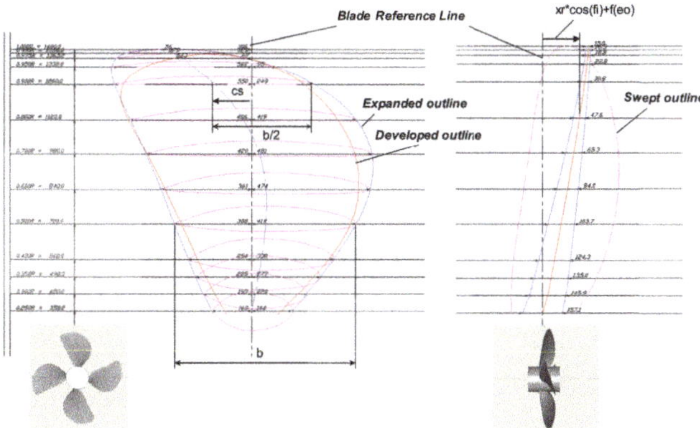

Figure A2. AKPD-AKPA geometrical definitions.

References

1. Young, Y.L. Fluid-structure interaction analysis of flexible composite marine propellers. *J. Fluids Struct.* **2008**, *24*, 799–818. [CrossRef]
2. Young, Y.L.; Garg, N.; Brandner, P.A.; Pearce, B.W.; Butler, D.; Clarke, D.; Phillips, A.W. Load-dependent bend-twist coupling effects on the steady-state hydroelastic response of composite hydrofoils. *Compos. Struct.* **2018**, *189*, 398–418. [CrossRef]
3. Mouritz, P.; Gellert, E.; Burchill, P.; Challis, K. Review of advanced composite structures for naval ships and submarines. *Compos. Struct.* **2001**, *53*, 21–423. [CrossRef]
4. Atkinson, P.; Glover, E.J. Propeller hydroelastic effects. In Proceedings of the Propellers 88 Symposium, Virginia Beach, VA, USA, 20–21 September 1988.
5. Savio, L. Measurements of the deflection of a flexible propeller blade by means of stereo imaging. In Proceedings of the SMP2015—The Fourth International Symposium on Marine Propulsors, Austin, TX, USA, 31 May–4 June 2015.

6. Maljaars, P.; Bronswijk, L.; Windt, J.; Grasso, N.; Kaminski, M. Experimental Validation of Fluid-Structure Interaction Computations of Flexible Composite Propellers in Open Water Conditions Using BEM-FEM and RANS-FEM Methods. *J. Mar. Sci. Eng.* **2018**, *6*, 51. [CrossRef]
7. Young, Y.L. Dynamic hydroelastic scaling of self-adaptive composite marine rotors, *Compos. Struct.* **2010**, *92*, 97–106. [CrossRef]
8. Achkinadze, A.S.; Krasilnikov, V.I. A Numerical Lifting-Surface Technique for Account of Radial Velocity Component in Screw Propeller Design Problem. In Proceedings of the 7th International Conference on Numerical Ship Hydrodynamic NSH7, Nantes, France, 19–22 July 1999.
9. Achkinadze, A.S.; Krasilnikov, V.I.; Stepanov, I.E. A Hydrodynamic Design Procedure for Multi-Stage Blade-Row Propulsors Using Generalized Linear Model of the Vortex Wake. In Proceedings of the Propellers/Shafting'2000 SNAME Symposium, Virginia Beach, VA, USA,20–21 September 2000.
10. JCGM. *JCGM 100:2008. GUM 1995 with Minor Corrections-Valuation of Measurement Data—Guide to the Expression of Uncertainty in Measurement*; JCGM: Paris, France, 2008.
11. Hou, G.; Wang, J.; Layton, A. Numerical Methods for Fluid-Structure Interaction—A Review. *Commun. Comput. Phys.* **2012**, *12*, 337–377. [CrossRef]
12. Bazilevs, Y.; Calo, V.M.; Hughes, T.J.; Zhang, Y. Isogeometric fluid-structure interaction: Theory, algorithms, and computations. *Comput. Mech.* **2008**, *43*, 3–37. [CrossRef]
13. Degroote, J.; Haelterman, R.; Annerel, S.; Bruggeman, P.; Vierendeels, J. Performance of partitioned procedures in fluid–structure interaction. *Comput. Struct.* **2010**, *88*, 446–457. [CrossRef]
14. Siemens PLM Software. *STAR-CCM+ v12.04 User Guide*; Siemens PLM Software: Plano, TX, USA, 2017

Journal of
Marine Science and Engineering

Article

Numerical Analysis of Full-Scale Ship Self-Propulsion Performance with Direct Comparison to Statistical Sea Trail Results

Wenyu Sun [1,2,*], Qiong Hu [1,2], Shiliang Hu [1,2], Jia Su [1,2], Jie Xu [1,2], Jinfang Wei [1,2] and Guofu Huang [1,2,3]

1 China Ship Scientific Research Center, Shanghai 200011, China; huqiong@702sh.com (Q.H.); hushiliang@702sh.com (S.H.); sujia@702sh.com (J.S.); xujie@702sh.com (J.X.); weijinfang@702sh.com (J.W.); huangguofu@702sh.com (G.H.)
2 Jiangsu Key Laboratory of Green Ship Technology, Wuxi 214082, China
3 CSSC Shanghai Marine Energy Saving Technology Development Co. Ltd., Shanghai 200011, China
* Correspondence: sunwenyu@702sh.com; Tel.: +86-0510-8555-5635

Received: 15 December 2019; Accepted: 2 January 2020; Published: 5 January 2020

Abstract: Accurate prediction of the self-propulsion performance is one of the most important factors for the energy-efficient design of a ship. In general, the hydrodynamic performance of a full-scale ship could be achieved by model-scale simulation or towing tank tests with extrapolations. With the development of CFD methods and computing power, directly predict ship performance with full-scale CFD is an important approach. In this article, a numerical study on the full-scale self-propulsion performance with propeller operating behind ship at model- and full-scale is presented. The study includes numerical simulations using the RANS method with a double-model and VOF (Volume-of-Fluid) model respectively and scale effect analysis based on overall performance, local flow fields and detailed vortex identification. The verification study on grid convergence is also performed for full-scale simulation with global and local mesh refinements. A series of sea trail tests were performed to collect reliable data for the validation of CFD predictions. The analysis of scale effect on hull-propeller interaction shows that the difference of hull boundary layer and flow separation is the main source of scale effect on ship wake. The results of the fluctuations of propeller thrust and torque along with circulation distribution and local flow field show that the propeller's loading is significantly higher for model-scale ship. It is suggested that the difference of vortex evolution and interaction is more pronounced and has larger effects on the ship's powering performance at model-scale than full-scale according to the simulation results. From the study on self-propulsion prediction, it could be concluded that the simplification on free surface treatment does not only affect the wave-making resistance for bare hull but also the propeller performance and propeller induced ship resistance which can be produced up to 5% uncertainty to the power prediction. Roughness is another important factor in full-scale simulation because it has up to an approximately 7% effect on the delivery power. As a result of the validation study, the numerical simulations of full-scale ship self-propulsion shows good agreement with the sea trail data especially for cases that have considered both roughness and free surface effects. This result will largely enhance our confidence to apply full-scale simulation in the prediction of ship's self-propulsion performance in the future ship designs.

Keywords: hull-propeller interaction; full-scale CFD; scale effect; self-propulsion; statistical sea trails

---

1. Introduction

An effective design of the ship hull and propulsion system is based on the full knowledge of fluid dynamics under the ship's real transportation condition. Accurate prediction of the self-propulsion performance is one of the most important factors for the energy-efficient design of a ship. In general, it is believed that the evaluation of the ship's hydrodynamic performance with high accuracy and confidence level can be achieved at model-scale through towing tank experiments or numerical simulations. Then, full-scale ship performance prediction can be obtained by extrapolating the model-scale results according to the law of similarity and the extrapolating method recommended by ITTC (The International Towing Tank Conference). This way is reliable for classical ships and typical propulsion systems, but it remains questionable for new types of hull form, propeller and energy-saving devices, especially on very large ships with high Reynolds number. Since the full-scale sea trial test is unavailable during the design stage, the numerical method is an effective and efficient way to investigate full-scale ship's performance and local flow characteristics.

Currently, the most state-of-art numerical tools in marine hydrodynamic analysis are based on Reynolds-averaged Navier–Stokes (RANS) equations and realized by the finite-volume method. For model-scale ship, the workshop on CFD in Ship Hydrodynamics [1] hold every five years in Gothenburg and Tokyo witnessed the development of Computational Fluid Dynamics (CFD) methods. Time-averaged RANS is able to accurately predict the mean flow field and force coefficients for ships with no drift angle [2]. However, the practice cases for full-scale ship is relatively less. In early 2000, Dubigneau et al. [3] also applied the full-scale CFD method to optimize hull form and found a significant difference in design results. At the same time, Leer–Andersen and Larsson [4] performed an experimental and numerical investigation on full-scale ship to evaluate the effect of different surface topographies on skin friction. As a result, a modification on friction resistance calculations were added to the CFD code SHIPFLOW. Wall roughness is the most challenging issue in full-scale simulation. Bhushan et al. [5] studied the modeling of wall roughness and demonstrates the versatility of a two-point, multilayer wall function in computing model- and full-scale ship flows. A systematic analysis including friction resistance prediction, self-propelled simulation, and seakeeping calculation were performed in their work and the result shows that rough-wall simulation predicted better results.

With the development of novelty ship forms, propeller, and ESDs, scale effect becomes the most challenging part in predicting ship resistance, propulsion performance, and energy-saving effect. Full-scale performance estimation with model-scale results uses friction correlation line and form factor and refers to the ITTC guidelines [6]. The ITTC guidelines assume that the form factor is the same for model- and full-scale ship and is independent of ship speed. Min and Kang [7] questioned these assumptions and performed towing tank tests for a series of Reynolds number. As form factor increased with the Reynolds number, they suggested a new extrapolating procedure for the prediction of the full-scale ship resistance. Katsui et al. [8], Eca and Hoekstra [9] also proposed their recommendations on friction correlation lines. Park [10] studied the scale effect of form factor depending on change in the Reynolds number with the CFD method and made a comparison with three kinds of friction resistance curves. He concludes that the self-propulsion components were sensitively influenced by large and small correlations owing to the different the revolution, thrust and torque of propellers and will cancel each other by analysis step.

For the prediction of ship's powering performance, scale effect on stern flow field and hull/propeller interaction is more significant [11]. Starke and Bosschers [12] discussed the scale effects in ship powering performance with a hybrid RANS-BEM method. Their result shows a maximum difference of 3% in thrust, which is higher than the difference in hull resistance. Lin and Kouh [13] focused on the scale effect of the thrust deduction factor. They assumed that the thrust deduction factor is not the same for model and full scales which is in contrast to the ITTC documents. Their research proposed an innovative self-propulsion balancing procedure and derived a simplified model for full-scale prediction. The large discrepancies between the wake fields of model-scale and actual ships are assumed to be the main source of the scale effect of ship performance. Wang et al. [14] resolved the viscous flow fields of ship at different scales by the RANS method and analyzed the scale effect that relies on the axial nominal wake field. They concluded that the reciprocal of mean axial wake fraction of propeller disc exhibits a near liner dependence on Reynolds number in logarithmic scale and the wake width in medium and outer radius reveals negative exponent power function dependence on Reynolds number in logarithmic scale. Guo et al. [15] studied the method to correct scale effect in the design procedure, they proposed a non-geometrically similar model to produce a closer wake field to that of an actual ship. Their method could help to access the full-scale flow field with model-scale experiments or simulations and has been successfully applied to the KRISO container ship (KCS).

Park et al. [16] combined the model-scale and full-scale computation to predict wake fraction for full-scale ship, and proposed a reliable and efficient propulsive performance prediction method for full scale ships with ESDs. Shen et al. [17] studied the scale effects for rudder bulb and rudder thrust fin on propulsion efficiency based on a numerical approach. Their result shows that the model scale simulations predicted 4.85% gains in terms of propulsive efficiency while full scale simulations indicated 2.28% efficiency gains.

For the numerical methods of self-propulsion simulation, several approaches have been developed by previous researchers, including fully discretized propeller approaches and some RANS/potential coupling approaches. The former one will consider the real propeller geometry and solve the rotation region and stationary region within a unique RANS model, such as multiple reference frame method [18], sliding mesh method and dynamic overset method [19]. These approaches are certainly time-consuming due to the large discrepancy between propeller rotation period and ship's traveling wave period. The later ones introduced actuator disk method or body force method to simulate the effect of propeller rotation, many attempts such as momentum theory [20], boundary element method [12,21] or optimal circulation lifting line theory [22,23] are performed to produce propeller's body force in ship wake. Free surface treatment is another important issue in simulating the self-propulsion. Usually, double-model approach is the first choice considering the computational cost, this is conducted by replacing the free surface with a slip wall [24]. A more realistic method is the VOF method, Wang et al. [25] analyzed the interaction of hull-propeller-rudder system considering free surface by RANS method and VOF model at model and full scale.

The series workshops on CFD in ship hydrodynamics have successfully promoted the world-wide validation of model-scale simulations. However, the accessible validation data is extremely rare for full-scale simulations. Studies on full-scale CFD always validate their results by the extrapolation from towing tank test data. The most well-known project for the validation study of full-scale CFD method is the EU cooperative project EFFORT (European Full-scale Flow Research and Technology) funded by the European Framework 5 program [26] which developed an appropriate physical modelling for full scale flows and validated the numerical method with direct full-scale measurements. More recently, a workshop on ship scale hydrodynamic computer simulations has been organized by the Lloyd's register to provide a basic platform for worldwide CFD comparison and validation [27]. Ponkratov and Zegos [28] directly compared their full -scale self-propulsion simulation of a ship at the same conditions recorded at the sea trial. Mikkelsen [24] made a comparison between sea trial measurements, ship scale CFD, model tank experiments and in-service performance.

In this article, we first focus on the flow field in the stern region to analyze the scale effect in the interaction of hull-propeller and free surface. Simulations were performed for a commercial bulk carrier propelled by a five-bladed, right-handed propeller at light ballast condition. A verification study on grid convergence is performed for full-scale simulation with global and local mesh refinements. Two sets of free surface treatments are included to analyze the free surface effect on the propulsion performance and local flow field evolution. Scale effect on ship wake, propeller bearing force and vortex evolution will be analyzed. Then, the method for full-scale self-prolusion balancing will be discussed for different free treatments and their effect on powering performance prediction will be analyzed. The most important part relies on the validation of CFD methods. Differ from the previous researchers who utilizing only towing tank experiments or single sea trail data, we introduce statistical sea trail results for the first time from nine times sea trail test for nine new-built ships with the same hull form, propeller and appendages. The uncertainty among the data acquisition of full-scale ship was thus reduced to a minimum up to now.

This article is organized as follows: Section 2 gives the mathematical models used in this study including governing equations and computation setup, and Section 3 presents the sea trail method and the analysis procedure of sea trail data. Then the simulation results will be analyzed and discussed in Section 4, including scale effect analysis and self-propulsion performance predictions. Detailed comparison of statistical sea trail results, towing tank experiments and CFD simulations is presented in this section to validate the powering predictions in a direct and reliable way. In Section 5, a short conclusion is presented and plan for future research is given.

2. Mathematical Model

2.1. Governing Equations

RANS method was used to predict the flow field around the ship at model- and full-scale. The governing equation for solving the time-averaging physical field can be written as:

$$\frac{\partial \overline{u_i}}{\partial t} + u_j \frac{\partial \overline{u_i}}{\partial x_j} = -\frac{1}{\rho}\frac{\partial \overline{p}}{\partial x_i} + v\frac{\partial}{\partial x_j}\left(\frac{\partial \overline{u_i}}{\partial x_j} + \frac{\partial \overline{u_j}}{\partial x_i}\right) + \frac{1}{\rho}\frac{\partial \tau_{ij}}{\partial x_j} \tag{1}$$

$$\frac{\partial \overline{u_i}}{\partial x_i} = 0, \tag{2}$$

where $\overline{u_i}$ refers to the time-averaged velocity components and $\overline{p}$ is the time-averaged pressure. ρ and v is the density and kinematic viscosity of fluid respectively. The unclosed item τ_{ij} is the Reynolds stresses tensor and requires an additional turbulence model to close the system.

The turbulence model selected in our simulations is shear stress transportation (SST) $k-\omega$ model, which is proved to have a good capacity for ship wake prediction [2].

Two free surface boundary treatments were adopted and compared in this paper. One is the double-model method, using symmetry boundary condition at water level (Z = 0), another is Volume of Fluid (VOF) scheme which introduces an additional transport equation for unknown variable α that represents the volume fraction of water inside each finite volume cell.

$$\frac{\partial \alpha}{\partial t} + \nabla \cdot (\alpha \mathbf{U}) = 0. \tag{3}$$

2.2. Computational Setup

The computational domain is shown in Figure 1, which consists of two computational regions: the outer region for the ship and the inner region for the propeller. The cartesian mesh was adopted for the most part of the computational region with the prism layer created in the near wall region. The targeted y+ is below 5 for model-scale simulations to avoid the use of wall function and is about 200 for full-scale simulations to avoid excessive computational cost. Special mesh refinements were applied near the free surface, bow and stern region as shown in Figure 2a. A sliding mesh strategy was applied to simulate the propeller's rotation behind the ship. In this article, the direction of the positive propeller phase angle is defined clock-wise while looking behind 0 degree relies on 12 o'clock position, as shown in Figure 2b.

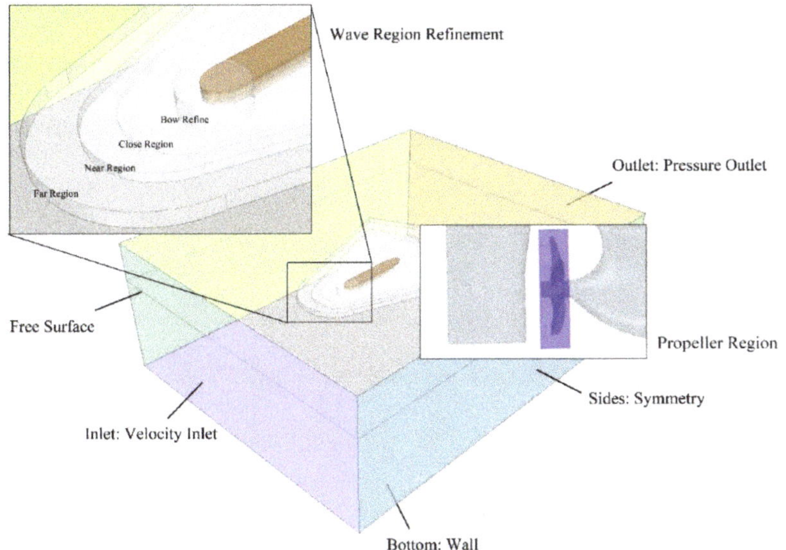

Figure 1. Computation domain, boundary condition and grid refinement strategy.

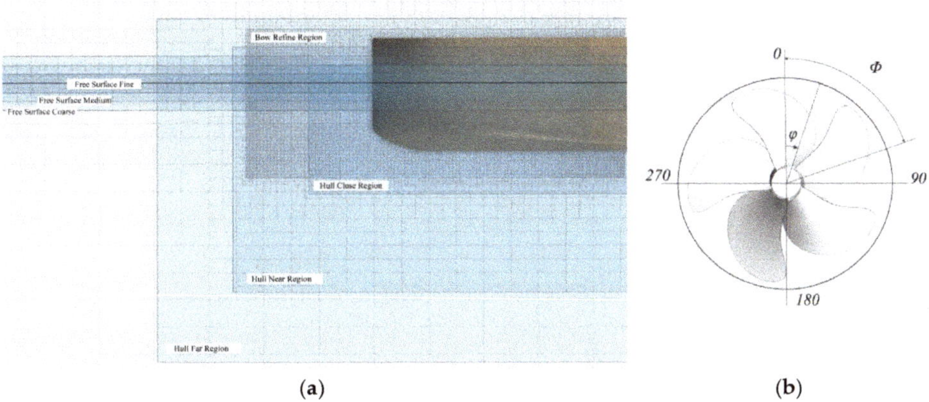

Figure 2. (**a**) Illustration of the bow region and free surface refinement; (**b**) diagram for the direction of positive propeller phase angle.

2.3. Grid Convergence and Sensitivity

It is important to perform grid sensitivity analysis for full-scale simulation with both global and local refinements. As the computational domain was discretized by Cartesian hexahedral mesh with octree-based local refinements, three meshes were made for global grid size study with cell number ranging from 4.27 M to 21.46 M. Simulations were then performed for the entire hull to identify the grid convergence based on drag force, which consists of skin-friction (F_v) and pressure (F_p). The force coefficients are defined as:

$$C_v = \frac{F_v}{0.5\rho V_s^2 S}. \tag{4}$$

$$C_p = \frac{F_p}{0.5\rho V_s^2 S} \tag{5}$$

The characteristic velocity is defined by ship speed (V_s) and S refers to the wetted surface area of the hull. For global refinements, the surface mesh size on the ship hull was selected as the base size for grid convergence study, as listed in Table 1, where Δ is the mesh size on the hull surface. The time step was set to a fixed value equals to 0.0047 Lpp/Vs, which is slightly lower than the recommendation of ITTC. From Table 1, the total coefficient of ship resistance achieved a monotonic convergence with convergence ratio $R_i = 0.048$ and the viscous coefficient and pressure coefficient received slight divergence with convergence ratio −1.33 and 1.44. Compared to the Richardson extrapolation result of C_t, the maximum discrepancy is about 1.27% for C_v and 0.83% for C_p, which is acceptable for numerical analysis. For the total coefficient C_t, the discrepancy between solution from fine mesh and RE result is 0.6%, so the fine mesh was selected and benchmark for local refinement study.

Table 1. Global grid convergence for force coefficients.

	Cell Number	Δ/Lpp	C_v ($\times 10^3$)	C_p ($\times 10^3$)	C_t ($\times 10^3$)
Coarse	4.27 M	0.0130	1.461	0.281	1.742
Fine	8.77 M	0.0091	1.452	0.269	1.721
Finer	21.5 M	0.0065	1.466	0.255	1.720
R_i	-	-	−1.33	1.14	0.048
RE	-	-	-	-	1.720

Local grid convergence study only focuses on the stern region as the wake field of the hull is mainly determined by the stern boundary layer. It is important to analyze the influence of stern mesh refinement on the velocity profile at the propeller plane. The fine mesh of global grid study was selected as the benchmark 1 (B1) and another benchmark grid (B2) was generated by extending the range of stern refinement. The grid size was changed based on two benchmark grids to achieve two levels of stern refinements separately. The contoured axial velocity of the results from four mesh sets are shown in Figure 3. All of these mesh sets have similar flow field structures with some differences in details. The main discrepancy mainly occurs in the inner region of the propeller plane and soon vanished with stern region refinement.

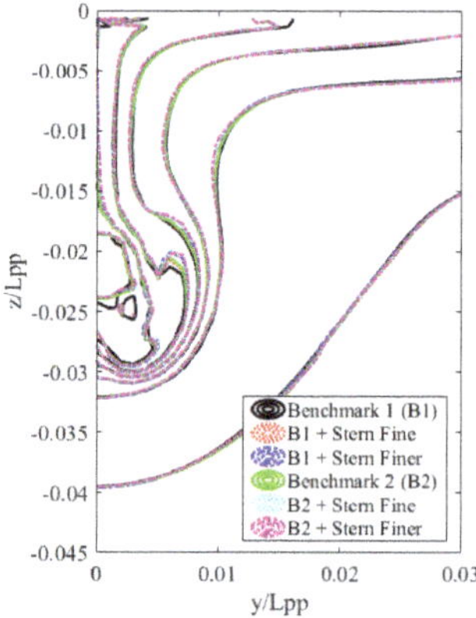

Figure 3. Grid convergence for nondimensional axial velocity contour.

3. Sea Trails and Data Analysis

Full-scale simulation could avoid the scaling errors induced by the assumptions in extrapolating the speed prediction. But CFD method still suffers doubts on its accuracy and reliability due to the lack of validation. In this article, sea trials especially speed tests were carried out for 9 bulk carriers with the same hull form, propeller, and rudder. All these 9 bulk carriers were built in the same shipyard and the sea trail tests were carried out by the same team from China Ship Scientific Research Center. Then, the statistical results after correction and analysis were used to validate the full-scale simulation.

The sea trail tests followed the International Organization for Standardization requirement ISO 15016:2015 [29], including ship heading, ship speed, shaft torque, shaft power, shaft revolution and trial environmental conditions such as water depth, relative wind speed and direction, wave height, wave period and wave direction. All trails were conducted under various engine settings at ballast draught. Prior to every trial, water temperature and density, air temperature, fore draught, midship draught, aft draught were measured at the trial site. For each engine setting, the speed trial is carried out using Double Runs, i.e., each run is followed by a return run in exactly the opposite direction and at the same engine settings.

3.1. Ship Heading and Speed Measurement

For all double runs, ship heading and ship's speed over ground were measured by an independent Differential Global Position System (DGPS) installed on the ships. Figure 4 shows an illustration of the ship's track during the Double Runs. The steady approach was long enough to ensure that the tested ships are in a steady condition prior to the commencement of each speed run. The test duration was 10 min for all speed runs.

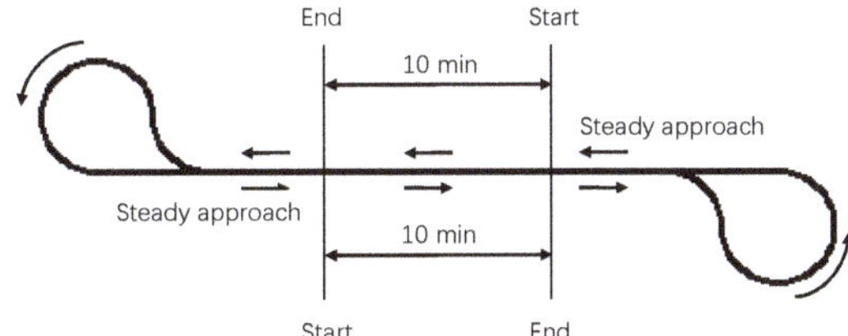

Figure 4. Ship's track during the double run.

3.2. Shaft Torque, Power and Revolution Measurements

Shaft torque and shaft power were measured by the torque telemetry system installed by CSSRC on the intermediate shaft. The system consists of the strain gauge, terminal block, TT10K torque telemetry system, magneto-resistive sensor, data acquisition card, and measurement software. Figure 5 shows the system composition proposed by Yun et al. [30].

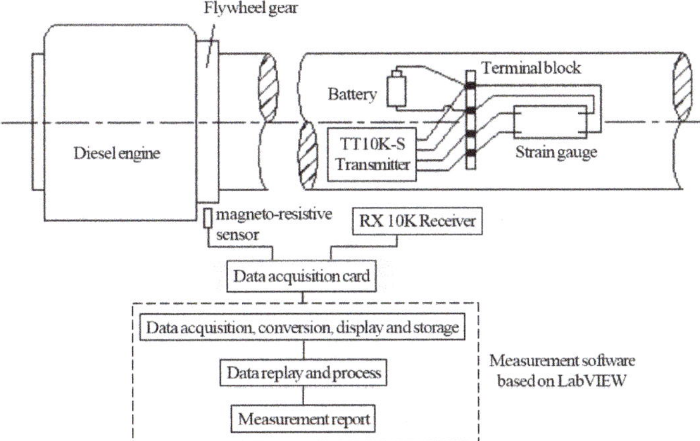

Figure 5. Composition of torque telemetry system.

Shaft torque could be obtained by measuring the surface shear strain force. Then, shaft power could be calculated by shaft torque and shaft revolution. As shown in Figure 5, the strain gauge was used to measure the surface shear strain force. One strain gauge is composed of two sets of resistance strain foils along the axis with an angle of 45° and 135°, the four-strain foils are connected by a full-bridge circuit used for converting the change in electric resistance value into a voltage signal. TT10K torque telemetry system (Binsfeld Engineering Inc., Maple City, MI, USA) is composed of 9 V battery, TT10K-S transmitter, and RX 10K receiver. The TT10K-S transmitter is fixed on the shaft and rotates with the shaft. Providing the bridge voltage for the full-bridge circuit of the strain gauge, the measured voltage signal is amplified and transmitted via an antenna. The RX 10K receiver receives and outputs signals in voltage with the range of 0–10 V. The magnitude of the output voltage is proportional to the shaft torque, as shown in the following formula:

$$M_e = \frac{V_{out}}{10} \times M_{FS} = \frac{V_{out}}{10} \times \frac{V_{FS}\pi E \times 4\left(d_o{}^4 - d_i{}^4\right)}{16000 V_{EXC} k_{GF} N(1+\mu) G_{XMT} d_o}, \tag{6}$$

where M_e is shaft torque in $N{\cdot}m$, M_{FS} is the full scale torque in $N{\cdot}m$, V_{FS} is the full scale output (10 V) of system, E is the elasticity modulus of shaft in $N{\cdot}mm^2$, d_i and d_o is the inner and outer diameter of shaft in mm, V_{EXC} is the bridge excitation voltage in V, k_{GF} is the gage factor which is specified on strain gage package, N is the number of active gages which is 4, μ is the Poisson's ratio of shaft material, G_{XMT} is telemetry transmitter gain.

Shaft revolution was measured by a magneto-resistive sensor that can perceive the movement of objects within a distance of 3 mm. The magneto-resistive sensor was installed directly to the flywheel and was about 2 mm away from the tooth surface. When the flywheel rotating through a tooth, the magneto-resistive sensor could output an inductive pulse, and the shaft revolution can be calculated based on the number of pulses accumulated in a specified time period as follows:

$$n = \frac{60N}{Mt}, \tag{7}$$

where n is the shaft revolution in *r/min*, t is the counting time period in seconds; N is the number of pulses accumulated in the counting time period, M is the number of flywheel teeth.

After measuring the output torque and shaft revolution under a certain engine setting, shaft power can be calculated according to the following formula:

$$P_S = \frac{M_e \times n}{9550}. \tag{8}$$

3.3. Trail Environment Measurement and Data Analysis

Directly simulating the sea trail environment is difficult and time-consuming with current CFD methods and could introduce more unknown uncertainties. Thus, it is better to compare and validate the CFD simulation results with ship performance data in still water. As it is difficult to carry out sea trials under ideal conditions, in practice, certain corrections for environmental conditions, such as water depth, wind, waves, current and deviating ship draught, are performed to achieve the corresponding still water performance of target ships. Therefore, during the speed trials, not only ship speed, shaft power, and shaft revolution, but also the relevant environmental conditions were measured at the same time. Water depth was measured by the ship's echo-sounder, relative wind speed and direction were measured by the ship's anemometer and wave data were observed by multiple experienced mariners.

In order to obtain ship's powering performance under ideal conditions, i.e., no wind, no wave, no current and deep water at 15 °C, the speed trial data were analyzed following the ISO 15016:2015 standard [29]. The analysis procedure includes corrections to power and speed considering the environmental influences. The analysis procedure could be basically divided into several steps, as shown in Figure 6.

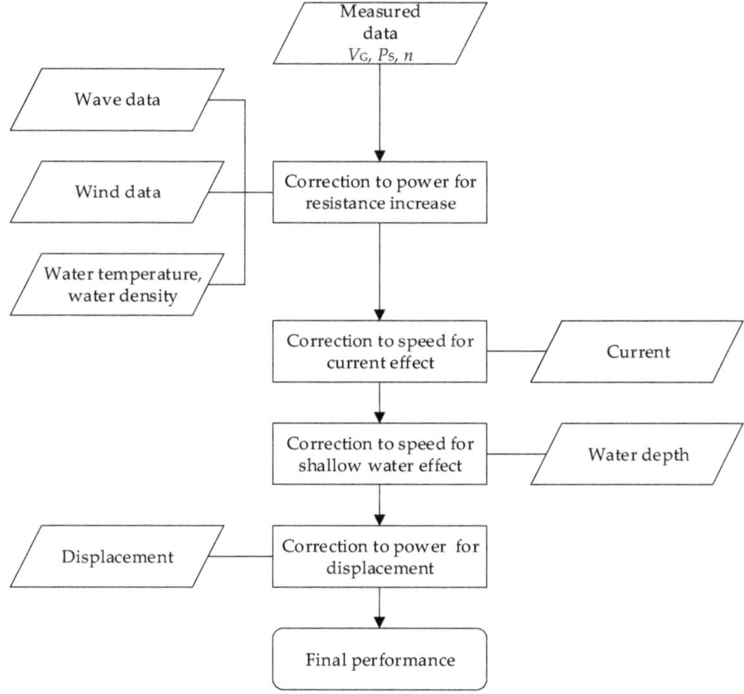

Figure 6. Flow chart of speed trail data analysis.

For each run, the total resistance increase associated with the measured power is calculated. The total resistance increase comes from relative wind, waves, and deviation of water temperature and water density. Then correction to power due to the total resistance increase is conducted based on the 'direct power' method. Additionally, the current effect is corrected by the 'iterative' method, the shallow water effect is corrected by the 'Lackenby' method and the displacement effect is corrected by Admiral-formula which applied to the power values. Finally, the ship's performance in terms of ship speed, power and shaft revolution under ideal conditions could be obtained.

4. Results and Analysis

4.1. Scale Effect of Hull-Propeller and Free Surface Interaction

Both model- and full-scale simulations were performed to analyze the scale effect of the ship's resistance, hull/propeller interaction and free surface's influence. The scale ratio for the ship model is 25.255 with full scale length of 195 m.

Ahead of the analysis of free surface, scale effect of hull/propeller interaction is presented using double-model simulation results. Four test cases including full scale and model scale simulations with/without propeller were performed without real free surface considered, as listed in Table 2. A fixed propeller speed was selected at ship speed 14.5 kn, which means that self-propulsion balancing was not carried in this sub-section. That is, the advance coefficient based on ship speed is $J_s = 0.861$ for both model- and full-scale simulation.

To analyze the scale effect, the proportions of pressure force and viscous force in total resistance are defined as:

$$R_p = \frac{F_p}{F_p + F_v}; R_v = \frac{F_v}{F_p + F_v}. \tag{9}$$

Without propeller, the proportion of viscous force and pressure force shows basically the same for model- and full-scale. The suction effect of the propeller can extremely increase the total force on the hull surface and then changed the ratio of two resistance components. Full-scale simulation leads to higher thrust deduction and has a higher effect on viscous force.

Table 2. Test cases using double-model.

Case	Scale	Propeller	w	R_v	R_p
1	full	without	0.2494	85.37%	14.63%
2	full	with	-	67.66%	32.34%
3	model	without	0.3402	84.73%	15.27%
4	model	with	-	70.43%	29.57%

As the boundary layer is relatively thinner (compared to L_{pp}) for full-scale ship, the decrease of wake fraction (deficit of axial velocity) can also be verified by numerical results. Details of axial and transversal velocity distribution at the propeller plane are shown in Figure 7. At full-scale, the velocity deficit region appears primarily in inner radius and shows a relatively obscure hook-like flow field structure. From the comparison of circumferential averaged axial velocity shown in Figure 8, propeller at full-scale runs in a relatively higher velocity field which has lower thrust and torque coefficients but higher efficiency, as listed in Table 3.

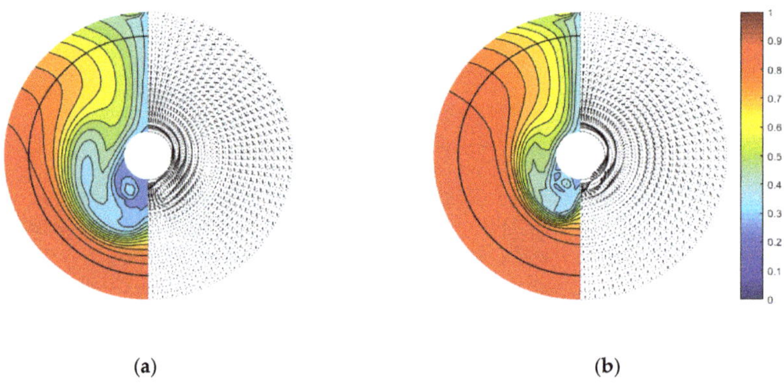

Figure 7. Non-dimensional axial velocity contour and transversal velocity vector at propeller plane: (**a**) model-scale; (**b**) full-scale.

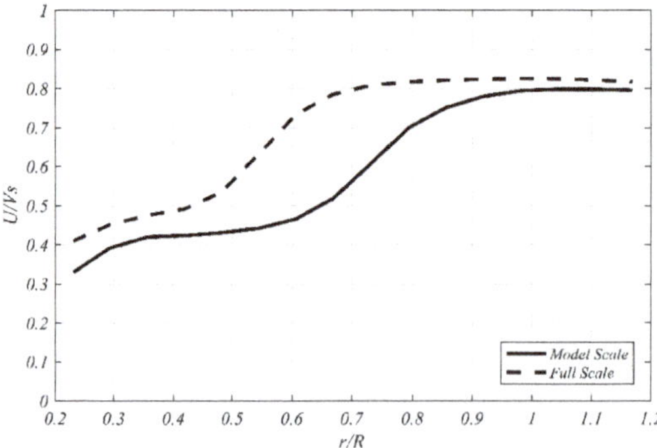

Figure 8. Non-dimensional circumferential averaged axial velocity at propeller plane.

Table 3. Thrust, torque, and efficiency based on Js.

Scale		Double-Model	VOF
full	K_T	0.1414	0.1547
	$10\ K_Q$	0.2087	0.2232
	η_B	0.6129	0.5907
model	K_T	0.1855	0.1916
	$10\ K_Q$	0.2777	0.2830
	η_B	0.5964	0.5900

To consider the real free surface effect when a propeller is operating behind the ship, simulations with the VOF model were made for both model- and full-scale ship. Results of propeller performance in Table 3 show that the symmetry boundary condition used in double-model underestimate the propeller force and also the efficiency. For full-scale ship, this discrepancy is relatively larger.

In consideration of the very difference in wake field and its effect on propeller loading, the thrust and torque variations of each propeller blade in its revolution period are plotted for better understanding the scale effect on propulsion. As shown in Figure 9, thrust coefficients for model- and full-scale ship are almost identical with little difference. The maximum single-blade thrust and torque appear at 60° for both model- and full-scale ship (due to the highly skew of propeller), but the minimums appear at 285° for model-scale, which is 15° later than full-scale condition. Also, the discrepancy of thrust and torque between model- and full-scale is small when propeller operating in the region from 285° to 60°, this is due to the smaller magnitude of inflow for propeller blades in portside. The underestimation by double-model can be figured out in Figure 9 especially around the peak of the propeller bearing force.

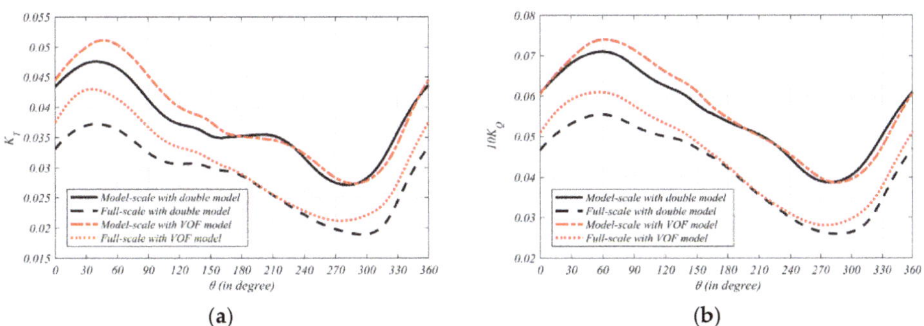

Figure 9. Single blade unsteady force coefficients in one rotation period: (**a**) thrust coefficients; (**b**) torque coefficients.

Additionally, the averaged loading of propeller can be achieved by circumferentially averaging the tangential velocity immediately after the propeller, as follows:

$$G(r) = \frac{\Gamma(r)}{\pi VD} = \frac{1}{\pi VDZ}\oint U_t \cdot r d\theta. \tag{10}$$

Higher circulation can be found for most radius at model-scale, which corresponding to higher propeller loads. Discrepancy of curve form only appears at inner radius; circulation shows a sharper drop for model-scale ship close to the hub and a small peak appears at 0.3 R for full-scale condition due to the intensive high-wake region. The red dashed lines in Figure 10 refer to the circulation distribution obtained by the VOF model, both model- and full-scale results show slightly higher circulation along the whole radius of propeller compared to the results from the double-model. But the circulation distributions remain almost the same for two free surface treatments except that real free surface effect can smoothing the load distribution for model-scale propeller at 0.85 R.

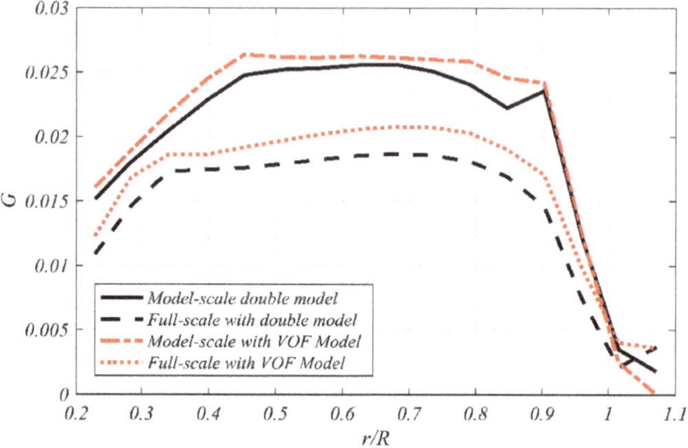

Figure 10. Radial distribution of propeller circulation.

The difference of propeller loading distribution can also be figured out by comparing the time-averaged axial velocity immediately behind the propeller. From the left column (model-scale) and right column (full-scale) of Figure 11, we can find out that in full-scale condition, the propeller has a larger high loading region and can produce a more uniform wake. The low-velocity zone near the hub is also smaller for full-scale ship, which is in coincidence with the ship wake. By comparing the results from two free surface treatments, the free surface boundary effect only appears in the upper half of the propeller wake and will not change the flow structure.

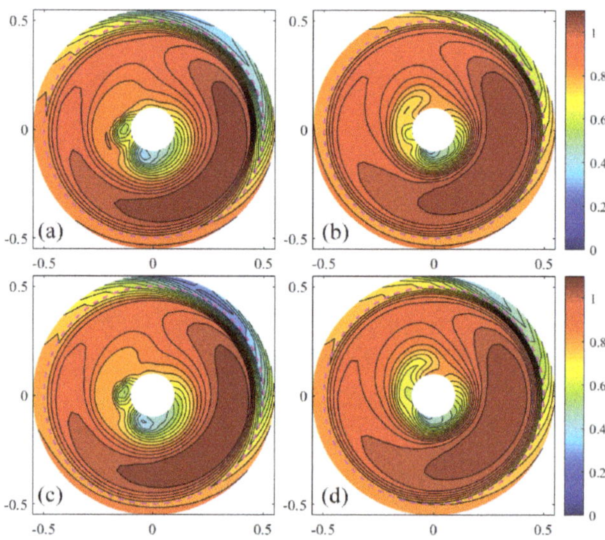

Figure 11. Nondimensional time-averaged axial velocity behind propeller: (**a**) model-scale with double-model; (**b**) full-scale with double-model; (**c**) model-scale with VOF model; (**d**) full-scale with VOF model.

The time-averaged axial velocity on the central longitudinal section (Y = 0) is shown in Figure 12. Propeller at full-scale can produce a large range of high velocity region compared to the model-scale results. But the low velocity region after the boss cap is larger for full-scale case. The difference of two free surface treatments appears when the VOF model can produce lower axial velocity near the free surface in Y = 0 plane.

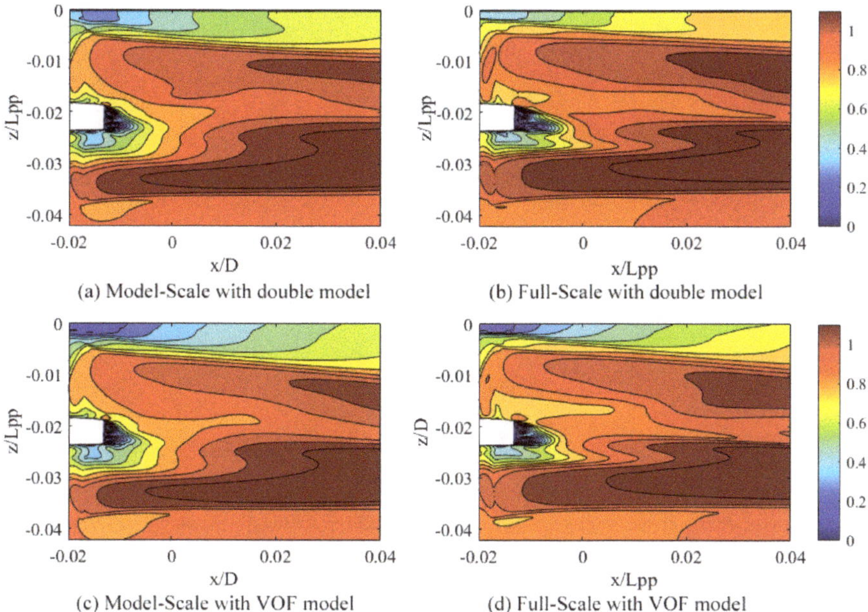

Figure 12. Time-averaged axial velocity at central longitudinal plane (Y = 0). (**a**) model-scale with double-model; (**b**) full-scale with double-model; (**c**) model-scale with VOF model; (**d**) full-scale with VOF model.

The instantaneous flow field is characterized in Figure 13 using nondimensional Q-criterion defined as:

$$\overline{Q} = \frac{Q}{(U/L)^2}. \tag{11}$$

After nondimensionalization, vortex structures at the model- and full-scale become comparable with a same value of $\overline{Q}$. For both model- and full-scale ship, vortex sheet in upper half of propeller wake undergoes flow fields with higher velocity gradients and are more difficult to main stable vortex structure. This phenomenon is also affected by the free surface. When the tip vortex reaches the low velocity region close to the free surface, it is more likely to break up to small vortex structures. This progress will certainly facilitate the turbulence dissipation and then speed up the mixing of the velocity field in the propeller wake. The breakup of tip vortex for full-scale ship comes relatively later and can maintain a more stable structure when translating to the intermediate wake. One explanation is that, for full-scale ship, the velocity deficit region locates limitedly in the inner radius region. So that the tip vortex undergoes a more uniform velocity field when translating to downstream. Another reason is the interaction between the propeller vortex sheet and ship shedding vortex. For full-scale ship, the shedding vortex from the flow separation of the hull boundary layer locates in the inner and upper region of the propeller disk which is relatively far from the propeller's main loading area and has little interaction with propeller tip vortex.

A comparison between Figure 13a,c reflects the difference of free surface boundary conditions on propeller vortex evolution. The tip vortex resulting from the double-model is more likely to break up, and this may be caused by the reflection effect of the symmetry boundary. But the VOF model can produce more real interaction between free surface and vortex transportation, thus results in more stable structures. For full-scale cases, a similar difference can also be figured out.

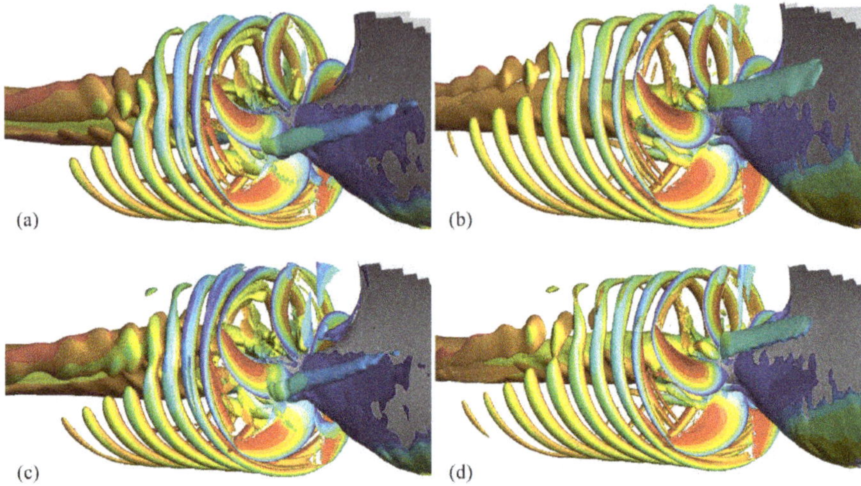

Figure 13. Instantaneous iso-surface of nondimensional Q-criterion, colored by axial velocity ratio: model-scale with double-model (**a**), full-scale with double-model (**b**), model-scale with VOF model (**c**), full-scale with VOF model (**d**).

The wave profiles along the Y = 0 plane at model- and full-scale conditions are plotted in Figure 14. The wave profiles are almost the same for two conditions with a slight discrepancy at crests and troughs. The nondimensional wave height is greater for the model-scale ship. The propeller's effect on wave profile is negligible compared to the wave-making effect of ship stern.

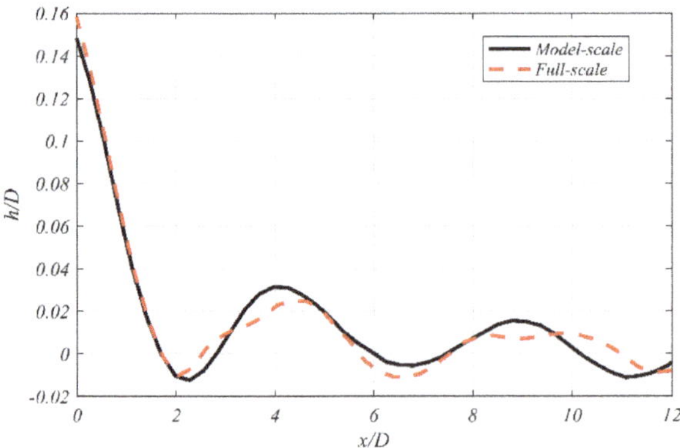

Figure 14. Comparison of the wave profiles at Y = 0 plane at model- and full-scale conditions.

4.2. Self-Propulsion Performance Prediction Compared to Sea Trail Results

In the previous sub-section, the scale effect of the ship's resistance and hull/propeller interaction considering different free surface treatments were analyzed at a fixed propeller rotation speed without a rudder. In this sub-section, we provide the results of full-scale self-propulsion performance prediction at the same condition with sea trail tests with a rudder installed. Due to the large amount of computational cost of the VOF method, it preferred to use double-model method when the free surface effect is negligible. So, both double-model and VOF methods were adopted in this sub-section to quantify the effect of free surface treatment on powering performance prediction.

4.2.1. Self-Propulsion Balance Condition

In order to achieve the self-propulsion balance, a series of propeller speeds were evaluated. For full-scale case, there is no friction force correction, so the self-propulsion balance can be achieved by equalizing propeller thrust T and ship resistance with propeller R_{SP}.

$$T = R_{SP}. \tag{12}$$

With the VOF method, T and R_{SP} could be obtained directly from the integration of force on propeller and hull surface respectively. However, an additional balance correction is needed for double-model method. From the results from the previous sub-section, it could be concluded that the free surface treatments have influence on self-propulsion balance from three aspects:

- Ship resistance
- Propeller performance
- Ship added resistance induced by propeller

The influence on bare hull resistance, nominal wake and propeller loading at a fixed propeller speed has been introduced in the previous section, while the effect on propeller performance at different revolutions is shown in Figure 15a. In accordance with the result of the previous sub-section, the propeller thrust and torque resulted from the VOF method is always larger than that from double-model treatment. This kind of difference may come from two sources: one is the difference in propeller inflow; another is the free surface boundary condition itself.

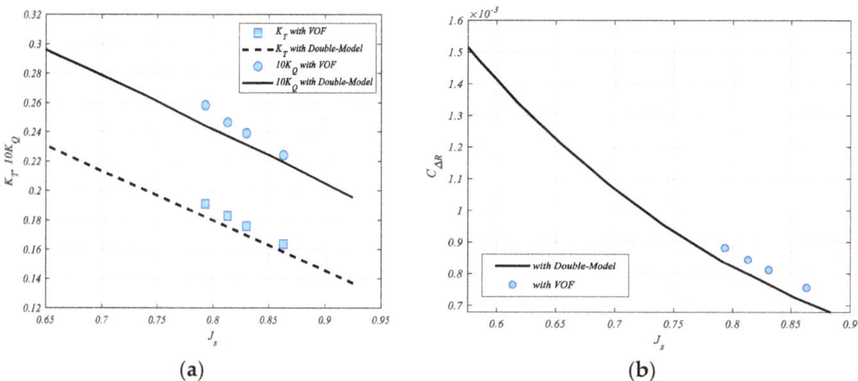

Figure 15. Effect of free surface treatment on full-scale propeller performance (**a**) and propeller induced resistance (**b**).

The influence of propeller induced resistance, as defined in Equation (13), is shown in Figure 15b, where R_t refers to ship resistance in calm water without propeller. The resultant value of ΔR is relatively higher with the VOF method. Traditionally, the effect of free surface treatment on ΔR is assumed negligible when performing the self-propulsion balance [21] for cases in the design draft. However, as sea trails test is always performed in a light ballast draft, where propeller rotates close to the free surface, this discrepancy should not be ignored.

$$\Delta R = R_{SP} - R_t. \tag{13}$$

Considering all the possible influence from free surface treatments, a modified self-propulsion balance correction method in ship scale is proposed, as shown in Equation (14).

$$\begin{aligned} T^{VOF} &= R_{SP}^{VOF} = R_T^{VOF} + \Delta R^{VOF} \\ T^{DM} + \Delta_T^{VOF} &= \left(R_T^{DM} + \Delta_{R_T}^{VOF}\right) + \left(\Delta R^{DM} + \Delta_{\Delta R}^{VOF}\right), \\ T^{DM} + \Delta_T^{VOF} &= R_{SP}^{DM} + \Delta_{R_T}^{VOF} + \Delta_{\Delta R}^{VOF} \end{aligned} \tag{14}$$

where the superscript DM refers to double-model treatment and VOF means the VOF treatment. Δ_φ^{VOF} refers to the discrepancy between VOF and double model treatment on quantity φ, that is $\Delta_\varphi^{VOF} = \varphi^{VOF} - \varphi^{DM}$.

Equation (14) represents the effect of free surface treatment on full-scale self-propulsion balance by three components: Thrust Δ_T^{VOF}, bare hull resistance $\Delta_{R_T}^{VOF}$, and propeller induced hull resistance $\Delta_{\Delta R}^{VOF}$. Then, free surface correction factors $\varepsilon_T, \varepsilon_R, \varepsilon_{R_T}, \varepsilon_{\Delta R}$ are introduced in Equation (15), which represent the correction to the calculated thrust and resistant of ship respectively.

$$\varepsilon_T = \frac{T^{VOF} - T^{DM}}{T^{DM}}, \varepsilon_{R_{SP}} = \frac{R_{SP}^{VOF} - R_{SP}^{DM}}{R_{SP}^{DM}}, \varepsilon_{R_T} = \frac{R_T^{VOF} - R_T^{DM}}{R_T^{DM}}, \varepsilon_{\Delta R} = \frac{\Delta R^{VOF} - \Delta R^{DM}}{\Delta R^{DM}}. \tag{15}$$

The influence of free surface treatments on propeller trust, torque and propeller induced resistance are relatively small compared to the effect on bare hull resistance, but still has non-negligible effect in this test case. The effect of free surface treatment increases with propeller speed and the correction factors are not constants for different advance ratio. For the cases where there is no detailed comparison between different free surface treatments, some easy to use regression models could be adopted to predict the correction factors. With known correction factors, the self-propulsion balance condition for double-model method could be simplified to Equation (16).

$$\begin{aligned} T^{DM}(1+\varepsilon_T) &= R_{SP}^{DM}\left(1+\varepsilon_{R_{SP}}\right) = R_T^{DM}\left(1+\varepsilon_{R_T}\right) + \Delta R^{DM}(1+\varepsilon_{\Delta R}) \\ &= R_T^{DM}\left(\varepsilon_{R_T} - \varepsilon_{\Delta R}\right) + R_{SP}^{DM}(1+\varepsilon_{\Delta R}) \end{aligned} . \tag{16}$$

4.2.2. Powering Performance Prediction

Results of the powering performance prediction from full-scale CFD simulations, towing tank tests, and sea trails are presented in this sub-section. As stated above, when self-prolusion balance was achieved by interpolating propeller speed to satisfy Equation (13) or Equation (16), the delivery power at a given ship speed could be calculated by

$$P_D = 2\pi n Q, \tag{17}$$

where n is the propeller speed and Q is the torque.

Usually, the hull roughness was ignored in model-scale ship resistance and self-propulsion prediction. However, in this sub-section, the full-scale simulation will be compared to the statistical sea trail results according to the analysis procedure proposed in Section 4. Thus, the roughness effect plays an important role to the increase of friction resistance and ship's shaft power. There have two approaches to take roughness effect into consideration according to the most recent work of Niklas and Pruszko [31]. One is the most widely used Bowden–Davison formula [6] as shown in Equation (18), where k_s is the roughness height and L_{WL} is the length of waterline. Therefore, the increase of friction resistance could be obtained by multiplying ΔC_F by $\frac{1}{2}\rho V_s^2 S$.

$$\Delta C_F = 105\left(\frac{k_s}{L_{WL}}\right) - 0.64. \tag{18}$$

Another one is to introduce the roughness effect directly into the wall function of the turbulence model. In this way, it is assumed that roughness could increase the local turbulence near the wall, thus could change the velocity profile in the log-law region. Recently, this method was widely used in many full-scale simulation practices, like Demirel et al. [32], Niklas and Pruszko [31]. According to their reported results, the selection of the turbulence model, grid arrangement and roughness model have a significant effect on the prediction results. According to the conclusion of Niklas and Pruszko [31], it is difficult to assess to what extent the discrepancies result from the roughness model and what is the influence of different turbulence models applied. As the roughness effect is not the focus of this article, we prefer the Bowden–Davison formula to avoid introducing more complicated uncertainty sources to CFD simulation. According to our historical data, the increase friction resistance coefficient was equal $\Delta C_F = 0.1779{\cdot}10^{-3}$ with roughness height assumed equal to $k_s = 90$ µm.

It is well known that the treatment of hull resistance, propeller performance, and hull-propeller interaction will significantly affect the result of self-propulsion balance and thus the resulted powering performance of a ship. So, a detailed comparison of the consideration of roughness and free surface correction was performed to quantify the influence. Table 4 shows all the tested cases included with different roughness and free surface corrections. When the free surface was treated by the VOF method, there have no other free surface corrections. When the free surface was treated by a double-model method, free surface corrections are considered in self-propulsion balance according to Equation (16). The correction on bare hull resistance is mandatory while corrections on trust and propeller induced resistance are optional.

Table 4. Tested cases.

Case NO.	Free Surface Treatment	Roughness	$\Delta_{R_T}^{VOF}$	Δ_{T}^{VOF}	$\Delta_{\Delta R}^{VOF}$
1	VOF	No	-	-	-
2	VOF	Yes	-	-	-
3	Double-Model	No	Yes	No	No
4	Double-Model	Yes	Yes	No	No
5	Double-Model	Yes	Yes	Yes	No
6	Double-Model	Yes	Yes	No	Yes
7	Double-Model	Yes	Yes	Yes	Yes

Self-propulsion points along with ship resistance, propeller trust and torque obtained by different approaches are presented in Figure 16. Figure 17 presents a comparison of the balanced propeller speed, torque, thrust and delivery power by the relative percentages of the test case 2 (VOF method with roughness). It could be figured out that the roughness effect has a significant effect on ship resistance and will change the self-propulsion point by 1.6% for RPS. Thus, it has a larger effect on propeller torque (5.8%) and the resultant delivery power (7%) prediction. For the double-model method, only correcting the bare hull resistance leads to a slight underestimation of balanced propeller speed (about 0.6%) and the delivery power (3%). From the comparison of cases 6 and 7, the leave-out-of-thrust-correction (case 6) overestimates the balanced RPS. Because the increase of propeller speed compensates the correction to propeller torque, the resultant delivery power only shows 1.6% higher than case 7.

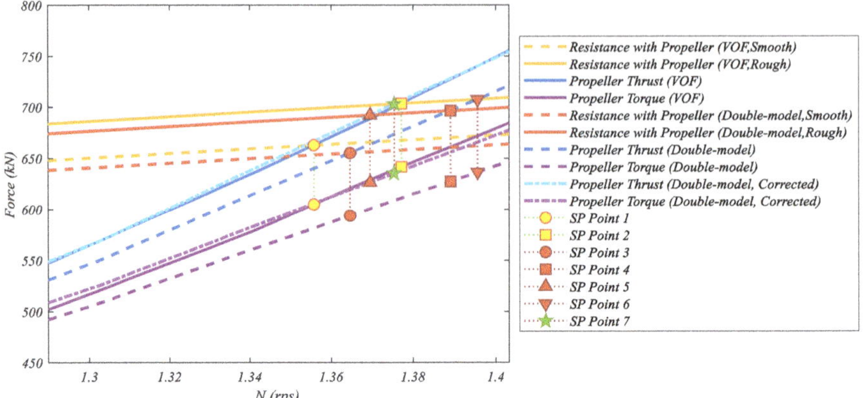

Figure 16. Self-propulsion balance interpolation.

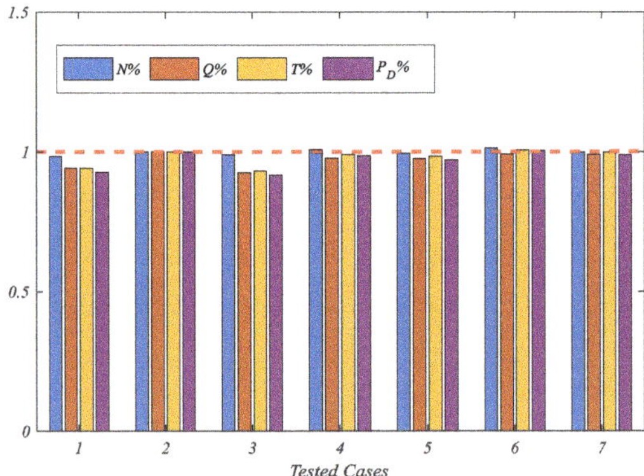

Figure 17. Comparison of the balanced propeller speed, torque, thrust, and delivery power.

The results of powering performance prediction obtained by several approaches mentioned above were compared to the extrapolated self-propulsion result from model test and statistical sea trail results. In this article, sea trail results were achieved by collecting nine times of sea trial test for nine new-build ships with the same hull form, propeller and appendages. The analysis of all the sea trails data followed the same procedure as stated in Section 3 to avoid any additional uncertainties. Figure 18 shows the uncertainty of sea trail results at different ship speeds, the uncertainty here is defined by the standard deviation of power-speed prediction result of the 9 tested ships. The overall statistical uncertainty along all speeds does not exceed 2% of the mean value and the estimated uncertainty due to weather effect could also be controlled under 2% according to the ITTC guidelines.

Figure 19 presents the model test results, sea trail results with confidence interval and the seven predictions. The extrapolated towing tank result overestimates the delivery power by approximately 2.4% compared to the statistical sea trail results. At service speed, all five CFD predictions with roughness effect included show good agreement (between −1.1% and 2.4%) with sea trail results especially for Case 2 and Case 7, where both roughness effect and free surface effect were fully considered. Approximately 6–7% discrepancy was found for cases that did not take roughness effect into consideration. The very little difference between the VOF method and the corrected double—model method mainly comes from the interpolating error and regression procedure of correction factors. Also, the correction method was applied to low speed prediction to validate its extensibility. As a result, the numerical method with double-model and free surface correction slightly overestimates the delivery power by about 2.5–4.8% compared to sea trails result in the consideration of hull roughness. That is, at low speed, the overestimation of numerical prediction is higher than that at design speed. Unfortunately, the genuine source of the modeling or numerical uncertainties is difficult to access with the present study, more systematic and sophisticated efforts should be made to quantify the uncertainties in the future. It is one of the most important issues regarding both ship CFD methods and full scale performance predictions.

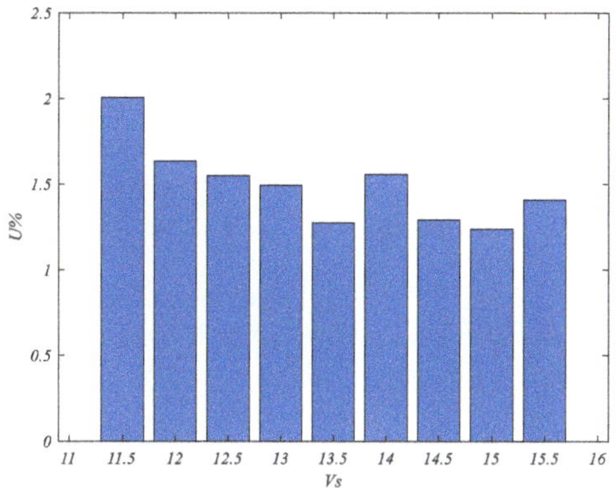

Figure 18. Uncertainty distribution of sea trail results.

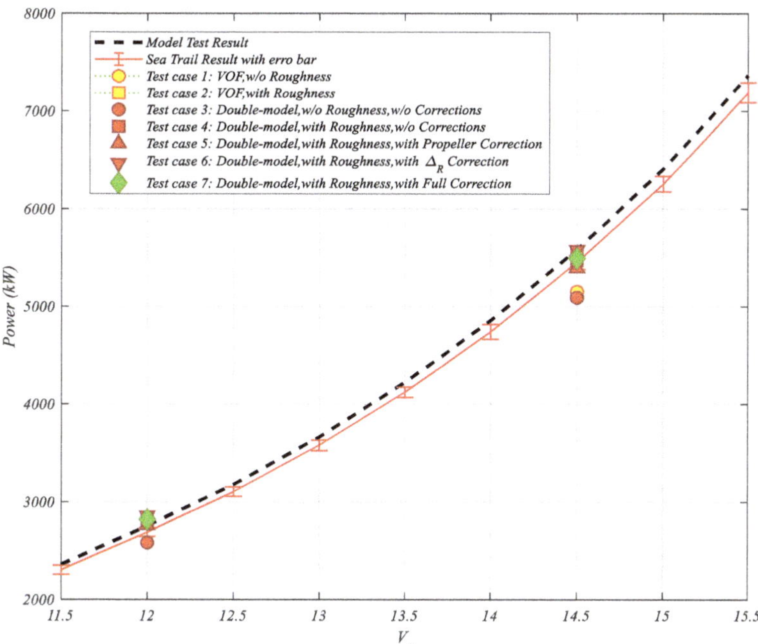

Figure 19. Speed-power correlation predicted by full-scale simulations with comparison to statistical sea trial result.

5. Discussion

Numerical simulations are performed for a commercial bulk carrier in light ballast condition to investigate the complicated interaction around the propeller and to predict the self-propulsion performance. Free surface effects are considered with double-model and VOF model respectively to quantify the difference between the two methods. Scale-effect of hull-propeller interaction under the free surface is detailed presented by wake field, propeller unsteady force and trailing vortex transportation.

The most significant scale effect appears in the ship wake. Full-scale simulation can produce a smaller high wake region around the shaft, which is the main source of the difference in propeller hydrodynamic performance. The overall circumferential axial velocity is higher for full-scale ship and makes propeller working with a high advance ratio. The propeller loading is significantly higher for model-scale ships according to the analysis of circulation distribution and downstream flow fields. Free surface treatments have non-negligible effect on propeller loading. Compared to the most realistic modeling of the free surface by the VOF model, the double-model treatment apparently underestimates the propeller loading. This phenomenon should be considered in the prediction of self-propulsion performance especially for ballast condition.

A non-dimensional vortex recognition criterion is introduced to make the vortex structure comparable at different Reynolds numbers. Full-scale and model-scale ship can produce similar trailing vortex sheet, which is more stable at full-scale condition due to the more favorable inflow. The interaction of ship shedding vortex, tip vortex and free surface impact the evolution of the vortex sheet. So, scale effect also originates from ship's boundary layer separation. Free surface treatments have a small effect on the propeller vortex sheet, the symmetry free surface boundary condition could make propeller tip vortex more vulnerable to break up in downstream.

It is well known that considering free surface with VOF in self-propulsion simulation requires a large amount of computational resources. Therefore, we compared the effect of free surface treatments on ship's self-propulsion balance to offer some suggestions for engineering practice with a more time-saving double-model approach. It could be concluded that the simplification of free surface treatment does not only affect the wave-making resistance for bare hull but also the propeller performance and propeller induced ship resistance. In this case, ignore the free surface correction on propeller performance and propeller induced resistance brings up to 5% more uncertainty to the result of power prediction. According to the simple treatment in this article, roughness is another important factor in full-scale simulation because it has up to approximately 7% effect on the delivery power. As have been discussed by other researchers, the appropriate direct modeling of roughness in CFD method still remains questionable and more experiments, numerical studies and modeling method developments should be made in the future.

The validation data provided by 9 times of sea trail test shows good quality with low uncertainty. For this ship, the towing tank prediction overestimates the power at service speed. Because the detailed information of the extrapolating method is unknown to us in this research, we prefer the statistical sea trial results to validate the CFD predictions. As a result, the numerical simulations of full-scale ship self-propulsion show good agreement with the sea trail data especially for cases that have considered both roughness and free surface effects. Predictions without the consideration of hull surface roughness significantly underestimates the delivery power.

In the future, more scientific studies should be performed to deeply investigate the mechanism of these mutual interactions. Quantitative, statistical or analytical models should be established in the future for better application to industrial design activities. Numerous efforts shall be made in the future to investigate more influence factors like wall roughness modeling, weather condition, and ship motions effect. Also, more detailed and systematical analysis of uncertainty in full-scale simulation and sea trails should be performed in the future to identify the confidence level of currently used predicting methodology.

Author Contributions: Conceptualization, W.S., J.S. and G.H.; methodology, W.S., S.H. and Q.H.; validation, J.S., J.X.; data curation, J.X. and W.S.; writing—original draft preparation, W.S., Q.H. and J.S.; writing—review and editing, W.S. and Q.H.; visualization, W.S.; supervision, J.W.; project administration, J.S.; funding acquisition, J.S. All authors have read and agreed to the published version of the manuscript.

Funding: This research was funded by the Key Project Funds from China Ship Scientific Research Center, grant number J1603 and the project "Research on the Intelligentized Design Technology of Hull Form" granted by Ministry of Industry and Information Technology of China, grant number [2018] 473.

Acknowledgments: Thanks are due to Shi-tang Dong and Leiping Xue for their valuable guidance and discussion and to those who helped the measurements in every sea trail test.

Conflicts of Interest: The authors declare no conflict of interest.

References

1. Larsson, L.; Stern, F.; Visonneau, M. *Numerical Ship Hydrodynamics—An Assessment of the Gothenburg 2010 Workshop*; Springer: Dordrecht, The Netherlands, 2014.
2. Pereira, F.; Vaz, G.; Eca, L. Verification and Validation Exercises for the Flow Around the KVLCC2 Tanker at Model and Full-Scale Reynolds Numbers. *Ocean Eng.* **2017**, *129*, 133–148. [CrossRef]
3. Duvigneau, R.; Visonneau, M.; Gan, B.D. On the role played by turbulence closures in hull shape optimization at model and full scale. *J. Mar. Sci. Technol.* **2003**, *8*, 11–25. [CrossRef]
4. Leer-Andersen, M.; Larsson, L. An experimental/numerical approach for evaluating skin friction on full-scale ships with surface roughness. *J. Mar. Sci. Technol.* **2003**, *8*, 26–36. [CrossRef]
5. Bhushan, S.; Xing, T.; Carrica, P.; Stern, F. Model- and Full-Scale URANS Simulations of Athena Resistance, Powering, Seakeeping, and 5415 Maneuvering. *J. Ship Res.* **2009**, *53*, 179–198.
6. ITTC. *1978 ITTC Performance Prediction Method*; ITTC: Zürich, Switzerland, 2014.
7. Min, K.-S.; Kang, S.-H. Study on the form factor and full-scale ship resistance prediction method. *J. Mar. Sci. Technol.* **2010**, *15*, 108–118. [CrossRef]
8. Katsui, T.; Asai, H.; Himeno, Y.; Tahara, Y. The proposal of a new friction line. In Proceedings of the Fifth Osaka Clolloquium on Advanced CFD Applications to Ship Flow and Hull from Design, Osaka, Japan, 14–15 March 2005.
9. Eca, L.; Hoekstra, M. The numerical friction line. *J. Mar. Sci. Technol.* **2005**, *13*, 328–345. [CrossRef]
10. Park, D.-W. A study on the effect of flat plate friction resistance on speed performance prediction of full scale. *Int. J. Nav. Archit. Ocean Eng.* **2015**, *7*, 195–211. [CrossRef]
11. Castro, A.M.; Carrica, P.M.; Stern, F. Full scale self-propulsion computations using discretized propeller for the KRISO container ship KCS. *Comput. Fluids* **2011**, *51*, 35–47. [CrossRef]
12. Satrke, B.; Bosschers, J. Analysis of scale effects in ship powering performance using a hybrid RANS-BEM approach. In Proceedings of the 29th Symposium on Naval Hydrodynamics, Gothenburg, Sweden, 26–31 August 2012.
13. Lin, T.-Y.; Kouh, J.-S. On the scale effect of thrust deduction in a judicious self-propulsion procedure for a moderate-speed containership. *J. Mar. Sci. Technol.* **2015**, *20*, 373–391. [CrossRef]
14. Wang, Z.Z.; Xiong, Y.; Wang, R.; Shen, X.-R.; Zhong, C.-H. Numerical study on scale effect of nominal wake of single screw ship. *Ocean Eng.* **2015**, *104*, 437–451. [CrossRef]
15. Guo, C.; Zhang, Q.; Shen, Y. A non-geometrically similar model for predicting the wake field of full-scale ships. *J. Mar. Sci. Appl.* **2015**, *14*, 225–233. [CrossRef]
16. Park, S.; Oh, G.; Hyung Rhee, S.; Koo, B.-Y.; Lee, H. Full scale wake prediction of an energy saving device by using computational fluid dynamics. *Ocean Eng.* **2015**, *101*, 254–263. [CrossRef]
17. Shen, H.-L.; Enock Omweri, O.; Su, Y.-M. Scale effects for rudder bulb and rudder thrust fin on propulsive efficiency based on computational fluid dynamics. *Ocean Eng.* **2016**, *117*, 199–209.
18. Carrica, P.M.; Mofidi, A.; Martin, E. Progress toward direct CFD simulation of Maneuvers in waves. In Proceedings of the VI International Conference on Computational Methods in Marine Engineering (Marine 2015), Rome, Italy, 15–17 June 2015; pp. 327–338.
19. Shen, Z.; Wan, D.; Carrica, P.M. Dynamic overset grids in OpenFOAM with application to KCS self-propulsion and maneuvering. *Ocean Eng.* **2015**, *108*, 287–306. [CrossRef]
20. Jasak, H.; Vukcevic, V.; Gatin, I.; Lalovic, I. CFD validation and grid sensitivity studies of full scale ship self propulsion. *Int. J. Nav. Archit. Ocean Eng.* **2019**, *11*, 33–43. [CrossRef]

21. Gaggero, S.; Villa, D.; Viviani, M. An extensive analysis of numerical ship self-propulsion prediction viaa coupled BEM/RANS approach. *Appl. Ocean Res.* **2017**, *66*, 55–78. [CrossRef]
22. Sánchez-Caja, A.; Martio, J.; Saisto, I.; Siikonen, T. On the enhancement of coupling potential flow models to RANS solvers for the prediction of propeller effective wakes. *J. Mar. Sci. Technol.* **2015**, *20*, 104–117. [CrossRef]
23. Sun, W.; Yang, L.; Wei, J.; Chen, J.; Huang, G. Numerical analysis of propeller loading with a coupling RANS and potential approach. *Proc. Inst. Mech. Eng. Part C J. Mech. Eng. Sci.* **2019**, *233*, 6383–6396. [CrossRef]
24. Mikkelsen, H.; Steffensen, M.L.; Ciortan, C.; Walther, J.H. Ship Scale Validation of CFD Model of Self-Propelled Ship. In Proceedings of the VIII International Conference on Computational Methods in Marine Engineering (Marine 2019), Gothenburg, Sweden, 13–15 May 2019.
25. Wang, C.; Sun, S.; Li, L.; Ye, L. Numerical prediction analysis of propeller bearing force for full-scale hull–propeller–rudder system. *Int. J. Nav. Archit. Ocean Eng.* **2016**, *8*, 589–610. [CrossRef]
26. Visonneau, M. A step towards the numerical simulation of viscous flows around ships at hfull scale—Recent achievements within the European Union Project EFFORT. In Proceedings of the 4th International Conference on Marine Hydrodynamics (Marine CFD 2005), London, UK, 30–31 March 2005.
27. Ponkratov, D. *Proceedings of the 2016 Workshop on Ship Scale Hydrodynamic Computer Simulations*; Lloyd's Register: Southampton, UK, 25 November 2016; Available online: https://www.lr.org/en/events/ (accessed on 2 January 2020).
28. Ponkratov, D.; Zegos, C. Validation of Ship Scale CFD Self-Propulsion Simulation by the Direct Comparison with Sea Trials Results. In Proceedings of the Fourth International Symposium on Marine Propulsors, Asutin, TX, USA, 31 May–4 June 2015.
29. ISO. *Ship and Marine Technology—Guidelines for the Assessment of Speed and Power Performance by Analysis of Speed Trial Data*; ISO: Geneva, Switzerland, 2015.
30. Yun, Q.Q.; Hu, Q.; Chen, W.W. SK-01 shaft power measurement system for marine diesel engine. *Shipbuild. China* **2010**, *51*, 126–131.
31. Niklas, K.; Pruszko, H. Full-Scale CFD simulations for the determination of ship resistance as a rational, alternative method to towing tank experiments. *Ocean Eng.* **2019**, *190*, 106435. [CrossRef]
32. Demirel, Y.K.; Turan, O.; Incecik, A. Predicting the effect of biofouling on ship resistanc e using CFD. *Appl. Ocean Res.* **2017**, *62*, 100–118. [CrossRef]

Journal of
Marine Science and Engineering

Article

Experimental Study of Supercavitation Bubble Development over Bodies in a Duct Flow

Lotan Arad Ludar * and Alon Gany

Faculty of Aerospace Engineering, Technion—Israel Institute of Technology, Haifa 32000, Israel; gany@tx.technion.ac.il

* Correspondence: lotanludar@gmail.com

Received: 18 December 2019; Accepted: 6 January 2020; Published: 8 January 2020

Abstract: Understanding the development and geometry of a supercavitation bubble is essential for the design of supercavitational vehicles as well as for prediction of bubble formation within machinery-related duct flows. The role of the cavitator (nose) of a body within the flow is significant as well. This research studied experimentally supercavitation bubble development and characteristics within a duct flow. Tests were conducted on cylindrical slender bodies (3 mm diameter) within a duct (about 20 mm diameter) at different water flow velocities. A comparison of supercavitation bubbles, developing on bodies with different nose geometries, was made. The comparison referred to the conditions of the bubbles' creation and collapse, as well as to their shape and development. Various stages of the bubble development were examined for different cavitators (flat, spherical, and conical nose). It was found that the different cavitators produced similar bubble geometries, although at different flow velocities. The bubble appeared at the lowest velocity for the flat nose, then for the spherical nose, and at the highest velocity for the conical cavitator. In addition, a hysteresis phenomenon was observed, showing different bubble development paths for increasing versus decreasing the water flow velocity.

Keywords: supercavitation; supercavities; cavitator; hysteresis

---

1. Introduction

The geometry of supercavitation bubbles has been studied for decades, using diverse tools and methods, because of its significance for the design of supercavitation vehicles and applications. In particular, comprehensive studies have examined axisymmetric supercavitation bubbles, proposing methods of calculation and presenting relations between the bubble dimensions and the flow conditions [1–5]. Other studies have analyzed the wall effect in channels on the bubble dimensions as well as the values of the cavitation number of fully developed cavitation bubbles [6]. Some experimental works showed the development of axially symmetric supercavities in bounded and unbounded flows and examined the bubble characteristics relating to natural and artificial supercavitation [7,8]. These studies are significant for hydraulic machinery in which the appearance and development of cavitation play a key role in the performance of the system. One of the most significant topics in the field is the role of the cavitator in determining the bubble shape and size. This has been examined analytically for 2D flows [9] and numerically for 3D flows for more accurate results and for specific geometries [10]. Numerical results have shown that the bubble dimensions are dependent on the cavitator shape [11]. In addition, some estimations and predictions have been done, mainly for unbounded bubbles [12–18]. Other experimental investigations have described the bubbles formed on different bodies, examining the gravitation effect, the closing modes, the separation point, etc. Many of them were summarized by Franc and Michel (2004) [19].

The objective of the present study was to investigate supercavitation bubble development over bodies in a duct water flow. We examined experimentally the influence of different cavitator geometries on axisymmetric cylindrical bodies, finding the relations between the dimensions of the bubble and the cavitation number. We also revealed and investigated a hysteresis phenomenon in naturally developed supercavitation bubbles. Previous studies have discussed this phenomenon for ventilated (artificial) bubbles. Semenenko (2001) [3] examined hysteresis found in bubbles closing on a solid body, investigating the angle of the bubble closure on the solid surface as well as the required gas supply to the bubble for maintaining its size. Wosnik and Arndt (2009) [20] examined hysteresis for different bodies in ventilated bubbles. In our research, we observed hysteresis in naturally formed bubbles that do not close on a solid body.

2. Problem Description

We considered axisymmetric supercavitation bubbles developing along a cylindrical object in a uniform flow of water within a convergent-divergent nozzle. Slender cylindrical bodies with different nose (cavitator) geometries were examined. The front edge nose causes velocity change, flow separation, and pressure drop. When the pressure decreases below the equilibrium vapor pressure of the liquid, the water starts to evaporate, developing a supercavitation bubble over the body. As the flow velocity increases, the pressure decreases, and the bubble grows and can envelope the entire body (see Figure 1). The geometry of the body, and especially its front edge, is the main factor that determines the flow field and the supercavitation bubble creation and development.

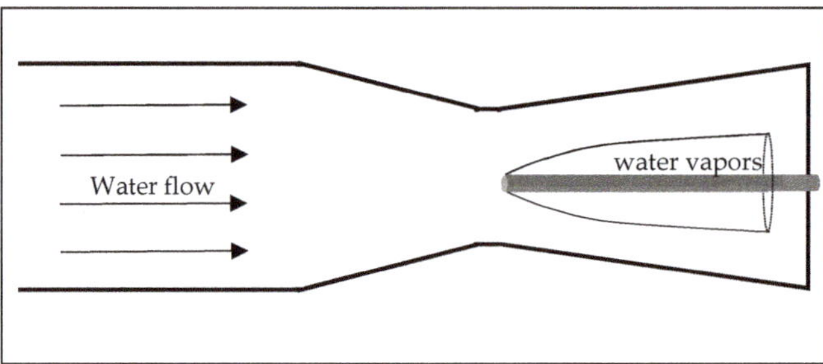

Figure 1. A scheme of the physical problem.

3. The Experimental System and Flow Conditions

The experimental system included a converging–diverging nozzle (converging angle 6.5°, diverging angle 2°), connected to a straight pipe from which the water flowed uniformly in a constant temperature of about 20 °C (Figure 2). The nozzle accelerated the flow to the appropriate speed in which cavitation could be created. A valve was used to adjust the water flow rate. The supercavitation body was placed right after the nozzle throat (Figure 3). Three slender cylindrical bodies of 2.97 mm diameter with different cavitators (noses) were tested: a flat cavitator, a spherical cavitator, and a conical cavitator with an angle of 15° (Figure 4). The bubble created over the body was examined while increasing the water flow rate. Seven pressure measurement taps were placed along the wall: one at the wall of the pipe before the entrance to the nozzle; two at the converging section of the nozzle; one at the nozzle throat, where the cavitator nose was placed (the location where the initial creation of the cavitation bubble was expected); and three additional gauges at the diverging section, along the supercavitation bubble (see again Figure 3). The Reynolds number of the flow, calculated with a relation to the object as well as the duct, was higher than 30,000 for all flow-rates tested.

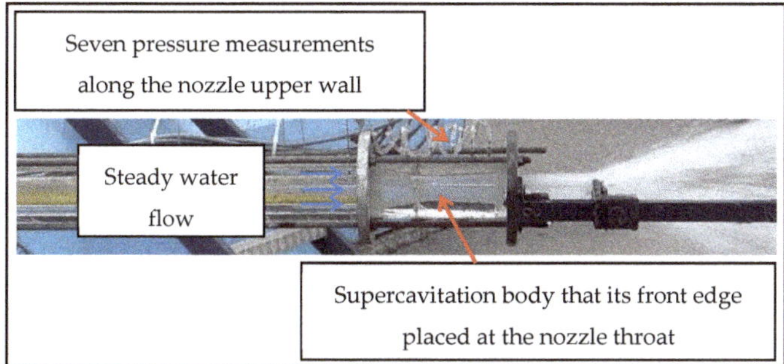

Figure 2. The experimental system.

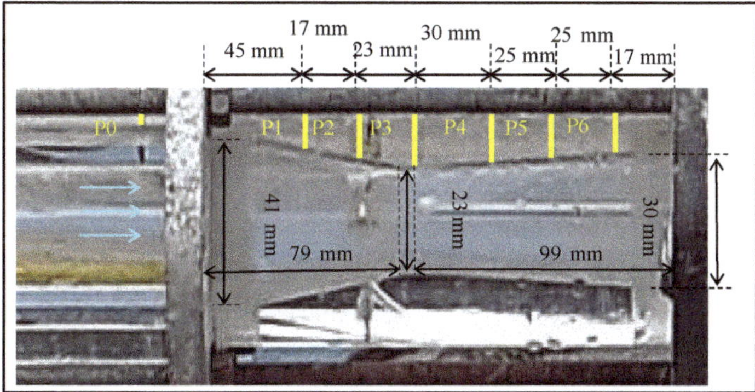

Figure 3. The nozzle geometry and the pressure measurement locations.

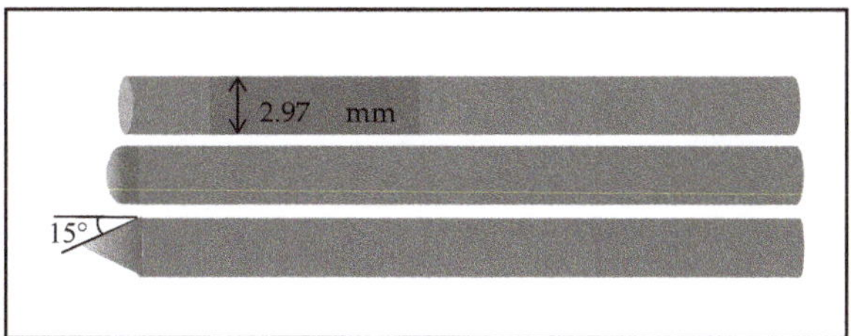

Figure 4. The supercavitation bodies.

4. Results

4.1. Stages of Bubble Development

The different stages in the bubble development along the body with increasing flow velocity are presented in Figure 5 for the flat cavitator. The observed geometries of the bubbles were practically similar for the three different cavitators. In every stage of the flow, the bubble did not close on the body, and it remained open in its back edge. At low velocities, the geometry of the bubble surface could be expressed by the function described in Equation (1):

$$f_a = \frac{d}{2}\left(1 + \frac{6}{d}z\right)^{\frac{1}{3}} \tag{1}$$

where d is the cavitator diameter. z is the coordinate of the axis of symmetry expressed with the same length units as d. This expression of the bubble geometry was used by Semenenko (2001) [3] based on Logvinovich empiric results (1973) [16].

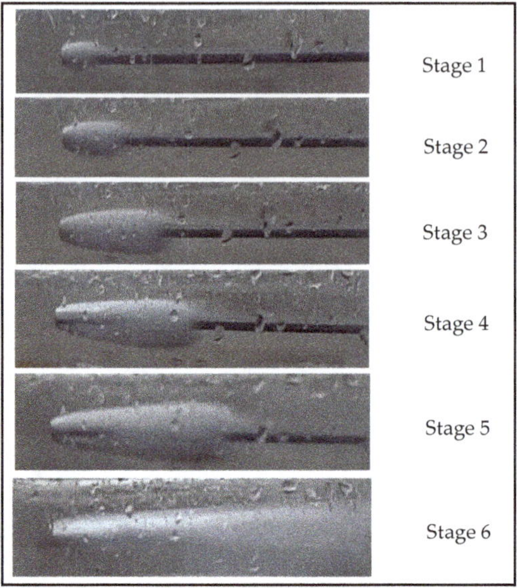

Figure 5. Stages of the bubble development for a flat cavitator.

When increasing the velocity, the function convexity changed, and the geometry of the bubble surface could be expressed by the function described in Equation (2):

$$f_b = \frac{d}{2}\left(1 + \frac{a}{d}z\right)^{\frac{1}{2}} \tag{2}$$

where a is a constant factor, depending on the geometry of the nozzle and of the body. When the velocity was further increased, the geometry of the bubble surface could almost be described as an open cone, with the function in Equation (3):

$$f_c = \frac{d}{2}\left(1 + \frac{a}{d}z\right) \tag{3}$$

In general, the convexity of the bubble decreases as the velocity increases in its front and in its back edges together. The two last functions, describing the bubble geometries in the later stages of development, are presented in Figure 6, revealing practically the same shape for all three cavitators.

Although the stages of development are similar for all three cavitators, the velocity at which the bubble starts as well as the characteristic velocity for each stage of development are different for the different cavitators. Table 1 presents the velocities of the flow at the front edge of the cavitator for the three cavitators in the six stages of the bubble development presented in Figure 5. The velocities were calculated from the water volume flow rate divided by the flow cross-section just before the cavitator.

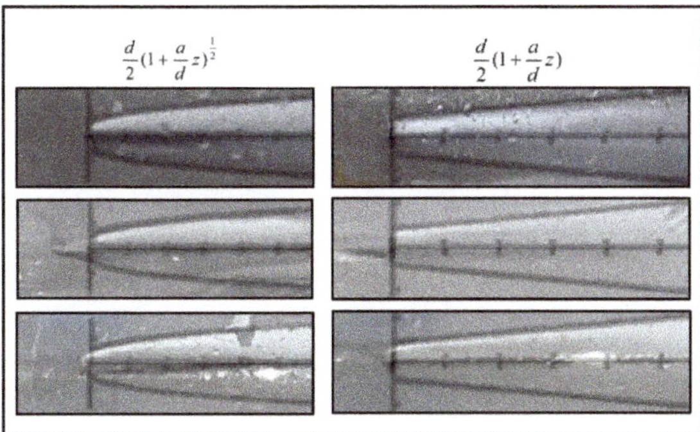

Figure 6. The geometries of the supercavitation bubble for three different cavitators.

Table 1. Flow velocity at the front edge of the cavitator in Stages 1–6 of the bubble development for the three cavitators.

Stage of Development	Conical Cavitator	Spherical Cavitator	Flat Cavitator
Stage 1	14.8 m/s	13.5 m/s	9.9 m/s
Stage 2	16.4 m/s	15.8 m/s	11.7 m/s
Stage 3	17.2 m/s	16.8 m/s	13.8 m/s
Stage 4	17.8 m/s	17.3 m/s	15.5 m/s
Stage 5	19.7 m/s	18.3 m/s	17.7 m/s
Stage 6	20.7 m/s	19.8 m/s	18.8 m/s

4.2. The Pressure Field

The change in the pressure was measured on the nozzle walls. The pressure decreases moderately with the increase of the flow velocity for each of the cavitators. A rapid decrease was detected when the cavitation bubble grew and extended, reaching the section of the gauge. Similar to the order that the beginning of the bubble creation appeared, the process took place at the lowest velocity for the flat cavitator, then for the spherical cavitator, and finally at the highest velocity for the conical cavitator (see measurements of P5 located at the divergent part of the nozzle where the bubble is extending in Figure 7).

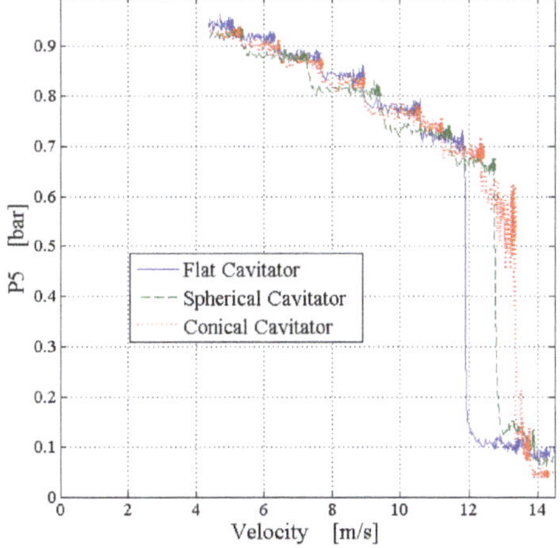

Figure 7. The wall pressure in P5 vs. the velocity in this section for the three cavitators [1].

The results of the measurements of P5 show that there are pressure fluctuations in the system. The largest fluctuations occur for the conical cavitator, where the system is the least stable; the slender body undergoes mechanical vibrations caused by the high flow rate in the pipe required for the creation and development of the supercavitation bubble. The lowest fluctuations in the graph appear for the flat cavitator, in which the bubble is created and developed at the lowest flow velocities. The lowest static wall pressure (the same for the three cavitators) measured by P4 and P5 was 0.065 and 0.07 bar (with an uncertainty of ±0.025 bar), respectively. These values were somewhat higher than the theoretical equilibrium vapor pressure (0.023 bar at 20 °C), presumably existing within the cavitation bubble. The magnitude of the pressure drop was similar for all cavitators. In addition, the pressure change was similar in all cavitators for every stage of the bubble development.

4.3. Comparison of Results with Theoretical Analysis

A control volume analysis was done (Figure 8) to calculate the pressure at a cross section i with relation to the conditions at the nozzle throat t (where the front edge of the cavitator is placed), assuming a one-dimensional flow with negligible viscosity effects. The theoretical calculation was compared with the experimental data. From conservation of mass, one can find the relation between the velocities at i and t cross-sections:

$$u_i = u_t \frac{A_t}{A_i} \tag{4}$$

where u_i and A_i are the flow velocity and cross-section area, respectively, at the location of gauge P_i, and u_t, A_t are the corresponding values at the nozzle throat. Using Bernoulli equation on a flow streamline at the wall between the throat and a cross section i of the gauge and substituting Equation (4), we derive the pressure difference between section i and the nozzle throat:

$$\Delta P = P_i - P_t = \frac{1}{2}\rho {u_t}^2 \left(1 - \left(\frac{A_t}{A_i}\right)^2\right) - \rho g h \tag{5}$$

where P_i, P_t are the pressures in section i and in the nozzle throat, correspondingly, g is the gravitational acceleration, ρ is the density of the water, and h is the height difference calculated in Equation (6) for the divergence angle of 2°:

$$h = z \tan 2^\circ \tag{6}$$

where z is the distance from the throat cross-section to section i of gauge P_i. As the height contribution is three orders of magnitude smaller than the dynamic pressure, $\rho g h / \frac{1}{2}\rho u^2 \sim O(10^{-3})$, this term in Equation (5) can be neglected.

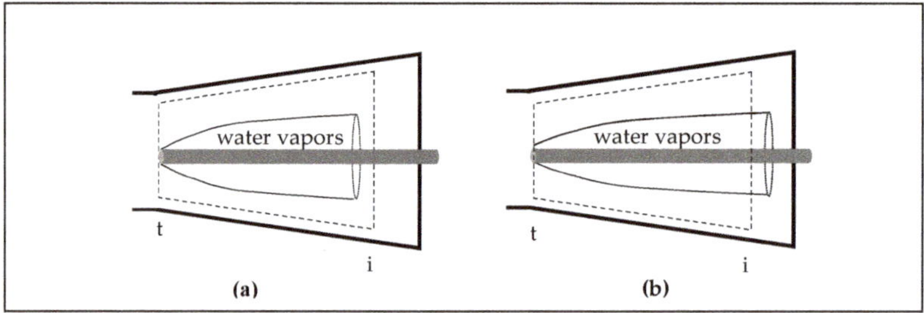

Figure 8. Control volume for theoretical analyses. (**a**) the cavitation does not reach the cross-section (**b**) the cavitation reaches the cross-section.

[1] Note that the velocities in Figure 7 describe the velocities in gauge P5 section, thus they are smaller than the velocities at the nozzle throat and those in Table 1 due to a mass conservation.

Figure 8a,b presents a situation where the cavitation bubble has not reached cross-section i and a situation where cross-section i is already within the bubble domain, respectively. In the former, the cross-section for the water flow at i is $A_i = \pi\left(d_i{}^2 - d^2\right)/4$, whereas the cross section for the water flow at t is $A_i = \pi\left(d_t{}^2 - d^2\right)/4$. The theoretical pressure difference between i and t cross-sections, $\Delta P = P_i - P_t$, should be derived from Equation (5). Figure 9 shows the theoretical line and experimental points versus the water flow velocity at cross section t for such a case, revealing very good agreement to within ±0.05 bar.

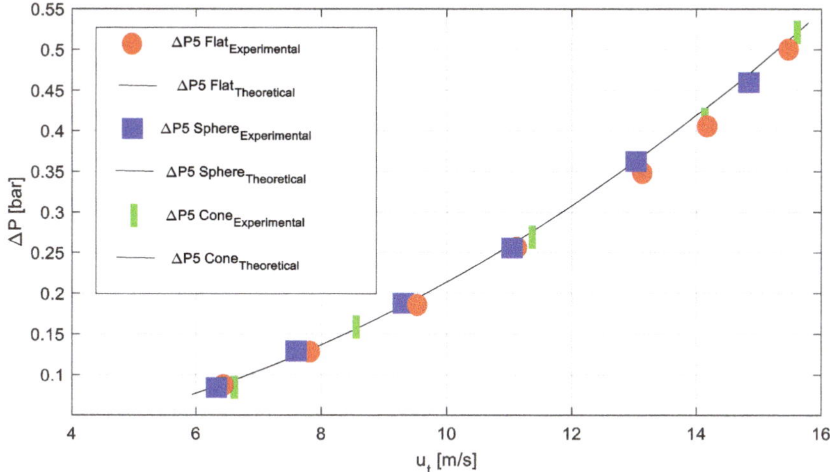

Figure 9. Comparison between the theoretical and experimental values of $\Delta P = P_5 - P_t$ vs. the velocity at the nozzle throat for the different cavitators.

Figure 8b shows a further developed bubble, exceeding section i. One can assume that the pressure within the naturally developed supercavitation bubble is equal to the equilibrium vapor pressure throughout the bubble. We can further assume that the pressure in a cross-section is more or less constant (meaning that, along the bubble, the pressure field is uniform and practically equal to the vapor pressure). Bernoulli equation implies that, to keep a uniform pressure along the flow path, the flow velocity should be constant, meaning the available cross section for the water flow should be constant. With the assumption that the water bypasses the contour of the vapor bubble, we obtain:

$$d_c = \sqrt{d_i{}^2 - d_t{}^2 + d^2} \tag{7}$$

where, in our case, the nozzle divergence angle is 2°, d_c is the cavitation bubble diameter, and

$$d_i = d_t + 2z\tan 2^\circ \tag{8}$$

Equation (7) means that the shape of the bubble in a bounded duct (nozzle) is dictated by the shape of the nozzle to practically keep a constant cross-section for the water flow between the supercavitation bubble boundary and the nozzle walls. It explains the finding in this research that the different cavitators (of different nose shape, but the same diameter) exhibited the same bubble shape. The function describing the bubble geometry was obtained as:

$$f(z) = \frac{d_c}{2} = \frac{d}{2}\left(1 + \frac{(2m)^2/d}{d}z\left(\frac{d_t}{m} + z\right)\right)^{1/2} \tag{9}$$

where $m = \tan 2°$. For small z, Equation (9) narrows to Equation (2) with the geometric factor:

$$a = 4m\left(\frac{d_t}{d}\right) \tag{10}$$

A comparison between the bubble geometry predicted in Equation (9) and the experimental results for all three cavitators is described in Figure 10, showing agreement within about 3%. The experimental results are related to Stage 5 of the bubble development.

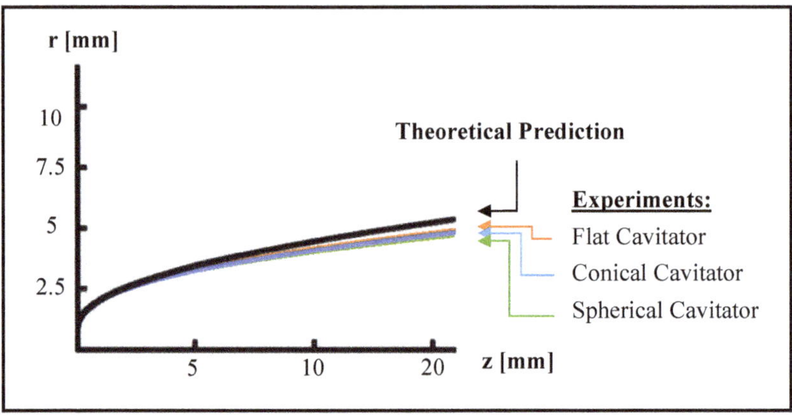

Figure 10. Comparison between the theoretical prediction and experimental results of the bubble geometry for the three cavitators.

4.4. The Bubble Dimensions

Based on the experiments, a relation between the supercavitation bubble dimensions and the cavitation number of the flow was deduced for all three cavitators (Figure 11), according to Equation (11):

$$l/c = A\sigma^n \tag{11}$$

where l is the bubble length; c is the diameter of the body; A, n are constants depending on the flow conditions, bubble position, and form; and σ is the cavitation number of the flow calculated according to Equation (12):

$$\sigma = \frac{p_a - p_v}{\frac{1}{2}\rho v^2} \tag{12}$$

where p_a is the atmospheric pressure, p_v is the vapor pressure of the water, ρ is the water density, and v is the water velocity.

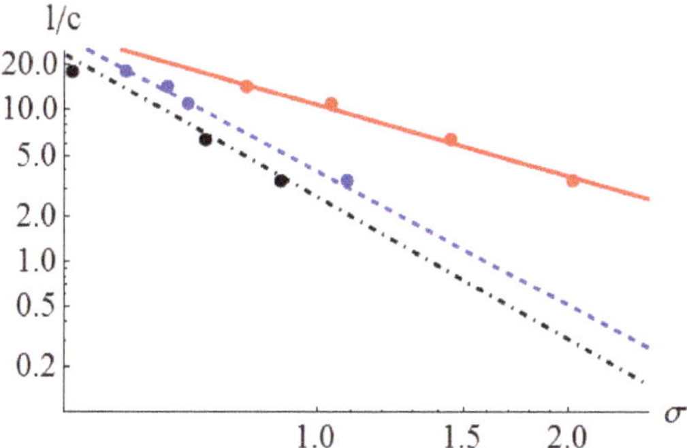

Figure 11. The measured supercavitation bubble dimension ratio vs. the cavitation number for the three cavitators.

All curves reveal the same general trend of decreasing the ratio of the bubble length to body diameter when increasing the cavitation number. The two cavitators with the gradually changing front shape, namely, the conical and spherical cavitators, exhibit very similar curves, both in slope and in magnitude. This could be expected as they generated similar bubble shapes at only slightly different flow velocities. The flat cavitator, however, showed bubble formation and development at substantially lower flow velocities, causing different slope and level of the curve. One may assume that the reason should be flow separation due to the abrupt change from the flat nose to the cylindrical body shape.

5. "Hysteresis"

Increasing the flow rate, the bubble grew and extended along the body until it reached the rear edge of the nozzle, which was open to the atmosphere. When opening to the atmosphere, the pressure measured at the wall grew rapidly becoming equal to the outside pressure (one atmosphere). Further increase of the flow rate did not change the pressure. Examining the situation when gradually decreasing the water flow rate revealed that initially the pressure stayed constant and the bubble had the same size and shape. Further decrease in the flow rate, dropping the flow velocity much below the value that initially caused the bubble creation, led to a sudden collapse of the bubble. Detecting the bubble size and shape as well as the wall pressure during the process, one could see a hysteresis-like phenomenon, although the physical processes were not utterly reversed. Figure 12 shows the different paths of P5 and P6 wall pressure variation for the spherical cavitator, for the two parts of the experiment (a velocity increase followed by a velocity decrease). The first part of the experiment revealed a gradual pressure decrease followed by a substantial drop, when the bubble reached the gauge section (as described in Section 4.2), and then an abrupt increase to the atmospheric pressure when reaching the nozzle exit. In the second part of the experiment, when decreasing the flow velocity, the bubble and the pressure remained steady until the bubble collapsed (with only a slight change in pressure). The hysteresis that we observed is for a naturally developed vapor bubble, which was influenced by the opening of the bubble to the atmosphere. At that stage, the bubble contained a mixture of water vapors and atmospheric air. It showed a similar behavior to that of an artificial cavity resulting from a gas supply.

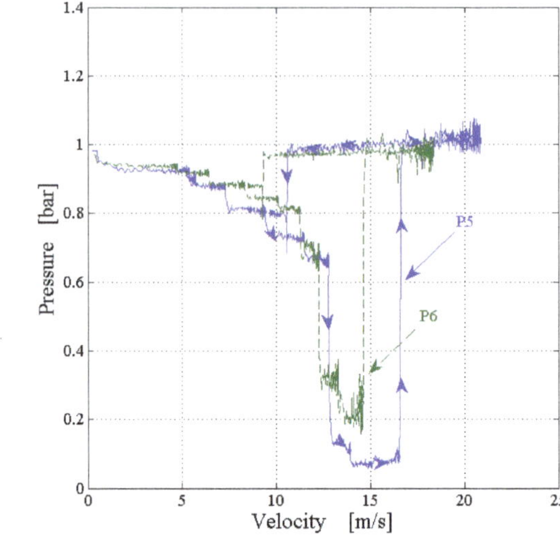

Figure 12. Wall pressure at the diverging part of the nozzle vs. the velocity for the spherical cavitator during a cycle including an increase followed a decrease of the flow velocity.

6. Conclusions

In a duct water flow, the wall has a significant influence on the flow regime and on the development of a supercavitation bubble on an object within the flow. The cavitator role in determining the bubble shape and dimensions becomes significantly smaller and its main impact is on the conditions (flow rate and velocity) of the bubble creation and collapse. It was found that different cavitators could produce similar bubble geometries, although at somewhat different flow velocities. Testing three different cavitators at the same flow conditions and examining the bubble development for each of the cavitators, one observes that a supercavitation bubble is created at the lowest flow velocity by a cavitator generating the largest disturbance in the flow, implying a more rapid change in the flow regime. This was the flat cavitator for which the bubble was created at a flow velocity of only 9.9 m/s. A bubble was created at the highest flow velocity for the cavitator causing the least and more moderate disturbance in the flow. This was the conical cavitator, for which the bubble was created at a velocity of 14.8 m/s (50% larger than for the flat cavitator). Another phenomenon observed was hysteresis. When increasing and decreasing the flow rates, the bubble as well as the pressure variations of growth and decrease underwent different paths. Supercavitation is a fluid dynamics phenomenon, depending mainly on the flow and pressure fields. The water quality, such as salt concentration (seawater vs. tap water), may have a small effect through the dependence of the vapor pressure. The vapor pressure of saline water is slightly lower than for pure water. For instance, at 20 °C, vapor pressure of pure water is 0.0234 bar, and that of seawater is 0.0229 bar. This small difference would practically not change the results but might cause a slight change in the velocity value for each stage of supercavitation bubble development.

Author Contributions: This academic research was conducted in full collaboration and involvement of both authors, as is commonly done by a PhD student (L.A.L.) and thesis Advisor (A.G.), with regards to conceptualization, evaluation, and presentation of the results. All authors have read and agreed to the published version of the manuscript.

Funding: This research received no external funding.

Conflicts of Interest: The authors declare no conflict of interest.

References

1. Logvinovich, G.V.; Serebryakov, V.V. On methods of calculating a shape of slender axisymmetric cavities. *Hydromechanics* **1975**, *32*, 47–54.
2. Serebryakov, V.V. Asymptotic solutions of axisymmetric problems of the cavitational flow under slender body approximation. In *Hydrodynamics of High Speeds*; Chuvashian State University: Cheboksary, Russia, 1990; pp. 99–111.
3. Semenenko, V.N. Artificial Supercavitation. Physics and Calculation. In Proceedings of the RTO Lecture Series 005 on Supercavitating Flows, Brussels, Belgium, 12–16 February 2001.
4. Serebryakov, V.V. The models of the supercavitation prediction for high speed motion in water. In Proceedings of the International Scientific School: HSH-2002, Chebocsary, Russia, 16–23 June 2002.
5. Savchenko, Y. Supercavitation-problems and perspectives. In Proceedings of the 4th International Symposium on Cavitation (CAV 2001), California Institute of Technology, Pasadena, CA, USA, 20–23 June 2001.
6. Wu, T.Y.T.; Whitney, A.K.; Brennen, C. Cavity-flow wall effects and correction rules. *J. Fluid Mech.* **1971**, *49*, 223–256. [CrossRef]
7. Ahn, B.K.; Lee, T.K.; Kim, H.T.; Lee, C.S. Experimental investigation of supercavitating flows. *Int. J. Nav. Archit. Ocean Eng.* **2012**, *4*, 123–131. [CrossRef]
8. Ahn, B.K.; Jeong, S.W.; Kim, J.H.; Shao, S.; Hong, J.; Arndt, R.E. An experimental investigation of artificial supercavitation generated by air injection behind disk-shaped cavitators. *Int. J. Nav. Archit. Ocean Eng.* **2017**, *9*, 227–237. [CrossRef]
9. Fridman, G.M.; Achkinadze, A.S. *Review of Theoretical Approaches to Nonlinear Supercavitating Flows*; Report No. ADP012079, Ship Theory Department, Saint Petersburg State Marine Technical University: St Petersburg, Russia, 2001.
10. Kirschner, I.I.; Chamberlin, R.; Arzoumanian, S.A. simple approach to estimating three-dimensional supercavitating flow fields. In Proceedings of the 7th International Symposium on Cavitation (CAV2009), Ann Arbor, MI, USA, 16–20 August 2009.
11. Kwack, Y.K.; Ko, S.H. Numerical analysis for supercavitating flows around axisymmetric cavitators. *Int. J. Nav. Archit. Ocean Eng.* **2013**, *5*, 325–332. [CrossRef]
12. Brennen, C. A numerical solution of axisymmetric cavity flows. *J. Fluid Mech.* **1969**, *37*, 671–688. [CrossRef]
13. Kinnas, S.A. The prediction of unsteady sheet cavitation. In Proceedings of the 3rd International Symposium on Cavitation, Grenoble, France, 7–10 April 1998.
14. Scardovelli, R.; Zaleski, S. Direct numerical simulation of free-surface and interfacial flow. *Annu. Rev. Fluid Mech.* **1999**, *31*, 567–603. [CrossRef]
15. Vasin, A.D. *The Principle of Independence of the Cavity Sections Expansion (Logvinovich's principle) as the Basis for Investigation on Cavitation Flows*; Central Aerodynamics Institute (TSAGI): Moscow, Russia, 2001.
16. Shi, H.H.; Wen, J.S.; Zhu, B.B.; Chen, B. Numerical simulation of the effect of different object nose shapes on hydrodynamic process in water entry. In Proceedings of the 10th International Symposium on Cavitation (CAV2018), Baltimore, MD, USA, 14–17 July 2018.
17. Logvinovich, G.V. *Hydrodynamics of Flows with Free Boundaries*; Naukova Dumka Publishing House: Kiev, Ukraine, 1973.
18. Ahn, B.K.; Jeong, S.W.; Park, S.T. An experimental investigation of artificial supercavitation with variation of the body shape. In Proceedings of the 10th International Symposium on Cavitation (CAV2018), Baltimore, MD, USA, 14–17 July 2018.
19. Franc, J.P.; Michel, J.M. *Fundamentals of Cavitation*; Kluwer Academic Publishers: Dordrecht, The Netherlands, 2004.
20. Wosnik, M.; Arndt, R.E.A. Control experiments with a semi-axisymmetric supercavity and a supercavity-piercing fin. In Proceedings of the 7th International Symposium on Cavitation, Ann Arbor, MI, USA, 17–22 August 2009.

Journal of
Marine Science and Engineering

Article

Towards a Realistic Estimation of the Powering Performance of a Ship with a Gate Rudder System

Noriyuki Sasaki [1], S. Kuribayashi [2], M. Fukazawa [3] and Mehmet Atlar [1,*]

[1] Department of Naval Architecture, Ocean and Marine Engineering, University of Strathclyde, Glasgow G4 0LZ, UK; noriyuki.sasaki@strath.ac.uk
[2] Kuribayashi Steamship Co. Ltd.,Tokyo 100-0004, Japan; A.Kotani@kuribayashi.co.jp
[3] Kamome Propeller Co. Ltd.,Yokohama 245-8542, Japan; m-fukazawa@kamome-propeller.co.jp
* Correspondence: mehmet.atlar@strath.ac.uk

Received: 11 December 2019; Accepted: 12 January 2020; Published: 15 January 2020

Abstract: This paper presents an investigation on the scale effects associated with the powering performance of a Gate Rudder System (GRS) which was recently introduced as a novel energy-saving propulsion and maneuvring device. This new system was applied for the first time on a 2400 GT domestic container ship, and full-scale sea trials were conducted successfully in Japan, in 2017. The trials confirmed the superior powering and maneuvring performance of this novel system. However, a significant discrepancy was also noticed between the model test-based performance predictions and the full-scale measurements. The discrepancy was in the power-speed data and also in the maneuvring test data when these data were compared with the data of her sister container ship which was equipped with a conventional flap rudder. Twelve months after the delivery of the vessel with the gate rudder system, the voyage data revealed a surprisingly more significant difference in the powering performance based on the voyage data. The aim of this paper, therefore, is to take a further step towards an improved estimation of the powering performance of ships with a GRS with a specific emphasis on the scale effect issues associated with a GRS. More specifically, this study investigated the scale effects on the powering performance of a gate rudder system based on the analyses of the data from two tank tests and full-scale trials with the above-mentioned sister ships. The study focused on the corrections for the scale effects, which were believed to be associated with the drag and lift characteristics of the gate rudder blades due to the low Reynolds number experienced in model tests combined with the unique arrangement of this rudder and propulsion system. Based on the appropriate semi-empirical approaches that support model test and full-scale data, this study verified the scale effect phenomenon and presented the associated correction procedure. Also, this study presented an enhanced methodology for the powering performance prediction of a ship driven by a GRS implementing the proposed scale effect correction. The predicted powering performance of the subject container vessel with the GRS presented an excellent agreement with the full-scale trials data justifying the claimed scale effect and associated correction procedure, as well as the proposed enhanced methodology for the practical way of predicting the powering performance of a ship with the GRS.

Keywords: gate rudder system; flap rudder; energy saving device; ducted propeller; maneuvring device; powering performance; scale effects; laminar separation

1. Introduction

1.1. Gate Rudder System

A gate rudder system (GRS) is a rather novel but straightforward arrangement of the ship rudder and propeller to act as an attractive and sound energy-saving propulsion and maneuvring device

(ESMD), for example, [1–3]. In this system, the classical single-rudder behind the propeller arrangement is replaced by twin-rudder blades with asymmetric cross-sections which are positioned on either side of the propeller. The two rudder blades encircle the propeller at the upper half of the propeller plane, as well as at both sides, such as a separated duct which is split into two sections with no bottom part. The arrangement of the gate rudder system (GRS) on the "Shigenobu" is shown in Figure 1a (left) in contrast to the conventional (flap-) rudder system (CRS) of the "Sakura" in Figure 1b (right). Shigenobu has the world's first GRS installed on it while her sister ship (Sakura) has a conventional rudder system (CRS) with the same hull and engine particulars. The principal dimensions of both vessels are presented in Table 1, Sasaki et al. (2019) [4].

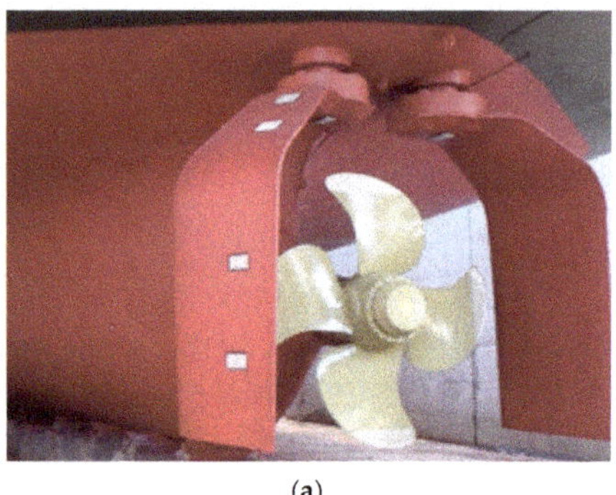

(a)

(b)

Figure 1. (**a**) Gate rudder system (GRS) on coastal container ship "Shigenobu." and (**b**) conventional flap-rudder system (CRS) on sister container ship "Sakura.".

Table 1. Principal characteristics of Shigenobu and Sakura.

Particulars	Sakura	Shigenobu
Length OA (m)	111.4	
Beam (m)	17.8	
Draft (m)	5.24	
Main Engine	3309 kW × 220 rpm	
Rudder	Flap rudder	Gate rudder
Delivery	August 2016	December 2017

Each rudder blade of the GRS can be controlled individually to affect the direction of the propeller's slipstream (i.e., to vector), and hence to steer the vessel with increased maneuvring and motion control capability. The GRS, therefore, takes advantage of additional thrust generated by the two rudder blades, in contrast to the extra resistance that results from the conventional rudder. This is somewhat similar to the ducted propulsor but with a much larger propeller diameter as compared with the traditional ducted propeller with less surface area, and hence improves the efficiency significantly. The GRS, therefore, can be categorized as a new "open-type ducted propeller" which is distinct from a conventional "closed-type ducted propeller" (e.g., Kort Nozzle) and a "front-type ducted propeller" (e.g., the Becker Mewis Duct or SILD and Sumitomo Integrated Lammern Duct), as shown in Figure 2.

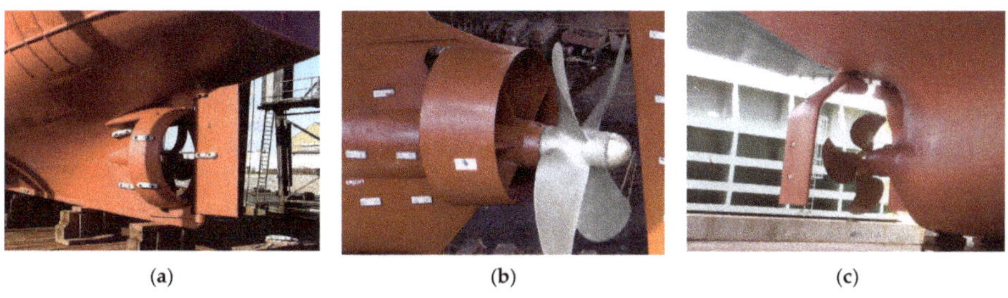

Figure 2. Kort nozzle (**a**), Becker Mewis Duct (**b**), and gate rudder system (**c**).

The foundation of a GRS is based on the originating activities, in Japan, to improve the maneuverability of coastal vessels which require tighter control of ships in their transverse motions at ports. This propulsion and maneuvering device was further developed by the recent R&D activities in the UK, e.g., Sasaki et al. (2015, 2018, 2017, 2019) [1,4–6]. The GRS was applied for the first time on a 2400 GT full-scale, new-built coastal container "Shigenobu", which was entered into service on November 2017. This was to demonstrate the vessel's performance, especially her excellent maneuverability performance. The comprehensive speed and maneuvering trials with this vessel and her sister ship "Sakura", indicated that the vessel with GRS was 14% more efficient at the design speed than her sister's while the gain in service can be as high as 30% in rough seas, as shown in Figure 3.

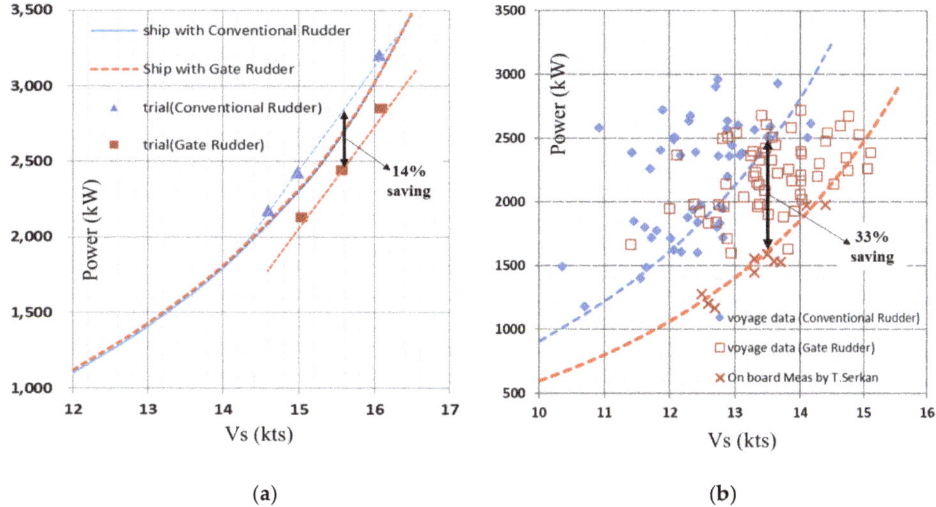

Figure 3. Powering performance comparisons of two sister ships, one with a gate rudder system and the other with a conventional flap-rudder system: In trials (**a**) and in-service (**b**).

The service performances of these sister ships, which operate on the same route and almost follow each other's path with the same mission in the northeast coast of Japan, are investigated based on the ship owner's standard logbook. This logbook continuously recorded the mean values of the important key performance indicators, such as ship speed, power, and fuel consumption for both vessels, since Nov 2017. As shown in Figures 3a and 4, the plot of the logbook data for powering and fuel consumption, respectively, indicates that the gain in performance for the ship with the GRS was as high as 30% or more than her sister ship with the conventional rudder system (CRS) [2,3].

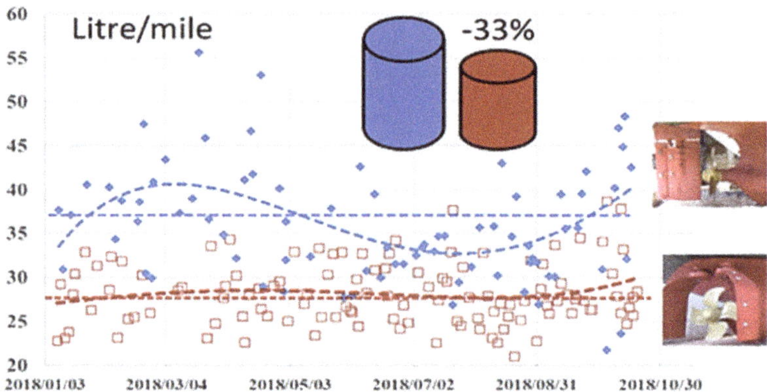

Figure 4. In-service fuel consumption comparisons of two sister ships, Shigenobu with GRS (red) and Sakura with CRS (blue).

1.2. Scale Effects

While the main remarkable features of a GRS are highlighted in the previous section (Section 1.1), it should be asked "why did no one think of the GRS idea before?" One reason appears to be that the conventional modelling techniques are not suitable for use in an unconventional propeller-rudder configuration. It is not easy to say, but it strongly relates to the testing technology and the analysis procedures. A GRS is not within the scope of the existing model testing and scaling technology. For example, the resistance of the gate rudder measured in model tests was found to be rather high, i.e., 5 to 10 times as compared with the full scale due to the suspected scale effects. This gives the wrong conclusion for the model test results. In fact, the decision to apply a GRS in full scale was taken in Japan even though the model test results achieved from existing model testing methodologies were unsupportive and the full-scale performance was remarkably above the predicted gain based upon the model tests. The comparative result of the power prediction for the world's first gate rudder system driven ship, "Shigenobu", with the trials data using the combined ITTC and Yazaki method [7], can be seen in Figure 3. As shown in Figure 3a, on the one hand, the conventional prediction method did not present any gain for the gate rudder system while the full-scale trials of Shigenobu presented a 14% energy savings. On the other hand, as also shown in Figure 3a, the same method predicted the trial performance of her sistership "Sakura" with the conventional flap-rudder system very well by raising the question for ships with the GRS. In fact, this was the major motivation of this paper to explore the potential scale effects, which may be the reason for the discrepancy between the predictions and the full-scale data for Shigenobu.

The power prediction of ships based on model tests is one of the main tasks for a towing tank facility. Within this context, the conduct of tank tests and their analysis procedures have been historically developed by taking into account not only theoretical approaches but also empirical model-ship correlation factors to achieve the accurate full-scale power predictions at sea trial. In fact, the introduction of the turbulence stimulators is the most well-known practice of the towing tanks by artificially tripping the flow to be turbulent on the model hull surface. However, it is also a well-known fact that the laminar flow can be experienced over the appendages in the stern region, even within the thick boundary layers which can be stimulated for turbulent flow by the above-mentioned procedure of using a turbulence stimulator. It is also true that the turbulence stimulator is not applied to the models of the conventional rudders because they operate in the propeller slipstream with the accelerated flow which can suppress the presence of laminar flow and its separation, whereas this may not be true for the GRS driven models due their different arrangements as will be discussed later.

In this paper, the reason for the above-mentioned scale effect for the GRS driven ships is explored and verified by using the model test results of two different size models with the GRS. The aim is to take a further step towards an improved estimation of the powering performance of ships with the GRS. The scale effect is associated with the drag and lift characteristics of the gate rudder blades and correction procedures were proposed based on the semi-empirical approaches supported with the available model test and full-scale trials data. The suggested scale effect correction was also implemented in the powering performance prediction method for ships with the GRS and results were validated with the full-scale trials data for Shigenobu.

2. Scale Effect Prediction for the Gate Rudder System

2.1. Effect of the Rudder Position

As shown in Figure 1a, the gate rudder blades are located on either side of a propeller with large clearances. Therefore, the flow field surrounding the gate rudder is quite different from that of a conventional rudder case (Figure 1b) with the following distinct features:

1. The flow field around the gate rudder is rather uniform without any strong disturbance from the propeller's slipstream;
2. The magnitude of the average flow speed at the gate rudder blades is close to the ship speed while the flow has a component in the transverse direction towards the ship's center plane;
3. The propeller accelerates the flow in the vicinity of the rudder blades and the top dead center position of the propeller where the high wake zone (or wake shadow) is observed, as in the case of a conventional rudder;
4. Larger transverse flow and, hence, the lift on the rudder blades are generated by the propeller's action which increases the thrust of the GRS.

The difference in the flow field between the model and full scale depends on the model size. If the model length (L_M) is not large enough (i.e., $L_M < 12$ m), the flow around the gate rudder blades has the possibility of being laminar and even developing a laminar flow separation because of their locations and that of the large thickness to chord ratio for the rudder blades. This is expected to result in the scale effect for the gate rudder blades. In fact, the scale effect of the rudder drag should be considered not only for the gate rudder but also for the conventional rudder. However, one can only appreciate the difference clearly if the comparisons of the flow characteristics are made based on the real case [5].

2.2. Drag Coefficients in Model and Full Scale

The resistance of a rudder blade, F_{RX}, and the side force (lift), F_{RY}, can be represented by Equations (1) and (2) as follows:

$$F_{RX} = F_{RD}\cos(\alpha) - F_{RL}\sin(\alpha) \quad (1)$$

$$F_{RY} = F_{RD}\sin(\alpha) + F_{RL}\cos(\alpha) \quad (2)$$

where, F_{RD} and F_{RL} is the contribution from the rudder drag and the rudder lift, respectively and α is the angle of attack of the rudder wing section to the flow. F_{RL} is negligibly small for the conventional rudder case except for the condition behind a rotating propeller.

F_{RD} can be predicted by the empirical formula given by Equation (3) when the flow velocity and rudder geometry is given.

$$F_{RD} = \frac{1}{2}\rho\int_0^H V_o(z)^2 C_F(z) * \left(1 + \frac{t(z)}{c(z)} + (\frac{t(z)}{c(z)})^2\right) \quad (3)$$

where V_o and C_F are the local flow velocity and frictional resistance coefficient, respectively, corresponding to flow direction and flow characteristics (laminar or turbulent, etc.) at z position and t/c is the thickness-chord ratio in the same position.

Here, the drag coefficient of a rudder can be represented by Equation (4) using ship speed and rudder area S_R.

$$C_{RD} = \frac{F_{RD}}{\frac{1}{2}\rho V_S{}^2 S_R} \tag{4}$$

It should also be noted that each of the resistance components obeys a different set of scaling laws and the problem of scaling can become more complicated because of the interaction between these components. Therefore, we explore the scale effect of the rudder drag by using data belonging to the gate rudder system of the Shigenobu, as described in the following:

Using Equation (3) the rudder drag was calculated for the gate rudder and the conventional rudder and results are presented as a comparison in Tables 2 and 3 for the model and full scale, respectively. The comparative results are also presented in Figure 5. The values in Figure 5 and Tables 2 and 3 are nondimensionalized by the measured ship resistance. To avoid confusion, it should be highlighted that the resultant rudder drag is different from these figures for the GRS because of the lift force acting on the rudder blades. The thrust (i.e., negative resistance) of gate rudder is quite often seen, even in the towing condition.

Table 2. Rudder drag calculations (in model scale).

Parameters	Flap Rudder	Gate Rudder
S_R	100%	157%
$mean V_0$ (model)	0.40	0.98
C_F (model)	0.00944	0.00700
C_{RD} (model)	0.0095	0.0405
% of ship resistance	0.8	5.3

Table 3. Rudder drag calculations (in full scale).

Parameters	Flap Rudder	Gate Rudder
S_R	100%	157%
$mean V_0$ (ship)	0.40	1.00
C_F (ship)	0.00281	0.0026
C_{RD} (ship)	0.0080	0.0068
% of ship resistance	0.2	1.7

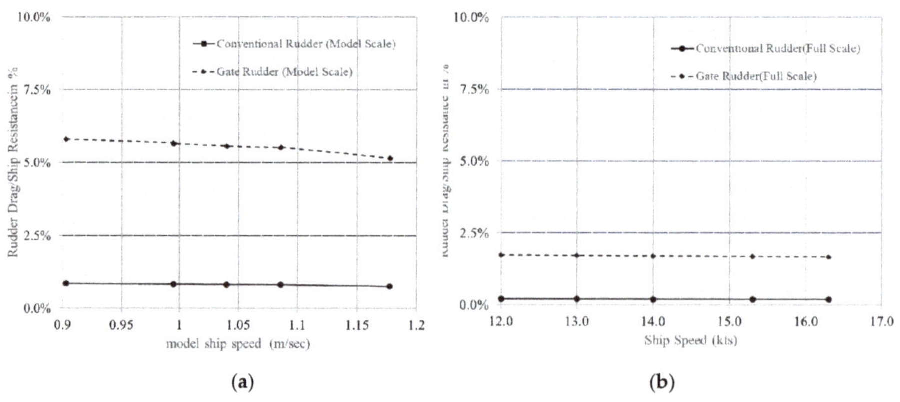

Figure 5. Rudder drag in % of ship resistance for model (**a**) and full scale (**b**).

As shown in Tables 2 and 3 and Figure 5, it is evident that the drag components of the two rudder cases are somewhat contrasting; the model scale drag of the gate rudder is six to eight times that of the conventional rudder, and both rudder drag components are decreased in full scale by a factor of one-third to one-quarter. Therefore, the scale effect of the gate rudder is relatively more significant as compared with that of the conventional rudder and the difference between the model and full-scale drag components of the gate rudder is 3.6% (5.3% minus 1.7%) of the ship resistance while it is only 0.6% (0.8% minus 0.2%) for the conventional rudder case. It is rather fortunate that the difference is six times, thanks to the flow speed difference around the rudder blades at each towing condition.

2.3. Rudder Drag Correction for the Scale Effect

As explained in Section 2.2, rudder drag cannot be measured directly during the resistance test of a hull with the gate rudder. This is because the blades of the gate rudder produce the lift force, and hence compensate for the rudder drag while a conventional rudder is simply a drag source contributing to the ship resistance. However, if we can measure the two rudder force components, F_{RX} and F_{RY}, in the ship fixed coordinate system, we can configure the rudder drag and lift components based on several assumptions within the wing theory. Hence, one can describe the lift coefficient as follows:

$$C_L = \kappa\left(\alpha + \alpha_g\right) \tag{5}$$

where C_L is the lift coefficient, α and α_g are attack angle and zero lift angle, respectively, and κ is the lift slope correction factor.

However, the rudder drag coefficient, C_{RD}, and angle of attack, α, can be estimated by the following equations;

$$C_{RD} = \frac{F'_Y - C_L\left[1 - \left(\alpha + \alpha_g\right)^2\right]^{0.5}}{\alpha + \alpha_g} \tag{6}$$

$$\alpha = \frac{\left(\kappa + F'_X\right) - \sqrt{(\kappa + F_{X'})^2 + 2F_{Y'}^2}}{-F'_Y} - \alpha_g \tag{7}$$

where, $F_{X'}$ and $F_{Y'}$ is nondimensional rudder force of F_{RX} and F_{RY}, respectively and given by Equations (8) and (9) as follows:

$$F_{X'} = \frac{F_{RX}}{\frac{1}{2}\rho V_S{}^2 S_R} \tag{8}$$

$$F_{Y'} = \frac{F_{RY}}{\frac{1}{2}\rho V_S{}^2 S_R} \tag{9}$$

We can observe a positive flow angle, α, for every kind of vessel when the rudder blades are off-centered, as sketched in Figure 6.

The lift slope correction factor, κ, can be expressed by Equation (10) as follows:

$$\kappa = \varepsilon \frac{6.13\lambda}{2.25 + \lambda} \tag{10}$$

where λ is the aspect of ratio, ε is the effect of low Reynolds number on the lift slope which will be explained in Section 2.4. C_L and C_D can be obtained from the measured F_X and F_Y during the resistance test directly.

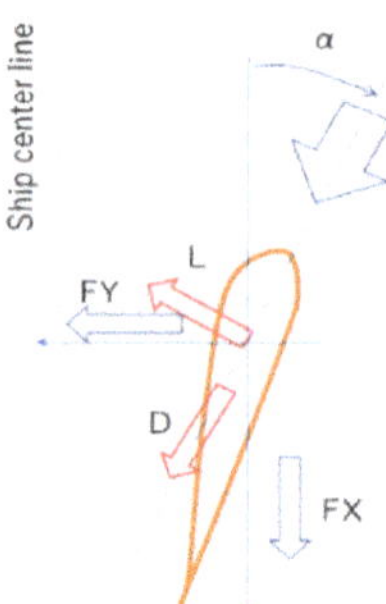

Figure 6. Rudder forces originated from lift and drag.

If we can express the rudder drag coefficient by Equation (11) which is combined with the drag coefficient C_{RD0} explained earlier:

$$C_{RD} = C_{RD0} + \delta C_{RD} \tag{11}$$

where δC_{RD} represents the additional resistance due to the rudder stock. Conservatively, we can use this additional resistance component without correction (i.e., no scale effect) such as wave resistance. Hence the effect on the rudder resistance:

$$\delta F_{RX} = \frac{1}{2}\rho V_S{}^2 \delta C_{RD0} S_R cos\alpha \tag{12}$$

$$\delta C_{RD0} = C_{RD0S} - C_{RD0M} \tag{13}$$

Figure 7 shows the difference between the obtained C_{RD} and calculated C_{RD0} (model). The δC_{RD0} is less than 1% of the total hull resistance, and the effect of this additional uncertain resistance on the full-scale performance is not as significant based on this figure.

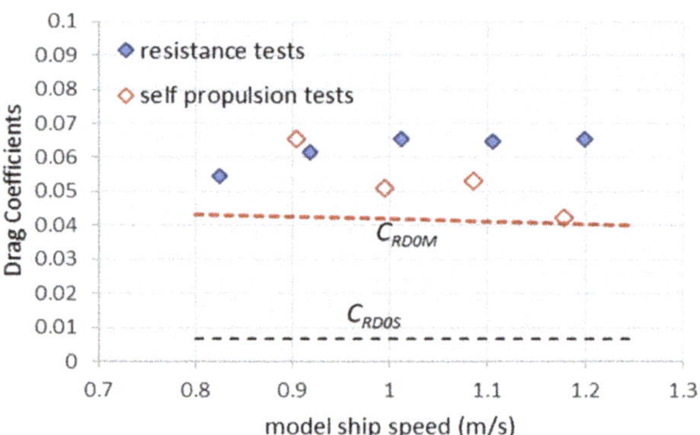

Figure 7. Obtained rudder drag coefficients.

2.4. Rudder Lift Correction for the Scale Effect

The lift coefficient of the gate rudder is also influenced by the model scale effect, and hence requires correction of the model test data. For this correction, the wind tunnel test data of NACA0012 section in low Reynolds numbers with the nonlinear lift slope curves are used, as shown in Figure 8 from McCormick (1995) [8].

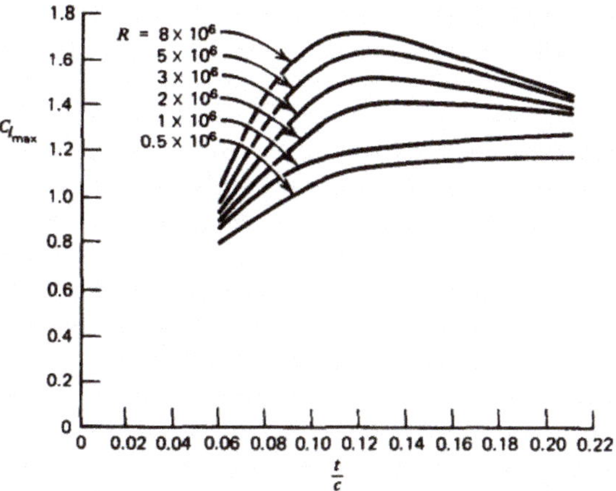

Figure 8. Effect of low Reynolds number on lift coefficients of typical wing section (NACA24xx series).

From Figure 8, the lift coefficient of the full-scale rudder (C_{LS}) can be estimated by using Equation (14) in conjunction with Equations (5) and (10) as follows:

$$\varepsilon = \frac{1.470{X_M}^3 - 3.109{X_M}^2 + 2.376X_M + 1.036}{1.470{X_S}^3 - 3.109{X_S}^2 + 2.376X_S + 1.036} \tag{14}$$

if $R_{NS} > 8 \times 10^6$, $X_S = 8 \times 10^6$
where

$$X_M = R_{NM} \times 10^{-7} X_S = R_{NS} \times 10^{-7}$$

Figure 9 also shows the Reynolds number effect on the lift coefficients of other typical NACA wing sections presented by the same study [8].

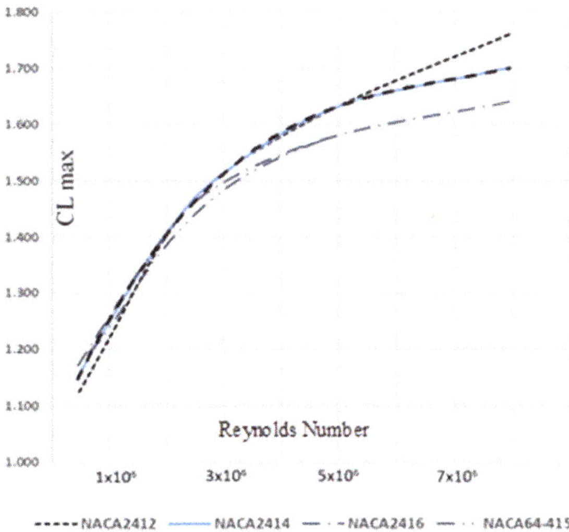

Figure 9. Reynolds number effect on the lift coefficients of typical NACA wing sections.

In order to show the significance of the resultant forces acting on the rudder in the model and full scale, which are evaluated earlier, Figure 10 is included. As shown in Figure 10, the effects of the rudder drag and lift correction on the gate rudder performance are rather significant, and therefore this phenomena explains the discrepancy between the model test and the full-scale results. In the Shigenobu case, two rudder blades seem to be generating a thrust of almost more than 10%. Therefore, the earlier derived two corrections are essential to assess the actual performance of the gate rudder system, which need to be implemented in the powering performance prediction and is discussed in Section 4.

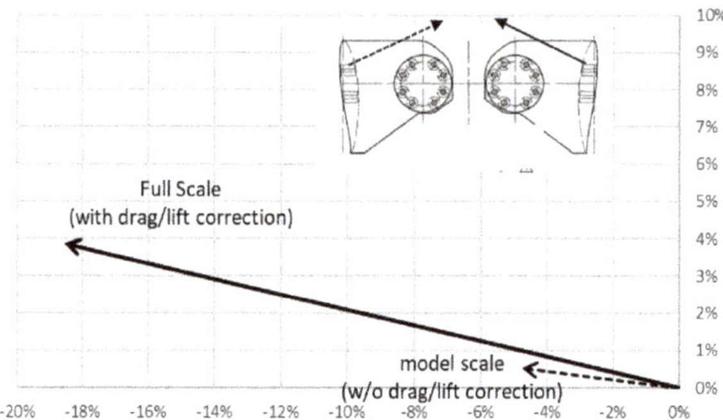

Figure 10. Effect of rudder drag and lift correction on rudder resultant force (model point case) to ship resistance.

3. Verification of Scale Effect Correction by Model Tests

In this section, the earlier proposed scale effect correction procedure is verified by the lift and drag coefficients of the gate rudder blades for two cargo ship models of different scales (2.0 m and 6.18 m long models). The verification is based on the data obtained from the resistance and self-propulsion tests of these models. Table 4 shows the principal dimensions of the cargo ship and its model size, and Figure 11 shows one of the models (6.18 m). Here, Lpp is the length between perpendiculars (m), B is the beam (m), d is the draft (m), and D_P is the propeller diameter (m). Table 5 presents the test conditions of the two models.

Table 4. Principal dimensions of ship and models.

Cargo Ship Particulars	
Lpp (m)	69.0
B (m)	12.0
d (m)	4.11
D_P (m)	2.3
Model Size (m)	6.18/2.00

Table 5. Model tests conducted with two scaled models.

Kind of Test	2.0 m Model	6.18 m Model
Resistance	with rudder	with rudder
Self-propulsion	with rudder	with rudder
Rudder force I	zero helm	zero helm
Rudder force II	−15 deg to +15 deg	−9 deg to + deg

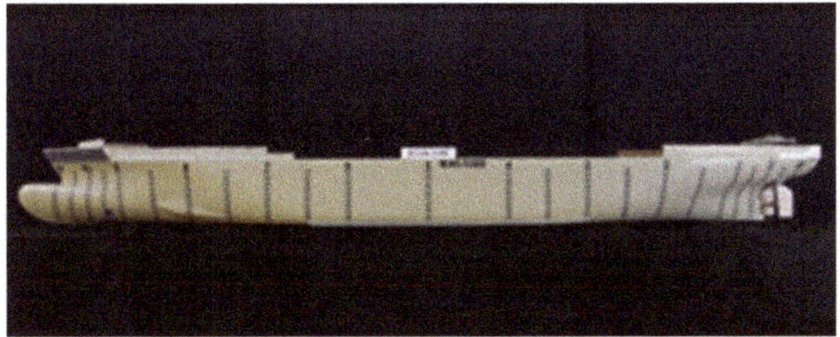

Figure 11. Large (6.18 m) ship model tested in 400 m towing tank.

From the model tests conducted, as shown in Table 5, the following verification data can be obtained:

(1) Scale effect on the rudder drag and lift during towing conditions;
(2) Scale effect on the rudder drag and lift during self-propelled conditions;
(3) Scale effect on the rudder normal force during steering conditions.

The verification data were calculated and are presented in Figure 12 in terms of the nondimensional drag and lift coefficients, (i.e., C_{DR} and C_L) of the rudders, as described earlier, using Equations (11) and (5), respectively. As shown in Figure 12, the proposed procedure shows good agreement with the two different scale model test data.

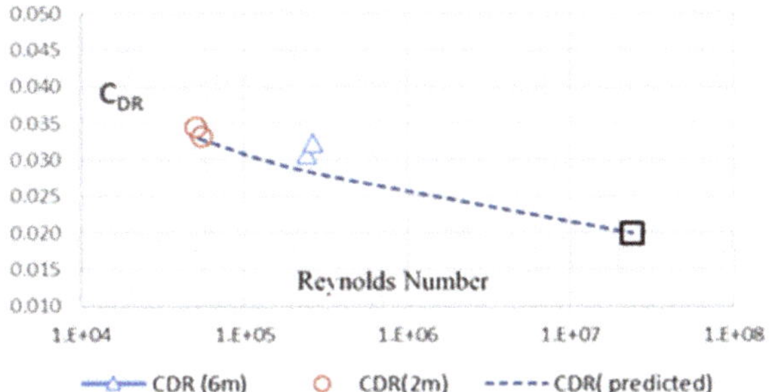

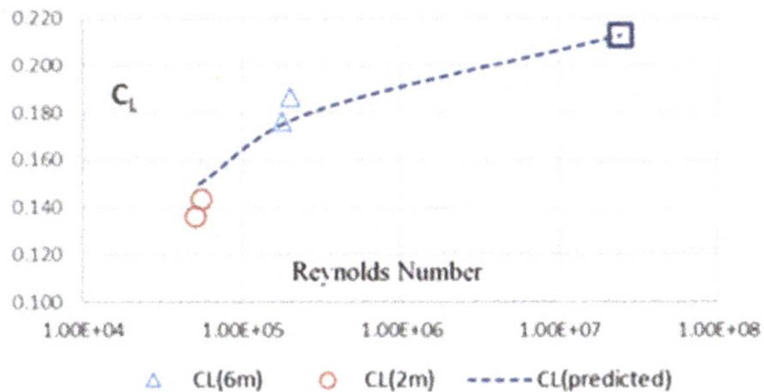

Figure 12. Gate rudder blades drag coefficient, C_{DR}, (top) and lift coefficient, C_L, (bottom) obtained with two different size models.

4. Enhanced Procedure for Powering Performance Prediction

Having established the scale-effect correction procedure in Section 2 and verified it with the model test data as described in Section 3, these corrections are introduced in the analysis of the resistance and self-propulsion test data to predict the powering performance of a ship with the gate rudder system in full scale as described in the following sections.

4.1. Effective Power

The effective horsepower (EHP) of a ship can be calculated based on the improved total resistance of the model by taking into account the scale effect correction on the rudder drag and rudder lift coefficients. Hence,

$$EHP = R_{TS}\ V_S \tag{15}$$

$$R_{TS} = C_{TS} 0.5 \rho {V_S}^2 S \tag{16}$$

where C_{TS} is the total resistance coefficient of the ship and is estimated based on the total model resistance coefficient according to the standard procedure, such as a recommended by the ITTC standard procedure.

Now, the total model resistance coefficient can be represented as:

$$C_{TM} = \frac{R_{TM} + \delta F_{RD} \cos\alpha + \delta F_{RL} sin\alpha}{0.5 \rho V^2 S_R} \tag{17}$$

4.2. Thrust Deduction Factor

The scale effect correction on the thrust deduction factor to be obtained from the self-propulsion tests can be included by introducing the drag and lift correction on the rudder blades, as described in Sections 2.3 and 2.4, and implemented in Equation (18):

$$1 - t = \frac{R_{TM} - F}{T + \delta F_{RD} \cos\alpha + \delta F_{RL} sin\alpha} \tag{18}$$

4.3. Effective Wake

The effective wake of the GRS is the most difficult item to analyze since the propeller advance speed is also accelerated by the gate rudder blades and this effect is not be considered as an integral part of the effective wake, since the effective wake, by definition, is originated from the boundary layer deformation due to a propeller's suction effect as the function of the propeller's thrust. Therefore, the actual effective wake of the GRS is less than that of a conventional rudder system because the propeller thrust is smaller than that of the conventional configuration.

Bearing in mind the above fact, to define the wake fraction of the gate rudder system, we need the mean flow speed at the propeller plane that can be represented in the nondimensional form, v_P, at the propeller plane of the gate rudder system, as in Equation (19):

$$v_P = C_1 v_{A0} + v_{inP} + v_{inR} \tag{19}$$

where v_{A0}, v_{inP} and v_{inR} are the propeller advance speed of the conventional rudder case, the propeller self-induced velocity, and the rudder induced velocity, respectively.

C_1 is a correction factor of the wake variation to account for the difference in the propeller diameter and the position and recommended value for this C_1 is <1.0 because of the smaller propeller diameter, and hence the resulting thrust of the gate rudder propeller as compared with the conventional type.

Then, the wake fraction for the gate rudder can be written as follows:

$$w_{GR} = C_1 * (w_0 - 0.04) + w_{inR} \tag{20}$$

where, w_{inR} is the rudder induced wake, as described earlier and its mean value can be estimated by the following formula:

$$w_{inR} = C_2 * C_T + w_{0R} \tag{21}$$

where C_2 is the correction factor for the thrust loading.

In Figure 13, a schematic representation of the scale effect for the gate rudder system is represented in comparison with the conventional rudder case. As shown in this figure, the dotted line presents the scale effect of a conventional rudder case, whereas the solid line shows the gate rudder case, respectively. One should also note in Figure 13 that, while point A′ can be predicted from point A by using the ITTC recommended procedure, point B′ cannot be predicted from point B which can be sometimes smaller, as shown in Figure 9, since point B′ does not follow the same principle.

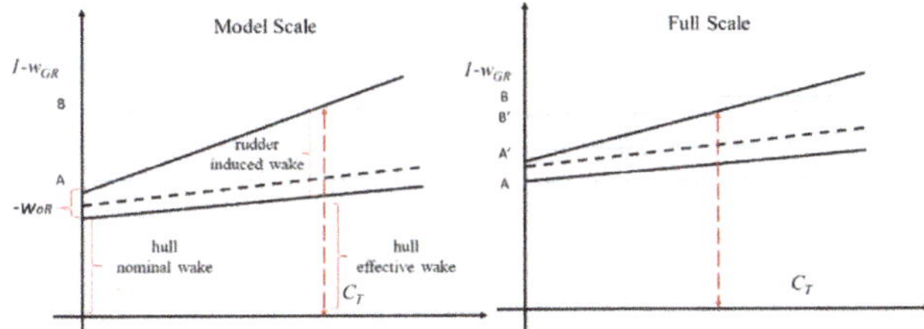

Figure 13. Schematic diagram of the scale effect of the wake of a gate rudder propulsion system.

Having represented the effective wake for the gate rudder system, Figure 14 is included to present the analysis results for Shigenobu (with GRS) and Sakura (with CRS-flap rudder) based on their respective sea trials and including the predicted results using the earlier described scale effects correction procedure.

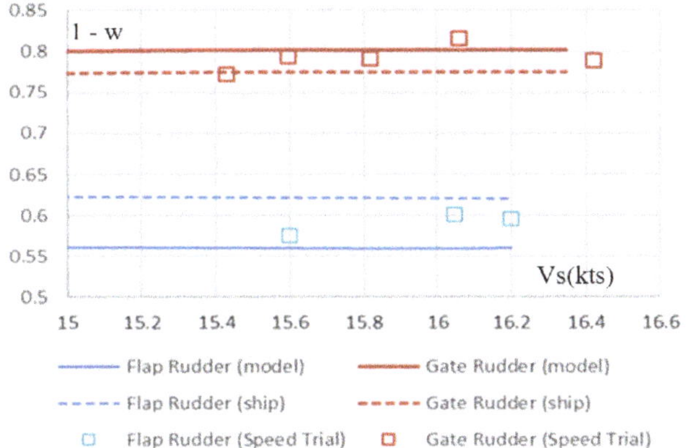

Figure 14. Effect of drag and lift correction on predicted and measured (speed trials) effective wake.

4.4. Powering Performance Prediction

In order to present an enhanced procedure for the powering performance prediction of a ship with the GRS, including the earlier described scale effect corrections, an algorithm is presented in Figure 15 in terms of a flow chart.

Finally, the above-described procedure is applied to predict powering performance of the Shigenobu (GRS), and results are shown in Figure 16 as compared with the prediction results for the Sakura (CRS) and including the trials data for both ships [9]. As one can see in Figure 16, the predicted power for Shigenobu is in excellent agreement with the trials data and, hence, justify the scale effect corrections applied on the gate rudder drag and lift characteristics.

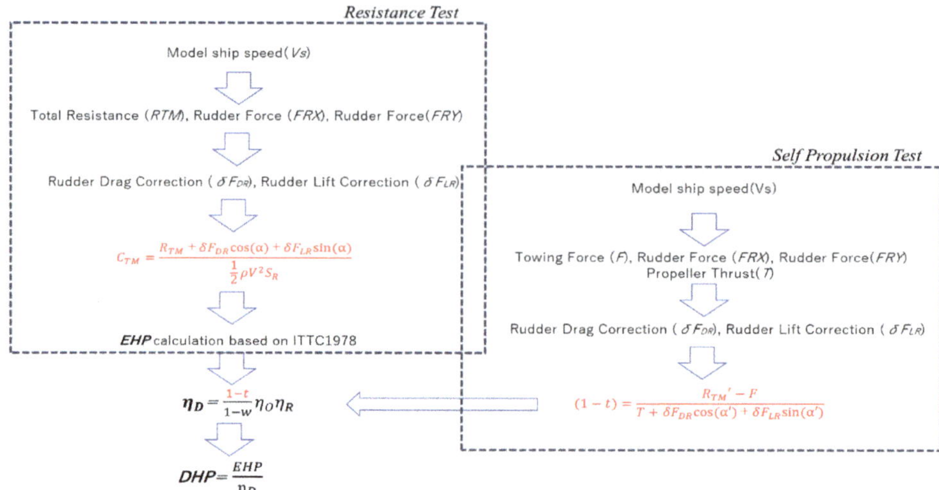

Figure 15. Algorithm for an enhanced powering performance procedure for a ship driven by a gate rudder system.

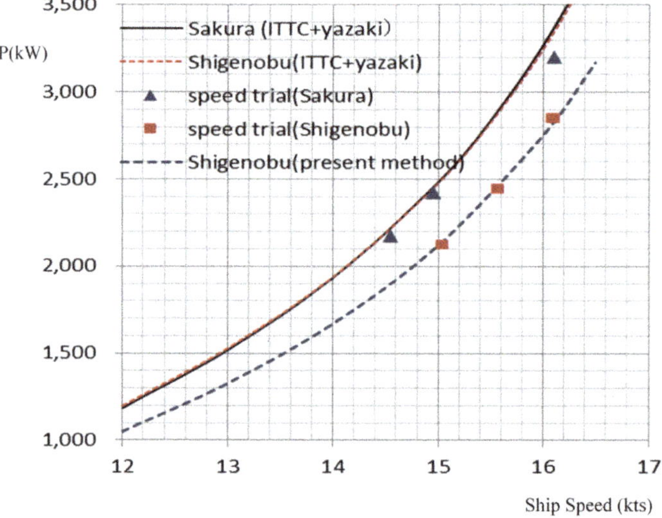

Figure 16. Comparison of trial power predictions for two ships using the ITTC and enhanced method for the gate rudder system.

It should be noted that the full-scale effective wake is estimated from the model test, and the same figures are used for the power prediction ($we_S = we_M$).

5. Conclusions

This study explored the scaling effect issues associated with the powering performance prediction of a ship fitted with a gate rudder system (GRS). The study aimed to take a further step towards a realistic estimation of the powering performance of ships with a GRS. The study focused on the corrections for the scale effects, which were believed to be associated with the drag and lift characteristics of the gate rudder blades due to the low Reynolds number experienced in model tests combined with the unique arrangement of this rudder and propulsion system. On the basis of the appropriate semi-empirical approaches and supporting model test and full-scale data, the study verified the scale effect phenomenon and presented the associated correction procedure. Together with this, the study also presented an enhanced procedure for the powering performance prediction of a ship driven by a gate rudder system implementing the proposed scale effect correction.

On the basis of the above, the following conclusions are obtained:

1. Scale effect of a gate rudder system can be considerably more significant than that of the conventional rudder system because of the unique arrangement of the gate rudder system behind the stern;
2. Flow characteristics around a gate rudder system in model scale can be laminar due to the low Reynolds numbers experienced, hence, the drag and lift coefficients of the gate rudder blades are strongly affected by this unfavorable scale effect;
3. The analysis presented in this study based on the semi-empirical procedure and the supporting data for two different models and full-scale ships clearly showed this scale effect and how to make corrections for the drag and lift coefficients of the gate rudder blades based on the verifications with the model test data;
4. The scale effect of the wake of a ship with a gate rudder system is not the same as that of a ship with a conventional rudder. The measured propeller advanced speed, based on the thrust identity, should be divided into two components, and the different scaling methods should be applied to each component;
5. An enhanced powering performance procedure which takes into account the subject scale effects presented in this study demonstrates excellent agreement with the full-scale trials data. This justifies the scale effect claim on the gate rudder drag and lift characteristics, and associated correction procedure, as well as the proposed enhanced methodology for the powering performance prediction.

This study focused on a rapid and practical solution to the powering problem associated with a gate rudder system. However, this study would greatly benefit from further in-depth studies involving CFD, systematic model tests, and full-scale trials which are underway with the increasing applications of this new attractive energy-saving device, Tacar et al. (2019) [10].

Author Contributions: N.S. conducted the study; S.K. introduced the original gate rudder conceptualization; M.F. contributed to the application and validation of the conceptualization; M.A. contributed further advancement of the concept, review, and editing. All authors have read and agreed to the published version of the manuscript.

Funding: In this paper, the Shigenobu full-scale data was obtained from a research project supported by the Nippon Foundation, whereas, the model test data of the cargo ship was obtained from another research project supported by the subsidy for a demonstration project for energy efficiency in commercial transport sector introduced by the Ministry of Economy, Trade and Industry of Japan.

Conflicts of Interest: The authors declare no conflict of interests.

References

1. Sasaki, N.; Atlar, M.; Kuribayashi, S. Advantage of twin rudder system with asymmetric wing section aside a propeller. *J. Mar. Sci. Technol.* **2015**, *21*, 297. [CrossRef]
2. The Naval Architect. *New Ducted Design Is a Boon for Propulsive Efficiency*; July/August issue; The Naval Architect, RINA: London, UK, 2019; pp. 32–36.
3. The Motorship. *New Ducted Propeller Design Offers Fuel Savings*; The Motorship: Fareham, UK, 2019; pp. 1–5. Available online: https://www.motorship.com/news101/ships-and-shipyards/new-ducted-propeller-design-offers-fuel-savings (accessed on 18 August 2019).
4. Sasaki, N.; Kuribayashi, S.; Atlar, M. GATE RUDDER®. In Proceedings of the 3rd International Symposium on Naval Architecture and Maritime (INT-NAM), Istanbul, Turkey, 24–25 April 2018.

5. Sasaki, N.; Kuribayashi, S.; Asaumi, N.; Fukazawa, M.; Nonaka, T.; Turkmen, S.; Atlar, M. Measurements and calculations of Gate rudder performance. In Proceedings of the 5th International Conference on Advanced Model Measurement Technology for Maritime Industry, AMT'17, Strathclyde University, Glasgow, UK, 11–13 October 2017.
6. Sasaki, N.; Atlar, M. Scale effect of gate rudder. In Proceedings of the Sixth International Symposium on Marine Propulsors, SMP'19, Rome, Italy, 26–30 May 2019.
7. Yazaki, A. A diagram to estimate the wake fraction for an actual ship from model tank test. In Proceedings of the 12th International Towing Tank Conference, Rome, Italy, 22–30 September 1969; p. 259.
8. McCormick, B.W. *Aerodynamics, Aeronautics, and Flight Mechanics*, 2nd ed.; John Wiley & Sons. Inc.: Hoboken, NJ, USA, 1995.
9. Fukazawa, M.; Turkmen, S.; Marino, A.; Sasaki, N. Full-Scale GATE RUDDER Performance obtained from Voyage Data. In Proceedings of the A. Yücel Odabaşı Colloquium Series: 3rd International Meeting-Progress in Propeller Cavitation and Its Consequences: Experimental and Computational Methods for Predictions, Istanbul, Turkey, 15–16 November 2018.
10. Tacar, Z.; Sasaki, N.; Atlar, M.; Korkut, E. Investigation of scale effects on Gate Rudder Performance. In Proceedings of the 6th International Conference on Advanced Model Measurement Technology for Maritime Industry, AMT'19, University of Rome, La Sapienza, Italy, 9–11 October 2019.

Journal of
Marine Science and Engineering

MDPI

Article

Numerical Analysis of Influence of the Hull Couple Motion on the Propeller Exciting Force Characteristics

Liang Li [1,2,*], Bin Zhou [1,2], Dengcheng Liu [1,2] and Chao Wang [3]

1. China Ship Scientific Research Center, Wuxi 214082, China
2. Jiangsu Key Laboratory of Green Ship Technology, Wuxi 214082, China
3. College of Shipbuilding Engineering, Harbin Engineering University, Harbin 150001, China

* Correspondence: heuliliang@foxmail.com; Tel.: +86-0510-8555-5635

Received: 31 August 2019; Accepted: 19 September 2019; Published: 23 September 2019

Abstract: The numerical calculation was performed for the KRISO Container Ship (KCS) hull-propeller-rudder system with different freedom hull motion by employing the Reynolds-Averaged Navier-Stokes (RANS) method and adopting the overset grid. Firstly, the numerical simulation of hydrodynamics for a bare hull with the heave and pitch motion is carried out. The results show that the space non-uniformity of a nominal wake in the disk plane with motion is comparable to the case without motion. However, the time non-uniformity increases sharply and it has a significant positive relationship with the motion amplitude. Then, the propeller exciting force is calculated in the case including single heave, single pitch and their couple motion. It was found that both the ship and propeller hydrodynamic performance deteriorated dramatically due to the hull motion. Furthermore, the spectrum peak at the motion frequency is dominant in all the peak values and the larger the amplitude is, the higher the motion frequency peak is expected to be. For the propeller bearing force, the effect of the different hull motions appears as linear superimposition. However, the superimposition of different hull motions enlarges the propeller-induced fluctuating pressure in a single motion.

Keywords: couple motion; propeller exciting force; free surface; wake non-uniformity; numerical simulation

1. Introduction

The high speed and large-scale ships mostly sail on the rough sea condition. With the effect of the wind, wave and wake, a significant six degrees of freedom motion to the hull is expected to be caused over a certain range of frequency. In addition, since the propeller is fixed on the stern, the hull motion also drives the rigid body motion of the propeller relative to the nearby fluid besides its rotational motion. As a result, the propeller works in the wake changing continuously and its efficiency decreases and at the same time, the exciting force of propeller increases sharply, which is unfavorable for the vibration and noise performance of ship. Therefore, the unsteady performance of the propeller under the influence of the hull motion, such as heave and pitch, has been one of concerns of researchers.

Researchers have done considerable work concerning the ship and propeller's unsteady hydrodynamic performance in the wind waves or motion condition. In the past, Sluijs [1] and Jessup [2] investigated the effect of unsteady contributions to the wake field caused by waves and wave-induced motions. They tried to explain how the propeller loads would change under these wave and wave-induced motion conditions. Sasajima [3] improved a quasi-steady way to predict the propeller bearing force. Breslin and Andersen [4] supported the adoption of a complete unsteady method to calculate the change of hydrodynamic force in the time domain. In the recent past, Politis [5] calculated the unsteady motion of a propeller in a fluid including free wake modeling. Xin Yu [6] analyzed the hydrodynamic performance of propeller in a wave by simplifying the problem into two aspects. One is the change of the submerged depth of the propeller axis due to the wave. The other is interference of the wave diffraction. Carrica [7] studied the KCS self-propulsion in a model scale free to sink and trim in head waves. The 0th and 1st harmonic amplitudes and 1st harmonic phase are computed for the total resistance coefficient CT, the heave motion z and the pitch angle θ. The comparisons between Computational Fluid Dynamics (CFD) and Experiment Fluid Dynamics (EFD) show that the pitch and heave are much better predicted than the resistance. Sharma [8] performed the numerical prediction of hydrodynamic performance for a hydrofoil or a marine propeller undergoing unsteady motion by adopting a panel method and RANS method. Kinnas [9] combined the vortex lattice method (VLM) with the boundary element method (BEM) to predict the unsteady hydrodynamic analysis of a propeller under a surge and heave motion. However, the effects of the turbulence and vortex separated flows are difficult to be handled by this method, resulting in some difference between the calculated results and the RANS simulation results. Recently, Tezdogan [10] carried out a numerical study of ship motions in shallow water for a full-scale large tanker model and obtained its heave and pitch response to head waves at various depths. The numerical results were found to be in good agreement with the experimental data. Lianzhou Wang [11] conducted a numerical simulation on a propeller impacted by heave motion in cavitating flow using the RANS method. The results show that the heave motion would aggravate the unsteady characteristics of the thrust and torque coefficient and lead to a non-uniform distribution of propeller sheet cavitation. Shuai Sun and Liang Li [12] calculated the propeller exciting force for the hull-propeller-rudder system in the oblique flow by the RANS method. The results show that the propeller thrust and torque fluctuation coefficient peak in the drift angle are greater than that in straight-line navigation, and the negative drift angle is greater than the positive. However, the calculation is conducted in the quasi-yaw condition with a different drift angle, which cannot show the change rule of propeller performance with time during the hull yaw motion. Liang Li [13] analyzed the influence of the hull heave motion on the propeller exciting the force characteristics. It was found that the spectrum peaks of the exciting force were richer compared with the condition without the heave motion after the fast Fourier transform. Moreover, the peak at the heave motion frequency is dominant in all the peak values. This work gives a good start to investigate the influence of a more complex hull motion on the propeller exciting force performance.

Overall, the literature surveys show that many of the previous studies have only studied the simple foil or single ship and single propeller. The mutual hydrodynamic interaction between the propeller and ship is not considered. On the other hand, the research is mainly conducted under the condition with a one degree of freedom hull motion [9,11]. The response of the propeller performance to the couple hull motion, which is more than one degree of freedom, is still unknown. Moreover, the propeller exciting force is not paid much attention to, which is the main source of the stern vibration and noise of the real ship. Hence, it is highly necessary and practically significant to perform a numerical analysis of influence of the hull couple motion on the propeller exciting force characteristics.

In the present study, the KCS hull-propeller-rudder system is employed to analyze the influence of the hull couple motion on propeller exciting force characteristics employing the RANS method. The hull couple motion includes the heave and pitch, both of which were simplified to a periodic motion based on a sinusoidal function and were achieved by an overset grid method. The change rule and characteristics of the hull resistance, wake field and propeller exciting force were obtained successfully, and the comparison was made between the couple motion and single motion. The results can provide an important reference for the prediction of a real ship's hydrodynamic performance in the motion condition.

2. Mathematic Base

2.1. Governing Equations

Fluid flow is governed by physical conservation laws. Basic conservation laws include the law of conservation of mass, the law of conservation of momentum and the law of conservation of energy [14].

As the medium in the calculation, water, is an incompressible fluid whose heat exchange is little enough to ignore, only the mass conservation equation and the momentum conservation equation are solved.

$$\frac{\partial \rho}{\partial t} + \frac{\partial(\rho u_i)}{\partial x_i} = 0 \tag{1}$$

$$\frac{\partial(\rho u_i)}{\partial t} + \frac{\partial}{\partial x_j}(\rho u_i u_j) = -\frac{\partial p}{\partial x_j} + \frac{\partial}{\partial x_j}(\mu \frac{\partial u_i}{\partial x_j} - \rho \overline{u'_i u'_j}) + S_j \tag{2}$$

Here, u_i and u_j is the averaged Cartesian components of the velocity vector (i, j = 1,2,3). p is the mean pressure. ρ is the fluid density and μ is the dynamic viscosity. $\rho \overline{u'_i u'_j}$ is the Reynolds stresses. S_j is the generalized source term of the momentum equation.

2.2. Turbulence Model and Free Surface Model

The governing equations are solved using the segregated method based on pressure-velocity, in which the second upwind scheme is used for the discretization of convective term and the second central differencing scheme is used for the discretization of dissipation term. In order to simulate the flow separation and strong adverse pressure gradients well, the shear stress transport (SST) $\kappa - \omega$ turbulence model is adopted [15]. This model is the one of the most advanced two-equation turbulence models currently, which has a good advantage in calculating the viscous flow around bodies. However, this model needs a certain range of Y plus value is needed for this model, in general, the Y plus being set to 30–200 is proper. The free surface is modeled by the volume of fluid (VOF) method [16], whose essential purpose is to determine the free surface by investigating the fluid-grid volume fraction function in the grid cells and trace the variation of the fluid, rather than the particle movement on the free surface. As long as the value of the function on each grid of the flow field is known, the movement interface can be traced.

2.3. Overset Grid

The hull motion is simulated by the overset grid method [17]. Compared to the general dynamic mesh, the overset grid is more adaptable and efficient in dealing with the large-amplitude motion. The overset grid method divides the computational domain into a number of subdomains whose grids are generated independently. Information transmission is implemented through grid nesting and overlap in these subdomains. In the boundary flow fields of the overset grids, information is coupled by interpolation. The overset grid generation involves two main steps: A hole-cutting operation on the grids of the background domain to shield the area inside the hole, mark those grids within the hole, and abandon them in the subsequent CFD computation; a point-searching operation to interpolate information transmitted at the hole boundaries for the subsequent numerical calculation. The schematic of overset grid is shown in Figure 1.

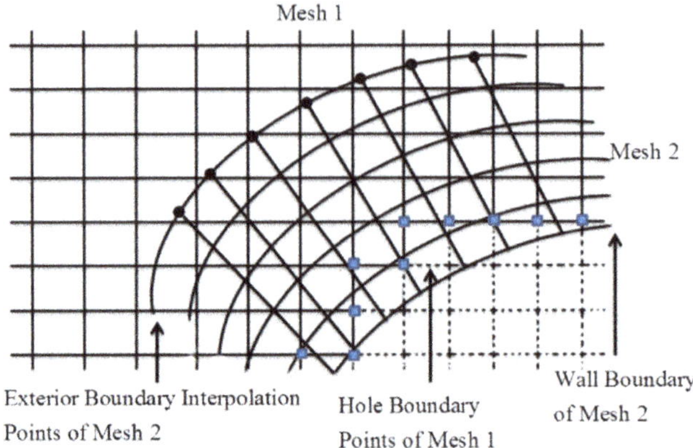

Figure 1. Schematic of overset grid.

3. Calculation Modeling

3.1. Calculation Object

The standard model KCS container ship with the scale factor of 31.6 is used as the study objects as Figure 2 shows, whose principle parameters is listed in Table 1. The notable bulbous bow and stern extension may result in a complex wake and wave, which can provide a good wake environment to study the propeller exciting force characteristics. The propeller that goes with the ship is a KP505 propeller. The propeller principle parameters are given in Table 2.

Figure 2. KCS ship and KP505 propeller geometry model.

Table 1. Principle parameters of KCS model.

Lpp (m)	**7.2786**
Draught (m)	0.3418
Wetted surface (m^2)	9.438
Reynolds No.	1.4×10^7
Froude No.	0.26

Table 2. Principle parameters of KP505 propeller model.

Diameter (m)	**0.250**	**Area Ratio**	**0.70**
No. of blades	5	P/D (0.7R)	1.00
Hub ratio	0.167	Skew angle(°)	12.66

3.2. Computational Domain and Boundary Condition

In order to simulate the hull motion, the computational domain is divided into the background domain and the overset grid domain as Figure 3 shows. The background domain is set as a cuboid water basin referring the towing tank. The inlet is 2 *Lpp* from the bow to ensure the inflow is uniform. The domain side and bottom from the hull surface are both 2 *Lpp* to avoid the hull flow field affected by the basin wall. Regarding the full development of the hull wake, the outlet from the stern is 3 *Lpp*.

The inlet is set as a velocity inlet. The outlet is set as a pressure outlet. The domain top plane is set as symmetry and the other boundaries are set as a wall. The overset grid domain is nested in the background domain, which includes the hull-propeller-rudder model and can realize the simulation of the hull heave motion.

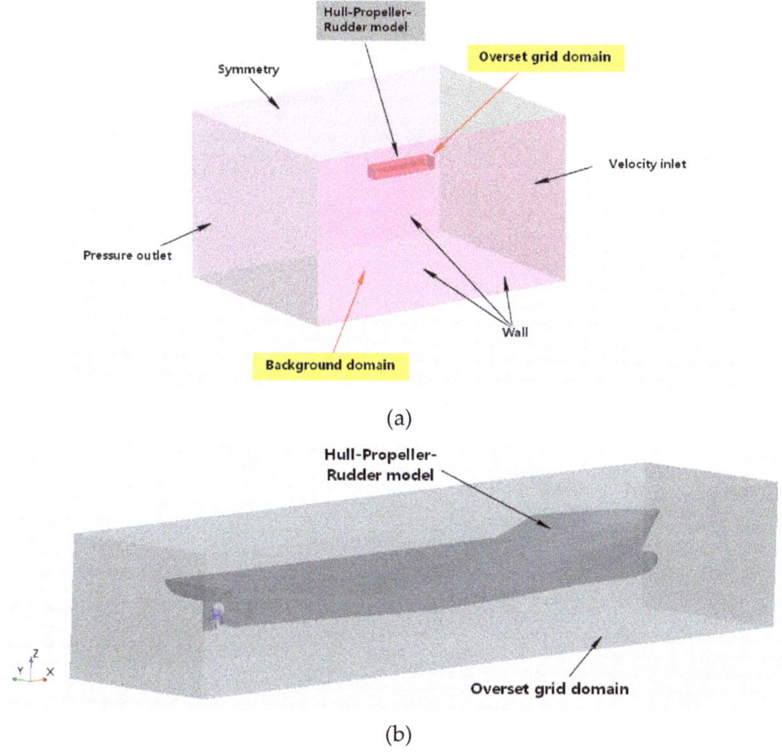

Figure 3. Computational domain and boundary condition: (**a**) Background domain and boundary condition; (**b**) Overset grid domain and hull-propeller-rudder model.

3.3. Grid Division

In our calculation, the trimmer mesh is used. It can capture the boundary layer flow effectively and control the total grid number of calculations. The grid division detail is shown in Figure 4. When performing the grid division, the grid size of the background domain around the overset area at best keeps the same with the grid size at the overset grid boundary. Furthermore, the motion range of the overset grid should be limited within the overset area. Regarding the hull surface mesh, more grids should be given to the bow and stern of the hull where the flow field varies dramatically. To capture the free surface better, the grids near the free surface and within the range of Kelvin waves are properly refined. The first layer of grids is 0.8–1 mm, corresponding to the Y plus which is approximately 60. Based on the previous analysis results of grid sensitivity [12], the total grids number is set as 4.5 million to save the computational time, but a satisfied calculation accuracy can also be reached.

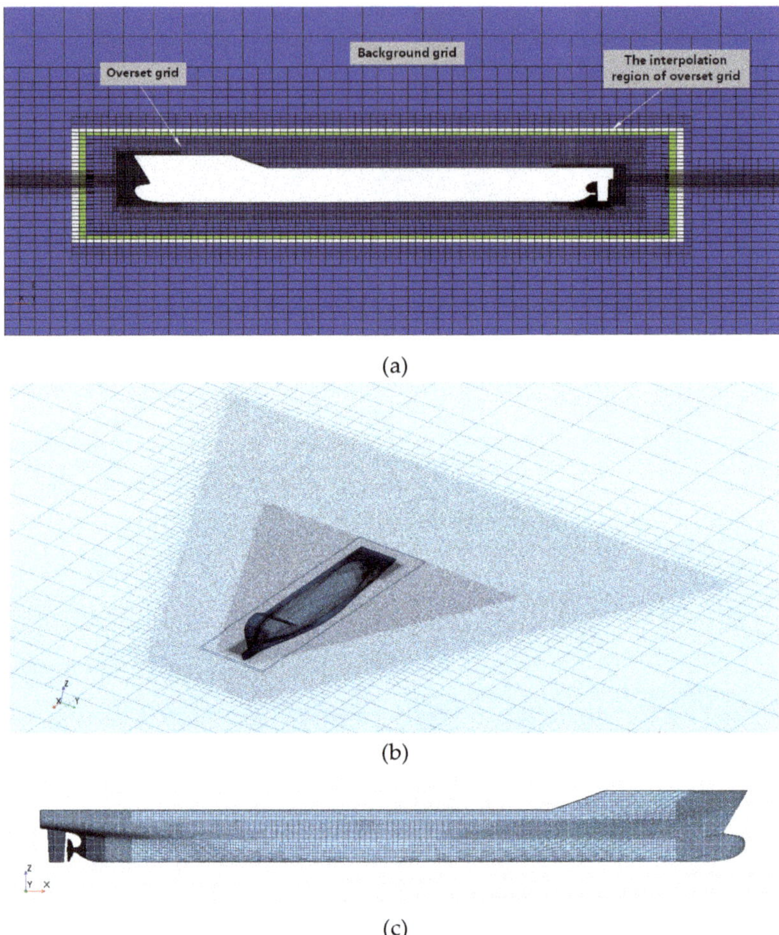

(a)

(b)

(c)

Figure 4. Computational grid distribution: (**a**) The distribution of overset grid; (**b**) The grids near the free surface; (**c**) Hull surface grids.

3.4. Calculation Conditions and Setting of Motion Model

The research shows that the heave and pitch motion have the dominant influence on the propeller performance among the hull six degrees of freedom motion. It can even lead to propeller emergence and the racing phenomenon when the hull is in the large amplitude motion condition. Next, only the heave and pitch motion will be investigated in detail. The calculation is performed by two steps. At the first step, only the hull and rudder model are used to research the change rule of the propeller inflow in the heave and pitch motion. Then, the propeller is taken into consideration as a whole system to get the exciting force data in the heave, pitch motion and their couple motion. In this study, the hull motion is defined independently using the sinusoidal function as follows.

(1) Heave motion

$$y(t) = A_p \sin(\omega(t - \Delta t)) \tag{3}$$

Here, y is the moving distance of hull in the vertical direction, corresponding with the moving distance at z direction in the present calculating coordinate system and upward is positive. A_p is the amplitude of the heave motion and is selected as $0.25T_m$ and $0.125\ T_m$ in the calculation by taking the amplitude of the forced oscillation model test as a reference. Tm represents the draft depth in the static water. ω is the frequency of heave motion and is dependent on the heave period T_e. T_e is selected as 2 s in this work. Δt is the delay time to ensure the numerical stability, being selected as 26 s, at which the hull system is without the heave motion. The schematic plan of the heave motion rule is shown as Figure 5.

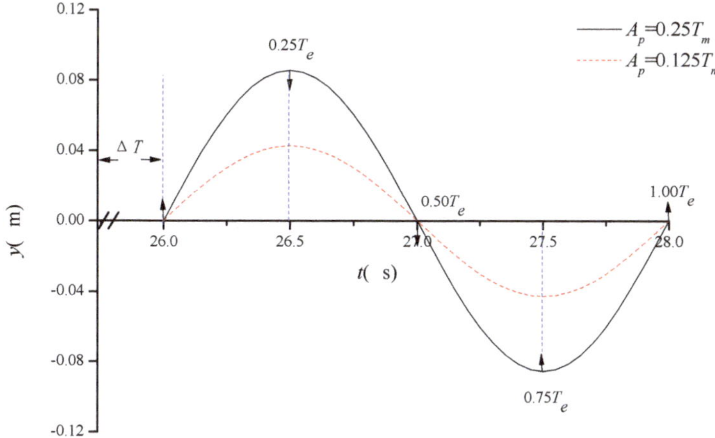

Figure 5. Schematic plan of heave motion rule.

(2) Pitch motion

$$\theta(t) = A_p \sin(\omega(t - \Delta t)) \quad (4)$$

Here, θ is the pitch angle of the hull and it is defined as positive when the pitch is forward. A_p is the amplitude of the pitch motion and is selected as 2 degrees and 1 degree in the calculation. ω is the frequency of the pitch motion and is dependent on the pitch period T_e. T_e is selected as 2 s that is the same with the heave period. Δt is also the delay time to ensure the numerical stability, being selected as 26 s. The schematic plan of the heave motion rule is shown as Figure 6.

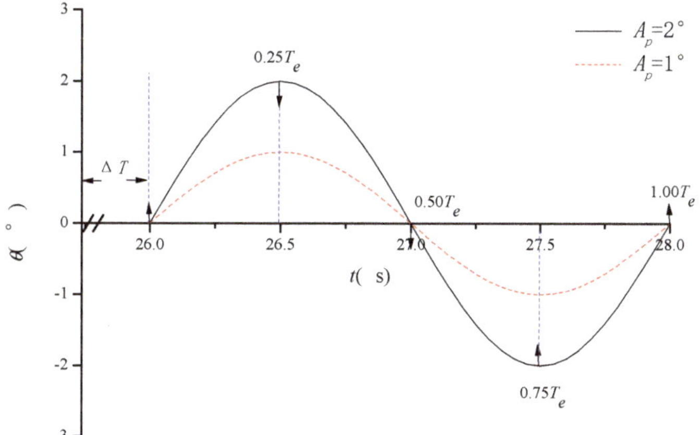

Figure 6. Schematic plan of pitch motion rule.

The difficulty to simulate the motion of the hull-propeller system is how to realize the superimposition of the hull motion on the propeller rotating motion without the hull-propeller model separation showing in the process of the simulation. In order to handle this problem, the superimposed coordinate system method is applied. Therefore, three different coordinate systems are created, including the initial coordinate system, a new local hull coordinate system and a new local propeller coordinate system. The original point of the initial coordinate system is located in the crossing point of the after-perpendicular and free surface. The original point of the local hull coordinate system is located in the mass center of the hull. Its coordinates in its initial coordinate system are [3.532 m, 0.0, −0.111 m]. The original point of the local propeller coordinate system is located in the crossing point of the propeller axis and propeller disk. Its coordinates in the local hull coordinate system are [−3.404 m, 0.0, −0.113 m]. The initial direction of XYZ in these three different coordinate systems is the same. The details are shown in Figure 7. It can be seen that it is convenient to define the hull motion in the local hull coordinate system, as it is with the propeller's rotation motion in the local propeller coordinate system. Then, the superimposed motion can be realized by attaching the locale propeller coordinate system to the hull motion in the local hull coordinate system.

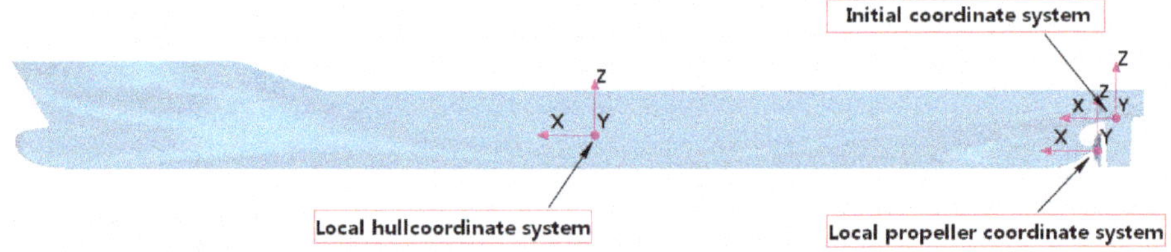

Figure 7. Schematic plan of different coordinate systems.

4. Calculation Result Analysis

4.1. Validation of Calculation Method

Firstly, the numerical calculation for both the bare hull and hull-propeller-rudder system are performed without the hull motion to validate the calculation method with the condition [18] that the flow speed is 2.196 m/s, the propeller revolution speed is 9.5 rps and the draft is 0.3418 m. The time step is set as the time in which the propeller rotates for 4 degrees.

The Table 3 shows the hydrodynamic calculation results. In general, it shows good agreement with the experimental data [19]. The error of the propeller thrust is the biggest, but it is still under the 5% that can be accepted. The possible reasons for this are neglecting the hull posture change during the calculation and the presence of the rudder in this work. Figure 8 shows the comparison between EFD and CFD for the wave contour. It can be seen that the values and position of the peaks and troughs are consistent between EFD and CFD. The previous problems that the wave on the hull's both sides dissipate too quickly and the details for the broken wave at the stern are not caught well and have been solved effectively by the logical grid division strategy. Figure 9 shows the comparison between EFD and CFD for the velocity contour. The calculated values also show good agreement with EFD's. However, the shrinkage of the velocity contour of EFD towards the mid-ship section is more serious, that means the boundary layer is thinner in the experimental condition. The use of the boundary layer trip in test may cause this phenomenon happen. Overall, the calculation method used in this work is accurate and satisfactory.

Table 3. Hydrodynamic calculation results of hull-propeller-rudder system.

Description	Resistance/N	Thrust/N	Moment/N.m
CFD (with rudder)	90.5	57.22	2.62
EFD (without rudder)	90.0	59.9	2.53
Error	+0.56%	−4.47%	+1.58%

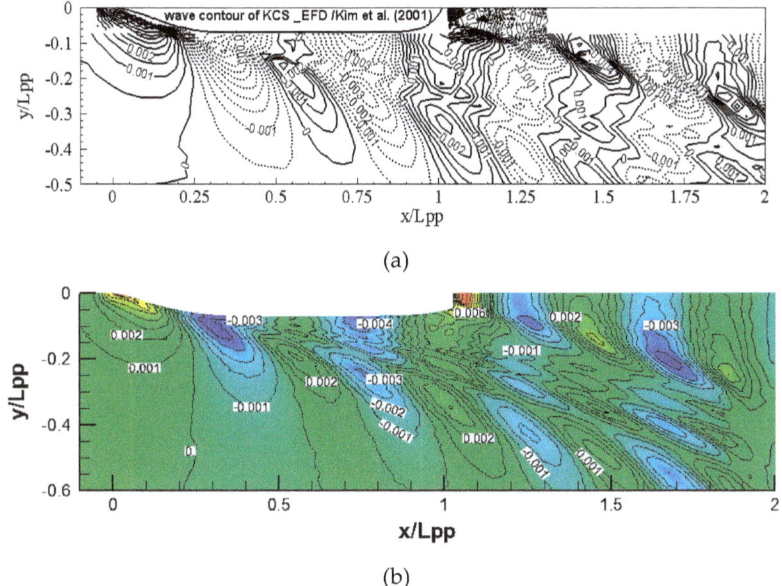

Figure 8. Comparison for wave contour: (**a**) Model test; (**b**) CFD calculation.

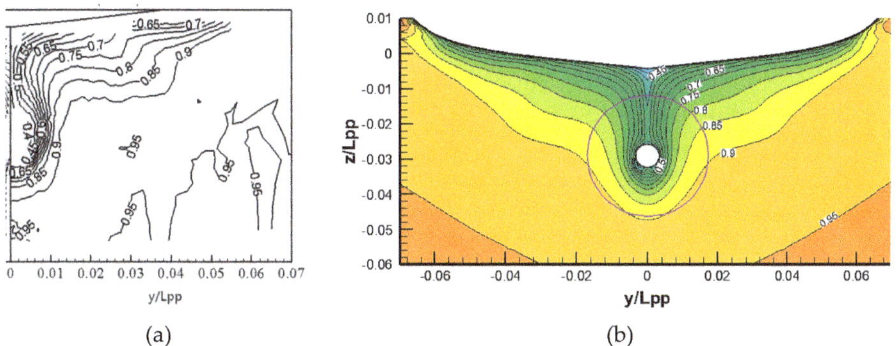

Figure 9. Comparison for velocity contour ($x/L_{pp} = 0.9825$): (**a**) Model test; (**b**) CFD calculation.

4.2. Free Surface and Wake Analysis

4.2.1. Free Surface in Motion Condition

Figures 10 and 11 show the wave patterns of the free surface at different times within a heave and pitch motion period. The wave pattern changes greatly with the time compared with it in the condition without the hull motion. The wave is generating and dissipating continuously due to the hull motion. It can be seen that, especially when the hull is heaving down or pitching up, the stern part becomes the main source to produce the wave because of the phenomenon of stern slapping water. From the viewpoint of energy, the wave costs the energy of the hull, therefore it will become the main reason to the growth of hull resistance in motion condition. As the hull moves periodically, the wave pattern on the hull's two sides gradually spread out and finally form an obvious transverse wave alternating with peaks and troughs. Compared with the heave condition, the transverse wave in the pitch condition seems to develop more fully and distribute more regularly. Hence, it can be considered that the pitch case in our calculation influences the hull resistance more powerfully than the heave case.

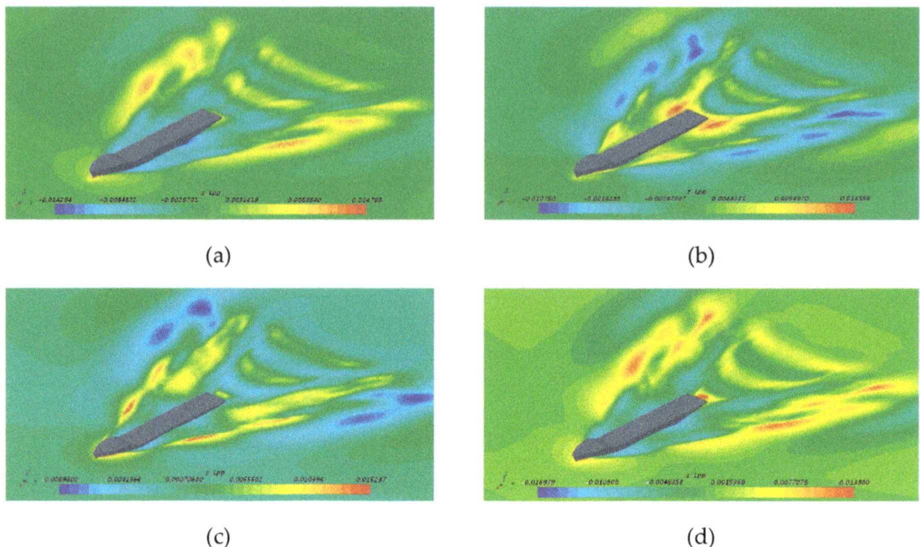

Figure 10. Free surface in a heave period ($A_p = 0.25T_m$): (**a**) t = 0.25 Te; (**b**) t = 0.50 Te; (**c**) t = 0.75 Te; (**d**) t = 1.0 Te.

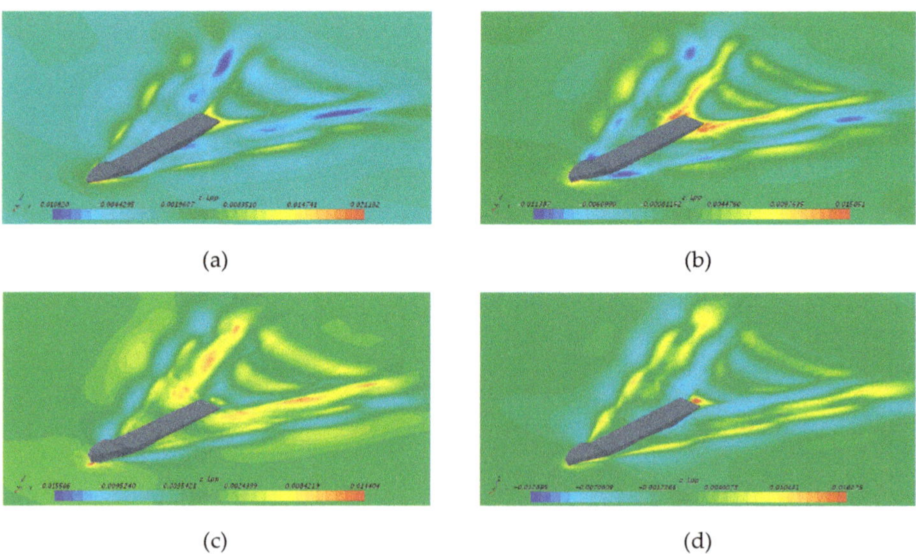

Figure 11. Free surface in a pitch period ($A_p = 2°$): (**a**) t = 0.25 Te; (**b**) t = 0.50 Te; (**c**) t = 0.75 Te; (**d**) t = 1.0 Te.

4.2.2. Wake in Motion Condition

The propeller exciting force performance is strongly associated with the wake non-uniformity. When the ship sails normally without a large amplitude motion, the wake temporal non-uniformity induced by the flow turbulence characteristics is slight. The spatial wake temporal non-uniformity is the dominant factor affecting the propeller exciting force. However, when the ship sail in rough sea with a significant heave or pitch motion, the wake temporal non-uniformity caused by the motion period is non-negligible. The priority is given to the axial wake in the three directional wake. To analyze the non-uniformity of the axial wake, the axial velocity is non-dimensionalized by the inlet velocity (2.196 m/s). The spatial non-uniformity is defined as:

$$\overline{u}'_x(r) = \frac{1}{2\pi}\int_0^{2\pi} u'_x(r,\theta)d\theta \tag{5}$$

$$\Delta' u'_x(r) = \frac{(u'_x(r))_{\max} - (u'_x(r))_{\min}}{\overline{u}'_x(r)} \tag{6}$$

Here, u'_x is the dimensionless axial velocity. $\overline{u}'_x(r)$ is the circumferential average of dimensionless axial velocity in a specific radius. $(u'_x(r))_{\max}$ is the peak value in a specific radius and $(u'_x(r))_{\min}$ is the valley value. $\Delta' u'_x(r)$ is the spatial non-uniformity. The bigger the spatial non-uniformity is, the worse the propeller exciting force performance is expected to be. The temporal non-uniformity is defined as:

$$\overline{u'_x}(t) = \frac{1}{S}\iint_S u'_x(r,\theta,t)rd\theta dr \tag{7}$$

$$\overline{u'_x} = \frac{1}{T_e}\int_0^{T_e} \overline{u'_x}(t)dt \tag{8}$$

$$\Delta' u'_x = \frac{(u'_x(t))_{\max} - (u'_x(t))_{\min}}{\overline{u'_x}} \tag{9}$$

Here, $\overline{u'_x}(t)$ is the disk surface average of the dimensionless axial velocity in a specific time. $\overline{u'_x}$ is the time average of the disk dimensionless axial velocity. $\Delta' u'_x$ is the temporal non-uniformity in a motion period which is defined similarly with the spatial non-uniformity.

Figure 12 shows the distribution of the nominal wake fields in the heave motion condition. At the time t = 0.25 Te, the velocity decreases dramatically as the hull rises to the highest position, corresponding with the propeller load becoming heavy. There is a rather strong vortex structure occurring on both sides of the hull bilge. At the time t = 0.75 Te, the velocity increase and its contour shrink to the hull obviously as the hull sinks to the lowest position, corresponding with the propeller load becoming light. Figure 13 shows the distribution of the nominal wake fields in the pitch motion condition. The wake is also strongly associated with the stern draught. The deeper the stern draught is, the bigger the velocity in the propeller plane can be.

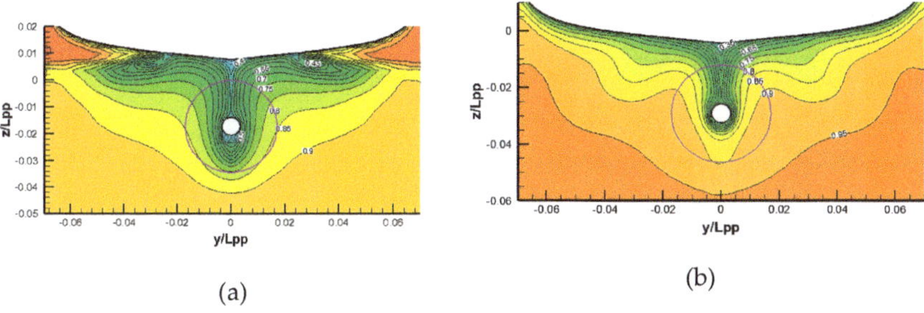

Figure 12. *Cont.*

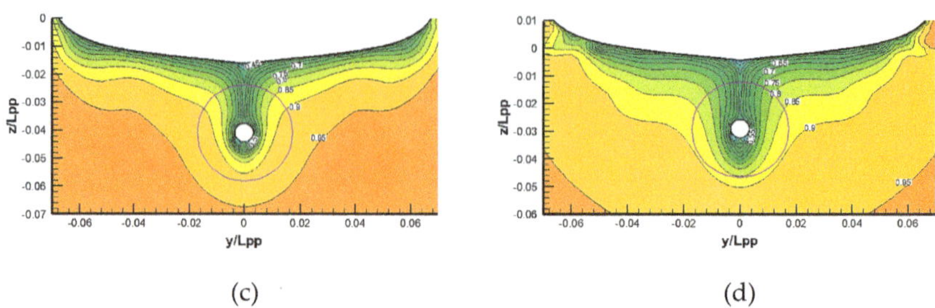

(c) (d)

Figure 12. Distribution of the nominal wake fields in heave motion condition ($x/L_{pp} = 0.9825$, $A_p = 0.25T_m$): (**a**) t = 0.25 Te; (**b**) t = 0.50 Te; (**c**) t = 0.75 Te; (**d**) t = 1.0 Te.

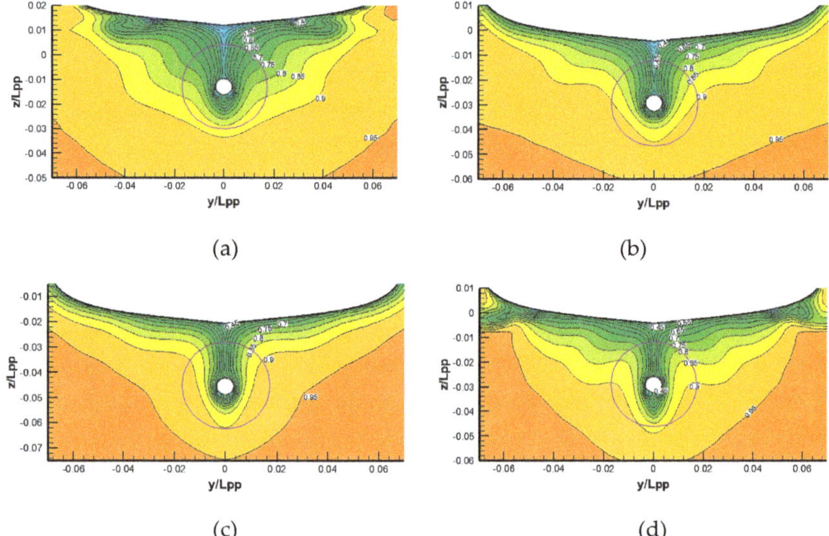

(a) (b)

(c) (d)

Figure 13. Distribution of the nominal wake fields in pitch motion condition ($x/L_{pp} = 0.9825$, $A_p = 2°$): (**a**) t = 0.25 Te; (**b**) t = 0.50 Te; (**c**) t = 0.75 Te; (**d**) t = 1.0 Te.

Since the flow field at 0.7 R radius has a dominant influence on the propeller hydrodynamic performance, the spatial non-uniformity of the nominal wake is analyzed by taking the axial velocity at 0.7 R radius as a representation. The circumferential angle is defined as Figure 14. Figure 15 shows the circumferential distribution of axial velocity at different times. The analysis results of the spatial non-uniformity are listed at Table 4. In general, the hull heave or pitch motion does not make the spatial non-uniformity become worse. Reversely, there is some improvement at the time that the hull draft is deeper, for example when t = 0.75 T_e.

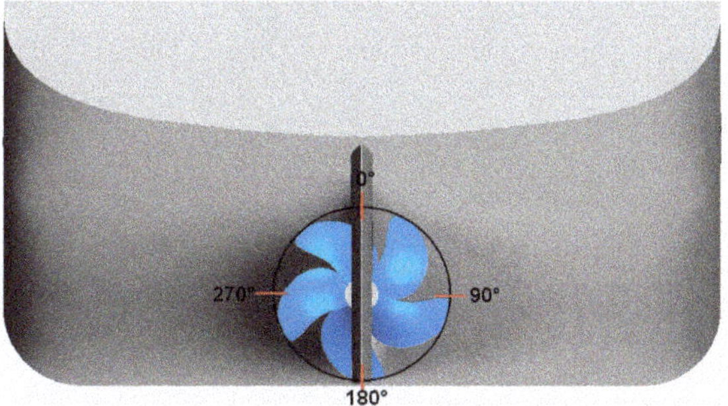

Figure 14. Definition of circumferential angle.

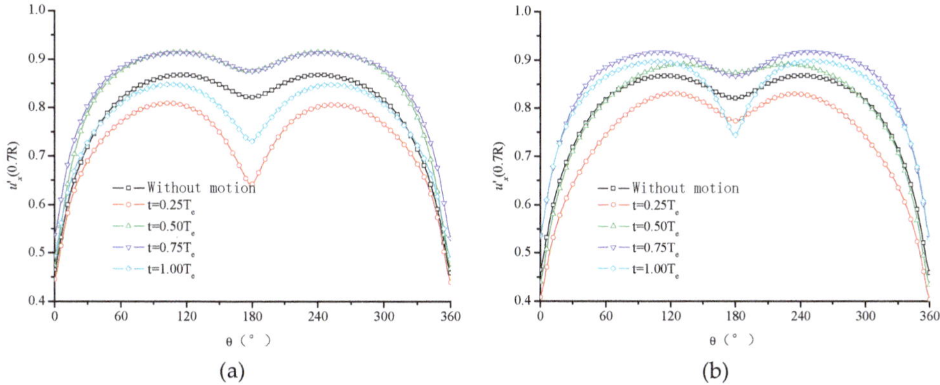

Figure 15. Circumferential distribution of axial velocity at different time for 0.7 R radius: (**a**) Heave motion $A_p = 0.25T_m$; (**b**) Pitch motion $A_p = 2°$.

Table 4. Spatial non-uniformity of axial velocity at different time for 0.7 R radius.

Condition	Without Heave	0.25Te		0.50Te		0.75Te		1.0Te	
		Heave	Pitch	Heave	Pitch	Heave	Pitch	Heave	Pitch
Average value	0.796	0.731	0.740	0.846	0.807	0.855	0.855	0.777	0.830
Peak value	0.868	0.808	0.830	0.915	0.891	0.913	0.916	0.847	0.897
Valley value	0.459	0.439	0.403	0.467	0.433	0.530	0.531	0.486	0.529
Spatial non-uniformity (%)	51.37	50.45	57.76	52.85	56.77	44.76	45.07	46.44	44.34

Figure 16 shows the average axial velocity curve at the disk plane in a motion period. From the analysis results listed in Table 5, compared with the condition without motion, the temporal non-uniformity of the heave increases sharply from 0.11% to 9.28% and 17.78% and the temporal non-uniformity of the pitch increases sharply from 0.11% to 7.28% and 14.17%. It also shows that the temporal non-uniformity has a significant positive relationship with the motion amplitude. The wake uniformity analysis is helpful to understand the change rule of the propeller exciting forcing in the motion condition later.

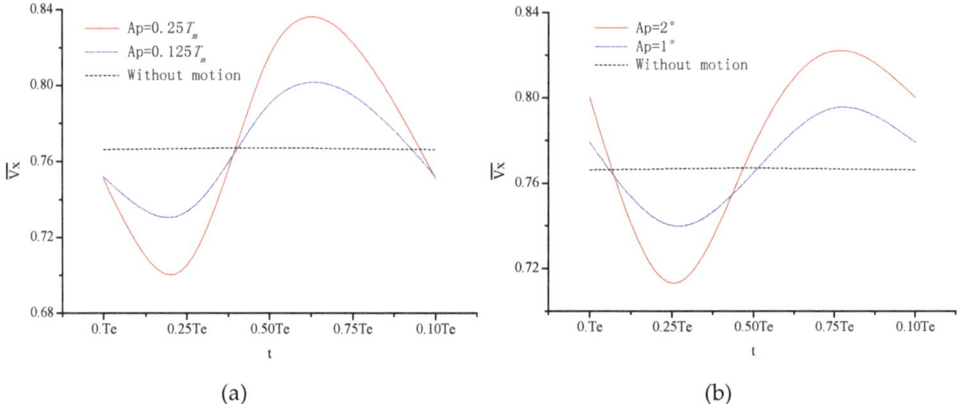

Figure 16. Average axial velocity curves at disk plane in a motion period: (**a**) Heave motion; (**b**) Pitch motion.

Table 5. Temporal non-uniformity of axial velocity in a motion period.

Condition	Without Motion	Heave		Pitch	
		A_p=0.25T_m	A_p=0.125T_m	A_p=2°	A_p=1°
Average value	0.7666	0.7661	0.7677	0.7704	0.7689
Peak value	0.7671	0.8361	0.8014	0.8221	0.7954
Valley value	0.7663	0.6999	0.7301	0.7129	0.7394
Temporal non-uniformity (%)	0.11	17.78	9.28	14.17	7.28

4.3. Hydrodynamic Performance Analysis

Table 6 lists the calculation results of the resistance. In the heave condition, the resistance increase percent is 54.23% when the heave amplitude is 0.25Tm and the resistance increase percent is 18.80% when the heave amplitude is 0.125Tm. In the pitch condition, the resistance increase percent is 69.66% when the pitch amplitude is 2° and the resistance increase percent is 20.68% when the pitch amplitude is 1°. As the motion amplitude become bigger, the resistance increases sharply. The resistance time average value in the couple motion condition is comparable with the single motion, but from the Figure 17, it can be seen that its amplitude of fluctuation is larger, which is unfavorable for the match of the ship-engine-propeller system. The growth of resistance mainly comes from the increase of the wetted surface area and residual resistance, which are induced by green water and wave making. Figure 18 has given the wave contour on the hull surface at a specific moment to show the change of the wetted surface area in different motion conditions.

Table 6. Time average resistance for hull-propeller-rudder system.

Condition	Without Motion	Heave		Pitch		Couple
		A_p=0.25T_m	A_p=0.125T_m	A_p=2°	A_p=1°	
Time average resistance (N)	90.50	139.58	107.52	153.54	109.22	136.39
Resistance increase percent (%)	-	54.23	18.80	69.66	20.68	50.70

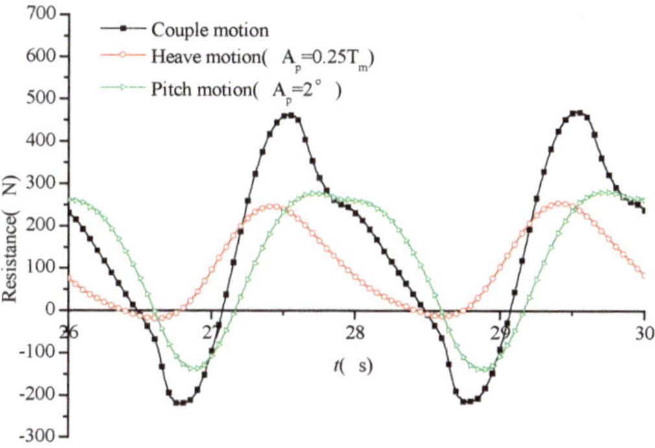

Figure 17. Time domain curves of resistance for different motion condition.

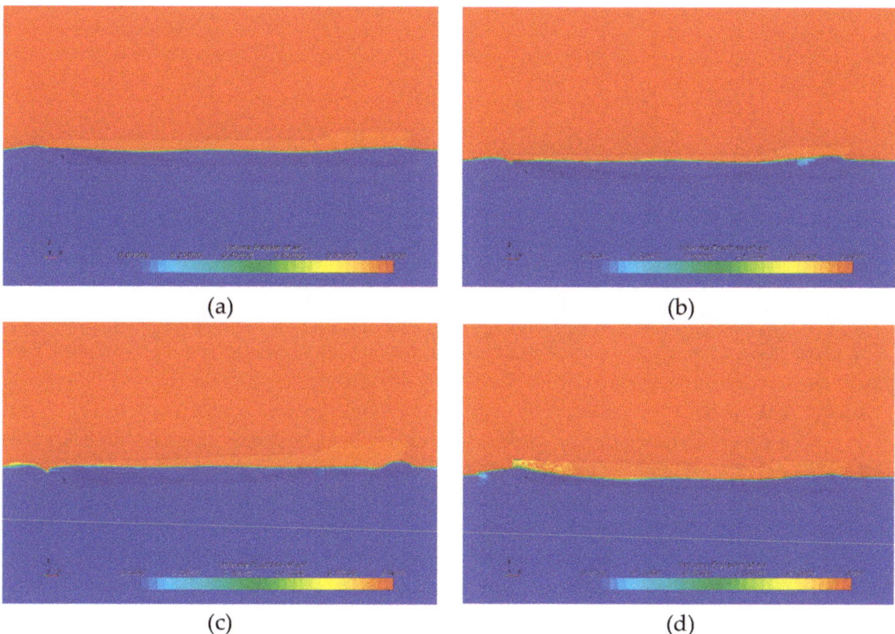

Figure 18. Wave contour on the hull surface at the specific moment: (**a**) Without motion; (**b**) Heave motion; (**c**) Pitch motion; (**d**) Couple motion.

Table 7 lists the calculation results of the propeller thrust. The drop of propeller thrust in the motion condition is remarkable. However, unlike the hull resistance, the thrust drop is not so sensitive with the motion amplitude. The difference between the time average thrust with different motion amplitude is very small. The decrease percent with the couple motion is comparable with a single motion, which is similar with the change rule of resistance. However, the amplitude of the thrust fluctuation in the couple motion is also larger, so the excitation from the propeller to the hull stern vibration is expected to be stronger. Figure 19 has shown the thrust change curves in the time domain.

Table 7. Time average thrust for propeller.

Condition	Without Motion	Heave		Pitch		Couple
		A_p=0.25T_m	A_p=0.125T_m	A_p=2°	A_p=1°	
Time average thrust (N)	57.22	33.31	34.50	32.94	34.35	33.76
Thrust decrease percent (%)	-	−41.79	−39.71	−42.43	−39.97	−41.00

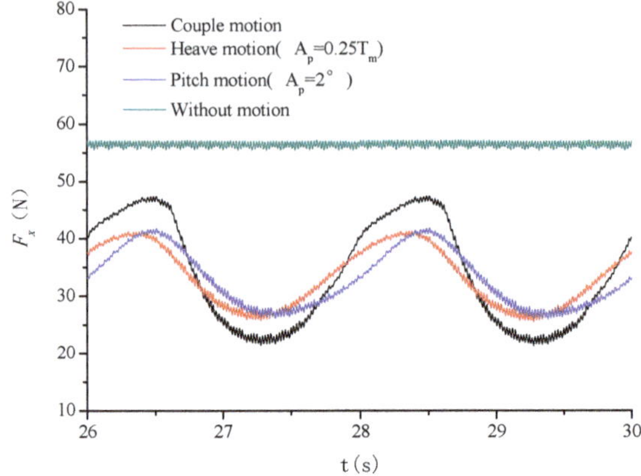

Figure 19. Time domain curve of thrust for different motion condition.

4.4. Propeller Exciting Force Analysis

4.4.1. Propeller Bearing Force

The propeller exciting force in the frequency domain was obtained after performing the FFT transform for the time domain signal. Figure 20 shows the frequency domain curves of the propeller exciting force without motion. Here, the Fx is the thrust, Fy is the horizontal force, Fz is the vertical force. It can be concluded that the propeller thrust and the side force have the same fluctuation frequency. The peaks appear at the axial frequency (9.5 Hz), the blade frequency (47.5 Hz), the double blade frequency (95.0 Hz) and the triple blade frequency (142.5 Hz), with the peak being the largest at the BPF and quickly attenuating afterwards. After 3 BPF, these forces can be ignored. The peaks between the thrust and side force do not show the obvious differences and that is because the propeller works in the non-uniform wake, the blades cannot balance the force in the Y, Z direction well. Although the time average of the side force is not very large, the fluctuating amplitude is comparable with the thrust.

Since the frequency domain characteristics of the side forces are similar with the thrust, only the domain curves of the thrust in the motion condition are given from Figures 21–23. The other calculation results are listed in Tables 8 and 9. Compared with the results without motion, the following rules can be concluded.

(1) Spectrum peaks are richer in the motion condition. The peaks show in motion frequency (0.5 Hz), double motion frequency (1.0 Hz) and triple motion frequency (1.5 Hz) to some degree besides the original blade frequency and axis frequency. (2) The peaks at the blade frequency that are induced mainly by the spatial non-uniformity at disk plane, is comparable with its value in the condition without motion. This rule corresponds with the wake analysis results that the hull motion does not have much influence on the spatial non-uniformity at a disk plane. (3) The peaks at a motion frequency which is induced mainly by the temporal non-uniformity at disk plane, is much bigger than its at blade frequency and it has become the main fluctuating quantity except for horizontal force. It means that temporal non-uniformity, compared with spatial non-uniformity, is the main factor that affects the propeller exciting force performance in the motion condition. Further, the side force peak at motion

frequency is related with the hull motion direction. (4) The peak at the motion frequency shows linear dependence on motion amplitude and the larger the amplitude is, the higher the motion frequency peak can be. However, the peak at blade frequency is almost the same for different motion amplitudes. (5) The motion frequency peaks in couple motion is almost equal to the sum of the peaks in the heave and pitch condition. It means that the effect of the different hull motion acting on the propeller bearing force appears as linear superimposition.

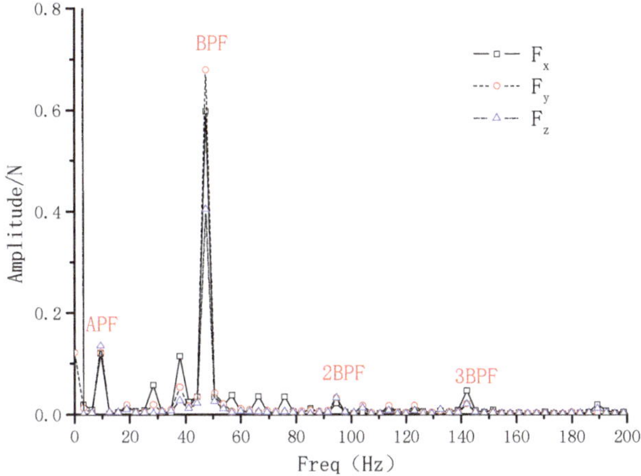

Figure 20. Frequency domain curves of propeller exciting force without motion.

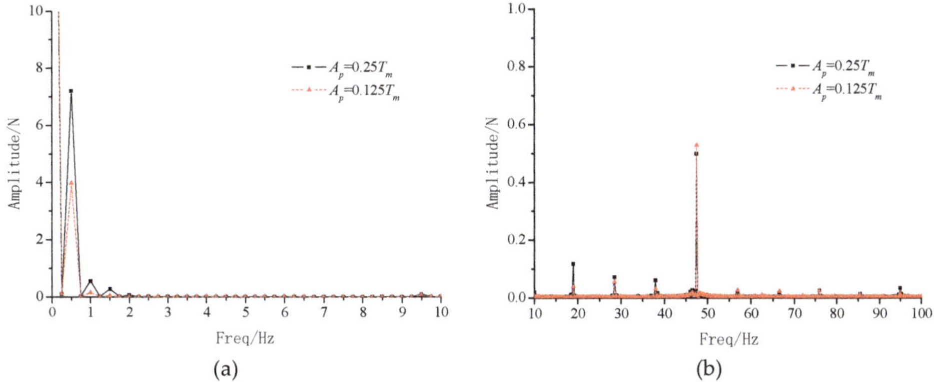

Figure 21. Frequency domain curves of thrust with heave motion: (**a**) 0–10 Hz; (**b**) 10–100 Hz.

Table 8. Peaks of horizontal force for different motion condition (unit: N).

Condition		0.5 Hz	1.0 Hz	1.5 Hz	9.5 Hz	19.0 Hz	28.5 Hz	47.5 Hz
Without motion		-	-	-	0.12	0.02	0.05	0.68
Heave	A_p = 0.25Tm	0.62	0.2	0.06	0.34	0.02	0.04	0.75
	A_p = 0.125Tm	0.33	0.06	0.01	0.36	0.04	0.04	0.73
Pitch	A_p = 2°	1.04	0.16	0.05	0.35	0.02	0.04	0.76
	A_p = 1°	0.51	0.08	0.02	0.37	0.03	0.05	0.71
Couple		1.43	0.53	0.23	0.35	0.02	0.04	0.78

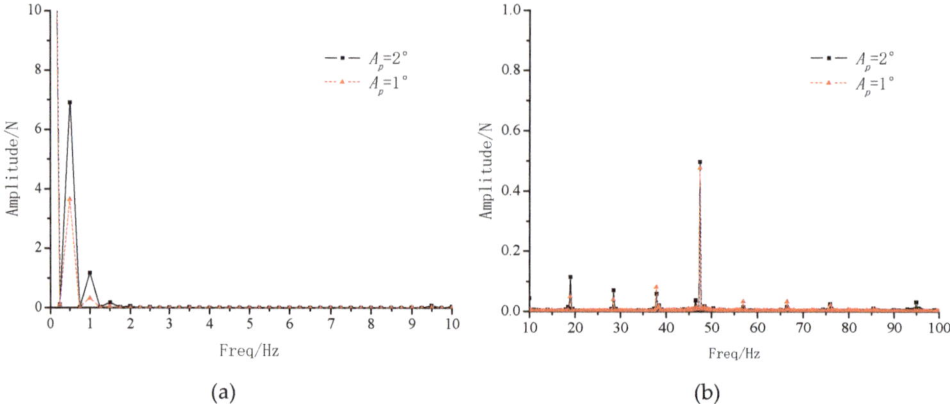

Figure 22. Frequency domain curves of thrust with pitch motion: (**a**) 0–10 Hz; (**b**) 10–100 Hz.

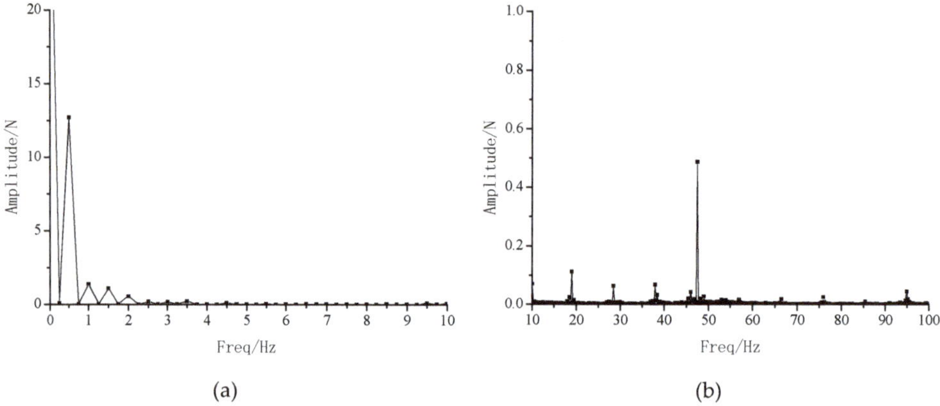

Figure 23. Frequency domain curves of thrust with couple motion: (**a**) 0–10 Hz; (**b**) 10–100 Hz.

Table 9. Peaks of vertical force for different motion condition (unit: N).

Condition		0.5 Hz	1.0 Hz	1.5 Hz	9.5 Hz	19.0 Hz	28.5 Hz	47.5 Hz
Without motion		-	-	-	0.13	-	-	0.40
Heave	A_p = 0.25Tm	2.44	0.11	0.08	0.36	0.03	0.02	0.46
	A_p = 0.125Tm	1.265	0.03	0.01	0.38	0.01	0.01	0.58
Pitch	A_p = 2°	3.00	0.18	0.07	0.35	0.03	0.02	0.50
	A_p = 1°	1.43	0.06	0.02	0.40	0.01	0.01	0.47
Couple		5.73	0.65	0.46	0.37	0.02	0.02	0.50

4.4.2. Propeller-Induced Fluctuating Pressure

The fluctuation pressure measurement points are arranged referring to Figure 24 in the plane Z −0.04 m above the propeller. An analysis is undertaken of the characteristics of the propeller-induced hull surface fluctuating pressure in the motion condition by monitoring the time signal data of these five points from P0 to P4. An explanation is required in that the absolute coordinates of monitor point is changing during the hull motion, but its relative position to the hull and propeller always keeps the same.

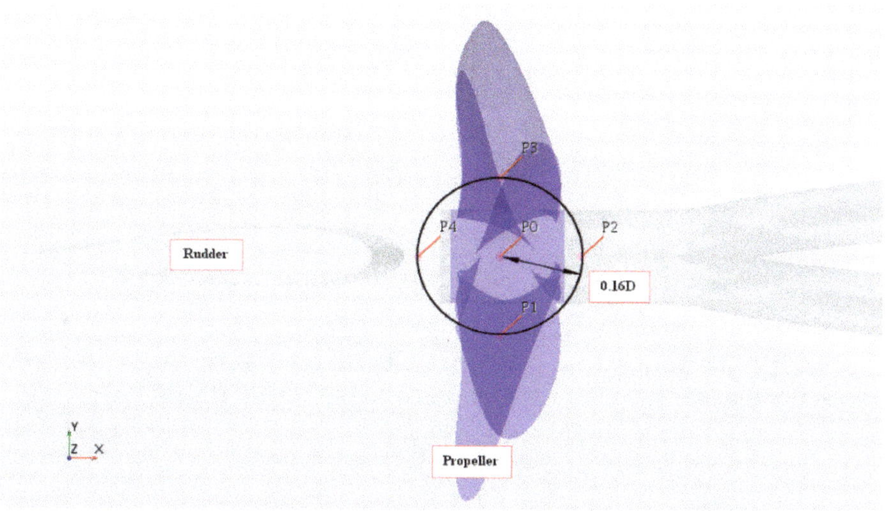

Figure 24. Arrangement of fluctuation pressure measurement points.

The authors discovered that the characteristics of fluctuation pressure at different points is almost the same and only the peak values are different after analyzing the time signal data by FFT method. Another point that needs to be mentioned is that the distance from P0 point to the propeller tip is the shortest among these five points. There is no doubt that the peak value of fluctuating pressure in P0 point is maximum, which can show the fluctuating pressure level of the whole stern. Therefore, the figures from Figures 25–28 just show the frequency domain curves of P0 point as representative to compare the characteristics of fluctuation pressure in different motion conditions.

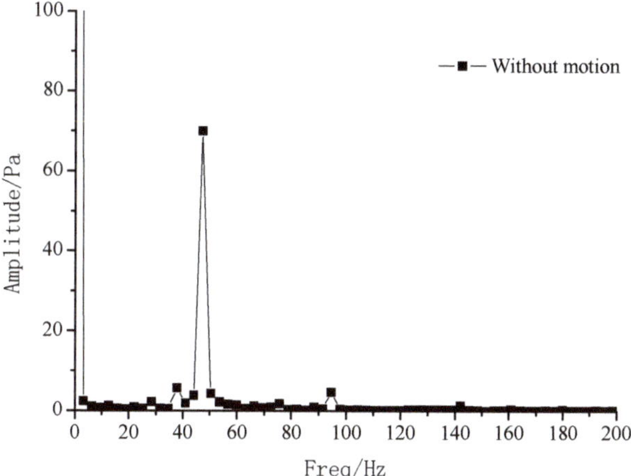

Figure 25. Frequency domain curves of P0 point without motion.

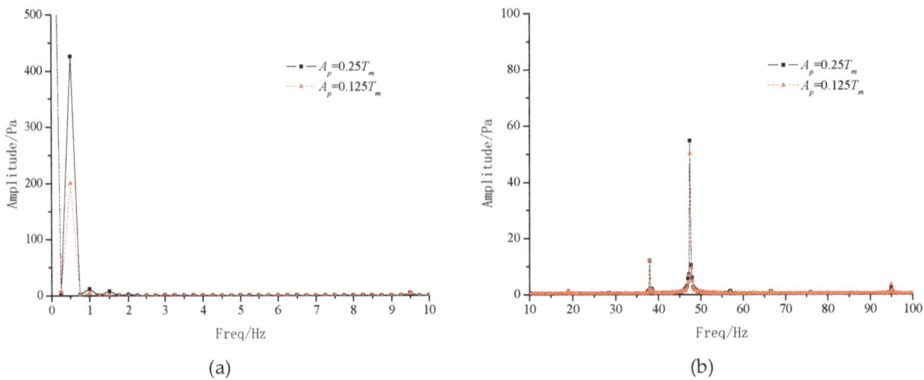

Figure 26. Frequency domain curves of P0 point in heave motion condition: (**a**) 0–10 Hz; (**b**) 10–100 Hz.

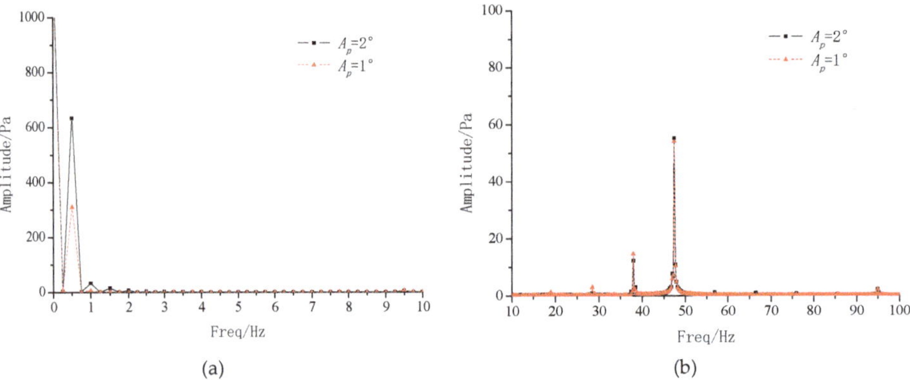

Figure 27. Frequency domain curves of P0 point in pitch motion condition: (**a**) 0–10 Hz; (**b**) 10–100 Hz.

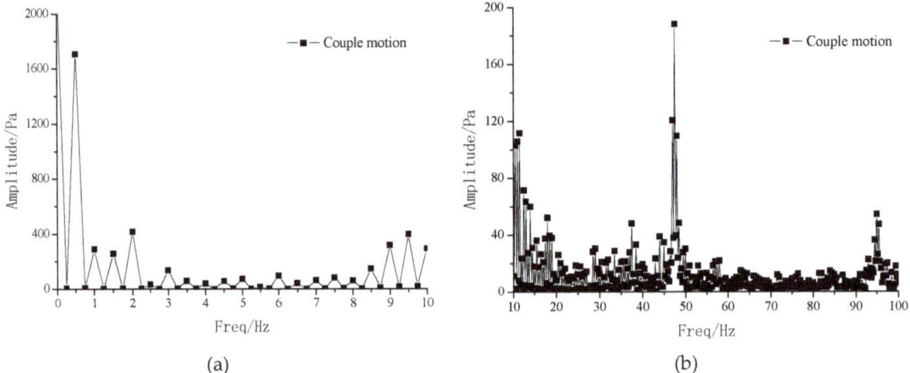

Figure 28. Frequency domain curves of P0 point in couple motion condition: (**a**) 0–10 Hz; (**b**) 10–100 Hz.

The spectrum peaks of P0 in the condition without motion mainly show in blade frequency (47.5 Hz), which is single compared with the condition with motion. The hull heave and pitch motion also excite the peaks in motion frequency (0.5 Hz) which is dominant among all the peaks. However, the value of the peaks in blade frequency does not show much difference during the motion. More attention should be paid to the peaks in the condition with the couple motion, unlike the single motion, in which the peaks associated with the motion frequency are aroused in a big range of frequency that is a quite difficult thing for the hull structure or equipment to avoid these characteristic frequencies. The fluctuating pressure is not only related to the excitation source of the propeller, but also highly related to the state of water medium between the propeller and hull surface. From the analysis results of the bearing force, the unsteady force of the propeller in the condition with couple motion mainly has a motion frequency component. Therefore, the possible reason for this phenomenon is that the superimposition of the heave and pitch motion has greatly changed the flow condition between the propeller and hull surface by different vortex structures and the generation of vortex structures are strongly related to the hull motion frequency.

In order to analyze the peaks of fluctuating pressure quantitatively, Figure 29 has given the peaks of every point at motion frequency and blade frequency for different conditions. It shows that when the motion amplitude is doubled, the peak value at motion frequency is also nearly doubled. The motion amplitude mainly influences the peaks associated with motion frequency rather than the peaks at blade frequency. This conclusion is consistent with the results given by the analysis of the wake and bearing force. However, unlike the bearing force, the peaks in the condition with the couple motion, which is at motion frequency or blade frequency, are greater than the sum of the peaks in the condition with the heave motion and pitch motion. It means that the couple motion can enlarge the fluctuating pressure in a single motion. Of course, it remains to be investigated how the fluctuating pressure will show if the motion frequency of the heave and pitch is not the same.

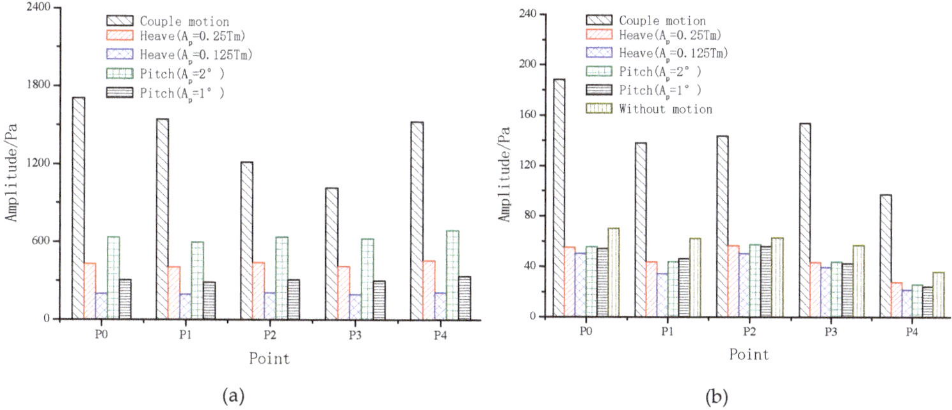

Figure 29. Fluctuation peaks at motion frequency (0.5 Hz) and blade frequency (47.5 Hz): (**a**) 0.5 Hz; (**b**) 47.5 Hz.

5. Conclusions

The numerical simulation has been conducted on the hull-propeller-rudder system of KCS model with the single and couple motion by employing the RANS method and overset grid method. Through the comparative analysis of hydrodynamic performance and the exciting force performance, the conclusions are obtained as follow.

(1) The calculation results without the hull motion are basically consistent with the experimental results, including the hydrodynamic performance of the hull-propeller–rudder system, the free surface wave contour and velocity contour in the propeller plane, which indicates that the calculation results are accurate and reliable.

(2) As the hull is in motion periodically, the wave pattern on the hull's two sides gradually spread out and finally form an obvious transverse wave. When the hull is in the heaving down or pitching up stage, the stern part becomes the main source to produce the wave as it slaps against the water. The change of wave contour can become the main reason for the resistance increase. From the

analysis of the wake field, it shows that the hull motion leads little changes to the disk wake spatial non-uniformity. However, the temporal non-uniformity in motion increases dramatically compared with case without motion.

(3) The increase of resistance and the drop of thrust induced by the hull motion are significant which can lead to obvious speed loss. In addition, when the motion amplitude becomes bigger, the resistance increases sharply. However, the thrust drop is not so sensitive. The time average value of resistance and thrust in the couple motion condition is comparable with the single motion, but their fluctuation value is much bigger.

(4) The spectrum peaks of the propeller bearing force are richer in the motion condition. Some peaks related to the motion frequency and its integral multiples are shown. The peak at heave frequency (0.5 Hz) is much bigger than the peak at blade frequency (47.5 Hz) and it has become the main fluctuating quantity, except for horizontal force. Furthermore, the motion frequency peak becomes bigger with the increase of motion amplitude. The motion frequency peaks in the couple motion are almost equal to the sum of the peaks in the heave and pitch condition. It means that the effect of the different hull motion acting on the propeller bearing force appears as linear superimposition.

(5) The characteristics of the propeller-induced fluctuating pressure in the frequency domain are similar with the bearing force. The different peaks associated with motion frequency are aroused in a big range of frequencies in the condition with the couple motion. The peaks, no matter which is at motion frequency or blade frequency, are greater than the sum of the peaks in the condition with the heave motion and pitch motion. It means that the couple motion can enlargement the fluctuating pressure in a single motion.

Our work only investigates the influence of the motion amplitude on the hydrodynamic performance of the hull-propeller-rudder system and propeller exciting force. The influence of the frequency and phase in the couple motion condition remain to be studied further. This is the point that our subsequent research will focus on.

Author Contributions: L.L. conducted CFD calculations and wrote original draft; B.Z. conducted the calculation results analysis; D.L. edited the paper; C.W. conducted some part of CFD calculations.

Funding: This research received no external funding.

Acknowledgments: At the beginning of this research, the help from Xiaoxing Peng to formulate the research idea is gratefully acknowledged.

Conflicts of Interest: The authors declare no conflicts of interest.

References

1. van Sluijs, M.F. *Performance and Propeller Load Fluctuations of a Ship in Waves*; TNO 163S; Netherlands Ship Research Centre: The Netherlands, 1972.
2. Jessup, S.D.; Boswell, R.J. The effect of hull pitching motions and waves on periodic propeller blade loads. In Proceedings of the 14th Symposium on Naval Hydrodynamics, Ann Arbor, MI, USA, 23–27 August 1982.
3. Sasajima, T. Usefulness of quasi-steady approach for estimation of propeller bearing forces. In Proceedings of the Propellers Symposium Transactions SNAME, Virginia Beach, VA, USA, 1978.
4. Breslin, J.P.; Andersen, P. *Hydrodynamics of Ship Propellers*, 1st ed.; Cambridge University Press: Cambridge, MA, USA, 1994.
5. Politis, G.K. Simulation of unsteady motion of a propeller in a fluid including free wake modeling. *Eng. Anal. Bound. Elem.* **2004**, *28*, 633–653. [CrossRef]
6. Yu, X. Research on Hydrodynamic Performance of Propeller in Heaving Condition. Master's Thesis, Harbin Engineering University, Harbin, China, 2008.
7. Carrica, P.M.; Fu, H.; Stern, F. Computations of self-propulsion free to sink and trim and of motions in head waves of the KRISO Container Ship (KCS) model. *Appl. Ocean Res.* **2011**, *33*, 309–320. [CrossRef]
8. Sharma, A. *Numerical Modeling of a Hydrofoil or a Marine Propeller Undergoing Unsteady Motion Via a Panel Method and RANS*; Tech. Rep. 11-02; Department of Civil Engineering, UT Austin: Austin, TX, USA, 2011.

9. Kinnas, S.A.; Tian, Y. Numerical modeling of a marine propeller undergoing surge and heave motion. *Int. J. Rotating Mach.* **2012**, *2012*, 257461. [CrossRef]
10. Tezdogan, T.; Incecik, A. Full-scale unsteady RANS simulations of vertical ship motions in shallow water. *Ocean Eng.* **2016**, *123*, 131–145. [CrossRef]
11. Lianzhou, W.; Chunyu, G. Numerical analysis of a propeller during heave motion in cavitating flow. *Appl. Ocean Res.* **2017**, *66*, 131–145.
12. Sun, S.; Li, L.; Wang, C.; Zhang, H. Numerical prediction analysis of propeller exciting force for hull-propeller-rudder system in oblique flow. *Int. J. Nav. Archit. Ocean Eng.* **2017**, *10*, 69–84. [CrossRef]
13. Li, L.; Zhou, B.; Liu, D.; Zheng, C. The numerical analysis of influence of the hull heave motion on the propeller exciting force characteristics. In Proceedings of the Sixth International Symposium on Marine Propulsors Smp'19, Rome, Italy, 27–30 May 2019.
14. Wang, F. *Principle and Application of CFD*; Tsinghua University Press: Beijing, China, 2004; pp. 7–11.
15. Menter, F.R. Two-equation eddy-viscosity turbulence models for engineering applications. *AIAA J.* **1994**, *32*, 1598–1605. [CrossRef]
16. Karim, M.D.; Bijoy, P.; Nasif, R. Numerical simulation of free surface water wave for the flow around NACA 0015 hydrofoil using the volume of fluid (VOF) method. *Ocean Eng.* **2014**, *78*, 89–94. [CrossRef]
17. Carrica, P.M.; Wilson, R.V. Ship motions using single-phase level set with dynamic overset grids. *Comput. Fluids* **2007**, *36*, 1415–1433. [CrossRef]
18. Hino, T. *Proceedings of the CFD Workshop*; NMRI Report 2005; National Maritime Research Institute: Tokyo, Japan, 2005.
19. Larsson, L.; Stern, F. CFD in Ship Hydrodynamics-Results of the Gothenburg 2010 Workshop. MARINE 2011. In *IV International Conference on Computational Methods in Marine Engineering*; Springer: New York, NY, USA, 2013; pp. 237–259.

Journal of
Marine Science and Engineering

Article

Multi-Degree of Freedom Propeller Force Models Based on a Neural Network and Regression

Bradford Knight * and Kevin Maki

Department of Naval Architecture and Marine Engineering, University of Michigan, Ann Arbor, MI 48109, USA; kjmaki@umich.edu
* Correspondence: bgknight@umich.edu

Received: 14 December 2019; Accepted: 26 January 2020; Published: 2 February 2020

Abstract: Accurate and efficient prediction of the forces on a propeller is critical for analyzing a maneuvering vessel with numerical methods. CFD methods like RANS, LES, or DES can accurately predict the propeller forces, but are computationally expensive due to the need for added mesh discretization around the propeller as well as the requisite small time-step size. One way of mitigating the expense of modeling a maneuvering vessel with CFD is to apply the propeller force as a body force term in the Navier–Stokes equations and to apply the force to the equations of motion. The applied propeller force should be determined with minimal expense and good accuracy. This paper examines and compares nonlinear regression and neural network predictions of the thrust, torque, and side force of a propeller both in open water and in the behind condition. The methods are trained and tested with RANS CFD simulations. The neural network approach is shown to be more accurate and requires less training data than the regression technique.

Keywords: neural network; machine learning; regression; propeller; body force; propeller-hull interaction

---

1. Introduction

Accurately and efficiently determining the multi-degree of freedom forces on a propeller for a maneuvering vessel is challenging. Potential flow methods are inexpensive but can be less accurate than viscous methods when the propeller operates off-design. Viscous flow methods, like RANS, DES, or LES, can be more accurate than potential flow methods, but require a very small time step and require extra mesh discretization for the propeller. Common ways of modeling a discretized propeller are with the overset method or sliding mesh methods [1–4]. The cost can be prohibitive for CFD of a maneuvering vessel, especially in waves and at full scale [5]. Maintaining the accuracy of a viscous CFD method while decreasing the computational cost is desirable. One way to minimize computational cost is to determine the body force using a model. The body force can be applied to the flow and to the vessel equations of motion. This approach can be used with a constant body force [6], a database driven model [7], Boundary Element Models [8], semi-empirical methods [9,10], or by using Blade Element Momentum Theory [5,11–13]. Methods that ignore viscous effects may not be as accurate as modeling a fully discretized propeller with CFD in off design conditions. Despite potential loss in accuracy in off-design conditions, Boundary Element Models and Blade Element Momentum Theory are less expensive than modeling a discretized rotating propeller with CFD [8].

This paper explores the use of data-driven models that are trained with RANS CFD. The intent is for the model to be as accurate as the CFD method used to train the model. When implemented as a body force in a maneuvering analysis the model has negligible cost. There are various methods for creating data-driven models for dynamical systems ranging from regression to neural networks. Hornik et al. [14] showed that neural networks can predict any function to any requisite level of accuracy, given adequate training and inputs. Neural networks have been applied to fluid dynamics problems ranging from improved turbulence modeling [15], to determining propeller forces for different advance coefficients [16].

This paper compares the ability for regression and a neural network to predict unsteady propeller forces due to prescribed motions. The thrust T, torque Q, and side force S of a propeller with unsteady forward velocity U, side velocity V, and propeller revolution rate n are examined in this paper. Both the open water and behind condition are considered. The objective of this paper is to evaluate the capabilities of regression compared to a neural network approach for predicting the unsteady three degree of freedom body force of a propeller undergoing unsteady motions. The goal is to train a model with a small amount of training data to limit the expense. Once the model is trained, it is tested on a variety of unsteady motions to evaluate the accuracy of the model. The intent of this method is to train an algorithm with limited data and implement the algorithm to apply the body force to the flow and the forces to the equations of motion for a vessel undergoing a maneuver. This method is especially useful when it is desired to model a vessel performing different maneuvers in a variety of sea states.

The propeller examined is the National Maritime Research Institute (NMRI) scale Kriso Container Ship (KCS) propeller. Details of the CFD setup as well as the derivation for a semi-empirical propeller force method can be found in the authors' previous work [9,17]. This paper examines both a nonlinear regression model and a neural network that are trained and tested on unsteady RANS CFD results to determine the three-degree of freedom force of the propeller in open water and the behind condition. This is an extension of the authors' prior work in which a neural network was used to determine the three-degree of freedom force of the propeller [18].

2. Methods

Three types of analysis are used in this paper. First, the RANS CFD setup and specification of unsteady motion is discussed. Second, the neural network framework is discussed. Third, the nonlinear regression modeling is discussed.

2.1. RANS CFD Setup & Specification of Unsteady Motion

The NMRI scale KCS propeller is examined in both open water and in the behind condition. The CFD used is very similar to that used by Knight & Maki [9,17], with the only difference of using an inlet condition on the sides of the domain when there is oblique flow. The OpenFOAM solver pimpleDyMFoam is used with the sliding mesh technique used for the unsteady rotating propeller simulations. The Moving Reference Frame approach is used to calculate the steady open water propeller forces using the OpenFOAM solver simpleFoam. The $k-\omega$ SST turbulence model is used with wall functions. For the behind condition case, a symmetry plane is used at the waterline. Therefore, a double body approximation is used. The mesh used in this study correlates to the coarse grid used by Knight & Maki [9,17]. The goal is for the model to be as accurate as the method that it is trained with. Therefore, the CFD used to train and test the model is treated as the truth with regards to evaluating the accuracy of the model. In this study, RANS CFD is used to train and test the models, but other techniques, such as experimental analysis or LES, could be used instead or in conjunction.

The motions of the vessel and the propeller are prescribed using a customized OpenFOAM six degree of freedom solver. The velocity is initially accelerated from rest to the average velocity over a time t_r. In all cases the t_r used is 0.0628 seconds. The general form of the velocity of the propeller after a time t_r can be given by Equation (1). The unsteady motions are prescribed to be harmonic about the average velocity for each respective degree of freedom. Thus, the motions depend upon the average velocity $\overline{\dot{x}}_j$, the amplitude of motion A_j, the frequency of oscillation $\omega_{osc,j}$, and time t. The velocity amplitude $\dot{x}_j'$ is the product of A_j and $\omega_{osc,j}$. In this form, the general motions are given with subscript j. The specific degrees of motion of forward speed U, side velocity V, and revolution rate ω can be given by Equations (2)–(4), respectively. The equations are functions of the average velocities ($\overline{U}$, $\overline{V}$, and $\overline{n}$), the velocity amplitudes (u', v', and n'), and time.

$$\dot{x}_j = \overline{\dot{x}}_j + A_j\omega_{osc,j}\cos\left(\omega_{osc,j}(t-t_r)\right) = \overline{\dot{x}}_j + \dot{x}_j'\cos\left(\omega_{osc,j}(t-t_r)\right) \tag{1}$$

$$U(t) = \dot{x}_1 = \overline{U} + u'\cos\left(\omega_{osc,1}(t-t_r)\right) \tag{2}$$

$$V(t) = \dot{x}_2 = \overline{V} + v'\cos\left(\omega_{osc,2}(t-t_r)\right) \tag{3}$$

$$\omega(t) = 2\pi n = \dot{x}_4 = 2\pi\overline{n} + 2\pi n'\cos\left(\omega_{osc,4}(t-t_r)\right) \tag{4}$$

2.2. Set-Up of the Neural Network for Unsteady Propeller Forces

A single hidden layer neural network is used with a sigmoid activation function. The inputs to the neural network are snapshots of the advance ratio, J, and the oblique flow angle, α. Thus, the inputs are instantaneous feature vectors, where the feature vector includes polynomials of J and α. The forces on the propeller depend upon the diameter of the propeller D, the density of water ρ, n, U, and α. The diameter of the propeller examined is 0.105 m, and the density of the water is 997.66 kg/m^3. Therefore, the output of the neural network is the instantaneous thrust coefficient, K_T, the instantaneous side force coefficient, K_S, and the instantaneous torque coefficient, K_Q. The equations for K_T, K_S, K_Q, J, and α are shown in Equations (5)–(9).

$$K_T = \frac{T}{\rho n^2 D^4} \tag{5}$$

$$K_S = \frac{S}{\rho n^2 D^4} \tag{6}$$

$$K_Q = \frac{Q}{\rho n^2 D^5} \tag{7}$$

$$J = \frac{U}{nD} \tag{8}$$

$$\alpha = \tan^{-1}\left(\frac{V}{U}\right) \tag{9}$$

The particulars of the neural network used to determine the open water propeller forces is discussed first. Second, the extension of the open water neural network to the behind condition is discussed.

2.2.1. Open Water Neural Network

A feedforward neural network with a single hidden layer is programmed in MATLAB. The input to the neural network is a feature vector, $\vec{x}$, for each snapshot in time. For each snapshot in time, the input feature vector, $\vec{x}$, is shown by Equation (10). The feature vector is comprised of permutations of the instantaneous J, up to second order, and the absolute value of α, up to sixth order. A single hidden layer with $k = 10$ hidden neurons is used. The output layer has $n = 3$ neurons correlating to K_T, K_S, and K_Q, which may be scaled and shifted to improve the accuracy of the neural network.

$$\vec{x} = [1, J, J^2, |\alpha|, \ldots, |\alpha|^6, J|\alpha|, \ldots, J|\alpha|^6, J^2|\alpha|, \ldots, J^2|\alpha|^6]^T \tag{10}$$

The snapshots for the training data are taken by incrementally sampling the time history of the observables of CFD simulations of an open water propeller with unsteady motions, U, V, and n. The unsteady motions are prescribed, and thus are known. Therefore, the feature vector comprised of J and α can be determined. The observables from the CFD simulations include the thrust, side force, and torque. These are processed to give instantaneous values of K_T, K_S, and K_Q. Similarly, J, α, K_T, K_S, and K_Q can all be preprocessed for the test data from CFD simulations. The force coefficients from the CFD are treated as the truth.

Based upon Equation (10), the input matrix has $m = 21$ rows and the number of columns equal to the number of snapshots examined. A schematic of a single hidden layer neural network is shown in Figure 1. The left hand column of neurons is the input, the middle column of neurons is the hidden layer, and the right hand column of neurons is the output layer. A sigmoid activation function is used, which is defined by Equation (11). Two weight matrices and two bias vectors are used. The first weight matrix $\mathbf{W_1}$ and bias vector $\vec{b}_1$ are used between the input layer and the hidden layer. The second weight matrix $\mathbf{W_2}$ and bias vector $\vec{b}_2$ are used between the hidden layer and the output layer. The values of the weight matrices and bias vectors are determined using back propagation. The accuracy for a single snapshot i can be evaluated by the loss function L_i defined in Equation (12), where $\hat{\vec{y}}_i$ is the prediction and $\vec{y}_i$ is the truth for that snapshot. $\hat{\vec{y}}_i$ is calculated from Equation (13) which depends on the output of Equation (14).

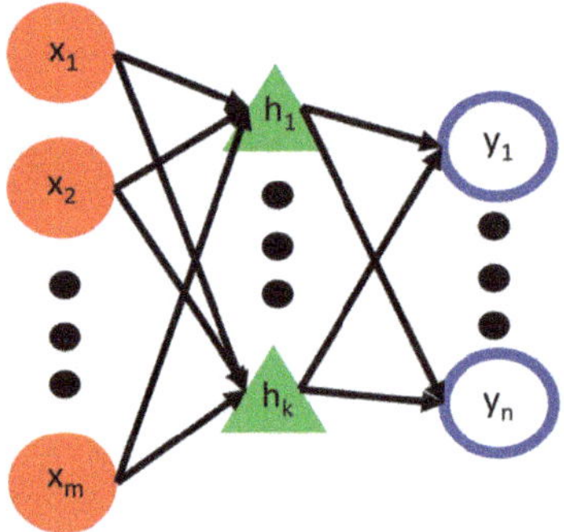

Figure 1. Schematic of neural network with single hidden layer.

$$\sigma(z) = \frac{1}{1+e^{-z}} \tag{11}$$

$$L_i = \frac{1}{2}\|\vec{y}_i - \hat{\vec{y}}_i\|_2^2 \tag{12}$$

$$\hat{\vec{y}}_i = \sigma(\mathbf{W_2}\vec{\eta}_i + \vec{b}_2) \tag{13}$$

$$\vec{\eta}_i = \sigma(\mathbf{W_1}\vec{x}_i + \vec{b}_1) \tag{14}$$

The average loss function L across all snapshots is most indicative of the error. Therefore, the loss function of each snapshot in time is calculated. The loss functions for each snapshot are summed and divided by the number of snapshots N as shown by Equation (15). The program iterates until the loss function is below a user defined tolerance. Stochastic gradient descent is used to improve the speed of convergence. The stochastic gradient descent algorithm works by randomly selecting one of the snapshots and calculating the gradients of the weight matrices based on that snapshot. Once the neural network is trained, it is tested with data from other unsteady propeller CFD simulations. The accuracy of the neural network is evaluated by how accurate it is compared to the force coefficients calculated from the CFD.

$$L = \frac{1}{N}\sum_{i=1}^{N} L_i \tag{15}$$

2.2.2. Extension of Neural Network to Predict Unsteady Behind Condition Propeller Forces

The neural network for the behind condition propeller forces is similar to the open water propeller neural network. For the behind condition neural network, the feature vector defined by Equation (10) is modified; the $|\alpha|$ is changed to α. Therefore, the oblique flow angle maintains its sign. In the open water propeller case, the direction of motion did not have an effect except to determine the sign of the side force. However, in the behind condition the hull interacts with the propeller. The direction of α affects the propeller forces since the propeller is rotating in the wake of the hull. Equation (16) shows the feature vector used as input to the neural network for the behind condition for each snapshot in time. Note that J and α are calculated based upon the prescribed vessel velocities and n, just like the open water method. The high dimensionality of the feature vector allows for faster training of the model. Early stopping is used to avoid overtraining of the model. A study is performed while training the behind condition model to determine when the neural network is sufficiently trained and not overtrained.

$$\vec{x} = [1, J, J^2, \alpha, \ldots, \alpha^6, J\alpha, \ldots, J\alpha^6, J^2\alpha, \ldots, J^2\alpha^6]^T \tag{16}$$

The second difference between the behind conditional neural network and the open water neural network is the manner in which they are trained. As aforementioned, the open water neural network uses stochastic gradient descent. The open water neural network is also trained with the results of one unsteady open water propeller CFD simulation. However, the behind condition neural network uses gradient descent instead. Therefore, the average gradient of the weights and bias vector is used to correct the current weights. The behind condition propeller neural network is trained with two different types of simulations. It uses one unsteady CFD simulation of the propeller operating in the behind condition. It also uses steady state open water propeller coefficients which are extended into the behind condition in accordance with [9,17]. Knight & Maki used the thrust identity [19] to extend steady state open water K_T and K_Q into K_T and K_Q coefficients for the propeller operating in the behind condition if the same propeller is run in the behind condition at a single U and n. The KCS propeller has been simulated using a Moving Reference Frame approach at different J values in open water and a rotating propeller simulation was performed for the propeller operating in the behind condition with constant U and n at one speed to determine the Taylor Wake fraction w using the thrust identity [19]. This parameter has been found in [9,17]. Therefore, the behind condition case is trained with the results of one unsteady behind condition propeller simulation and the open water K_T and K_Q values extended into the behind condition using the thrust identity.

2.3. Nonlinear Regression for Unsteady Propeller Forces

MATLAB's non linear regression function "fitnlm" [20] is used to develop a regression model for K_T, K_Q, and K_S. The same training data is used for the regression model and the neural network. Thus, snapshots of a feature vector of different permutations of J and α are used as input. To limit the propensity for overfitting a simpler feature vector than the neural network is used. The feature vector for the regression contains terms of J and α up to second order, as shown in Equation (17). In the case of the open water data the absolute value of α is used for the same reason as for the neural network. Equation (18) shows the form of the regression model to determine the force coefficients. "fitnlm" determines the a, b, and c coefficients. In the behind condition analysis, a study is performed in which different feature vectors are used to develop the regression model.

$$\vec{x} = [1, J, J^2, \alpha, \alpha^2, J\alpha]^T \tag{17}$$

$$\begin{aligned} K_T &= a_1 + a_2 J + a_3 J^2 + a_4\alpha + a_5\alpha^2 + a_6 J\alpha \\ K_S &= b_1 + b_2 J + b_3 J^2 + b_4\alpha + b_5\alpha^2 + b_6 J\alpha \\ K_Q &= c_1 + c_2 J + c_3 J^2 + c_4\alpha + c_5\alpha^2 + c_6 J\alpha \end{aligned} \tag{18}$$

3. Training and Testing the Neural Network and Regression Model to Predict Unsteady Open Water Propeller Forces

The models to predict unsteady open water propeller forces are trained with the results of an open water unsteady propeller CFD simulation. Discrete snapshots are examined for the training data.

Before testing the neural network on a second unsteady propeller motion, a validation step is taken. The validation step examines the full time history of the unsteady propeller data that was used for training the algorithm. Once the models are trained and validated, they are tested on an unsteady propeller which undergoes a different unsteady motion. Table 1 shows the parameters that define the unsteady motion for each of the cases.

Table 1. Open water parameters for unsteady motion.

Parameter	Train	Test
$\overline{U}$ (m/s)	1.68	1.68
$\overline{V}$ (m/s)	0.00	0.00
$\overline{n}$ (rev/s)	32.0	32.0
$u'/\overline{U}$	0.476	0.262
$v'/\overline{U}$	0.500	−0.476
$n'/\overline{n}$	0.156	0.156
$\omega_{osc,1}/(2\pi\overline{n})$	0.020	0.020
$\omega_{osc,2}/(2\pi\overline{n})$	0.020	0.020
$\omega_{osc,4}/(2\pi\overline{n})$	0.005	0.005

3.1. Training the Neural Network and Nonlinear Regression Model

Both the neural network and the nonlinear regression model are trained on a case that has unsteady velocity in all three degrees of freedom. The average forward velocity is $\overline{U} = 1.68$ m/s, the average side velocity is $\overline{V} = 0$ m/s, and the average propeller revolution rate is $\overline{n} = 32$ rev/s. The amplitudes for each degree of freedom are $A_1 = 0.2$ m, $A_2 = 0.213$ m, and $A_4 = 31.4$ rad. The frequencies of oscillation are $\omega_{osc,1} = 4$ rad/s, $\omega_{osc,2} = 4$ rad/s, and $\omega_{osc,4} = 1$ rad/s. The non-dimensionalized velocity amplitudes are $u'/\overline{U} = 0.476$, $v'/\overline{U} = 0.500$, and $n'/\overline{n} = 0.156$. The motion of the propeller in surge, sway, and rotational degrees of freedom is imposed in the customized OpenFOAM multi-degree of freedom solver. Similarly, since the motion is prescribed, Equations (1)–(4) can be used to extract the velocities for each degree of freedom. The first 0.1 seconds are neglected to account for the ramp time and the initial transient effects. After this time, the CFD forces are uniformly sampled to generate 70 snapshots. J and α for each of the snapshots are shown in Figure 2. The feature vector is generated as a function of these J and α values.

At each of these snapshots, the output matrix is extracted from the CFD results. As discussed earlier, each column of output from the neural network represents the predicted $[K_T\ K_S\ K_Q]^T$ of each snapshot. Similarly, the nonlinear regression model determines the values the values of K_T, K_S, and K_Q.

K_Q is much smaller than K_T. Additionally, as K_S is zero at $\alpha = 0$ rad, K_S can be very small. Furthermore, K_S oscillates about zero depending upon the direction of the sway velocity. Therefore, for both the training and testing of the open water unsteady propeller models, scaling is applied to the K_Q and K_S extracted from the forces calculated by the CFD. The value of K_S is also shifted to improve the scaling. K_S and K_Q are recast in Equations (19) and (20), where the subscript u denotes unscaled coefficients, correlating to the original K_S and K_Q defined in Equations (6) and (7) respectively. The absolute value of $K_{S,u}$ is used as the input matrix contains functions of the absolute value of the oblique flow angle; thus, directionality is not accounted for. It was found that the neural network was more robust when directionality was omitted in the neural network. The sign of the side force is corrected after the models calculate K_S. This is demonstrated when the models are tested.

$$K_S = 10|K_{S,u}| + 0.3 \tag{19}$$

$$K_Q = 10K_{Q,u} \tag{20}$$

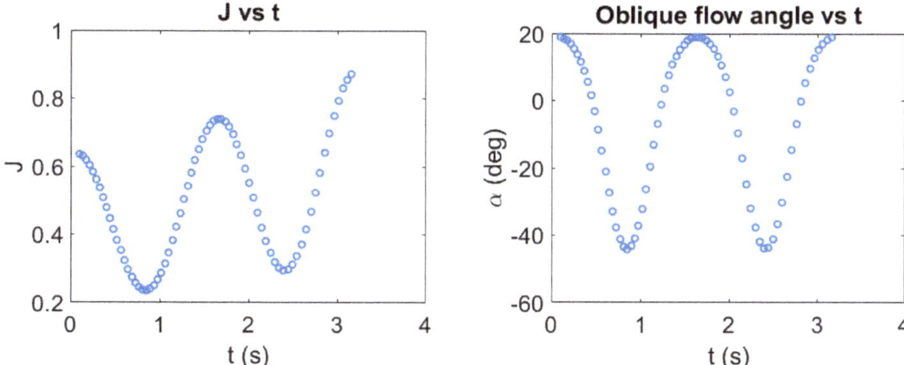

Figure 2. Discrete samples of J (**left**) and α (**right**) as a function of time for the unsteady open water propeller training case.

3.1.1. Training the Neural Network for Unsteady Open Water Propeller Forces

Figure 3 shows L versus the training iteration. As stochastic gradient descent is used, the loss function may increase on some iterations, but the loss function overall tends to decrease until reaching a floor. The training loss function is reduced to 3.1×10^{-5}. To note, the neural network was examined with the same second order feature vector used by the nonlinear regression method. Using this lower order feature vector leads to similar prediction of the forces, but requires twice as many training iterations compared to the higher order feature vector.

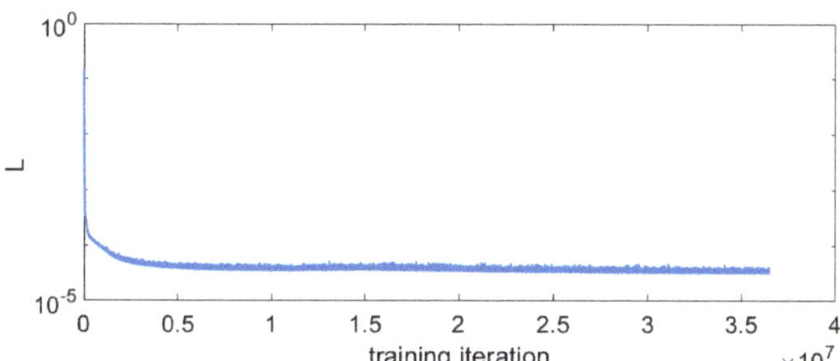

Figure 3. L as a function of training iteration for the open water neural network.

Figure 4 shows the neural network prediction for the training data. This demonstrates that the neural network is trained sufficiently, as it does a good job predicting the data that it was trained with. The neural network can further be validated by examining the neural network prediction of the coefficients versus the CFD predictions for the whole time series, instead of the 70 discrete snapshots used to train the neural network. Figure 5 shows that the neural network does a good job of predicting the force coefficients from the CFD for the whole time series. Note that the side force predicted by the CFD has a high frequency oscillation due to the side force varying as a function of the rotation angle of the propeller. However, this oscillation is not modeled by the neural network since it was not trained with the rotational position of the propeller as a feature, nor were enough snapshots used to train the neural network to discern this characteristic. This oscillation is small and since the intent of the neural network is for use in vessel maneuvering this small oscillation is not necessary to capture.

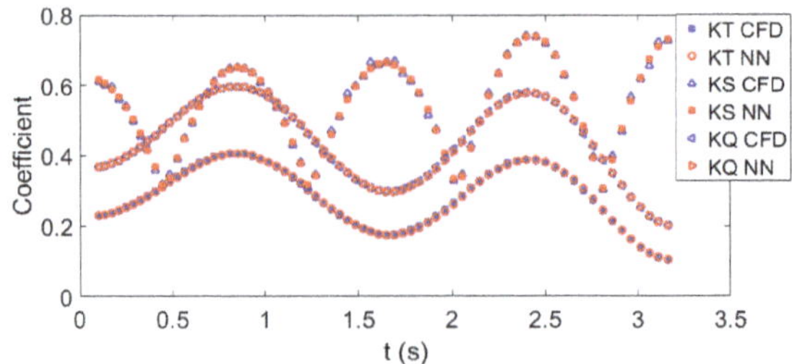

Figure 4. Coefficients as a function of time for open water training using the neural network.

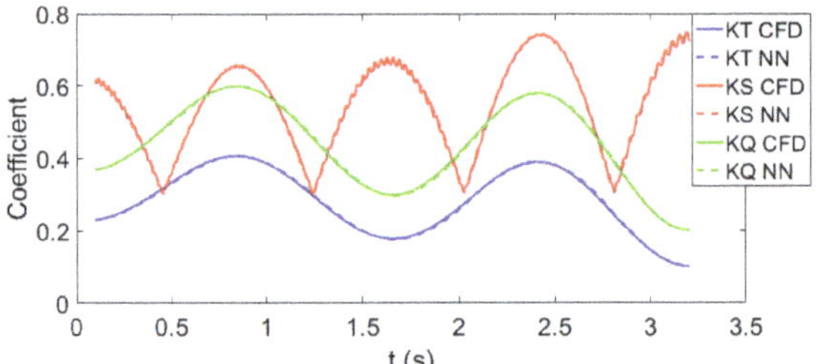

Figure 5. Coefficients as a function of time for the entire time series used for open water training. Seventy snapshots of this series were used to train the neural network.

3.1.2. Training the Nonlinear Regression Model for Unsteady Open Water Propeller Forces

The nonlinear regression fit determined using MATLAB for the prediction of K_T, K_S, and K_Q for the open water propeller is given by Equation (21). Figure 6 shows the nonlinear regression prediction for the training data. The left hand figure illustrates how well the regression model would predict the training data if only first order terms were used and the regression model was trained neglecting the J^2, α^2, and $J\alpha$ features. The right hand image shows an improvement if the second order terms are included. This demonstrates that the nonlinear regression model is trained sufficiently when the second order features are used.

$$\begin{aligned} K_T &= 0.477 - 0.281J - 0.061J^2 + 0.134|\alpha| - 0.204|\alpha|^2 + 0.047J|\alpha| \\ K_S &= 0.320 - 0.017J - 0.126J^2 + 1.611|\alpha| - 0.028|\alpha|^2 + 0.252J|\alpha| \\ K_Q &= 0.680 - 0.332J - 0.076J^2 + 0.162|\alpha| - 0.291|\alpha|^2 + 0.069J|\alpha| \end{aligned} \tag{21}$$

3.2. *Testing the Models for Unsteady Open Water Propeller Force*

The test case has unsteady velocity in all three degrees of freedom. The average forward velocity is $\overline{U}$ = 1.68 m/s, the average side velocity is $\overline{V}$ = 0 m/s, and the average propeller revolution rate is $\overline{n}$ = 32 revolutions/s. The amplitudes for each degree of freedom are A_1 = 0.107 m, $A_2 = -0.2$ m, and A_4 = 31.4 radians. The frequencies of oscillation are $\omega_{osc,1}$ = 4 rad/s, $\omega_{osc,2}$ = 4 rad/s, and $\omega_{osc,1}$ = 1 rad/s. This results in non-dimensionalized velocity amplitudes of $u'/\overline{U}$ = 0.262, $v'/\overline{U} = -0.476$, and $n'/\overline{n}$ = 0.156. The resulting J and α as a function of time are shown in Figure 7.

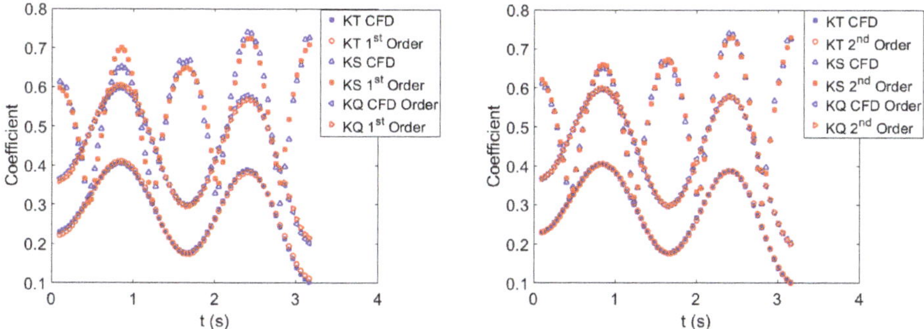

Figure 6. Coefficients as a function of time for open water training using nonlinear regression. (**Left**) Nonlinear regression using a first order feature vector. (**Right**) Nonlinear regression using the second order feature vector.

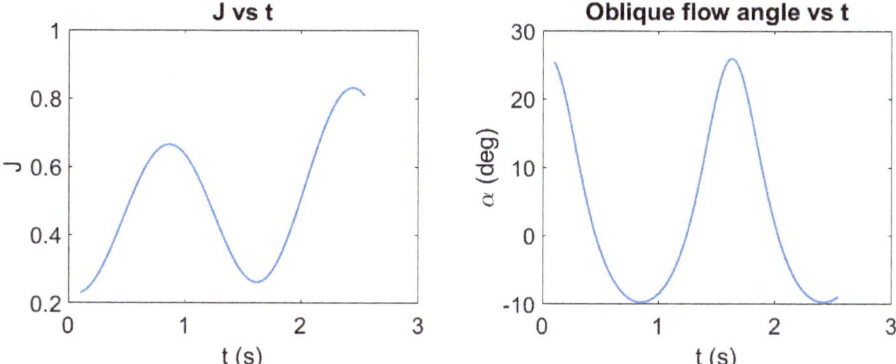

Figure 7. J and α as a function of time for the unsteady open water propeller test case.

3.2.1. Testing the Neural Network for Unsteady Open Water Propeller Force

Figure 8 shows the coefficients predicted by the neural network compared to those computed from the CFD. The average loss function is 6.56×10^{-5}, the average L2 norm error of the K_T is 0.0035, the average L2 norm error of the K_S is 0.01015, and the average L2 norm error of the K_Q is 0.0040. Thus, the neural network does a good job of predicting the coefficients for the case that it was not trained with. The coefficients can be expanded back into the forces using Equations (5)–(7), (19), and (20). The sign of the side force is determined with the knowledge that the force acts in the opposite direction that the vessel sways. The side force predicted by the CFD oscillates as a function of the azimuthal position of the propeller. This high frequency oscillation is ignored by the model, as the time scale of this oscillation and the low amplitude of this oscillation will have negligible effect on the maneuvering characteristics of a vessel. The thrust, side force, and torque are plotted in Figure 9. Good agreement is seen between the forces predicted by the CFD and the neural network.

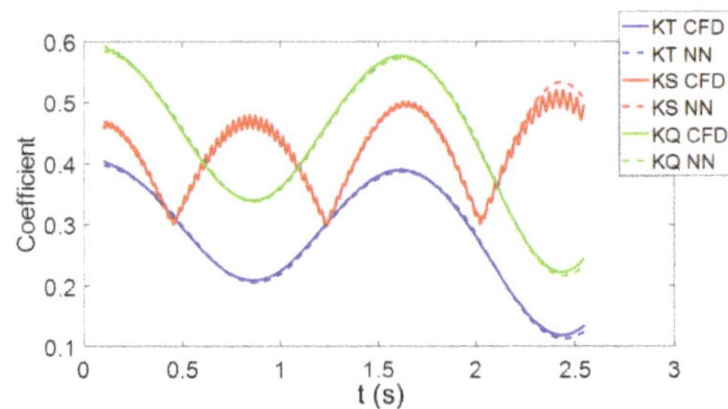

Figure 8. Open water coefficients for the test data set comparing CFD to the neural network.

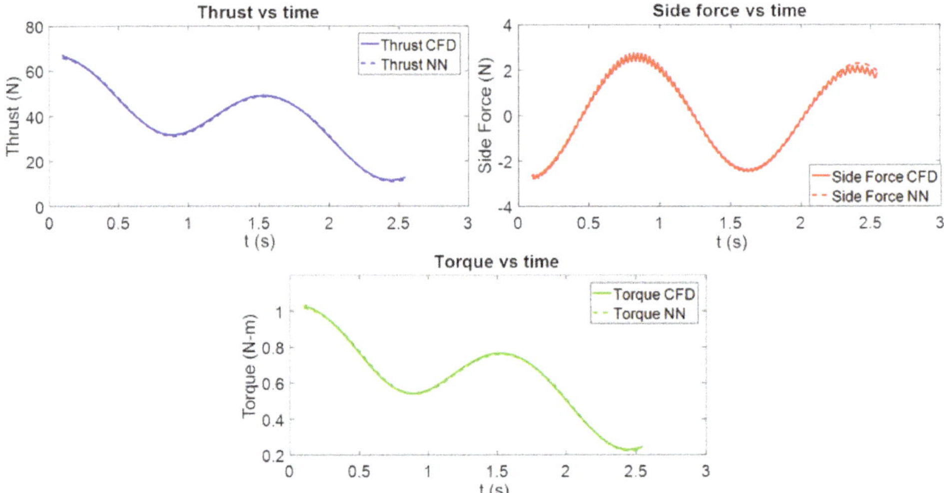

Figure 9. Open water thrust, side force, and torque for the test data set as a function of time comparing CFD to the neural network.

3.2.2. Testing the Nonlinear Regression for Unsteady Open Water Propeller Force

The coefficients predicted by the nonlinear regression model are quite accurate compared to those computed from the CFD. The average loss function is 5.13×10^{-5}, the average L2 norm error of the K_T is 0.0038, the average L2 norm error of the K_S is 0.00385, and the average L2 norm error of the K_Q is 0.0077. Thus, the nonlinear regression also does a good job of predicting the coefficients of the test data. The average loss function and K_S are predicted better using the nonlinear regression, but the K_T and K_Q are predicted better using the neural network. The differences in prediction between the neural network and the nonlinear regression are quite small in this case. The coefficients determined by nonlinear regression are expanded back into the forces using Equations (5)–(7), (19), and (20). The thrust, side force, and torque are plotted in Figure 10. Good agreement is seen between the forces predicted by the CFD and the nonlinear regression.

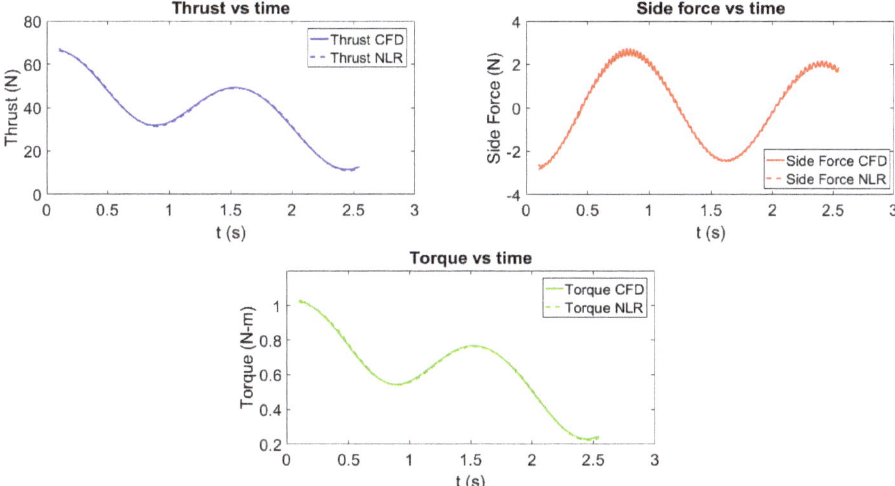

Figure 10. Open water thrust, side force, and torque for the test data set as a function of time using nonlinear regression.

4. Training and Testing the Neural Network and Nonlinear Regression to Predict Unsteady Behind Condition Propeller Forces

The neural network and nonlinear regression model are used to predict unsteady behind condition propeller forces. First, the models are trained with the results of a behind condition unsteady propeller CFD simulation and the steady state open water coefficients extended to the behind condition. Discrete snapshots are examined for the training data. Once the models are trained they are tested with several different unsteady behind condition propeller CFD cases that have various motions. These three different cases are denoted "Test 1", "Test 2", and "Test 3". Table 2 shows the parameters that define the unsteady motion for each of the cases. Figure 11 shows α-J space for the validation and test cases compared to the training space.

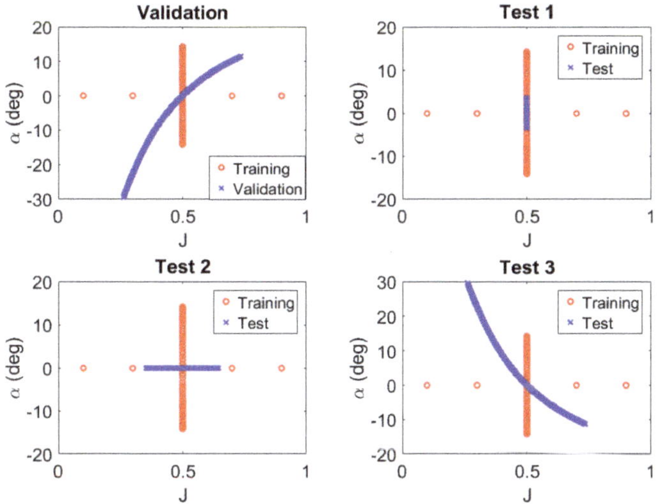

Figure 11. α-J space for training, validation, and test cases.

Table 2. Behind condition parameters for unsteady motion.

Parameter	Train	Validation	Test 1	Test 2	Test 3
$\overline{U}$ (m/s)	1.68	1.68	1.68	1.10	1.68
$\overline{V}$ (m/s)	0.00	0.00	0.00	0.00	0.00
$\overline{n}$ (rev/s)	32.00	32.00	32.00	20.95	32.00
$u'/\overline{U}$	0.00	0.48	0.00	0.31	0.48
$v'/\overline{U}$	0.25	0.30	0.07	0.00	−0.30
$n'/\overline{n}$	0.00	0.00	0.00	0.00	0.00
$\omega_{osc,j}$ (rad/s)	1.00	1.00	1.00	6.72	1.00

4.1. Training the Models to Predict the Unsteady Behind Condition Propeller Force

The training data is split into two parts. The first part is the unsteady behind condition propeller training case. The average forward velocity is $\overline{U}$ = 1.68 m/s, the average side velocity is $\overline{V}$ = 0 m/s, and the average propeller revolution rate is $\overline{n}$ = 32 rev/s or $\dot{x}_{4,avg}$ = 201.06 rad/s. The amplitudes for each degree of freedom are A_1 = 0 m, A_2 = 0.4265 m, and A_4 = 0 rad. The amplitude of sway is equal to the maximum waterline beam (BWL). Therefore, only the sway degree of freedom is unsteady. The resulting velocity amplitudes are u' = 0 m/s, v' = 0.42 m/s, and n' = 0 rev/s. The frequency of oscillation is $\omega_{osc,2}$ = 1 rad/s. The second half of the training data is the steady state open water propeller coefficients which are converted to the behind condition coefficients using the thrust identity. Therefore, the K_T and K_Q for steady-state open water J values of [0.1, 0.3, 0.5, 0.7 and 0.9] are expanded to the behind condition if the vessel has the same U and n used for those open water coefficients. The K_S is assumed to be zero, as without drift the side force is small and this eliminates the need to perform additional behind condition simulations. This series is repeated twenty times so that the unsteady data and the steady data have similar data sizes for the determination of the training of the models.

The behind condition coefficients are scaled differently than for the open water case. K_T is shifted up by 0.2 as shown by Equation (22). K_S is multiplied by five and is shifted up 0.5 as shown by Equation (23). K_Q follows the same scaling as in the open water modeling, shown by Equation (20).

$$K_T = K_{T,u} + 0.2 \tag{22}$$

$$K_S = 5|K_{S,u}| + 0.5 \tag{23}$$

4.1.1. Training the Neural Network for the Unsteady Behind Condition Propeller Force

As noted in the open water discussion, the feature vector defined by Equation (10) trains the neural network faster than a lower order feature vector. To avoid over training the neural network, different tolerances of L are examined and are evaluated by a single test case used to validate the model. As the tolerance of L decreases, it is expected that the accuracy of the model improves until it reaches a point where the neural network may become overtrained.

The "Validation" data set is used to validate the behind condition neural network to evaluate whether or not the model is overtrained. The average forward velocity is $\overline{U}$ = 1.68 m/s, the average side velocity is $\overline{V}$ = 0 m/s, and the average propeller revolution rate is $\overline{n}$ = 32 rev/s. The resulting velocity amplitudes are u' = 0.81 m/s, v' = 0.50 m/s, and n' = 0 rev/s. The frequency of oscillation is $\omega_{osc,1}$ = $\omega_{osc,2}$ = 1 rad/s. This data set is used to validate the model since it incorporates both unsteady J and unsteady α. The validation results for five different tolerances of L are presented: 1×10^{-3}, 5×10^{-4}, 1×10^{-4}, 8.3×10^{-5}, and 7.7×10^{-5}. The L as well as the L2 norm error of K_T, K_S, and K_Q are shown in Table 3. The validation L is used to determine whether the neural network is appropriately trained. For these five different levels of tolerance for the training L, the optimal corresponds to a training L of 8.3×10^{-5}. As L is a function of all three force coefficients, it can be seen that actually the K_T and K_S are better predicted if a lower L tolerance is used, but the K_Q gets worse.

Figure 12 shows the validation force coefficients, the thrust, side force, and torque as a function of time. Each row correlates to a different training L, with $L = 0.001$ in the top row, and the smallest L in the bottom row. This figure illustrates how the prediction of validation forces vary as a function of the tolerance on the L used for training. A better correlation between the neural network forces and the CFD forces can be seen as the training tolerance is reduced. However, both Table 3 and Figure 12 show that as the tolerance of the training L is reduced, some forces may be better predicted in the validation set at different levels of tolerance. One way to mitigate this would be to develop a neural network specifically to determine each force. However, for the purposes of implementation as a body force in a CFD simulation, this would be less desirable due to the increased training cost.

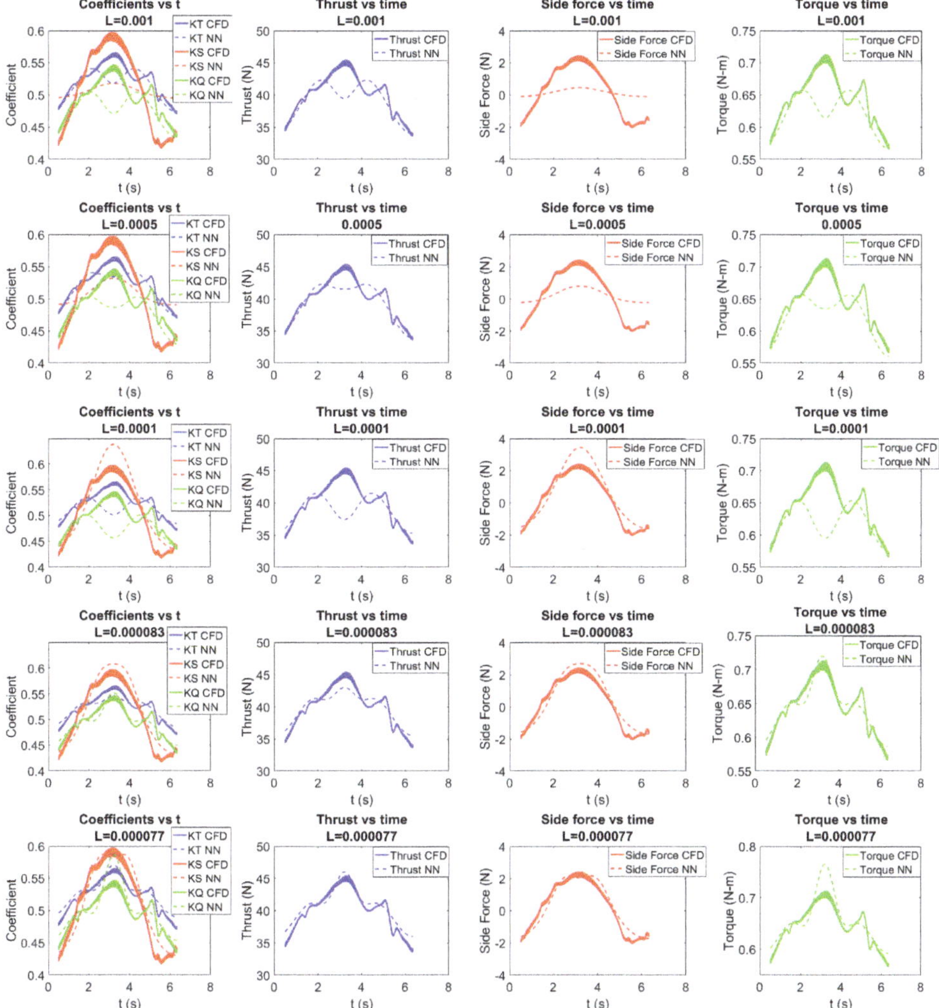

Figure 12. Coefficients, thrust, side force, and torque for behind condition neural network validation case. Top row: highest tolerance for L. Bottom row: lowest tolerance for L examined.

Table 3. L2 norm error for different tolerances of L for validation of neural network for behind condition forces.

Train L	L2 Error K_T	L2 Error K_S	L2 Error K_Q	Validation L
1.0×10^{-3}	2.8×10^{-2}	5.3×10^{-2}	1.0×10^{-1}	1.0×10^{-2}
5.0×10^{-4}	1.2×10^{-2}	4.7×10^{-2}	2.4×10^{-2}	1.5×10^{-3}
1.0×10^{-4}	2.4×10^{-2}	2.6×10^{-2}	3.4×10^{-2}	1.1×10^{-3}
8.3×10^{-5}	9.6×10^{-3}	1.9×10^{-2}	8.8×10^{-3}	2.58×10^{-4}
7.7×10^{-5}	8.3×10^{-3}	1.6×10^{-2}	1.7×10^{-2}	3.0×10^{-4}

As noted, the best tolerance found from this validation study is for $L = 8.3 \times 10^{-5}$. Figure 13 shows the coefficients predicted by the neural network compared to those extracted from the CFD for the behind condition training case. The plot on the left shows the neural network prediction of the unsteady training data. The plot on the right shows the neural network prediction of the steady training data. Thus, this setting of $L = 8.3 \times 10^{-5}$ leads to good results for both the training data as well as the validation test case and will be used to further test this neural network.

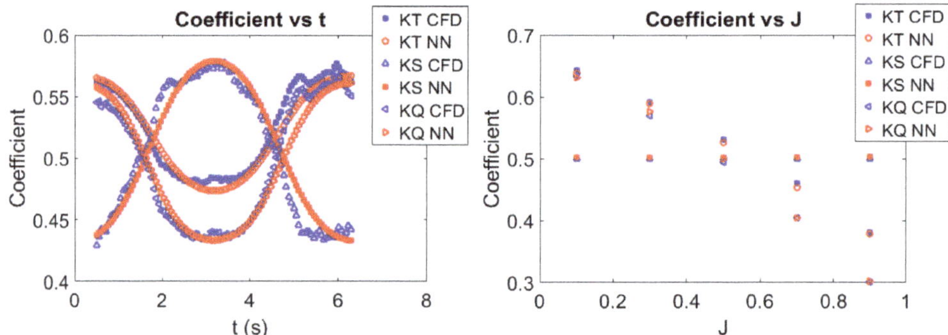

Figure 13. Coefficients for the behind condition training cases. Left: Unsteady training. Right: Steady training.

4.1.2. Training the Nonlinear Regression Model for the Unsteady Behind Condition Propeller Force

A study of how many features to incorporate into the regression model is performed. The validation case is used to verify the model is accurate and is used to select the number of features used to analyze the test cases. Four different forms of the regression are examined. The features examined in each of these include a second-order model, a third-order model, a fourth-order model, and a model using the neural network features. The feature vector for each of these is shown in Equation (24). Table 4 and Figure 14 are used to evaluate the different feature vectors for the regression. Table 4 shows the training L compared to the L2 error norms of the force coefficients and the L for the validation case. This table illustrates that even though the training loss function decreases as higher features are added, the model becomes overtrained and the accuracy of the regression model on the validation test case becomes worse. This phenomenon is shown graphically in Figure 14. The top row shows the force coefficients and the forces predicted by the second order regression, the second row shows the results for the third order regression, the third row shows the results for the fourth order regression, and the last row shows the results if the feature vector used by the neural network is used. As additional terms are added to the feature vector, the prediction of the validation test case become worse. Therefore, the second-order model is used to analyze the test cases. Note that Figure 14 also shows that even the second-order model does a poor job of predicting the thrust and torque for this validation case. More training data could be used to improve the model, but that would also increase the cost of implementing the model. Therefore, it can be seen that for this set of training data, the neural network is able to do a better job of predicting the validation test case than the regression approach.

$$\begin{aligned}
&Second\ Order : \vec{x} = [1, J, J^2, \alpha, \alpha^2, J\alpha]^T \\
&Third\ Order : \vec{x} = [1, J, J^2, J^3, \alpha, \alpha^2, \alpha^3, J\alpha, J\alpha^2, J^2\alpha]^T \\
&Fourth\ Order : \vec{x} = [1, J, J^2, J^3, J^4, \alpha, \alpha^2, \alpha^3, \alpha^4, J\alpha, J\alpha^2, J^2\alpha, J^2\alpha^2, J^3\alpha, J\alpha^3]^T \\
&Neural\ Network\ Features : \vec{x} = [1, J, J^2, \alpha, \ldots, \alpha^6, J\alpha, \ldots, J\alpha^6, J^2\alpha, \ldots, J^2\alpha^6]^T
\end{aligned} \tag{24}$$

Table 4. L2 norm error for different tolerances of L for validation of neural network for behind condition forces.

Regression Model	Train L	L2 Error K_T	L2 Error K_S	L2 Error K_Q	Validation L
Second Order	9.3×10^{-5}	5.0×10^{-2}	1.8×10^{-2}	5.1×10^{-2}	2.7×10^{-3}
Third Order	7.3×10^{-5}	2.7×10^{-2}	1.4×10^{-1}	3.9×10^{-2}	1.1×10^{-2}
Fourth Order	7.0×10^{-5}	9.9×10^{-2}	1.4×10^{-1}	6.4×10^{-2}	1.6×10^{-2}
Neural Network Features	6.9×10^{-5}	2.0×10^{-1}	7.3×10^{-1}	5.9×10^{-1}	4.6×10^{-1}

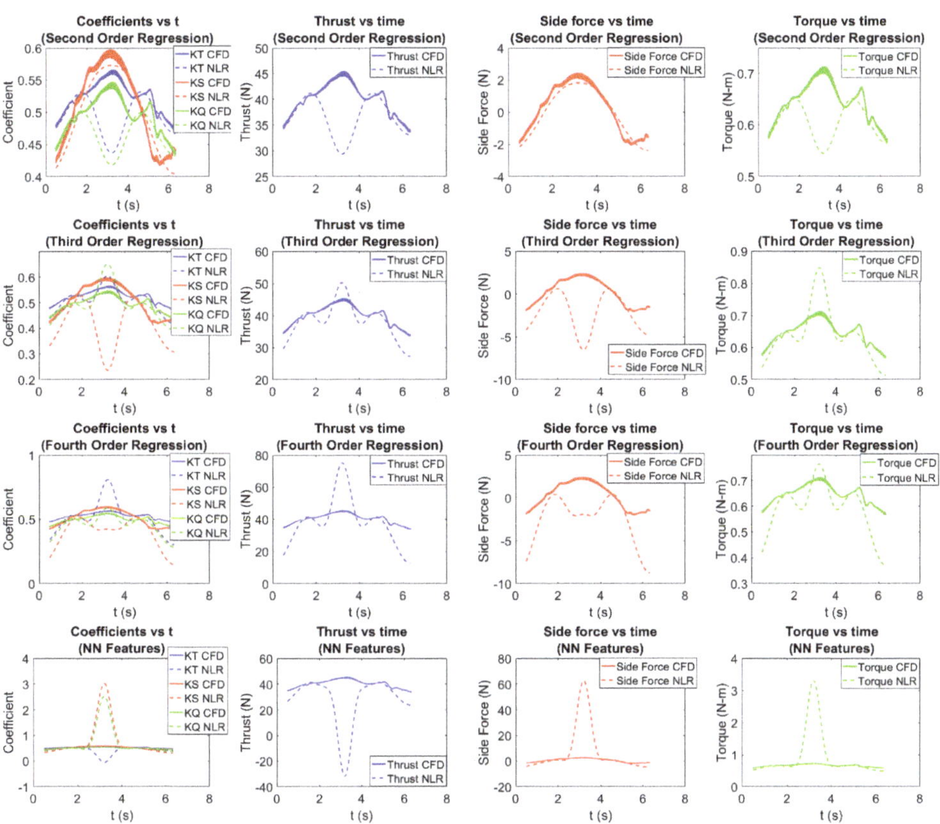

Figure 14. Coefficients, thrust, side force, and torque for behind condition nonlinear regression validation case. Top row: Lowest order regression. Bottom row: Highest order regression.

4.2. Testing the Neural Network for the Unsteady behind Condition Propeller Force

The behind condition neural network is tested against three different CFD simulations, each of which have different prescribed motions as shown by Table 2. Test 1 examines a lower amplitude sway motion than the training case with J held constant at 0.5. Test 2 examines a case where the oblique flow angle is held constant at 0 rad, but U, and thus J, vary with time. Test 3 examines a case in which both α and J vary in time. In each of these cases, the first 0.5 seconds are neglected to allow the flow to develop.

4.2.1. Behind Condition Test 1

The average forward velocity for Test 1 is $\overline{U}$ = 1.68 m/s, the average side velocity is $\overline{V}$ = 0 m/s, and the average propeller revolution rate is $\overline{n}$ = 32 rev/s. Only the sway degree of freedom is unsteady, with an amplitude of $A_2 = 0.11$ m and a frequency of oscillation of $\omega_{osc,2}$ = 1 rad/s. This correlates to a sway velocity amplitude of v' = 0.11 m/s. This case has a sway amplitude that is a quarter of the amplitude of the training case. The J is constant at 0.5.

Figure 15 shows the coefficients predicted by the neural network compared to those computed from the CFD for Test 1. The average loss function is 3.43×10^{-5}, the average L2 norm error of the K_T is 0.0032, the average L2 norm error of the K_S is 0.003, and the average L2 norm error of the K_Q is 0.0067. Figure 16 shows the thrust, side force, and torque as a function of time. As the error in the coefficients is low, the predictions of the force by the neural network are also close to that of the CFD. This demonstrates that the neural network can correctly predict the forces when the amplitude of sway is less than it was trained with.

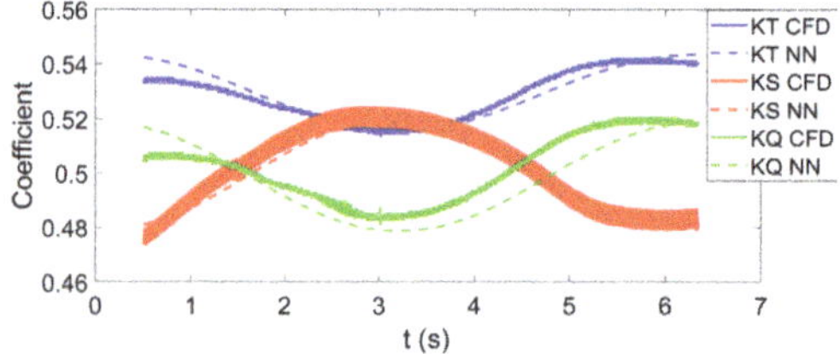

Figure 15. Behind condition Test 1 coefficients as a function of time predicted by neural network.

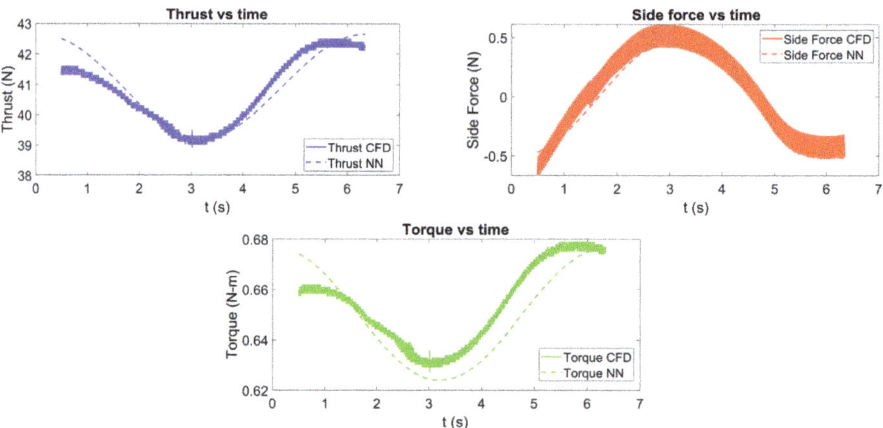

Figure 16. Behind condition neural network predictions of thrust, side force, and torque for Test 1 as a function of time.

Figure 17 shows the coefficients predicted by the nonlinear regression compared to those computed from the CFD for Test 1. The average loss function is 3.43×10^{-5}, the average L2 norm error of the K_T is 0.003, the average L2 norm error of the K_S is 0.004, and the average L2 norm error of the K_Q is 0.0067. Figure 18 shows the thrust, side force, and torque as a function of time for the nonlinear regression predictions. These results are very similar and nearly indistinguishable compared to those predicted by the neural network. Therefore, both the nonlinear regression and the neural network do a good job of predicting the forces for this case.

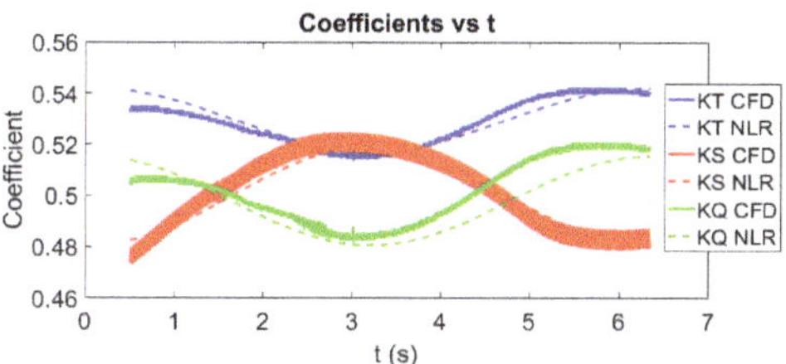

Figure 17. Behind condition Test 1 coefficients as a function of time predicted by nonlinear regression.

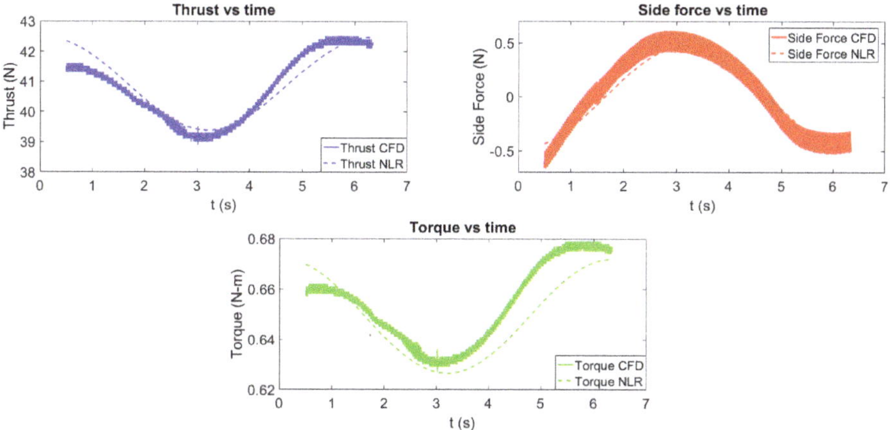

Figure 18. Behind condition nonlinear regression predictions of thrust, side force, and torque for Test 1 as a function of time.

4.2.2. Behind Condition Test 2

The average forward velocity for Test 2 is reduced to $\overline{U} = 1.1$ m/s, the average side velocity is $\overline{V} = 0$ m/s, and the average propeller revolution rate is $\overline{n} = 20.95$ rev/s. Only the surge degree of freedom is unsteady, with an amplitude of $A_1 = 0.0487$ m and a frequency of oscillation of $\omega_{osc,1} = 6.72$ rad/s. This frequency of oscillation was used in [9,17], and it represents the frequency of encounter if the vessel were to be in head seas with the wavelength equal to the length of the waterline. The α is constant at 0 rad. J as a function of time is shown in Figure 19.

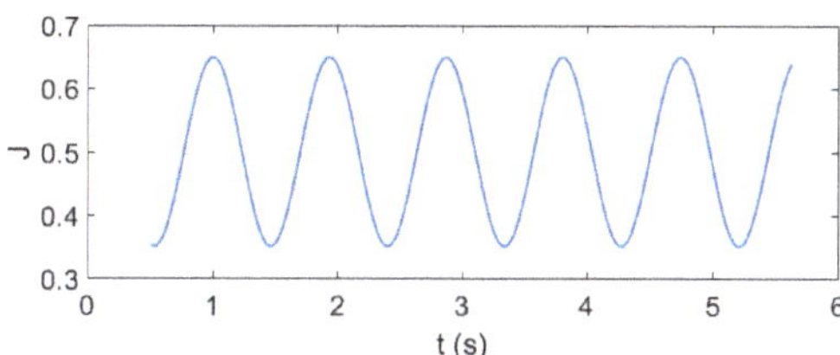

Figure 19. J as a function of time for behind condition Test 2.

The top left of Figure 20 shows the comparison between the neural network coefficient predictions and the CFD predictions. The average loss function is 1.18×10^{-4}, the average L2 norm error of the K_T is 7.25×10^{-3}, the average L2 norm error of the K_S is 4.58×10^{-3}, and the average L2 norm error of the K_Q is 1.27×10^{-2}. The oscillation in side force is caused by the azimuthal position of the propeller and the resulting force is around zero. The error for K_T and K_Q is higher than Test 1, but still quite accurate.

The top right and bottom plots of Figure 20 show the thrust and torque respectively as a function of time. This demonstrates that the neural network can predict the forces well when J oscillates.

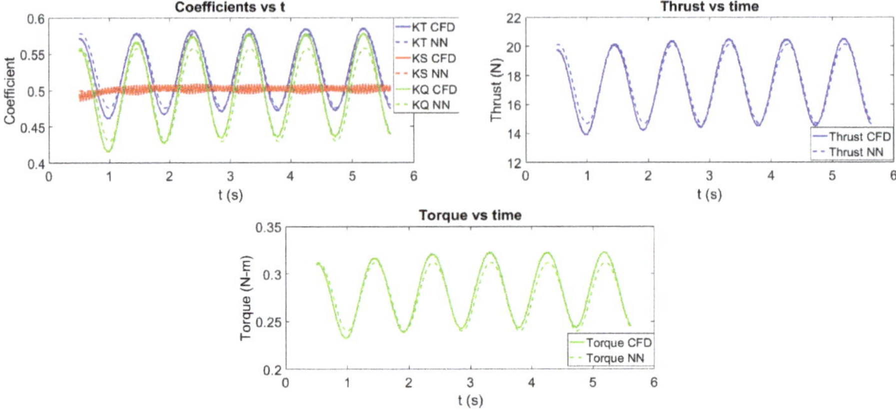

Figure 20. Behind condition coefficients, thrust, and torque for Test 2 as a function of time predicted by the neural network.

The top left of Figure 21 shows the comparison between the nonlinear regression coefficient predictions and the CFD predictions. The average loss function is 1.18×10^{-4}, the average L2 norm error of the K_T is 8.25×10^{-3}, the average L2 norm error of the K_S is 4.56×10^{-3}, and the average L2 norm error of the K_Q is 1.38×10^{-2}. Thus, the forces are predicted quite similarly to the forces predicted by the neural network. The top right and bottom plots of Figure 21 show the thrust and torque respectively as a function of time.

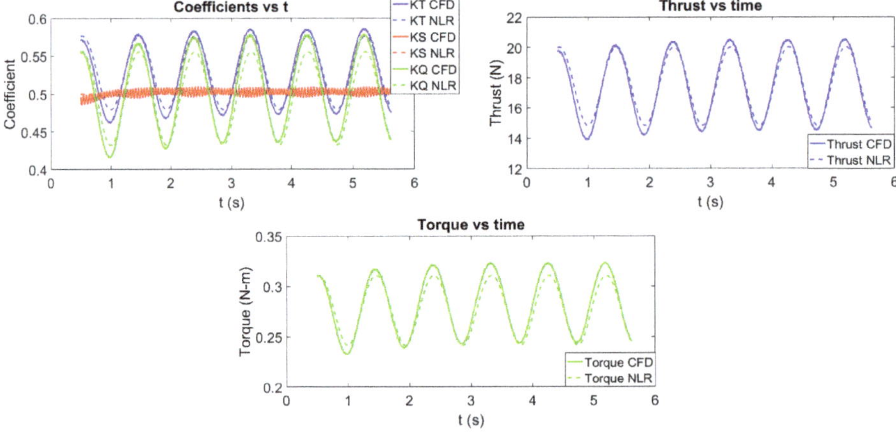

Figure 21. Behind condition coefficients, thrust, and torque for Test 2 as a function of time predicted by nonlinear regression.

4.2.3. Behind Condition Test 3

Test three has both unsteady forward velocity as well as unsteady sway velocity. The average forward velocity is $\overline{U} = 1.68$ m/s, the average side velocity is $\overline{V} = 0$ m/s, and the average propeller revolution rate is $\overline{n} = 32$ rev/s. Frequency of oscillation for both surge and sway is $\omega_{osc,1} = \omega_{osc,2} = 1$ rad/s. The amplitude of unsteady motion is $A_1 = 0.8$ m and $A_2 = -0.5$ m. Thus, the sway amplitude is larger than the training case and the sway direction is opposite of the Validation case. Figure 22 shows how J and α vary with time for Test 3.

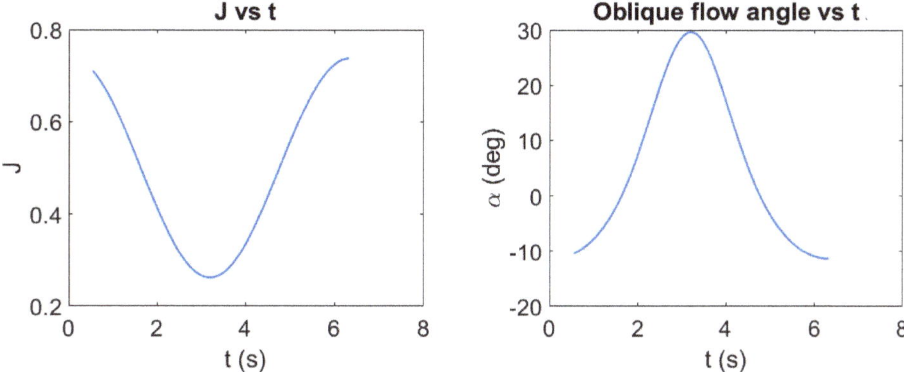

Figure 22. J (left) and α (right) as a function of time for Test 3.

The top left of Figure 23 shows the comparison between the neural network force coefficient predictions and the CFD predictions. The average loss function is 7.30×10^{-4}, the average L2 norm error of the K_T is 1.52×10^{-2}, the average L2 norm error of the K_S is 2.76×10^{-2}, and the average L2 norm error of the K_Q is 2.16×10^{-2}. The top right, bottom left, and bottom right plots of Figure 23 show the thrust, side force, and torque, respectively, as a function of time predicted by the neural network. Overall, good agreement is seen between the CFD and the neural network prediction. The largest error occurs in the prediction of the side force between three and four seconds when the J is low and the oblique flow angle is large.

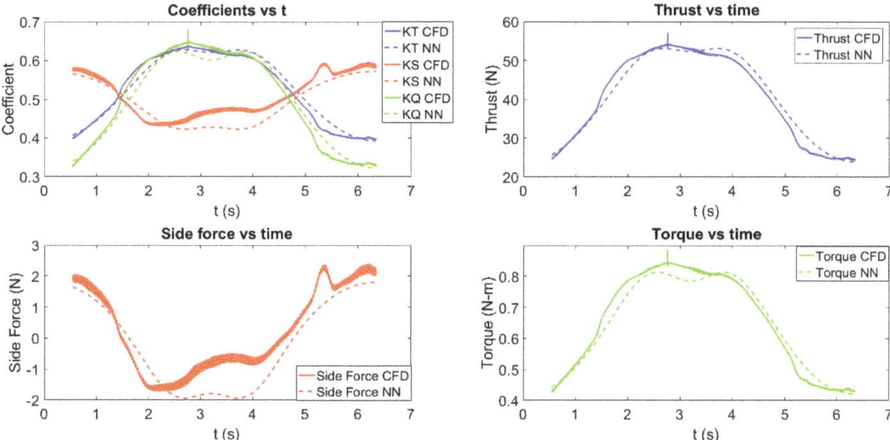

Figure 23. Behind condition coefficients, thrust, and torque for Test 3 as a function of time predicted by the neural network.

The top left of Figure 24 shows the comparison between the nonlinear regression force coefficient predictions and the CFD predictions. The average loss function is 2.66×10^{-3}, the average L2 norm error of the K_T is 4.32×10^{-2}, the average L2 norm error of the K_S is 1.72×10^{-2}, and the average L2 norm error of the K_Q is 5.63×10^{-2}. Thus, the forces predicted by the nonlinear regression method are not as accurate as those predicted by the neural network in terms of the L, K_T, and K_Q. However, the regression model does outperform the K_S neural network prediction. The top right, bottom left, and bottom right plots of Figure 24 show the thrust, side force, and torque, respectively, as a function of time.

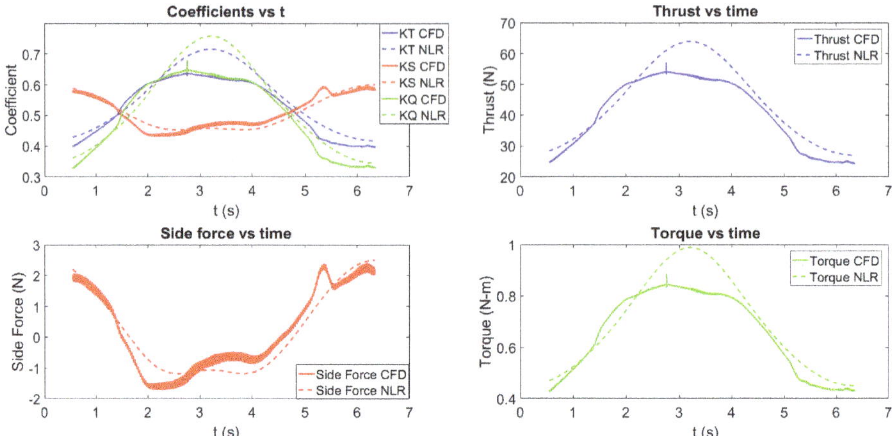

Figure 24. Behind condition coefficients, thrust, and torque for Test 3 as a function of time predicted by nonlinear regression.

5. Conclusions

A neural network and a nonlinear regression model have been trained and tested to determine the unsteady propeller forces due to unsteady motions. The first neural network examined is for open water and it takes a feature vector comprised of various permutations of the instantaneous J and $|\alpha|$ for each instant in time. A second-order nonlinear regression model was also trained using the instantaneous J and $|\alpha|$ for each instant in time. The models are extended to the behind condition. The neural network for the behind condition used a similar feature vector, but with the difference of using α instead of $|\alpha|$. An early stopping procedure is used for the neural network to avoid overtraining. Different feature vectors are examined for the behind condition nonlinear regression, but the second order regression model is shown to be more accurate than higher order regression models that overfit the training data. Both the neural network and the regression approaches are able to predict the unsteady propeller forces for unsteady motions, but the neural network is more accurate for the cases examined. This study is a demonstration of how a neural network and nonlinear regression model can be trained and tested to predict the correct propeller forces for unsteady motion.

The behind condition models could be further improved by incorporating other features in the input. For example, the feature vector could be expanded to incorporate history terms to account for memory effects in the flow. Furthermore, probes could be placed upstream of the propeller plane and the effects of shed vortices from the hull or incident velocities could be accounted for.

The regression model could be improved by supplying additional training to the model. Latin hyper cube sampling could be used to generate training data to cover the full J-α space. However, additional training would clearly lead to increased training cost. Thus, for the cases examined, as the neural network tends to be more accurate than the nonlinear regression, it may be preferable since it can be more accurate than the regression model with limited training.

The models, especially the behind condition neural network, could be very useful for analyzing a maneuvering vessel. Instead of having to discretize the propeller and use a rotating mesh to simulate the viscous effects on the propeller, the neural network or regression model could be used. This could dramatically reduce the cost of a maneuvering calculation using CFD. Furthermore, the models could be trained with any type of training data whether it be experiments, potential flow calculations, RANS

CFD, LES, or any combination of these. For example, LES could be used to train the neural network in off design circumstances and potential flow methods could be used to train the on design points.

Author Contributions: B.K. performed the analysis and wrote the paper. K.M. provided guidance for the paper and edited the paper. All authors have read and agreed to the published version of the manuscript.

Funding: This research was funded by the Office of Naval Research Grant N00014-16-1-2969.

Conflicts of Interest: The authors declare no conflicts of interest.

Abbreviations

The following abbreviations are used in this manuscript:

CFD	Computational Fluid Dynamics
DES	Detached Eddy Simulation
LES	Large Eddy Simulation
NLR	Nonlinear Regression
NN	Neural Network
RANS	Reynolds Averaged Navier–Stokes

References

1. Carrica, P.M.; Sadat-Hosseini, H.; Stern, F. CFD analysis of broaching for a model surface combatant with explicit simulation of moving rudders and rotating propellers. *Comput. Fluids* **2012**, *53*, 117–132. [CrossRef]
2. Mizzi, K.; Demirel, Y.K.; Banks, C.; Turan, O.; Kaklis, P.; Atlar, M. Design optimisation of propeller boss cap fins for enhanced propeller performance. *Appl. Ocean Res.* **2017**, *62*, 210–222. [CrossRef]
3. Wang, J.; Zou, L.; Wan, D. Numerical simulations of zigzag maneuver of free runnig ship in waves by RANS-Overset grid method. *Ocean Eng.* **2018**, *162*, 55–79. [CrossRef]
4. Shen, Z.; Wan, D.; Carrica, P.M. Dynamic overset grids in OpenFOAM with application to KCS self-propulsion and maneuvering. *Ocean Eng.* **2015**, *108*, 287–306. [CrossRef]
5. Ortolani, F.; Dubbioso, G.; Muscari, R.; Mauro, S.; Mascio, A.D. Experimental and Numerical Investigation of Propeller Loads in Off-Design Conditions. *J. Mar. Sci. Eng.* **2018**, *6*, 45. [CrossRef]
6. Mousaviraad, S. CFD Prediction of Ship Response to Extreme Wind and/or Waves. Ph.D. Thesis, University of Iowa, Iowa City, IA, USA, 2010.
7. Yao, J. On the Propeller Effect When Predicting Hydrodynamic Forces for Manoeuvring Using RANS Simulations of Captive Model Tests. Ph.D. Thesis, Technical University of Berlin, Berlin, Germany, 2015.
8. Gaggero, S.; Dubbioso, G.; Villa, D.; Muscari, R.; Viviani, M. Propeller modeling approaches for off-design operative conditions. *Ocean Eng.* **2019**, *178*, 283–305. [CrossRef]
9. Knight, B.; Maki, K. A Semi-Empirical Multi-Degree of Freedom Body Force Propeller Model. *Ocean Eng.* **2019**, *178*, 270–282. [CrossRef]
10. White, P.F.; Knight, B.G.; Filip, G.P.; Maki, K.J. Numerical Simulations of the Duisburg Test Case Hull Maneuvering in Waves. In *SNAME Maritime Convention*; The Society of Naval Architects and Marine Engineers: Tacoma, WA, USA, 2019.
11. Broglia, R.; Dubbioso, G.; Durante, D.; Mascio, A.D. Simulation of turning circle by CFD: Analysis of different propeller models and their effect on manoevring prediction. *Appl. Ocean Res.* **2013**, *39*, 1–10. [CrossRef]
12. Phillips, A.; Turnock, S.; Furlong. Evaluation of manoeuvring coefficient of a self-propelled ship using blade element momentum propeller model coupled to a Reynolds averaged Navier–Stokes flow solver. *Ocean Eng.* **2009**, *36*, 1217–1225. [CrossRef]
13. Winden, B. Powering Performance of a Self Propelled Ship in Waves. Ph.D. Thesis, University of Southampton, Southamptom, UK, 2014.
14. Hornik, K.; Stinchcombe, M.; White, H. Multilayer feedforward networks are universal approximators. *Neural Netw.* **1989**, *2*, 359–366. [CrossRef]
15. Singh, A.P.; Medida, S.; Duraisamy, K. Machine-Learning-Augmented Predictive Modeling of Turbulent Separated Flows over Airfoils. *AIAA J.* **2017**, *55*, 2215–2227. [CrossRef]

16. Abramowski, T. Prediction of Propeller Forces During Ship Maneuvering. *J. Theor. Appl. Mech.* **2005**, *43*, 157–178.
17. Knight, B.; Maki, K. Body Force Propeller Model for Unsteady Surge Motion. In Proceedings of the ASME 37th International Conference on Ocean Offshore and Arctic Engineering (OMAE), Madrid, Spain, 17–22 June 2018.
18. Knight, B.; Maki, K. A Data-Driven Multi-Degree of Freedom Body Force Propeller Model for Maneuvering. In Proceedings of the Sixth International Conference on Marine Propulsors (SMP 19), Rome, Italy, 26–30 May 2019.
19. ITTC. ITTC-Recommended Procedures and Guidelines. 1978 ITTC Performance Prediction Method. International Towing Tank Conference, 2011, 7.5-02, 03-01.4, pp. 1–9. Available online: https://ittc.info/media/8372/index.pdf (accessed on 2 February 2020).
20. Seber, G.A.F.; Wild, C.J. *Nonlinear Regression*; Wiley-Interscience: Hoboken, NJ, USA, 2003.

Journal of
Marine Science and Engineering

Article

Prediction of Unsteady Developed Tip Vortex Cavitation and Its Effect on the Induced Hull Pressures

Seungnam Kim * and Spyros A. Kinnas

Ocean Engineering Group, CAEE, The University of Texas at Austin, Austin, TX 78712, USA; kinnas@mail.utexas.edu

* Correspondence: naoestar@utexas.edu (S.K.); Tel.: +1-512-751-8829

Received: 20 January 2020; Accepted: 6 February 2020; Published: 13 February 2020

Abstract: Reducing the on-board noise and fluctuating pressures on the ship hull has been challenging and represent added value research tasks in the maritime industry. Among the possible sources for the unpalatable vibrations on the hull, propeller-induced pressures have been one of the main causes due to the inherent rotational motion of propeller and its proximity to the hull. In previous work, a boundary element method, which solves for the diffraction potentials on the ship hull due to the propeller, has been used to determine the propeller induced hull pressures. The flow around the propeller was evaluated via a panel method which solves in time for the propeller loading, trailing wake, and the sheet cavities. In this article, the propeller panel method is extended so that it also solves for the shape of developed tip vortex cavities, the effects of which are also included in the evaluation of the hull pressures. The employed unsteady wake alignment scheme is first applied, in the absence of cavitation, to investigate the propeller performance in non-axisymmetric inflow, such as the inclined-shaft flow or the flow behind an upstream body. In the latter case, the propeller panel method is coupled with a Reynolds-Averaged Navier–Stokes (RANS) solver to determine the effective wake at the propeller plane. The results, including the propeller induced hull pressures, are compared with those measured in the experiments as well as with those from RANS, where the propeller is also simulated as a solid boundary. Then the methods are applied in the cases where partial cavities and developed tip vortex cavities coexist. The predicted cavity patterns, the developed tip vortex trajectories, and the propeller-induced hull pressures are compared with those measured in the experiments.

Keywords: developed tip vortex cavity; propeller-induced hull pressures; hull-propeller interaction; boundary element method; unsteady wake alignment model

---

1. Introduction

From the time when people used propellers as a prime propulsion system for maneuvering ships, reducing the propeller-induced noise and the fluctuating hull pressures has been challenging and represents added value research. Among possible causes for the noise and vibrations, the propeller-induced hull pressures have been one of the main causes, and some efforts have been made to predict and reduce the nuisance through the numerical or experimental approaches [1–3]. Among those, Hallander & Lars (2013) [2] performed the experiments with a model-scale ship with pressure transducers mounted on the hull above the propeller plane to record the propeller-induced pressure signals and noise levels. Johan (2018) [4,5] introduced prediction methods for broadband noise, including the hull pressure fluctuations through developed tip vortex cavitation. Fundamental aspects of a cavitating vortex were investigated to understand the basic mechanisms of broadband noise through theoretical and computational studies of the kinematics, dynamics, and acoustics of tip vortex cavitation.

As for a numerical approach, Boundary Element Method (BEM) has been an alternative tool to predict the propeller-induced hull pressures, as it significantly reduces the computing time and omits the complexity of 3D meshing, compared to other commercial Raynolds-Averaged Navier–Stokes (RANS) solvers. Göttsche et al. (2019) [6] dealt with the prediction of the underwater noise induced by ship propellers with/without sheet cavitation on the blade. The seabed and the free surface effects are also considered in the calculations. The calculation was conducted using a hybrid method combining BEM with the Ffowcs–Williams–Hawkings equation. The numeric results and their comparisons with the full-scale measurements of the research vessel show in general good agreement with existing limitations due to the environmental aspects. Another test program was proposed by Tani et al. (2019) [7] as a means to better understand the accuracy and reliability of underwater radiated noise measurement by a round robin test for open water propellers. It was done to overcome limitations on predicting underwater radiated noise via model tests which contain uncertainty sources, thus requiring suitable postprocessing techniques. Gosda et al. (2018) [8] studied uncertainties due to Reynolds number scale effects when extrapolating model propeller to geometrically similar full-scale ship, considering that most assessments of the propeller's cavitation behavior is conducted via model tests. The scaling effects on tip vortex cavitation was investigated using an axisymmetric viscous line vortex. As results, isolated vortex cavities within the viscous core of the tip vortex effectively dampen the alternations of the cavity radius in both the amplitude and frequency as the core radius increases.

The BEM applications in this article focus on the effect of cavities on the hull pressures, as the cavity arising on or near the propeller surface is the main contributor to the high-level noise and vibrations. For cavitating propellers, the cavity source is the major excitation in the acoustic BEM model, so the influence of the propeller lifting force, blade thickness, and the blade trailing wake can be neglected. For a wetted propeller, however, the influence of propeller and its wake becomes dominant contribution to the propeller-induced hull excitation. In the previous works presented in SMP'17 by Su et al. (2017) [9] and in CAV'18 by Kim et al. (2018) [10], fluctuating hull pressures are predicted numerically using the acoustic BEM (HULLFPP), coupled with hydrodynamic BEM (PROPCAV [11]) in wetted and partially cavitating conditions, repectively. A sequence of numerical methods was applied to predict the propller-induced hull pressures given the hull/rudder/propeller geometries with emphasis on the unsteady wake alignment model in periodic non-axisymmetric inflow. The effects of tip vortex cavitation were considered later in the work presented in SMP'19 by Kim and Kinnas (2019) [12]. In the work, tip vortex radius was kept constant over the streamwise length, so the hull excitation with different tip vortex radii were mainly discussed. The fluctuating hull pressures start to capture a high-frequency as the propeller rotates with thicker vortex radius as wide as 4.0% of the blade diameter. Although omitting inception stage of the tip vortex cavitation is far from its physical development, as observed in the experiment by Hallander and Gustafsson (2013) [2], the predicted pressure signals followed overall trends of the experimental measurements at the eight different transducers although quantitative comparisons still maintain evident deviations from the measurements. Due to the discrete singularity arrangement used to represent the tip vortex cavitation at the wake tip, induced hull pressures are sensitive to the distance from the tip cavity. In the inclined shaft flow, for example, tip vortex cavitation has relatively significant effects on the hull pressures downstream because of the inclination in the wake, and is expected to be affected less when the uniform inflow is assumed, as it imposes relatively large distance between the hull surface and the tip vortex cavitation.

Depending on the time dependency of the problem, either steady or unsteady wake alignment scheme can be used in the hydrodynamic BEM. As for the testing cases, uniform inflow is assumed on a propeller with the two different alignment schemes, and the results were in good agreement for open propellers (Kim 2017) [13]. For steady problems, simplified wake alignment scheme was proposed by Greely and Kerwin (1982) [14], which limits geometrical variations of the streamlines. This wake shows efficient computing time and reasonable results at the low loading conditions but over predicts the propeller forces when high loading is imposed at the lower J_S, especially for ducted propellers as shown by Kinnas et al. (2015) [15]. To investigate the unsteady hull pressures induced by propeller and developed tip vortex cavitation in non-axisymmetric inflow, an unsteady wake alignment scheme is used in this article, and a sequence of numerical methods is applied as follows.

- For the effective wake prediction, BEM/Reynolds-Averaged Navier–Stokes (RANS) interactive scheme (PC2NS) is used. PC2NS is the interface that couples the BEM with RANS solver in an iterative manner for the prediction of the effective wake on the propeller plane [16].
- Hydrodynamic BEM is applied to open propellers subject to the periodic non-axisymmetric inflow. The unsteady wake alignment scheme is employed to allow the physical variation of the trailing wake in space and time.
- BEM models tip vortex cavitation at the tip of fully aligned wake. Trajectory of the tip vortex cavitation is assumed to be following the wake panels from wetted runs. This assumption is valid only for the limited amount of blade sheet cavity as large cavity volume affects the overall blade loading.
- The unsteady cavity problem, including both the sheet cavitation on the blade and developed tip vortex cavitation, is solved for the singularities (discrete vortex and source strength) on the wetted propeller surfaces and on the surfaces beneath the cavity.
- The induced potentials on a flat barge or ship hull due to the propeller, its wake, sheet cavity, and developed tip vortex cavity are determined by solving Green's 3rd identity in the acoustic BEM solver.
- Finally, the pressure BEM solver is implemented to predict the diffraction potentials which will lead to the solid boundary factor to estimate the fluctuating hull pressures.

Figure 1 shows the flow chart summarizing the above sequence.

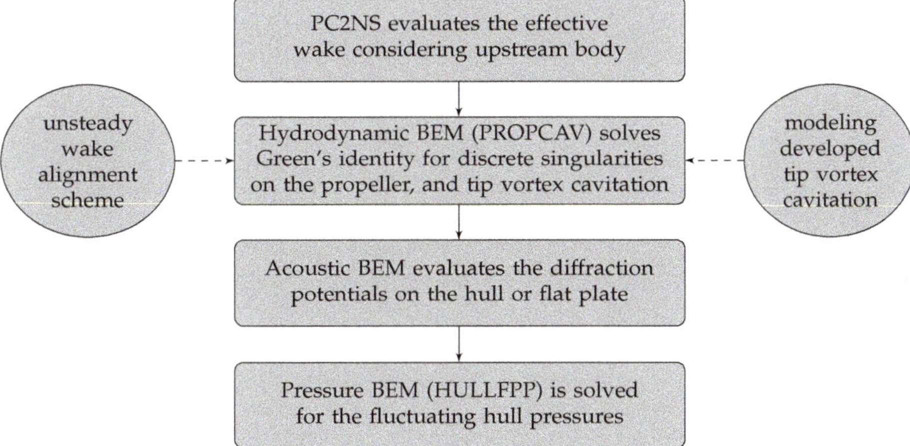

Figure 1. Flowchart of the numerical methods applied in a series to solve the unsteady hull pressure problems.

The predicted hull pressure signals from the present methods are compared to the experimental measurements to validate the numerical predictions. Qualitative comparisons of the sheet/tip vortex cavity patterns from the BEM are made with the observations in cavitation tunnel. The results from RANS simulations are also presented to validate the numerical predictions for the trailing wake, unsteady propeller forces, and the propeller-induced hull pressures. Before discussing the results, detailed descriptions on the unsteady wake alignment scheme and the tip vortex cavitation model are presented in the following sections.

2. Methodology

2.1. The Unsteady Wake Alignment Scheme

The unsteady wake alignment scheme was first introduced by Lee (2004) [17] and has been improved by Kim (2017) [13], such that the alignment scheme allows the variation of wake panels in non-periodic inflow conditions. Su et al. (2018) [18] first employed this alignment scheme in calculating the hull pressures, and it was shown that the unsteady wake behaved better when predicting the downstream hull pressures in fully-wetted case. In this article, the alignment algorithm is coupled with the Full Wake Alignment (FWA) scheme by Tian & Kinnas (2012) [19] and Kim et al. (2018) [20], which models the trailing vortices as material lines, following the local stream in steady state. FWA will be implemented on the initial helical wake at $t = 0$ time step, and its results will be used as inputs for the unsteady scheme at $t > 0$, during which the wake panels are updated at each time step as the propeller completes several revolutions.

2.1.1. Steady Regime ($t = 0$)

Alignment starts with the FWA algorithm (Equation (1)) in steady state using the 0th harmonic from the effective wake. Circumferential variation in the effective wake field is given in terms of the harmonic coefficients in the BEM. FWA updates the initial guess of helical wake with the dominant harmonic to prevent abrupt changes in the wake panels when proceeding to the unsteady alignment procedure $(t > 0)$ from the steady state $(t = 0)$. It improves the overall alignment stability.

$$\vec{X}_i^{N+1} = \vec{X}_i^N + \epsilon \left[\left(\vec{U}_{in,i} + \vec{u}_i \right) \Delta t - \frac{\Delta t}{\Delta t_i^*} \left(\vec{X}_i^N + \vec{X}_{i-1}^N \right) \right], \tag{1}$$

where $\vec{X}_i^{N+1}$ denotes the coordinate of the aligned wake panel i at $N+1$ iteration step. ϵ is an under-relaxation factor, which has different value depending on the given loading conditions. At conditions close to the design advance ratio, for example, ϵ is taken to be equal to 0.5, whereas at low-advance ratios, it is equal to 0.25. Δt is the time step size of the shedding vortex determined by the angular velocity (radian per unit second) of propeller and the angular increment of the discretized wake sheet, $\Delta\theta$.

$$\Delta t = \frac{\Delta\theta}{\omega} = \frac{\Delta\theta}{2\pi n} \tag{2}$$

Δt_i^* is the adjusted time step size, which is introduced to maintain the length of wake panel segment. Detailed description on the adjusted time step size is presented in Kim et el. (2018) [10]. $\vec{U}_{in,i}$ is the inflow velocity at the panel node i and $\vec{u}_i$ is the induced velocity which considers the effects of blade, its wake, hub, and both the sheet and tip vortex cavitation (if detected). The induced velocity in wake is obtained by solving Equation (3), which is Green's 3rd identity.

$$\begin{aligned} \vec{u}_i = & \frac{1}{4\pi} \iint_{S_B \cup S_T} \left[\phi_q \nabla \frac{\partial}{\partial n_q} \left(\frac{1}{R(p;q)} \right) - \frac{\partial \phi_q}{\partial n_q} \nabla \left(\frac{1}{R(p;q)} \right) \right] dS \\ & + \frac{1}{4\pi} \iint_{S_W} \Delta\phi_w(r_q) \nabla \frac{\partial}{\partial n_q} \left(\frac{1}{R(p;q)} \right) dS, \end{aligned} \tag{3}$$

where ϕ_q denotes the potentials at the variant point q, which is the distance $R(p;q)$ apart from the field point p. n_q is the unit normal vector at point q pointing out of the propeller surface S_B or tip vortex cavitation surface S_T, and $\Delta\phi_w$ denotes the potential jump across the wake surface S_W. The induced velocity from the tip vortex cavitation brings local effects on the wake panels, and the effects become significant as the wake panels move closer to the cavitation, as discussed in the result section.

BEM is solved for the solution potentials through the several iterations with the updated wake, and this process is repeated until the changes in the wake geometries fall within a certain tolerance, less than 1.0×10^{-4}.

2.1.2. Unsteady Regime ($t > 0$)

When $t > 0$, alignment procedure enters the unsteady regime where the time variation of the incoming flow is considered in evaluating the inflow velocity. Euler-explicit scheme is used with consideration of the full harmonics (up to 16th in the present method), as shown in Equation (4). The key wake is aligned and saved at the current time step to be used later when the following blade is located at the saved location. Different from the steady algorithm, the wake strength $\Delta\phi_w$ also needs to be saved at the current time step, as it convects along the wake surface. It advances to the downstream with time step Δt, as shown in Figure 2.

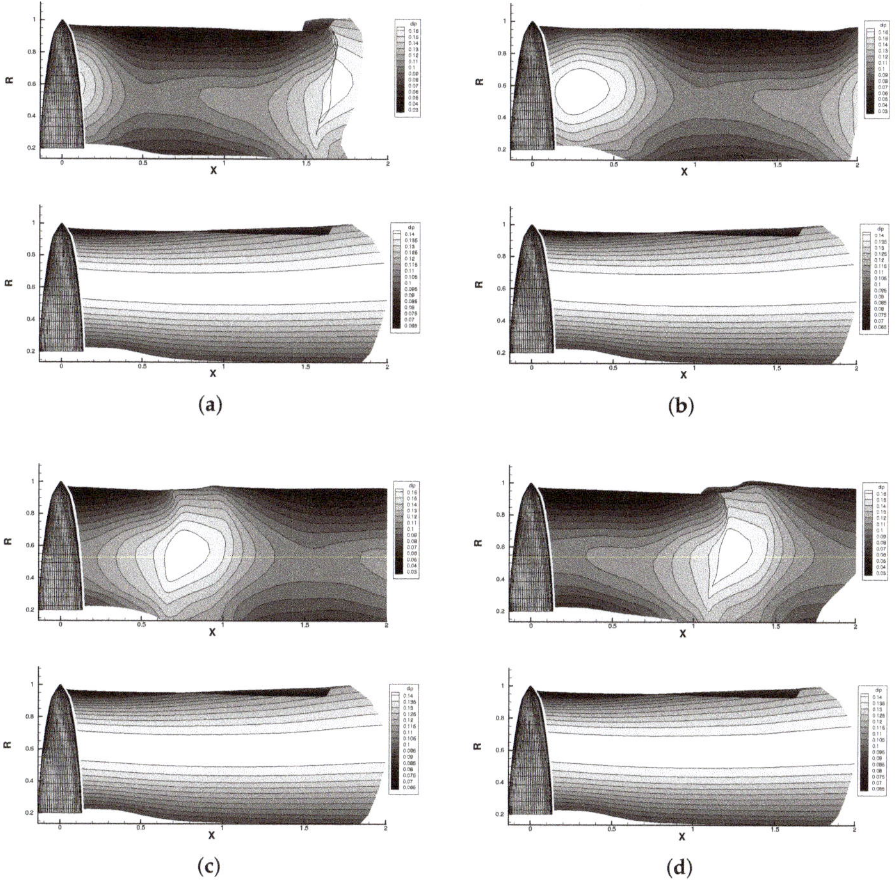

Figure 2. Convection of the wake strength $\Delta\phi_w$ along the wake surface, plotted on the x–r plane at the blade angle (**a**) $\theta = 0°$, (**b**) $\theta = 90°$, (**c**) $\theta = 180°$, and (**d**) $\theta = 270°$. Nonuniform wake field is assumed in the upper figure of every pair, and uniform inflow in the lower figure. Only the key blade and its wake are presented.

$$\vec{X}_{i+1}^{n+1} = \begin{Bmatrix} X_{i+1}^{n+1} \\ Y_{i+1}^{n+1} \\ Z_{i+1}^{n+1} \end{Bmatrix} = \begin{Bmatrix} X_i^n + \frac{1}{2}\left(U_{T,X_i}^n + U_{T,X_{i+1}}^n\right)\Delta t \\ Y_i^n + \frac{1}{2}\left(U_{T,Y_i}^n + U_{T,Y_{i+1}}^n\right)\Delta t \\ Z_i^n + \frac{1}{2}\left(U_{T,Z_i}^n + U_{T,Z_{i+1}}^n\right)\Delta t \end{Bmatrix}, \quad (4)$$

$$\vec{U}_{T,i}^n = \begin{Bmatrix} U_{T,X_i}^n \\ U_{T,Y_i}^n \\ U_{T,Z_i}^n \end{Bmatrix} = \begin{Bmatrix} U_{in,X_i} + u_{X_i} \\ U_{in,Y_i} + u_{Y_i} \\ U_{in,Z_i} + u_{Z_i} \end{Bmatrix}, \quad (5)$$

where i and n denote the ith nodal point at the nth time step. $\vec{U}_T$ is the total velocity in combination of the induced velocity $\vec{u}_i$ and the incoming flow $\vec{U}_{in,i}$. X, Y, and Z are the coordinates of the wake nodal points in the Cartesian coordinate system.

As this step aligns the wake in a progressive manner, calculations need to go over several revolutions until the predicted propeller thrust and torque converge. Given one revolution normally consists of 60 time steps ($\Delta\theta = 6°$) for reasonable predictions, applying FWA to unsteady problem might not be an appropriate option because of its multiple inner loops for each time step. Equation (4) predicts the aligned wake in good convergence when the time step Δt is small enough. With larger time step, however, it aligns the wake as expanding in radial direction far downstream. This is because Equation (4) aligns the wake along the straight line in the global X, Y, and Z axes, neglecting the segment of the actual curve that the shedding vortex should follow. Considering the dominant swirl component in the total velocity belongs to the propeller rotation, wake panels are first rotated by $\Delta\theta$ along the circular path with radius $R_i^n = \sqrt{(Y_i^n)^2 + (Z_i^n)^2}$, and the final coordinates are determined by the sum of the induced velocity from propeller, its wake, hub, cavities, and inflow velocity, as shown in Equation (6).

$$\vec{X}_{i+1}^{n+1} = \begin{Bmatrix} X_{i+1}^{n+1} \\ Y_{i+1}^{n+1} \\ Z_{i+1}^{n+1} \end{Bmatrix} = \begin{Bmatrix} X_i^n + \frac{1}{2}\left(U_{T,X_i}^n + U_{T,X_{i+1}}^n\right)\Delta t \\ R_i^n \cos\left(\theta_i^n + \Delta\theta\right) + \frac{1}{2}\left(U_{T,Y_i}^{n*} + U_{T,Y_{i+1}}^{n*}\right)\Delta t \\ R_i^n \sin\left(\theta_i^n + \Delta\theta\right) + \frac{1}{2}\left(U_{T,Z_i}^{n*} + U_{T,Z_{i+1}}^{n*}\right)\Delta t \end{Bmatrix}, \quad (6)$$

where $\theta_i^n = \tan^{-1}\left(Z_i^n / Y_i^n\right)$, U_{T,Y_i}^{n*}, and U_{T,Z_i}^{n*} are the total velocity components without the propeller angular velocity in Y and Z directions at the ith panel node.

2.1.3. Fully Unsteady Regime (Last Two Revolutions)

During the last two revolutions, wake alignment is no longer implemented, but the saved wake geometry and the wake strength $\Delta\phi_w$ from the previous revolutions are used based on the time step location of the key blade. Propeller performance normally converges after completing up to 5 or 6 revolutions in most unsteady problems. The BEM model for propeller blade, its wake, hub, tip vortex cavitation, and the indices for the wake nodal points are shown in Figure 3.

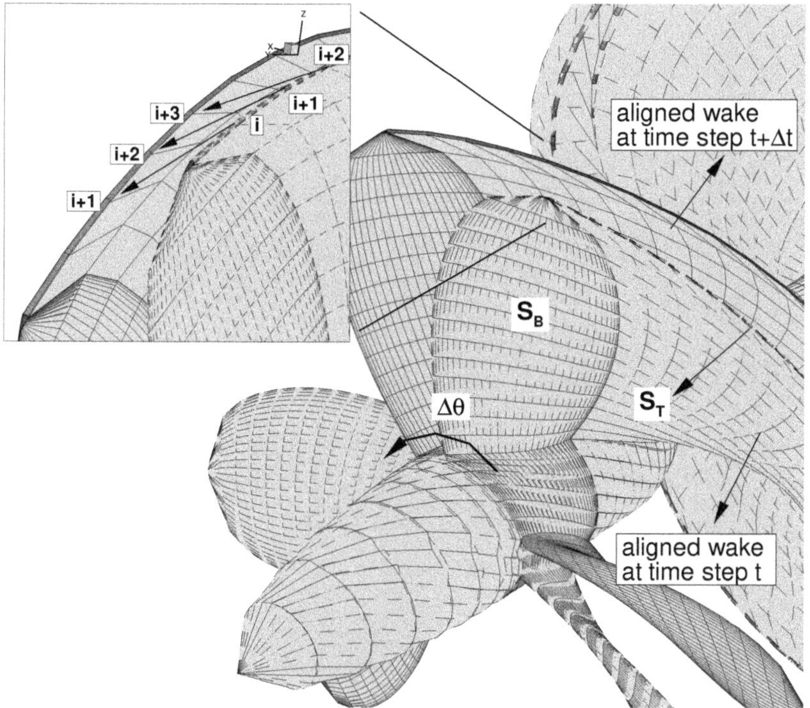

Figure 3. Schematics showing the wake alignment process from the time step t (dashed line mesh) to $t + \Delta t$ (solid line mesh). Wake panels convect downstream with time.

2.2. Tip Vortex Cavitation Modeling

2.2.1. Modeling Developed Tip Vortex Cavitation Using Jacobian Method

Flow around the blade tip, from the pressure side to suction side, produces the tip vortex, which flows from blade tip to downstream. The tip vortex cavitation usually starts downstream of the propeller in a detached form, and as the cavitation number σ_n decreases, the detached tip vortex cavity moves closer to the blade tip to become fully developed tip vortex. Developed tip vortex appears along with the sheet cavity on the blade and is known as the main source of the propeller induced hull excitations. Prediction of the developed tip vortex cavity is thus crucial not only for the assessment of the propeller performance but also for the prediction of the hull pressure fluctuations induced by the rotational motion of propellers.

H. Lee (2002) [21] conducted initial research on the tip vortex cavitation modeling. In his method, the initial shape of the tip vortex is assumed to be a solid cylinder with a circular cross section. Then, the BEM integral equation is solved for the unknown potentials on the tip vortex surface. After finding the solutions ϕ, a two-point Newton–Raphson scheme was applied to adjust the tip vortex radius in both the circumferential and streamwise directions, based on Equation (7).

$$\delta p \left(h; \sigma_n\right) = 0, \tag{7}$$

where $\delta p = -C_p - \sigma_n$. A two-point Newton–Raphson scheme solves for the circumferential grid at each section in streamwise direction by applying a M-dimensional Newton–Raphson scheme in case the circumferential panel number is M. The updated cavity heights h at the $(n+1)$th iteration are given as

$$[h]^{n+1} = [h]^{n} - [J]^{-1} [\delta p]^{n}, \tag{8}$$

where

$$[h]^{T} = [h_1, h_2, h_3, ..., h_M] \tag{9}$$

$$[\delta p]^{T} = [\delta p_1, \delta p_2, \delta p_3, ..., \delta p_M], \tag{10}$$

and the Jacobian matrix J is defined as

$$J_{ij} = \frac{\partial \delta p_i}{\partial h_j}. \tag{11}$$

A two-point finite difference scheme is used to evaluate the Jacobian, and to accelerate the scheme, the off-diagonal terms in the Jacobian matrix are discarded, therefore reducing the Newton–Raphson method into a two-point Secant method. Jacobian method is solved repeatedly to determine the tip vortex radius until the changes in the radius fall within a certain tolerance. Jacobian method modifies the entire tip vortex radius section by section throughout the iterations. The partial cavity algorithm introduced by Kinnas and Fine (1993) [22] also uses Newton–Raphson method to determine the sheet cavity length on the blade surface.

$$\delta_m(l_1, l_2, ..., l_M; \sigma_n)) = 0, \qquad m = 1, ..., M \tag{12}$$

where δ_m is the openness of the cavity trailing edge, and l_m is the cavity planform length at the mth spanwise strip. In this case, direct relation is established between the cavity planform length and the openness of the cavity trailing edge. Increase or decrease in one parameter directly causes corresponding changes on the other parameter, as shown in Figure 4. In the tip vortex modeling, however, the vortex radius h and δp do not have a direct correlation in the Newton–Raphson method, as the $-C_P$ is evaluated from the solution potentials. Also, due to the proximity between the panels surrounding vortex core, small changes in the panel geometry in turn cause large differences in the potential solutions. Moreover, in some cases under intense effective wake, wake panels rollup more than a panel method can handle, quite often leading to the singular behavior due to the twist on the tip vortex strand or the crossing of the tip vortex panels with the wake surface.

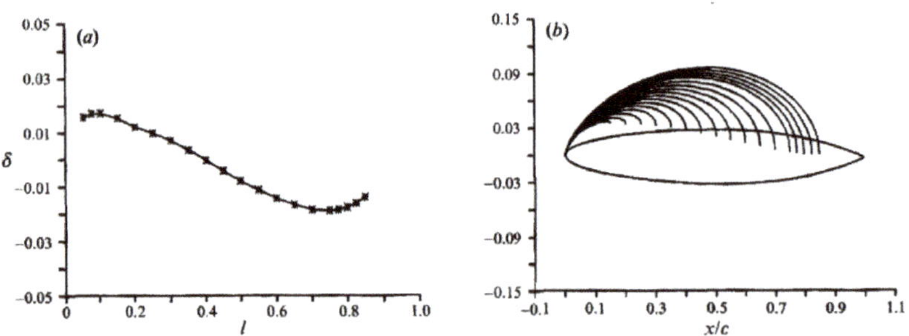

Figure 4. NACA16006 hydrofoil at $\alpha = 4°$ and $\sigma = 1.097$ (corresponding to the correct length $l = 0.4$). (**a**) The resulting behavior of δ vs. l; (**b**) The predicted cavity shapes in ratio of blade radius (y-axis) for different guesses of l, obtained from Figure 4 of [22], with permission from the prime author, Spyros A. Kinnas in 2019.

An approach via Direct method, on the other hand, can be a reliable alternative to Jacobian method. Direct method, as described in the following section, directly evaluates the solution potentials on the tip vortex surface from the Bernoulli equation, and the source strength is obtained by solving Green's identity.

2.2.2. Modeling Developed Tip Vortex via Direct Method

Direct method starts from the initial tip vortex geometry, which is a thin cylinder following the streamline of the aligned wake. To find the initial shape, a local coordinate system needs to be defined along the unit vectors $\mathbf{n_3}$ and $\mathbf{n_4}$ at the outer rim of the wake panels, as shown in Figure 5.

$$\mathbf{n_3} = \frac{\mathbf{n_1} \times \mathbf{n_2}}{|\mathbf{n_1} \times \mathbf{n_2}|} = (n_{3x}, n_{3y}, n_{3z}) \tag{13}$$

$$\mathbf{n_4} = \frac{\mathbf{n_3} \times \mathbf{n_1}}{|\mathbf{n_3} \times \mathbf{n_1}|} = (n_{4x}, n_{4y}, n_{4z}) \tag{14}$$

where $\mathbf{n_1}$ vector is an unit vector connecting the mid points of the wake panels placed between the $(i-1)$th to $(i+1)$th strips. $\mathbf{n_2}$ is also an unit vector pointing cross flow direction along the ith wake strip. With the two orthogonal unit vectors $\mathbf{n_3}$ and $\mathbf{n_4}$ as the local axes, the $(j+1)$th circumferential point on the tip vortex cavitation at the ith wake strip is determined as follows,

$$\begin{Bmatrix} X_{i,j+1} \\ Y_{i,j+1} \\ Z_{i,j+1} \end{Bmatrix} = \begin{Bmatrix} W_{X_i} + n_{3x}h_{i,j+1}(1-\cos\delta\theta) + n_{4x}h_{i,j+1}\sin\delta\theta \\ W_{X_i} + n_{3y}h_{i,j+1}(1-\cos\delta\theta) + n_{4y}h_{i,j+1}\sin\delta\theta \\ W_{X_i} + n_{3z}h_{i,j+1}(1-\cos\delta\theta) + n_{4z}h_{i,j+1}\sin\delta\theta \end{Bmatrix} \tag{15}$$

$$\delta\theta = \frac{2\pi}{M}, \tag{16}$$

where W_{X_i}, W_{Y_i}, and W_{Z_i} are the coordinates of wake tip at the end of the ith strip. M is the total panel number in circumferential direction surrounding the vortex core. The Bernoulli equation can be set up with respect to the propeller fixed system to satisfy the dynamic boundary condition of constant vapor pressure on the cavity surface. By connecting the two points, i.e., one on the tip vortex cavity surface and the other on the shaft axis far upstream, we have

$$\frac{P_0}{\rho} + \frac{1}{2}\left(|U_W|^2 + \omega^2 r^2\right) = \frac{\partial\phi}{\partial t} + \frac{P_v}{\rho} + \frac{1}{2}|q_t|^2 + g y_s, \tag{17}$$

where ρ is fluid density and r is the distance from the axis of rotation. q_t is the total cavity velocity. P_0 is the reference pressure far upstream, and g is the gravitational constant. y_s is the vertical distance from the horizontal plane through the axis of rotation. U_W is the inflow velocity and ω is the propeller angular velocity. Cavitation number σ_n is defined as

$$\sigma_n = \frac{P_0 - P_v}{\frac{\rho}{2}n^2 D^2}, \tag{18}$$

where n is revolution per unit second (rps) and D is the propeller diameter. From Equation (17), total cavity velocity on the tip vortex cavity surface can be derived as

$$|q_t|^2 = n^2 D^2 \sigma_n + |U_W|^2 + \omega^2 r^2 - 2 g y_s - 2\frac{\partial\phi}{\partial t}. \tag{19}$$

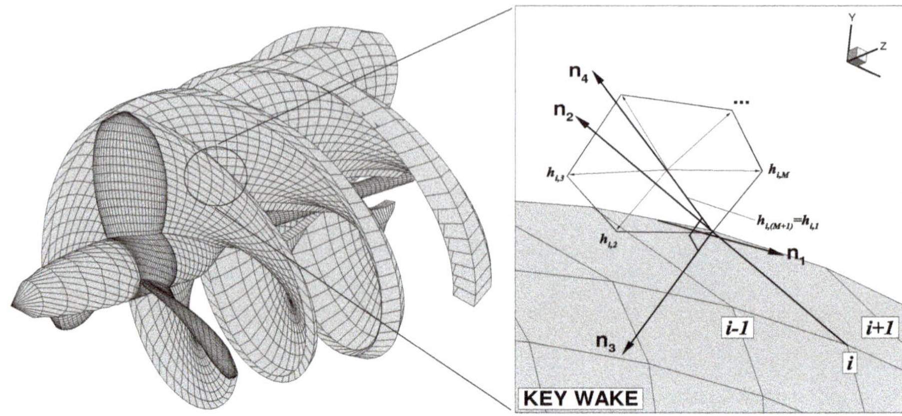

Figure 5. Schematics of tip vortex paneling at each section of the wake tip. Panel nodes are placed clockwise direction when viewed from downstream.

In addition to Equation (19), the total cavity velocity q_t is also represented in terms of the directional derivatives of the perturbation potential ϕ and the inflow components along the curvilinear coordinates system, which is in general not orthogonal. The coordinate system on the tip vortex surface consists of s (streamwise) and v (circumferential) axes, as shown in Figure 6.

$$q_t = \frac{\left(\frac{\partial\phi}{\partial s} + U_s\right)[\hat{s} - (\hat{s}\cdot\hat{v})\hat{v}] + \left(\frac{\partial\phi}{\partial v} + U_v\right)[\hat{v} - (\hat{s}\cdot\hat{v})\hat{s}]}{||\hat{s}\times\hat{v}||^2} + \left(\frac{\partial\phi}{\partial n} + U_n\right)\hat{n}, \tag{20}$$

$\hat{s}$, $\hat{v}$, and $\hat{n}$ are the unit vectors along the local s, v, and n directions, respectively. U_s, U_v, and U_n are the s, v, and n components of the inflow velocity, respectively. Equations (19) and (20) can be combined to produce a quadratic equation of the unknown total velocity in the s and v directions.

$$\begin{aligned}\left(\frac{\partial\phi}{\partial v} + U_v\right)^2 - 2\left(\frac{\partial\phi}{\partial v} + U_v\right)\left(\frac{\partial\phi}{\partial s} + U_s\right)\cos\psi + \left(\frac{\partial\phi}{\partial s} + U_s\right)^2 + \sin^2\psi\left(\frac{\partial\phi}{\partial n} + U_n\right)^2 \\ = \sin^2\psi\left(n^2D^2\sigma_n + |U_W|^2 + \omega^2r^2 - 2gy_s - 2\frac{\partial\phi}{\partial t}\right).\end{aligned} \tag{21}$$

$$\begin{aligned}V_s &= \frac{\partial\phi}{\partial s} + U_s \\ V_v &= \frac{\partial\phi}{\partial v} + U_v \\ V_n &= \frac{\partial\phi}{\partial n} + U_n.\end{aligned} \tag{22}$$

Compared to the streamwise perturbation velocity $\frac{\partial\phi}{\partial s}$, circumferential perturbation velocity $\frac{\partial\phi}{\partial v}$ is the dominant velocity component, as the swirl velocity is established around the tip vortex core. Equation (21), thus, can be solved for the unknown total velocity in the v direction on the tip vortex cavity surface. There are two solutions for $\frac{\partial\phi}{\partial v}$ as the roots of Equation (21). The solution with positive square root is chosen to have $\frac{\partial\phi}{\partial v}$ increase in the positive v direction.

$$\frac{\partial\phi}{\partial v} = -U_v + \left(\frac{\partial\phi}{\partial s} + U_s\right)\cos\psi + \sin\psi\sqrt{n^2D^2\sigma_n + |U_W|^2 + \omega^2r^2 - 2gy - 2\frac{\partial\phi}{\partial t} - Vs^2 - V_n^2}. \tag{23}$$

ψ is an angle between the s and v axes, as shown in Figure 6. V_v becomes a function of the unknown total velocity in the s and n directions, which are evaluated iteratively. Equation (23) is integrated along the positive v direction on the tip vortex surface to determine $\phi(s,v,t)$ using trapezoidal quadrature. The initial value of the integral is taken to be equal to the potential jump across the wake surface connected to the tip vortex. To make Equation (23) fully nonlinear, both the $\frac{\partial \phi}{\partial n}$ and U_n terms need to be considered for the normal velocity V_n pointing normal to the cavity surface. However, $\frac{\partial \phi}{\partial n}$, known by solving Green's identity (Equation (41)) is very sensitive to the relative distance between the panels surrounding the vortex core, and large change in the panel geometry due to the kinematic boundary condition produces very strong local $\frac{\partial \phi}{\partial n}$, therefore leading to the negative values inside the square root. To avoid the full nonlinearity and produce stable tip vortex model even in the irregular wake field, $\frac{\partial \phi}{\partial n}$ term is neglected and V_n is approximated to U_n, i.e., $V_n \simeq U_n$. Therefore,

$$\begin{aligned} \phi(s,v,t) &= -\Delta\phi(s) \\ &+ \int_0^v \left[-U_v + \left(\frac{\partial \phi}{\partial s} + U_s \right) \cos\theta + \sin\theta \sqrt{n^2 D^2 \sigma_n + |U_W|^2 + \omega^2 r^2 - 2gy - 2\frac{\partial \phi}{\partial t} - Vs^2 - U_{\tilde{n}}^2} \right] dS. \end{aligned} \tag{24}$$

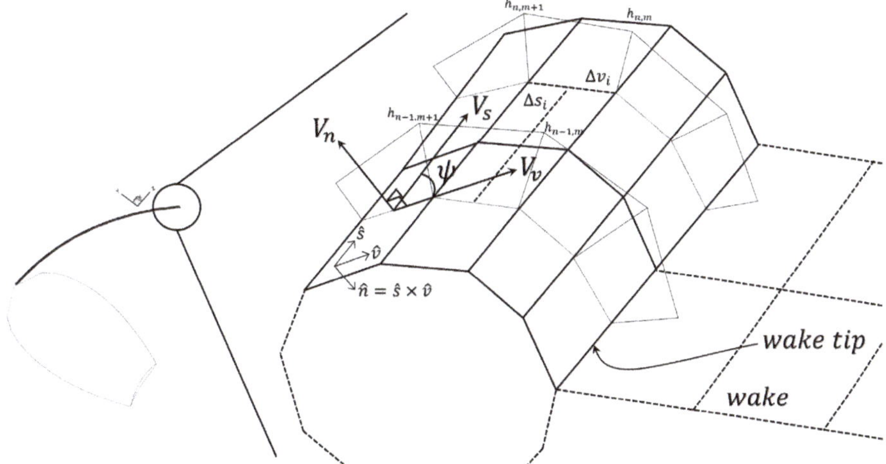

Figure 6. Schematics showing directions of the local velocities in s, v, and n directions on the curvilinear coordinate system. Integral in Equation (24) is computed along the positive v direction from the Mth panel which is attached to wake tip.

The kinematic boundary condition on the tip vortex cavitation requires that the total derivative of the cavity surface should vanish on the tip vortex surface.

$$\frac{D}{Dt}[n - h_{n,m}] = \left(\frac{\partial}{\partial t} + q_t \cdot \nabla \right) [n - h_{n,m}], \tag{25}$$

where $h_{n,m}$ means the tip vortex radius at the nth and mth streamwise and circumferential directions on the cavity surface, respectively. The gradient in the local curvilinear coordinate system is represented as

$$\nabla = \frac{[\hat{v} - (\hat{v}\cdot\hat{s})\hat{s}]\frac{\partial}{\partial v} + [\hat{s} - (\hat{v}\cdot\hat{s})\hat{v}]\frac{\partial}{\partial s}}{\|\hat{v}\times\hat{s}\|^2} + \hat{n}\frac{\partial}{\partial n}. \tag{26}$$

Taking this gradient on the total cavity velocity in Equation (20) to calculate Equation (25) can represent the derivative of tip vortex height in s and v directions. To discretize $\frac{\partial h}{\partial v}$ and $\frac{\partial h}{\partial s}$, two points backward scheme is used.

$$\frac{\partial h}{\partial v}\left[V_v - \cos\psi V_s\right] + \frac{\partial h}{\partial s}\left[V_s - \cos\psi V_v\right] = \sin^2\psi V_v\left(V_n - \frac{\partial \bar{h}}{\partial t}\right), \tag{27}$$

where

$$\begin{aligned} \frac{\partial h}{\partial v} &= \frac{h_{n,m} - h_{n,m+1}}{\Delta v_i} \\ \frac{\partial h}{\partial s} &= \frac{\bar{h}_{n,m} - \bar{h}_{n-1,m}}{\Delta v_i} \\ \bar{h} &= \frac{h_{n,m} + h_{n,m+1}}{2}. \end{aligned} \tag{28}$$

Equation (27) in combination with Equation (28) can be rearranged to represent the tip vortex height $h_{n,m}$ on the nth wake panel at the mth circumferential panel node on the tip vortex:

$$h_{n,m} = \frac{H_1 h_{n+1,m+1} + H_2\left(h_{n-1,m} + h_{n-1,m+1}\right) + V_n^* \sin^2\psi}{H_1}; (n = 1 \sim N_W, m = M \sim 1), \tag{29}$$

where N_W denotes the total panel number on the wake surface in streamwise direction,

$$\begin{aligned} H_1 &= \frac{V_v - \cos\psi V_s}{\Delta v_i} - \frac{V_s - \cos\psi V_v}{2\Delta s_i} - \frac{\sin^2\theta}{2\Delta t} \\ H_2 &= \frac{V_s - \cos\psi V_v}{2\Delta s_i} \\ V_n^* &= U_n + \frac{\partial \phi}{\partial n} + \frac{h_{n,m} + h_{n+1,m}}{\Delta t}. \end{aligned} \tag{30}$$

Computation of the tip cavitation height starts from the $M + 1$th panel node connected to wake tip and proceeds in counterclockwise direction when viewed from downstream. Cavity height from the previous time step is used as the initial guess for the first revolution, and the saved data at the same blade angle from the previous revolutions are used for the rest calculations. Tip vortex height at $m = M + 1$ is taken to be equal to the height at $m = 1$ after solving Equation (29) M through 1 to satisfy the periodic boundary condition at the point where the tip vortex is closed with wake tip.

The potential difference between the first ($m = 1$) and the last ($m = M$; if M panels are assumed circumstantially on the tip vortex cavitation) panel is equivalent to the potential jump across the wake surface. This is because the tip vortex is attached to the end of wake tip as shown in Figure 7. From Equation (24), the potentials on the first and last panels are known:

$$\begin{aligned} \phi_{1,1} &= -\Delta\phi_s^{MR} + \int_0^M \frac{\partial \phi}{\partial v} dv, (s = 1 \sim N_W) \\ \phi_{1,M} &= -\Delta\phi_s^{MR} + \int_0^1 \frac{\partial \phi}{\partial v} dv, (s = 1 \sim N_W), \end{aligned} \tag{31}$$

The difference of the two potentials—$\phi_{1,1}$ and $\phi_{1,M}$—is equal to the potential jump $\Delta\phi_1^{MR}$, therefore, we have

$$\phi_{1,1} - \phi_{1,M} = \Delta\phi_1^{MR} = \int_0^M \frac{\partial \phi}{\partial v} dv - \int_0^1 \frac{\partial \phi}{\partial v} dv = \int_M^1 \frac{\partial \phi}{\partial v} dv. \tag{32}$$

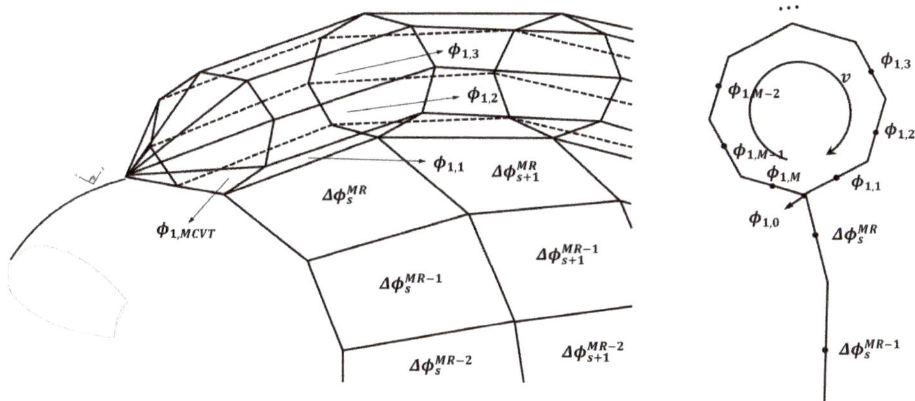

Figure 7. Potentials around the tip vortex cavitation in connection with the wake surface (**left**), and its cross section view with arrangement of the potentials calculated by Equation (24) along the v direction (**right**).

However, the potential difference is not always equal to the potential jump (or wake strength) as the potentials are from the direct method, which evaluates $\frac{\partial\phi}{\partial v}$ based on the initial guess on the vortex geometry. To resolve this, relative distance between the control points around the tip vortex needs to be adjusted iteratively as follows. Equation (29) is first applied to the cylindrical tip vortex shape to update the vortex radii in both the streamwise and circumferential directions. Then, the tip vortex radius is adjusted by α percentage of its original length to satisfy the potential jump condition (Equation (32)), as shown in Figure 8. Trapezoidal rule is used for the calculation of the integral in Equation (32). Satisfying the potential jump condition prevents the tip vortex geometry from expanding or contracting with abnormally large or negative radius during the calculations. The value of α can be evaluated as

$$\begin{aligned}\Delta\phi_s^{MR} &= \int_M^1 \frac{\partial\phi}{\partial v}\,dv \\ &= \frac{1}{2}\Delta\theta(1+\alpha)\left\{r_2\left(\left.\frac{\partial\phi}{\partial v}\right|_1 + \left.\frac{\partial\phi}{\partial v}\right|_2\right) + r_3\left(\left.\frac{\partial\phi}{\partial v}\right|_2 + \left.\frac{\partial\phi}{\partial v}\right|_3\right) + \cdots + r_M\left(\left.\frac{\partial\phi}{\partial v}\right|_{M-1} + \left.\frac{\partial\phi}{\partial v}\right|_M\right)\right\}.\end{aligned} \tag{33}$$

From Equation (33), α is reduced to

$$\alpha = \frac{2\Delta\phi_s^{MR} - \Delta\theta\mathcal{L}}{\mathcal{L}}, \tag{34}$$

where

$$\mathcal{L} = \left\{r_2\left(\left.\frac{\partial\phi}{\partial v}\right|_1 + \left.\frac{\partial\phi}{\partial v}\right|_2\right) + r_3\left(\left.\frac{\partial\phi}{\partial v}\right|_2 + \left.\frac{\partial\phi}{\partial v}\right|_3\right) + \cdots + r_M\left(\left.\frac{\partial\phi}{\partial v}\right|_{M-1} + \left.\frac{\partial\phi}{\partial v}\right|_M\right)\right\}. \tag{35}$$

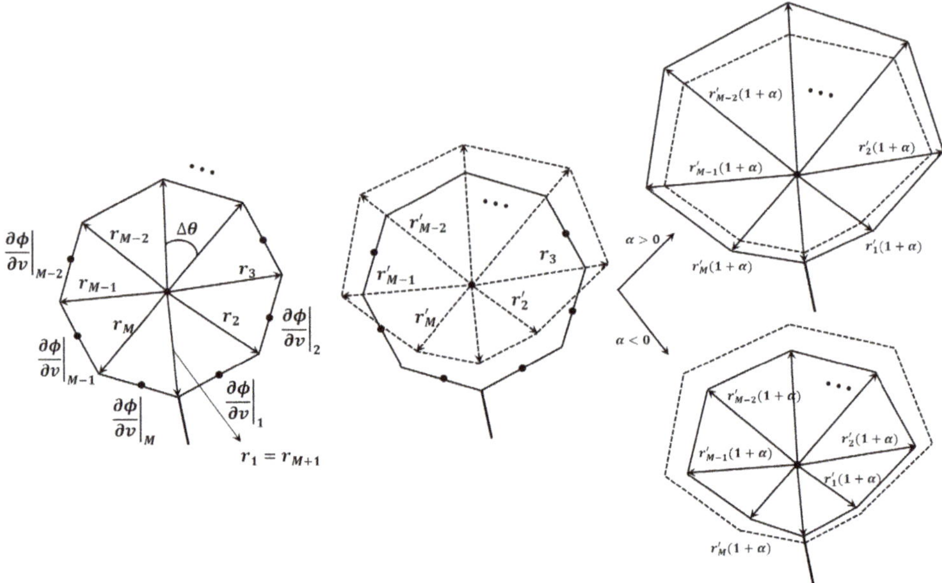

Figure 8. Adjusting tip vortex radii from the initial shape (**left**). Kinematic boundary condition (Equation (27)) updates the radius in radial direction (**middle**), and the potential jump condition expands or contracts the overall shape, depending on the sign and the magnitude of the α (**right**).

2.2.3. Treatments for the Tip Vortex Collapse and Panel Distortion

After the tip vortex height is updated, $\frac{\partial\phi}{\partial v}$ on the control points is reevaluated due to the changes in the panel coordinates, and thus in the influence functions between the panels. This process is conducted iteratively each time step until the α converges at different sections along the tip vortex. Convergence history shows that the convergence is normally achieved after five to six iterations. Figure 9 shows an example of thickness variation along a vortex strand after the adjustment. A weak wake strength Γ shortens the vortex radius to satisfy the potential jump condition (Equation (33)), and thus brings the panels close within a critical radius which leads to the singularities and consequently crash in the code (0.1 percentage of blade radius R in the present method). The section with small radius is considered as *no tip vortex cavity zone* where the tip vortex collapses. In the event the developed tip vortex collapses, the viscous core reappears and that cannot be captured by the present inviscid model. The effect of the cavity sources in the neglected region is therefore not included in the calculations making the actual matrix size that the BEM solves smaller than the fully wetted cases depending on the local Γ along the tip vortex. The matrix size is dynamically allocated in time and space to account for the sections along the tip vortex cavity which have or have not collapsed. The current method uses a semi-empirical criterion of the non-dimensional circulation G_R, below which the radius of the tip vortex cavity will become smaller than 0.3% of the propeller radius.

$$G_R(r = R, s, t) = \frac{100\Gamma(r = R, s, t)}{2\pi R U_R}, \tag{36}$$

where $U_R = \sqrt{V_S^2 + (0.7n\pi D)^2}$ is the reference velocity. By applying the Bernoulli equation between a point on the tip vortex cavity and another point on the propeller shaft, and by using the approach in Kinnas (2006) [23] where the unsteady terms are ignored, a relation between the tip vortex cavity radius r/R and the vortex core strength G_R can be established as follows,

$$P_v + \frac{1}{2}\rho \left|q_t\right|^2 + \rho g y_s = P_0, \tag{37}$$

where $q_t = \Gamma/2\pi r$, which leads to the minimum vortex core strength $G_{R,min}$ as a function of advance ratio J_S, cavitation number σ_n, Froude number F_n, and the minimum tip vortex radius $r_{t,min}$, above which the tip vortex cavity does not collapse.

$$G_{R,min}(r = R, s, t) = \frac{100 r_{t,min}}{R}\sqrt{\frac{\sigma_n - \frac{y_s}{F_n}}{J_S^2 + (0.7\pi)^2}}, \tag{38}$$

where $F_n = \frac{n^2 D}{g}$ and $y_s = y/R$, with y being the vertical distance from the shaft, taken as positive towards the free surface. To account for possible fluctuations in tip vortex radius during the calculations and to make sure that the radius is above the critical value, the current method assumes the $r_{t,min}/R$ to be ~0.003R, which will lead to the corresponding minimum vortex core strength $G_{R,min}$ depending on the vertical elevation of the tip vortex cavity and the propeller loading. The $G_{R,min}$ serves as the final criterion to determine the no tip vortex cavity zone[1] where the local G_R is less than $G_{R,min}$.

Another numerical issue, which happens almost every time trailing wake undergoes intense rollup, is the twisting in the vortex strand (Figure 9). When the non-axisymmetric effective wake is assumed with the effect of upstream body, for example, it makes the tip vortex panels twist irregularly by following the rotation of the local $\mathbf{n_3} - \mathbf{n_4}$ axes, as shown in Figure 10b. This kind of numerical issue is hard to avoid as long as the panel method models the tip vortex cavitation with quadrilateral panels, most likely leading to the crash in the code. To go around the twisting issue, the following *bridge technique* can be proposed. First, each section at s_{i-1}, s_i, s_{i+1} (Figure 10a) finds a plane containing the wake stream line (junction with the tip vortex) and the section diameter extended from the wake tip, thus along the local $\mathbf{n_4}$ axis (Figure 10c). The two adjacent planes α and β in Figure 10c are not parallel in general, so a non-zero dihedral angle θ can be calculated. Both the upstream and downstream sections rotate toward each other in the opposite direction by $\frac{\theta M_{total}}{4\pi}$ after truncating it to the integer part. This is to retain the control point still on the panel centroid (Figure 10d). Panel perimeters are redefined across the twisted panels like a bridge, and the geometrical data for the quadrilaterals are reevaluated based on the adjusted sections, as shown in the dashed lines in Figure 10e. The bridge technique helps the presented model complete several revolutions even with the intense rollup in the wake until the geometrical changes in the wake panels fall below the predefined tolerance.

The trajectory of a tip vortex in non-cavitating conditions was found to be close to that of the cavitating conditions by Arndt et al. (1991) [24]. Thus, the unsteady wake from fully-wetted runs can be used to predict the trajectory of the developed tip vortex cavitation.

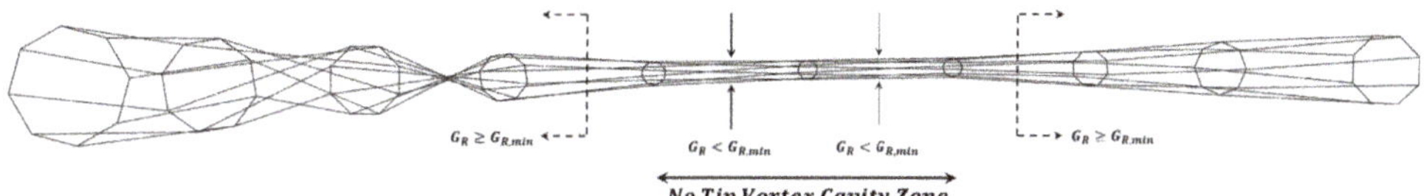

Figure 9. Twisted tip vortex cavity sections before adjustment. The zone over which a tip vortex cavity is not considered is also shown. The criterion to determine the zone with no tip vortex cavity is $G_R < G_{R,min}$.

1 Note that this criterion, due to its simplicity, might allow the tip vortex cavity radius to become less than 0.003R, but not smaller than 0.001R.

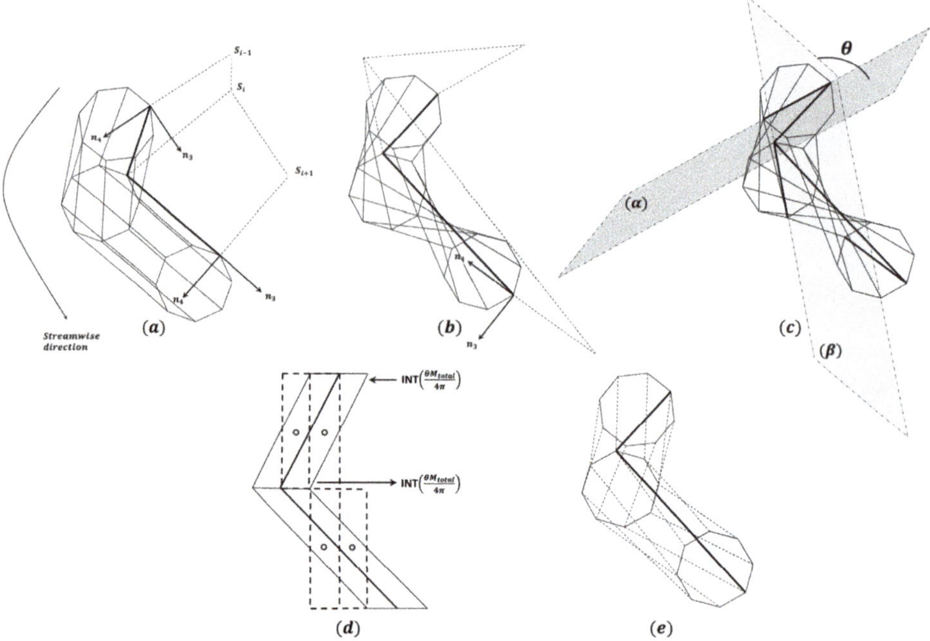

Figure 10. Initial guess of tip vortex cavity with $\mathbf{n_3}$ and $\mathbf{n_4}$ the local coordinate axes along the streamline in the trailing wake and the cross flow direction, respectively (**a**); distortions. of tip vortex cavity panels due to the rollup in the wake (**b**); a non-zero dihedral angle θ between the two adjacent planes α and β (**c**); adjusted panel perimeters (**d**) and updated cavity panels (**e**).

2.2.4. Discretized Green's Formula

In fully-wetted condition at time t, the perturbation potentials at any field point p on the propeller surface S_B and tip vortex surface S_T (Figure 3) satisfy Green's third identity as follows,

$$
\begin{aligned}
2\pi\phi_p(t) = & \iint_{S_B(t)\cup S_T(t)} \left[\phi_q(t)\frac{\partial}{\partial n_q}\left(\frac{1}{R(p;q)}\right) - \frac{\partial\phi_q(t)}{\partial n_q}\left(\frac{1}{R(p;q)}\right)\right] dS \\
& + \iint_{S_W(t)} \Delta\phi_w(r_q,\theta)\frac{\partial}{\partial n_q}\left(\frac{1}{R(p;q)}\right) dS.
\end{aligned}
\tag{39}
$$

Equation (39) expresses the perturbation potential as a superposition of the potential induced by a continuous source distribution $\frac{\partial\phi_q(t)}{\partial n_q}$ and a continuous dipole distribution $\phi_q(t)$ on $S_B(t)$ and $S_T(t)$, and a continuous wake strength $\Delta\phi_w(r_q,\theta)$ on the trailing wake surface $S_W(t)$. To uniquely determine the potentials, both the kinematic and dynamic boundary conditions are enforced on the propeller surface, along with the Kutta condition, which requires the fluid velocity be finite at the blade trailing edge. Classic kinematic boundary condition applied to the propeller problems in wetted condition provides the unknown source strength by setting it equal to the normal component of the inflow velocity on the propeller surface. It remains the dipole strength ϕ as the unknowns, which can be obtained by solving Green's identity.

In case the tip vortex surface $S_T(t)$ is included as the cavitating surface, the unknown perturbation potentials on the tip vortex surface are determined via the dynamic boundary condition (Equation (24)), and Green's formula is solved for the unknown source strengths instead. Solution potentials beneath the sheet cavity on the blade surface are also determined via the dynamic boundary condition if included [22]. To convert Equation (39) into the discretized Green's Formula, the key blade is divided into NC chordwise and MR spanwise quadrilateral panels covering the propeller surface S_B, and tip vortex cavitation surface into M circumferential and $N_W - n_{not}$ streamwise quadrilateral panels. n_{not} is the number of sections within the no tip vortex cavity zone which are skipped in Green's formula. Therefore, among the discreted dipoles and sources, we have

- $NC \times MR$ known source strengths, which are equal to $-\vec{U}_{in} \cdot \vec{n}$ on S_B.
- $NC \times MR$ unknown dipole strengths on S_B.
- $M \times (N_W - n_{not})$ known dipole strengths, which are evaluated by Equation (24) on S_T.
- $M \times (N_W - n_{not})$ unknown source strengths on S_T.

Expression for the discretized Green's formula in fully-wetted condition reads as Equation (40) and can be converted into Equation (41), which considers the developed tip vortex as cavitating surface. By solving Green's identity given as Equation (41), the unknown dipole strengths on the wetted blade surface and the source strengths on the tip vortex surface are obtained simultaneously.

$$\begin{aligned} &\sum_{m=1}^{MR}\left\{\sum_{n=1}^{NC} A_{inm}^{1}\phi_{nm}^{1}(t) + W_{i1m}^{1}\Delta\phi_{w_{im}}^{1}(r,t)\right\} = \sum_{m=1}^{MR}\left\{\sum_{n=1}^{NC} B_{inm}\frac{\partial\phi}{\partial n}_{nm}^{1}\right\} \\ &\quad - \sum_{K=2}^{N_{blade}}\left\{\sum_{m=1}^{MR}\sum_{n=1}^{NC}\left[A_{inm}^{K}\phi_{nm}^{K} - B_{inm}^{K}\frac{\partial\phi}{\partial n}_{nm}^{K}\right] + \sum_{l=2}^{NW_{panel}} W_{ilm}^{K}\Delta\phi_{w_{lm}}^{K}\right\}, \\ &(i = 1, \ldots, NC \times MR) \end{aligned} \tag{40}$$

$$\begin{aligned} &\sum_{m=1}^{MR}\left\{\sum_{n=1}^{NC} A_{inm}^{1}\phi_{nm}^{1}(t) + W_{i1m}^{1}\Delta\phi_{w_{im}}^{1}(r,t)\right\} + \sum_{m=MR+1}^{MR+M}\left\{\sum_{n=NC+1}^{NC+NW_{panel}-n_{not}} -B_{inm}^{1}\frac{\partial\phi}{\partial n}_{nm}^{1}\right\} \\ &= \sum_{m=1}^{MR}\left\{\sum_{n=1}^{NC} B_{inm}^{1}\frac{\partial\phi}{\partial n}_{nm}^{1}\right\} - \sum_{m=MR+1}^{MR+M}\left\{\sum_{n=NC+1}^{NC+NW_{panel}-n_{not}} A_{inm}^{1}\phi_{nm}^{1}(t)\right\} \\ &- \sum_{K=2}^{N_{blade}}\left\{\sum_{m=1}^{MR}\sum_{n=1}^{NC}\left[A_{inm}^{K}\phi_{nm}^{K} - B_{inm}^{K}\frac{\partial\phi}{\partial n}_{nm}^{K}\right] + \sum_{l=2}^{NW_{panel}} W_{ilm}^{K}\Delta\phi_{w_{lm}}^{K}\right\}, \\ &(i = 1, \ldots, NC \times MR + (N_W - n_{not}) \times M) \end{aligned} \tag{41}$$

where N_{blade} denotes the total blade number. $\Delta\phi_{wlm}^{K}$, ϕ_{nm}^{K}, and $\frac{\partial\phi}{\partial n}_{nm}^{K}$ on other blades are taken from the saved data by the key blade in the previous revolutions.

2.3. Acoustic Boundary Element Method: Solver for the Oscillating Hull Pressure

According to the Bernoulli equation, the small amplitude pressure oscillation $P^{(t)}$ can be represented by the steady velocity potential $\Phi^{(t)}$, as shown in Equation (42). In this equation, Φ_n is the total velocity potential magnitude at a certain frequency.

$$P^{(t)} = -\rho\frac{\partial\Phi^{(t)}}{\partial t} = -\rho\sum_{n=1,2,\ldots} nN_{blade}\omega\Phi_n i e^{inN_{blade}wt} \tag{42}$$

Instead of solving for the oscillating pressure field, the velocity potential field Φ_n is solved at different frequencies in the acoustic BEM solver. The potential field Φ_n is governed by the Helmholtz equation. For the near-field small-amplitude pressure fluctuations caused by marine propellers, an infinite sound speed can be assumed so that the Helmholtz equation is reduced to a Laplace equation (Equation (43)), which can be solved by the BEM solver.

$$\nabla^2 \Phi_n = 0 \tag{43}$$

Similar to the hydrodynamic BEM, the nth total velocity potential field Φ_n can be decomposed into a radiated potential field $\Phi_n^{(R)}$ (taken equal to the potential due to the propeller flow in the absence of the hull) and a diffraction potential field $\Phi_n^{(D)}$. The radiated potential field $\Phi_n^{(R)}$ is taken from the Fourier series of the unsteady perturbation potential field in the hydrodynamic BEM, which provides the singularities on the propeller, its wake, hub, and developed tip vortex. In the propeller problem, the lowest frequency for the Fourier decomposition is the blade-passing frequency (BPF),

$$BPF = \frac{N_{blade}\omega}{2\pi}. \tag{44}$$

Based on the BEM, the total potential field $\Phi_n^{(D)}$ and the unsteady pressure field on the hull surface can be solved by integral Equation (45), where S_H is the hull surface and S_I is the image of the hull. Constant strength dipole panels are placed on the ship hull surface, and the effect from the top tunnel wall is considered by the image model.

$$\frac{1}{2}\Phi_n^{(D)} = \Phi_n^{(R)} - \frac{1}{4\pi}\iint_{S_H+S_I} \Phi_n^{(D)} \frac{\partial G}{\partial n}\, dS. \tag{45}$$

$\Phi_n^{(D)}$ is a diffraction potential induced by a rotating flow-field of the propeller relative to the hull [1]. It is determined by subtracting the solution on the image model from that on the hull after solving Equation (45). Different from the hydrodynamic BEM model, the total potential, instead of the diffraction potential, is solved. This eliminates the source-induced potentials in the boundary integral equation and thus saves computational cost. However, the method cannot be used for the hydrodynamic BEM because the total velocity field can be vortical and therefore is not governed by the Laplace equation. Once the $\Phi_n^{(D)}$ is solved for, the pressure fluctuations on the hull $P(t)$ will be determined by:

$$P(t) = -\rho \frac{\partial \Phi_n^{(D)}}{\partial t}. \tag{46}$$

Section 3.3 will present applications of this procedure.

3. Numerical Results and Comparisons with Other Measurements

In this section, numerical applications presented in methodology section are tested with several propellers and a hull model, aligning with the items in Figure 1. In summary, a sequence of the numerical models will be applied as follows.

- BEM/RANS interactive method for the effective wake predictions .
- Application of the unsteady wake alignment (UWA) scheme in the inclined-shaft flow for an open propeller.
- Predictions of the developed tip vortex cavitation in the effective wake for an open propeller.
- Finally, predictions of the propeller-induced hull pressures in fully-wetted or cavitating conditions, considering hull/propeller/rudder interactions.

Among those, the BEM/RANS interactive method will be presented in Section 3.3, as this is not the main topic of this article but required for any application with propeller/hull interactions. After investigating the UWA and tip vortex cavitation model in Sections 3.1 and 3.2, Section 3.3 will apply both models and the pressure-BEM to the propeller, operating behind the hull in wetted/cavitating conditions. Table 1 summarizes the following sections with applied numerical models.

Table 1. Applications of the numerical models in Section 3.3.

Subsection	Propeller Geometry	Wake Field	UWA	Tip Vortex Cavitation Model	Pressure-BEM
Section 3.1	DTMB [1] 4661	10° inclined uniform flow	○	×	×
Section 3.2	DTMB N4148	effective wake from [25]	○	○	×
Section 3.3	SSPA P2772 [2]	effective wake from BEM/RANS coupling method	○	○	○

[1] David Taylor Model Basin (DTMB), U.S. Navy, Maryland, U.S.

3.1. Unsteady Wake Alignment in 10° Inclined Shaft Flow

To validate the results from the hydrodynamic BEM using the unsteady wake alignment scheme in periodic non-axisymmetric inflow, DTMB 4661 propeller, tested by Boswell et al. (1984) [26] is first used in 10° inclined shaft flow. DTMB4661 propeller is a five bladed propeller with a high skew distribution. Hub geometry has an elliptic shape that closes both upstream and downstream. The design advance ratio of this propeller is $J_S = 1.14$. As shown in Figure 11, the initial wake geometries do not show noticeable inclination in the geometry at the beginning, but it starts to incline toward the free surface as completing several revolutions. Direction of the fully aligned wake is toward the inflow direction, as shown in Figure 12.

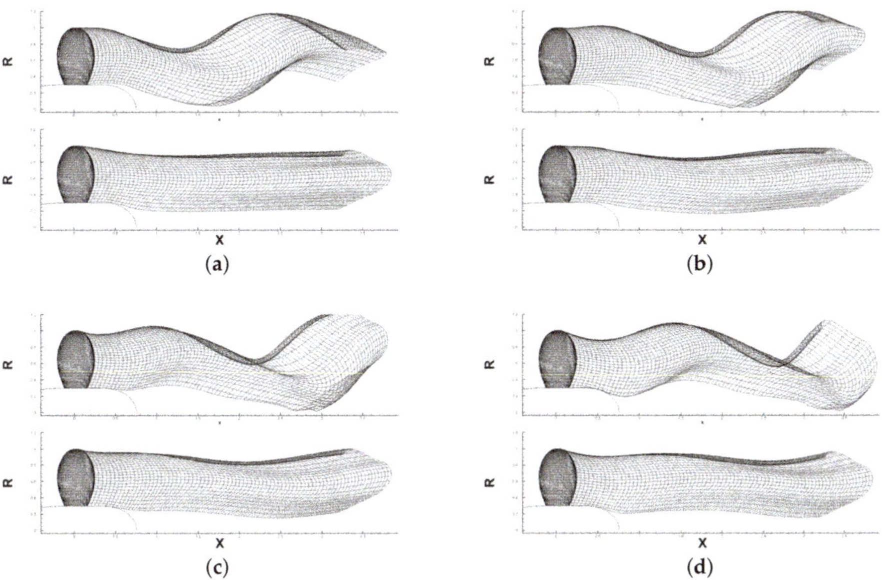

Figure 11. Projected view of the fully aligned wake (**up**) and its initial geometry (**down**), plotted on the $X-R$ plane at the blade angle of (**a**) $\theta = 0°$, (**b**) 72°, (**c**) 144°, and (**d**) 216°.

Propeller forces (KT and $10KQ$), predicted by the hydrodynamic BEM, are compared with the RANS results at four different advance ratios, i.e., 0.8, 1.0, 1.2, and 1.4, in Figure 13. Overall, the two methods seem to agree well with each other, especially around the design $J_S = 1.14$. The forces (torque coefficients and thrust coefficients) from RANS are evaluated with the time step $\delta t = 0.003$ during more than 6 revolutions until the unsteady forces converge. BEM completed 10 wetted revolutions, and the results converged after the 6th revolution. BEM takes several wetted revolutions to fully align the wake due to its gradual alignment process in non-axisymmetric effective wake.

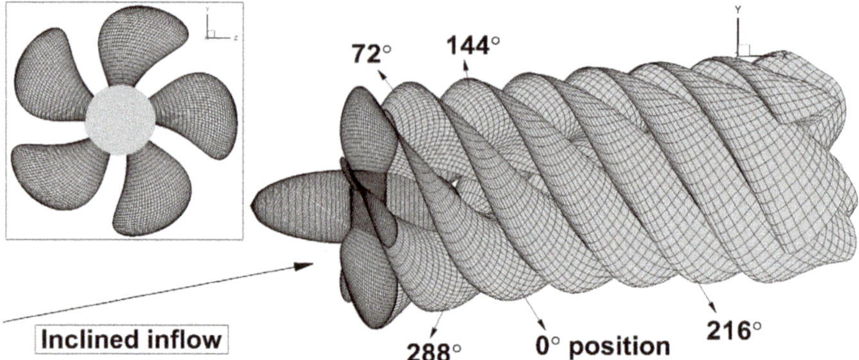

Figure 12. Projected view of the fully aligned wake of DTMB4661 propeller with hub, which has elliptical shape at both ends. Geometry of the 5-bladed propeller is shown in a separate frame.

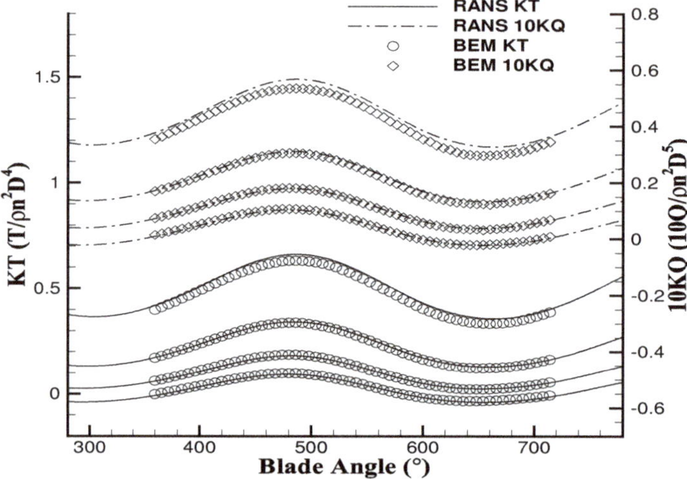

Figure 13. Predicted DTMB 4661 propeller forces by BEM using the unsteady wake alignment scheme, in comparisons with RANS results in 10° inclined shaft flow at the advance ratio of 0.8, 1.0, 1.2, and 1.4 from top to bottom lines for every pair of KT and $10KQ$.

Figure 14 shows the pressure coefficient on the blade surface at the four different blade angles, i.e., $\theta = 0°$, $\theta = 90°$, $\theta = 180°$, and $\theta = 270°$. The pressure coefficients are plotted along the chord-wise direction at the two blade sections $r/R = 0.56$ and $r/R = 0.75$. Comparisons are made between the results from RANS and BEM without the boundary layer correction. Good agreement is established between the two, and due to the high advance ratio used in this case $J_S = 1.14$, boundary layer thickness is not significant at the suction side near the trailing edge; therefore, viscous pitch correction is enough to account for the viscous effects in this case. The viscous pitch correction introduces an empirical correction to the blade pitch angle by applying a constant friction coefficient over the blade to consider the friction forces, as described in Kerwin and Lee (1978) [27].

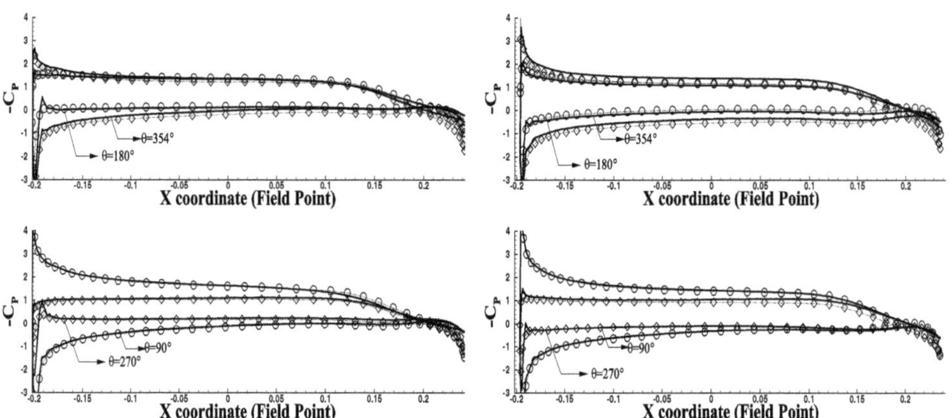

Figure 14. Key-blade surface pressures at the four different time steps at $r/R = 0.56$ (**left**) and $r/R = 0.75$ (**right**). Comparisons between RANS (solid line) with BEM (symbols) in 10° inclined shaft flow for DTMB 4661 propeller: $J_S = 1.14$.

Contour plots of the vorticity magnitude plotted on the $x - y$ and $x - z$ planes from RANS simulations are overlaid on the wake geometries predicted by BEM in Figure 15 for the advance ratios of $J_S = 0.8$ and $J_S = 1.0$. Vorticity in RANS simulation diffuses as it convects downstream, whereas the diffusion effect cannot be captured by BEM as it represents vorticity as the concentrated wake panels. The locations of the trailing wake from BEM is in good agreement with the locations of the diffused vorticity from RANS simulations. Note the rollup of the wake panels in BEM brings more panels around, and it also corresponds to the RANS results, which show strong vortex magnitude due to the tip vortex shed from the blade tip.

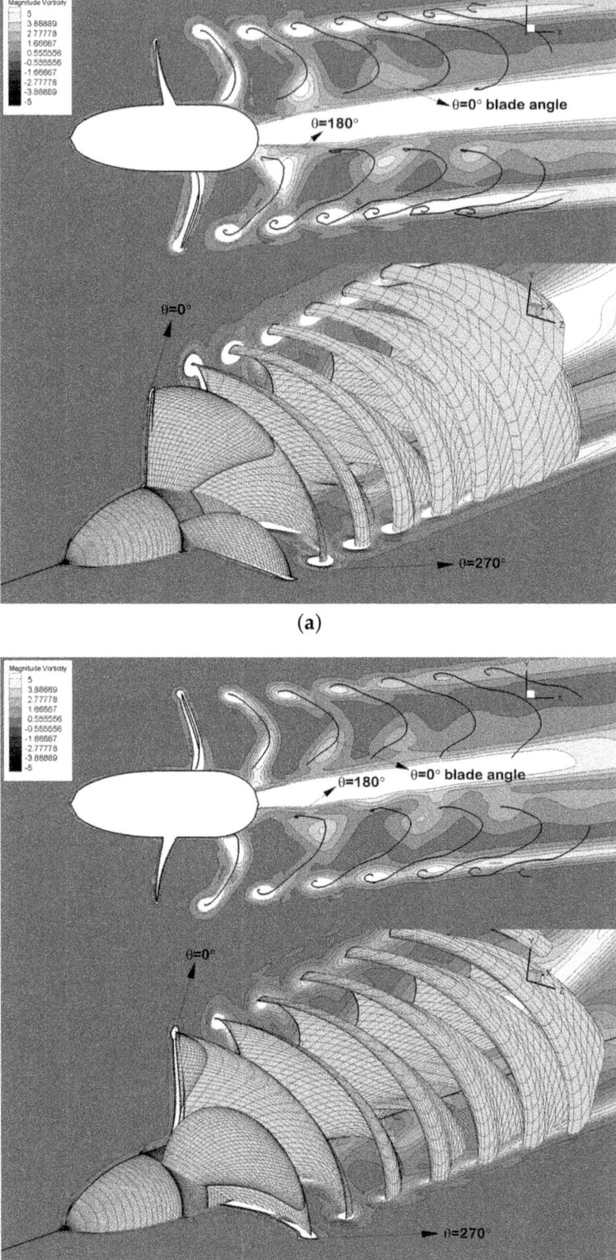

Figure 15. Comparisons of the contour plots showing vorticity magnitude predicted by RANS, and the trailing wake by BEM (black solid line) for the advance ratios of (**a**) 0.8 and (**b**) 1.0. All the shown quantities and wake shapes correspond to the intersection of a vertical and a horizontal planes through the propeller axis.

3.2. Partially Cavitating Propeller (DTMB N4148) in the Effective Wake

In this section, unsteady wake alignment scheme and the tip vortex cavitation model are applied to DTMB N4148 propeller in partially cavitating condition. Comparisons with the experiments conducted by Mishima et al. (1995) [28] are presented for the validation of the sheet/tip votex cavity patters. The geometry of N4148 propeller in the non-axisymmtetric effective wake is presented in Figure 16. The effective wake used in this article includes the effects of the tunnel walls and vortical propeller/inflow interaction, calculated by Choi and Kinnas (1998, 2000) [25,29]. The induced velocity from the tip cavitation brings additional swirl velocity on the local wake close to the cavity, thus causing additional rollup on the wake panels. Figure 17 addresses this curly behavior, showing that the aligned wake with consideration of the induced velocity from the tip cavitation rolls up further than the case without the tip cavitation effect. This behavior can be more addressed if additional panels are placed in the rollup region.

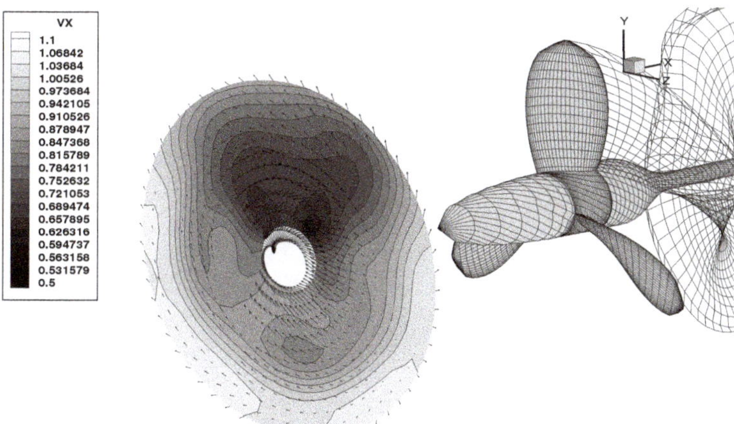

Figure 16. Geometry of DTMB N4148 propeller subject to the non-axisymmetric effective wake. Only the key wake is shown, and the wake disk is shifted upstream for clarity.

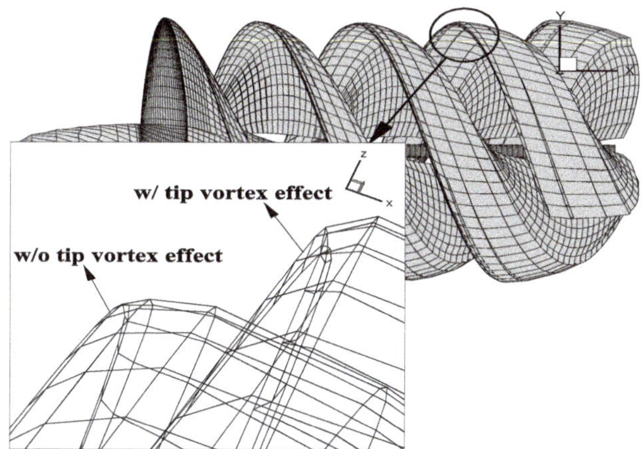

Figure 17. Rollup of the wake tip after the wake alignment with/without induced velocity from the tip vortex cavitation: DTMB N4148 propeller, uniform inflow. Tip cavitation geometry is not presented. The wake geometry w/o tip vortex effect is shifted by -0.5 in y direction for clarity in the lower frame.

Figure 18 shows the predicted tip vortex patterns during one propeller revolution with 6° discretized time step. As expected from the effective wake in Figure 16, tip cavitation volume increases around the 0° blade angle and decreases as the blade passes negative y region (note the propeller center is located at $y = 0$ in vertical direction) where the incoming axial velocity is relatively faster.

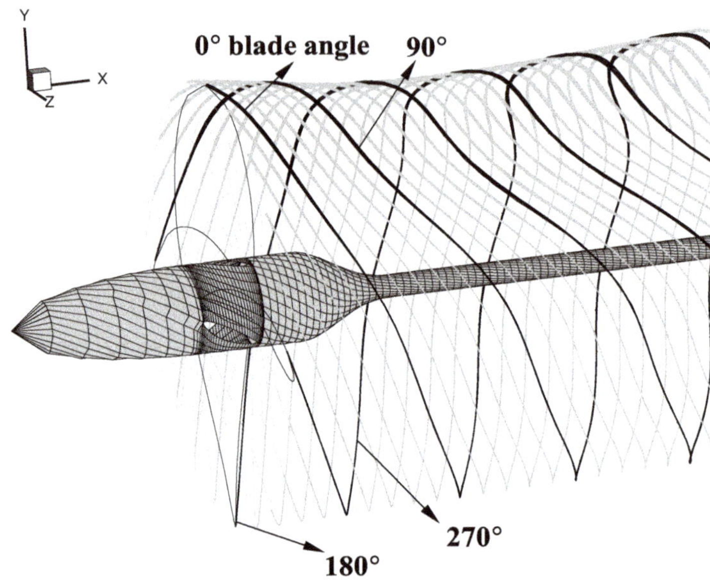

Figure 18. Developed tip vortex cavitation modeled by BEM during one period of propeller revolution: DTMB N4148 propeller, $\sigma_n = 2.576$, $J_S = 0.954$. Sheet cavity on the blade is not presented.

The predicted blade sheet and tip vortex cavities compared to the photographs taken during the experiments are presented in Figure 19. The observed angles are −30°, 6°, and 30°. Qualitative comparisons of the predicted cavity patterns by the BEM seem to be following the observations. The conical bulb is introduced at the blade tip to represent the tip vortex before the detachment from the blade tip. Radius of the conical bulb is increasing with the square root of the distance from zero at the leading edge to the radius of the first tip vortex section at the trailing edge. The kinematic boundary condition is not applied on this region, as small changes in the radius can produce large changes in the velocity due to the dense panels around tip vortex core.

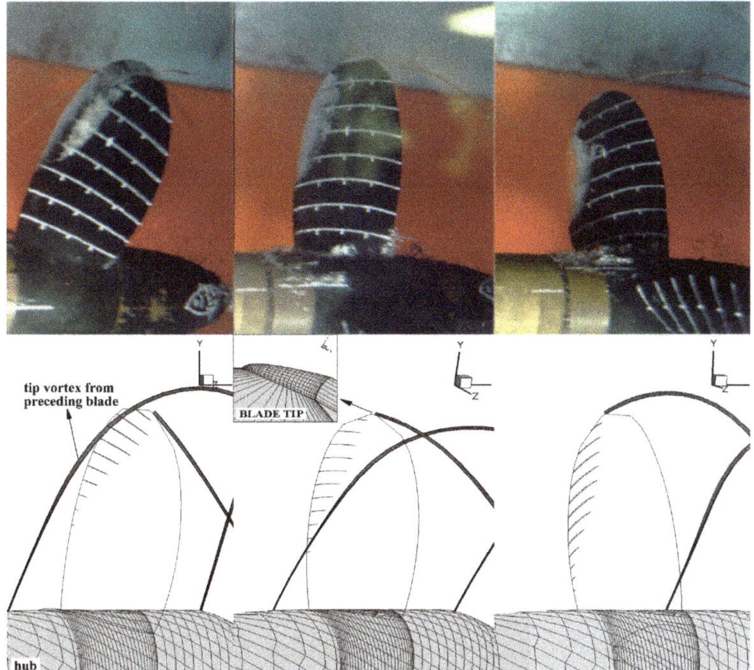

Figure 19. Comparisons of the blade sheet and tip vortex cavities predicted by the present method (**bottom**) with the measurements (**top**) for DTMB N4148 propeller: $J_S = 0.9087$, $F_n = 9.159$, and $\sigma_n = 2.576$.

Figures 20–23 (top) show the projected view of the tip vortex cavitation on the 2D $X - R$ plane at several time steps, i.e., $\theta = 0°$. $\theta = 90°$, $\theta = 180°$, and $\theta = 270°$. Cavitation volume is pulsating by satisfying both the kinematic boundary condition and the potential jump condition to adjust its sectional radius. Figures 20–23 (bottom) also show the predicted pressure $-C_P$ plotted on the tip vortex surface along the circumference before/after adjusting its shape at several axial locations. The initial $-C_P$ deviates from the cavity number $\sigma_n = 2.576$, whereas the $-C_P$ on the converged vortex clusters around the σ_n except some locations where the wake panels rollup due to the lower velocity magnitude. If the circumference of the tip vortex after the adjustment is still close to the circular shape, differences of the predicted pressure $-C_P$ from σ_n are minute, e.g., $x = 1.1; \theta = 0°$ and $x = 2.3; \theta = 180°$. Similarly, for the same propeller subject to uniform inflow, predicted $-C_P$ on the same tip vortex locations show relatively even distributions around the cavitation number. On the other hand, if the adjusted shape has a region with concentrated panel ($x = 1.7; \theta = 270°$) or does not fully converge ($x = 1.1; \theta = 270°$), the predicted $-C_P$ still shows some differences from the target value, although the difference is not so significant as in the case of initial guess. The difference can be avoided by having more panels on the tip vortex perimeter, but to many panels in the limited vortex region shortens the side length of each panel surrounding the tip vortex cavity, and it leads to the crash in the code due to the insufficient panel area in the BEM matrix. In the present method, the tolerance for the minimum panel side length is 1.0×10^{-6} and the blade radius equal to 1.0. Figures 20–23 (middle) show the convergence history of the tip vortex section at the denoted downstream locations. The initial tip vortex has circular cross section with radius of 0.005. At each iteration, adjusted tip vortex shape is saved at the current time step and later used as the updated initial when any blade is located on the saved position. The first panel on the tip vortex is connected to the wake tip and any changes in its radius moves the center of tip vortex during the iterations.

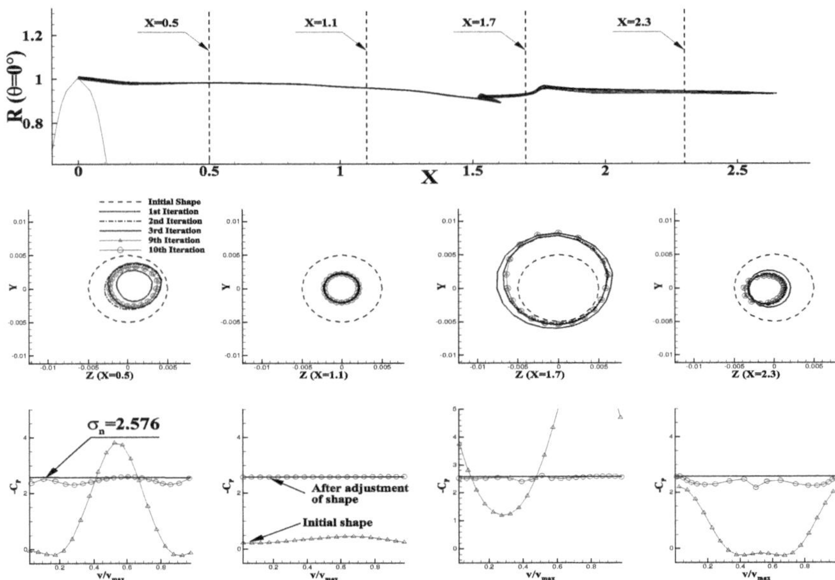

Figure 20. Projected view of the developed tip vortex cavitation from the BEM at blade angle $\theta = 0°$ on $X - R$ plane (**up**). Convergence history of the cavitation sections at the denoted downstream locations (**middle**). Pressure distributions on the tip cavity surface at the same locations from the initial and the converged shape (**bottom**).

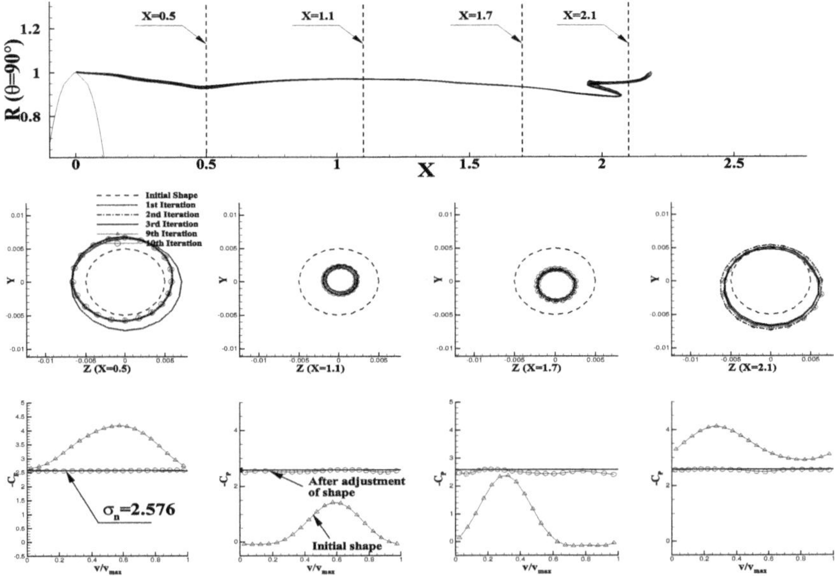

Figure 21. Projected view of the developed tip vortex cavitation from the BEM at blade angle $\theta = 90°$ on $X - R$ plane (**up**). Convergence history of the cavitation sections at the denoted downstream locations (**middle**). Pressure distributions on the tip cavity surface at the same locations from the initial and the converged shape (**bottom**).

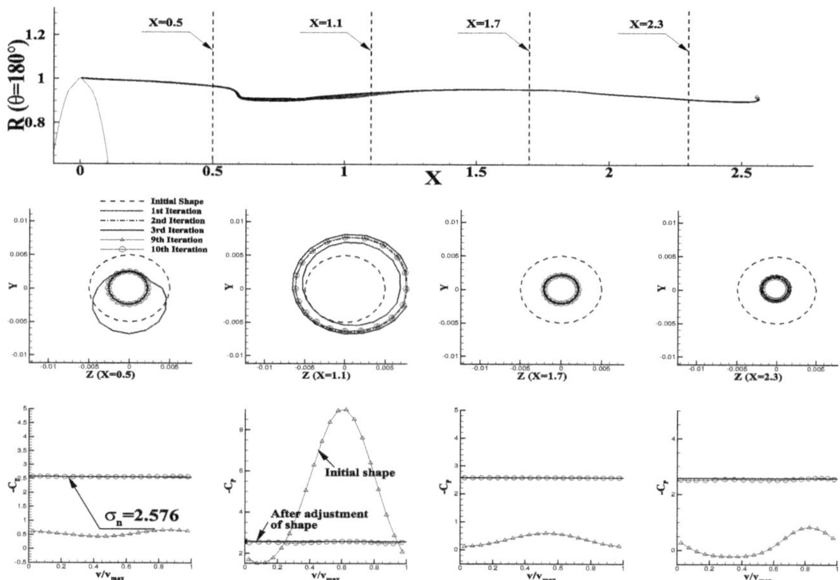

Figure 22. Projected view of the developed tip vortex cavitation from the BEM at blade angle $\theta = 180^{\circ}$ on $X - R$ plane (**up**). Convergence history of the cavitation sections at the denoted downstream locations (**middle**). Pressure distributions on the tip cavity surface at the same locations from the initial and the converged shape (**bottom**).

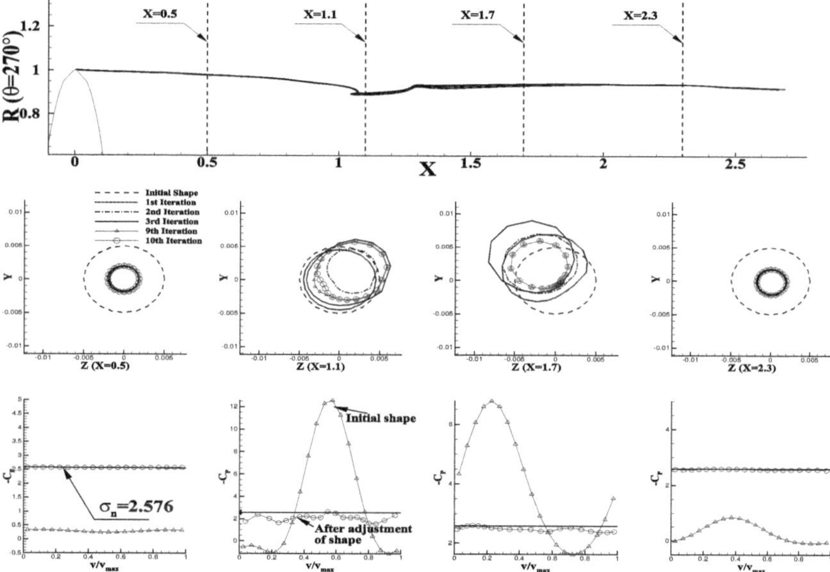

Figure 23. Projected view of the developed tip vortex cavitation from the BEM at blade angle $\theta = 270^{\circ}$ on $X - R$ plane (**up**). Convergence history of the cavitation sections at the denoted downstream locations (**middle**). Pressure distributions on the tip cavity surface at the same locations from the initial and the converged shape (**bottom**).

3.3. SSPA P2772 Propeller in Effective Wake Field and Hull Excitation

In this section, a series of numerical methods is applied to a propeller–hull interaction problem. The results are compared to the experiments conducted by Hallander and Gustafsson (2013) [2]. In the experiments, a 4-bladed SSPA P2772 propeller is tested with a 1:20 model hull positioned on the ceiling of the cavitation tunnel at its design draft. The geometrical arrangement of the eight transducers used to monitor the pressure signals and a model-scaled ship are shown in Figure 24. A left-handed propeller is used in the experiments and it is converted to the right-handed counterpart in the BEM model. The cavity observations and the pressure pulse measurements are performed with several loading conditions, and among those, design draft condition with $P/D(0.7R) = 0.87$ is chosen for the validation of the BEM results.

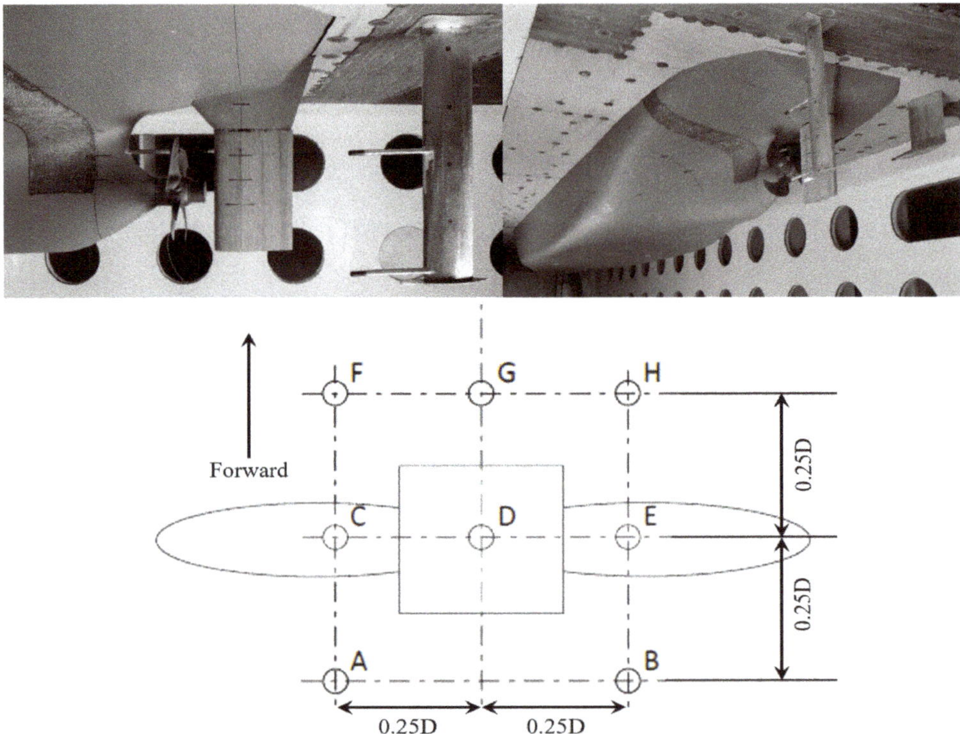

Figure 24. SSPA P2772 propeller mounted on a model-scaled ship in the cavitation tunnel (**up**). Arrangement of the pressure transducers on the hull surface viewed from tunnel ceiling (**down**).

The hydrodynamic BEM discretizes the propeller with 80 × 20 panels (chordwise × spanwise) to represent a single blade geometry. One-hundred-and-twenty panels in streamwise direction are used to align the trailing wake, which proceeds farther downstream beyond the rear end of the hull. Constant spacing and full cosine spacing are used to discretize the blade geometry in spanwise and chordwise directions, respectively. Although the hub geometry in the experiments has finite axial length, infinite hub with circular cross section is used in the numerical model to prevent the wake panels from intersecting the hub surface during the wake alignment procedure as the effective wake is unknown inside the hub.

The hydrodynamic BEM considers neither the hull geometry nor the rudder during the calculations. Instead, the BEM uses the effective wake (Figure 25) from the BEM/RANS coupling method, which accounts for the 3D effects of the upstream body (Figure 25) and vortical inflow/propeller interactions. Four million cells are used in RANS model. A slip-wall boundary condition is adopted on the propeller shaft, and a non-slip boundary condition is used to the ship hull and the rudder surface. It takes five iterations for the BEM/RANS coupling method to produce the converged forces and the effective wake field within 2 h of the wall-clock computing time. To study the effect of the rudder in the predicted propeller-induced hull pressure, a test was made by Su (2017) [18], in which the two cases are tested in the pressure BEM model: one includes only the upper part of the rudder, the other is totally neglecting the rudder geometry. According to the presented results, the difference is negligible except the transducers relatively close to the leading edge of the rudder. That is, including the upper part of the rudder locally affects the predicted hull pressures. He (2017) [30] introduced a numerical fence outside of the rudder surface to de-singularize the wake–rudder interaction. The numerical fence allows the wake sheet to wrap around the rudder surface, therefore prevents the wake from penetrating inside the rudder. In the present method, however, rudder geometry is ignored to avoid the numerical complexity in determining an artificial smoothing parameter for the numerical fence model, and to utilize the fact that the hull pressures on the given transducer locations have minor effect from the rudder geometry. Another test has been made on the rudder effects using 3D full-blown RANS, and the results are presented in Section 3.3.1.

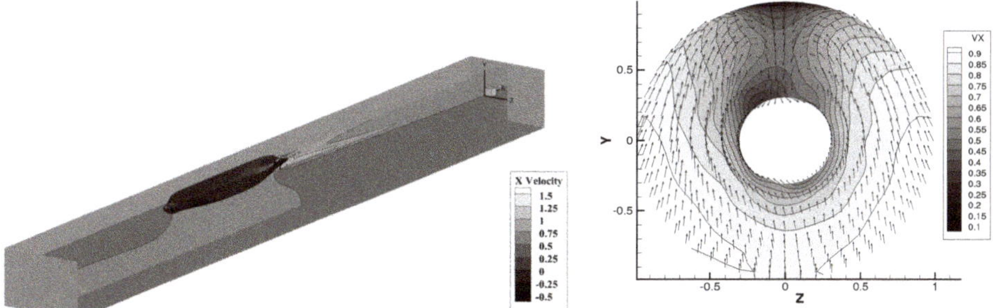

Figure 25. Computational domain of RANS model in the BEM/RANS coupling method (**left**). Non-axisymmetric effective wake predicted by the same method at $J_S = 0.7758$ (**right**). The effective wake is evaluated on the mid-chord disk; therefore, the actual effective wake may vary in the axial direction. Axial component is represented by contour plot, and the in-plane velocities are presented by arrows.

Figure 26 shows open water characteristics of P2772 propeller predicted by the BEM and the experiment in uniform flow. Good agreement is established between the two approaches over a wide range of advance ratios. In the experiment, the rate of revolutions is kept constant while the speed of advance is varied so that a loading range of the propeller can be investigated.

In the pressure-BEM model, an external flow problem is solved using BEM panels placed on the surface of a closed body. Practically, the fore part of the hull has very little influence on the transducer locations where the propeller-induced pressure is measured; therefore, only the rear part of the hull is included in the BEM model, as shown in Figure 27. An image hull model is included to consider the tunnel effect. If the hull geometry is a perfectly closed body, i.e., with consideration of the fore part and the image model, summation of the influence function on a BEM panel from other panels including itself is equivalent to 4π. In the present problem, ratio of the summation on the panel at the transducer location to 4π is found to be equal to 92.5%, therefore neglecting the fore hull geometry is a reasonable assumption for the pressure predictions, and also for the efficient calculations. The hull pressure P is non-dimensionalized by $\rho n^2 D^2$ in the pressure-BEM, where the n and D are the propeller revolution per second (rps) and propeller diameter, respectively.

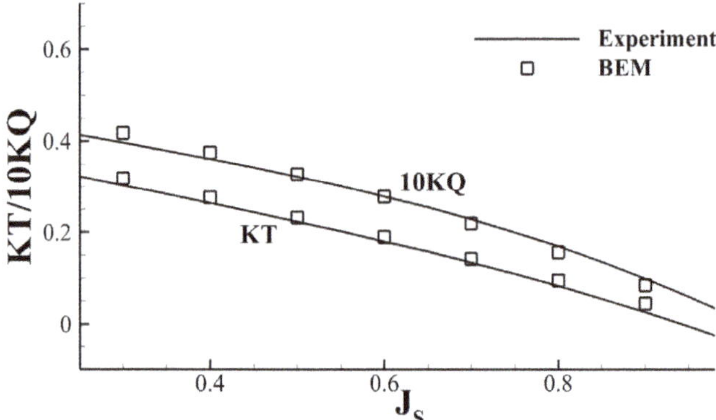

Figure 26. Open water characteristics of P2772 propeller predicted by BEM and the model scale experiment; $P/D = 0.87$ $(r/R = 0.7)$.

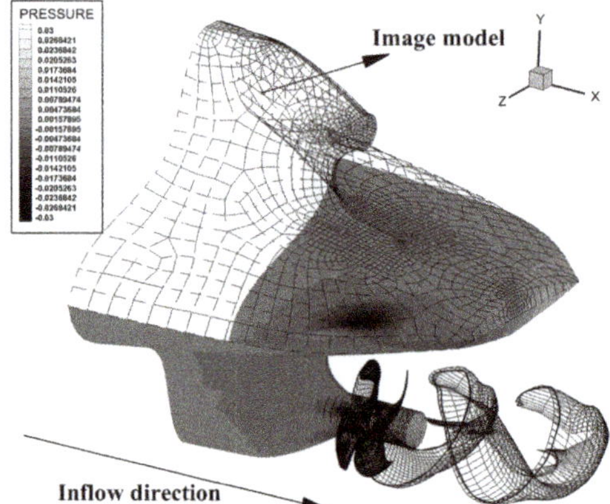

Figure 27. Pressure-BEM model without rudder geometry. Contour plot represents the dimensionless propeller-induced hull pressures $(P/\rho n^2 D^2)$ at time step $t = 0$. The fore part of the hull is neglected for the sake of the efficient computation. The panels on the image model are not shown for transparency.

Figure 28 shows the fully aligned wake for SSPA P2772 model propeller. When the blade angle is approximately $-15°$ to $+15°$, wake panels are legged behind its neighboring wake in the axial direction because of the relatively slower axial velocity around the blade tip and near the hub as opposed to the other angular positions. Unsteady wake alignment scheme considers the non-axisymmetric component of the effective wake, therefore the wake geometry varies in both the radial and tangential directions. More elaborate cases, such as non-periodic unsteady inflow, hull–propeller–rudder interaction problem, or the time-accurate problem with accelerating inflow, can be handled by coupling BEM with RANS solver as the BEM/RANS interactive method [16].

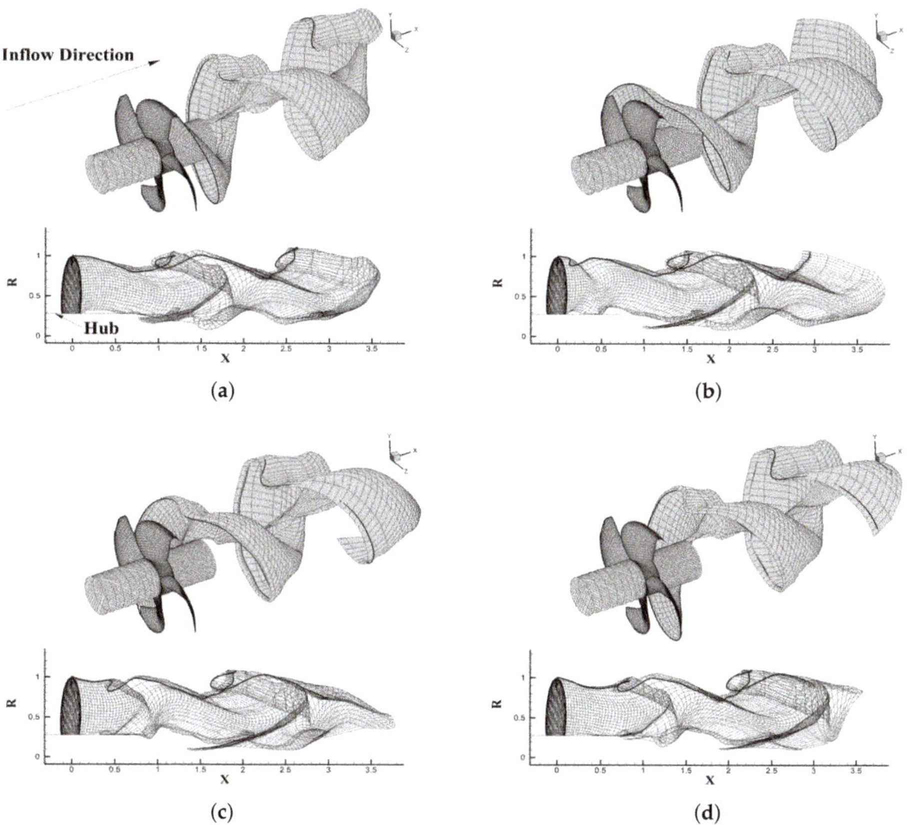

Figure 28. Fully aligned wake of SSPA P2772 propeller with infinite hub at the blade angle (**a**) 0°, (**b**) 90°, (**c**) 180°, and (**d**) 270° positions. Only the key wake is shown.

A cavitating case is investigated using the hydrodynamic BEM with the tip vortex cavitation model to validate the predicted cavity patterns. SSPA P2772 propeller is operated at $J_S = 0.7758$, and the cavity number $\sigma_n = 3.507$ is adopted as reported in the experiments. Figure 29 shows the predicted tip cavitation patterns during a propeller period with time step $\Delta\theta = 6°$. Figure 30 compares the cavity patterns with the photographs taken during the experiments at several blade angles. Both the BEM and the experiments detect a small amount of sheet cavity toward the blade tip when the key blade is passing 0° to 30° angular positions. Tip vortex cavitation detaches from the blade tip beyond 80° position due to the weak shedding vortex strengths Γ and is reattached to the blade tip as the blade enters 18° position. At 0° position, the key blade captures a small amount of sheet cavity from $0.9R$ to blade tip, and then the sheet cavity develops into the tip vortex cavitation afterwards. Sheet cavity also grows, but soon disappears as the propeller proceeds into the higher pressure zone. BEM shows the tip vortex inception around 20° blade angle, at which the measurement also captures the transition from the sheet cavitation to the tip vortex cavitation. As the key blade further rotates, tip vortex keeps growing and is eventually detached from the tip to follow the local stream.

Note that the BEM still has strong tip vortex cavitation downstream after the detachment, while the experiment shows very little amount at the same downstream location. It is because of the dissipation of the shedding vortex in the experiment, whereas the tip vortex panels in the BEM represent the concentrated vortex with its core inside. It is similar to the wake panels representing the concentrated shedding vortex by its panel edges.

The actual panel numbers constituting the BEM matrix can be adjusted at each time step allowing the growth or collapse of the tip vortex cavitation based on the vortex core strength. This task requires careful attention on the I/O operations such that the surplus data from the larger panel case are not overwritten or remaining in the smaller panel cases.

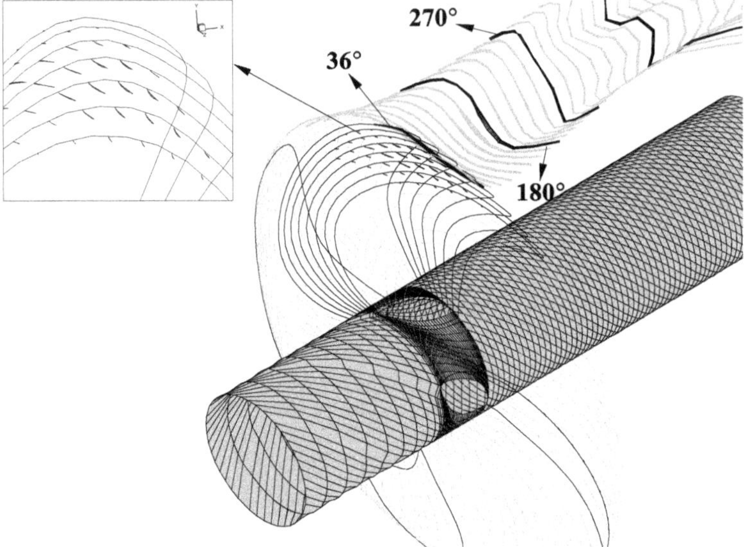

Figure 29. Developed tip vortex cavitation modeled by the BEM during one propeller revolution; SSPA P2772 propeller, $\sigma_n = 3.507$, $J_S = 0.7758$.

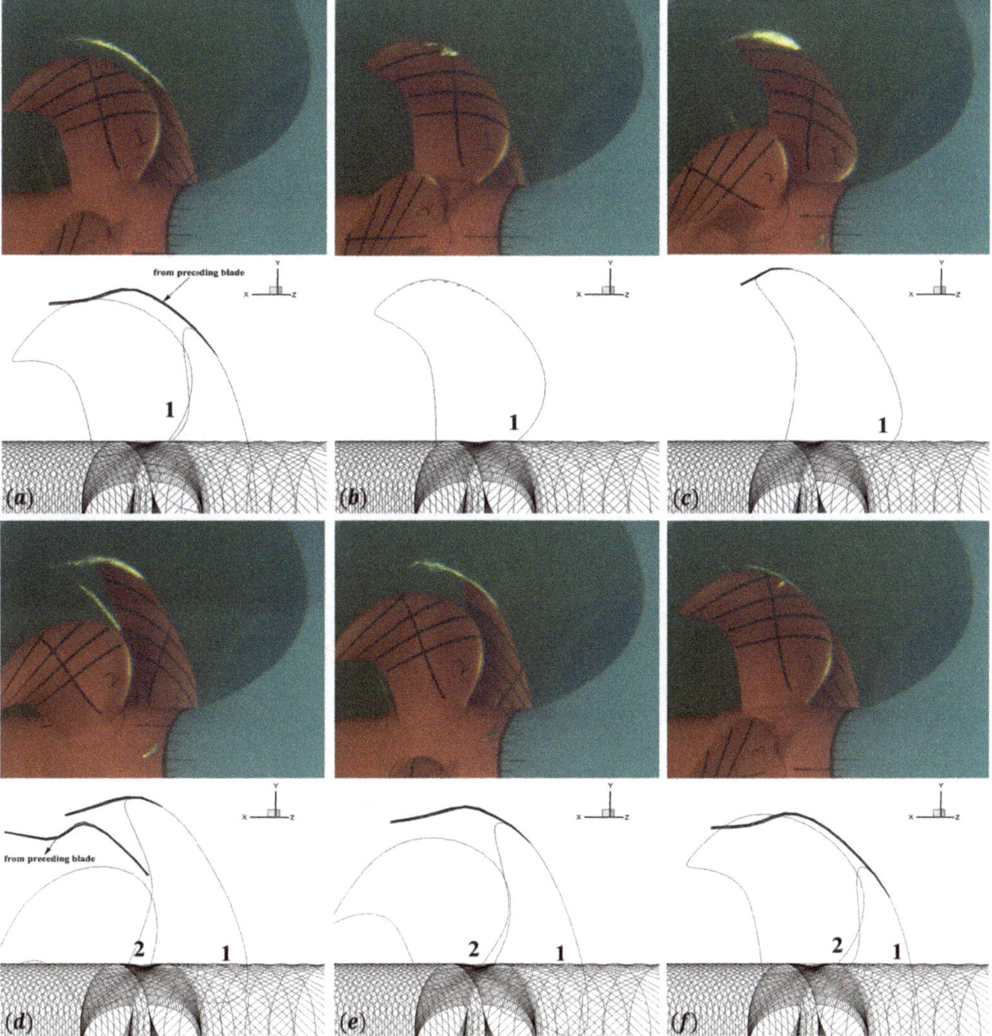

Figure 30. Comparisons of the tip vortex caivty predicted by BEM (bottom of every pair) with the experiments for SSPA P2772 propeller; Key blade angle: (**a**) 340°, (**b**) 0°, (**c**) 20°, (**d**) 40°, (**e**) 60°, and (**f**) 80°.

3.3.1. Fully-Wetted Condition: Effects of Rudder on the Hull Pressures

The predicted hull pressure fluctuations are compared to the experiments where the hull pressures are measured at the eight different transducers. Arrangement of the pressure transducers on the hull is shown in Figure 31. The pressure history is recorded under the draft load condition (the LC1 condition [2]). To investigate the rudder effects on the hull pressures especially near the rudder, 3D full-blown RANS simulations are conducted using ANSYS/Fluent (version 18.2) with/without rudder geometry. Parameters for loading conditions are shown in Table 2.

Table 2. Loading conditions used in RANS solver.

Advance Ratio ($J_S = V_S/nD$)	Propeller Diameter D (m)	Water Density ρ (kg/m^3)	Ship Speed V_S (m/s)
0.7758	2.0	1.0	1.0

Unstructured meshing model is used for both the inner zone where the blade rotates with sliding mesh motion and the outer zone containing hull and rudder. $k-\omega\ SST$ turbulent model is adopted with a Reynold's number of 1.0×10^6. QUICK and SIMPLEC schemes are used for the spatial discretization and the pressure correction, respectively. Over 1.76 million polyhedral meshes are used to discretize the fluid domain. Second-order implicit scheme is used for time stepping scheme in transient model. Non-slip boundary condition is applied on the ship hull, rudder, and propeller blades; slip boundary condition on the hub; inlet boundary condition is set at the upstream surface with 1.0 m/s inlet velocity; and zero pressure gradient is imposed at the downstream boundary. Calculation time takes over 16 h on 192 Intel Xeon Platinum 8160 2.1 GHz cores to reach the converged blade thrust after ~8200 iterations with time step $\delta t = 0.005$. Convergence residuals are set 4.0×10^{-5} for the continuity and the momentum terms. Figures 31 and 32 show the hull geometries used in RANS. When neglecting the rudder, flow separation occurs at the sharp corner of the hub end. To avoid convergence issue due to the strong swirls in the region, hub geometry is extended by $0.8R$ and closed with elliptic hub. Despite the change in the geometry, it has negligible effects on the hull pressures with far distance from the hull.

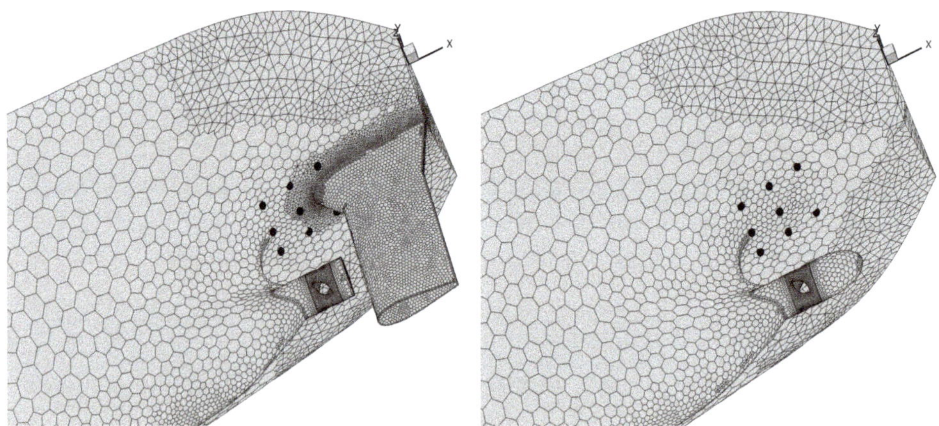

Figure 31. Grids on the hull geometry with (**left**) and without (**right**) rudder. Transducer locations are denoted in black scatters. Propeller mesh is omitted in the figure.

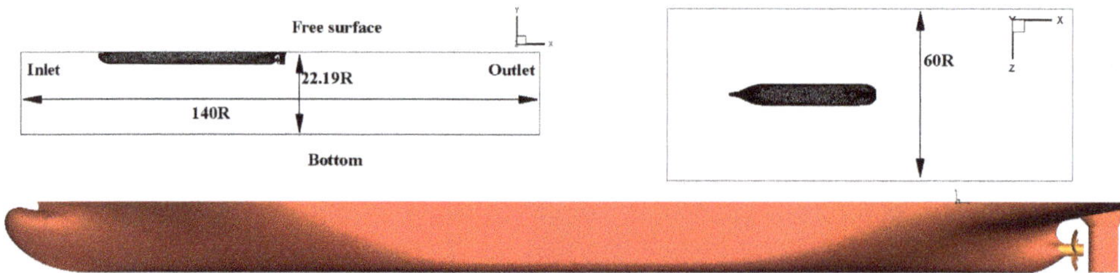

Figure 32. Boundaries in the finite volume mesh used in RANS simulations for the design draft condition.

Figure 33 shows the fluctuating hull pressures predicted by RANS during about one-fourth period. To measure the pressure fluctuations at the transducer locations, vertex average is used which sums the hull pressure of the vertex values and divides by the total vertex numbers.

Figure 34 first compares the pressure history predicted by BEM with the measurements in fully-wetted condition. Results are shown up to the three-fourths period with the pressures non-dimensionalized by $\rho n^2 D^2$. The current method can predict the peak amplitudes on the most transducers locations as close to the experiments except the transducers A, B, and D where the locations are right around the rudder. Importantly, effects of the rudder geometry on the hull pressures is very minor on the given measuring points. RANS results predict pretty close peak to peak pressure amplitudes regardless of the rudder geometry on all transducer locations (transducers A to H). As the propeller loading is one of the major contributors to the propeller-induced hull pressures in fully-wetted condition, the same loading conditions would predict the similar pressure pulses at atmospheric pressure.

In the downstream locations (transducers A and B), both the BEM and RANS capture relatively small pressure amplitudes compared to the experiments. According to the RANS predictions in Figure 33, minute changes in the locations of the transducers A and B bring them to the higher pressure zone around the rudder, where the downstream flow along the hull surface stagnates due to the upcoming flow from the rudder. In the region where the pressure distribution has a high gradient, slight changes in the measuring points might produce large differences.

RANS predictions on the propeller head and upstream points (transducers D and G) show small excitation amplitudes compared to the BEM and experiment. A possible explanation for the discrepancy could be in the effective wake. Unlike the inclined-shaft flow, which has constant inclination in the inflow direction, the effective wake in the propeller–hull–rudder interaction problem has upstream body effects which makes streamlines nonuniform in all directions, combined with the propeller-inflow interaction. 3D variation of the effective wake, and thus might affect the overall propeller loading and wake geometry during the alignment procedure. In the present method, the 2D wake field evaluated on the propeller mid-chord disk is used for evaluating the strength of the source singularities on the blade, and for the wake alignment regardless of downstream locations of shedding vortices. As for the comparisons, tests have been made on tip loaded propellers (TLPs) [31] laid under flat plates in uniform flow where no upstream body is assumed. In the cases, pressure pulses on the flat plate from the pressure BEM and the experiments show reasonable agreement with each other at the propeller head point (transducer D) at the design J_S, and also in the increasing or decreasing 20% on the J_S values. Therefore, consideration of the axial variation in the wake field and also the wake/rudder interaction is expected to improve the results, remaining as future research.

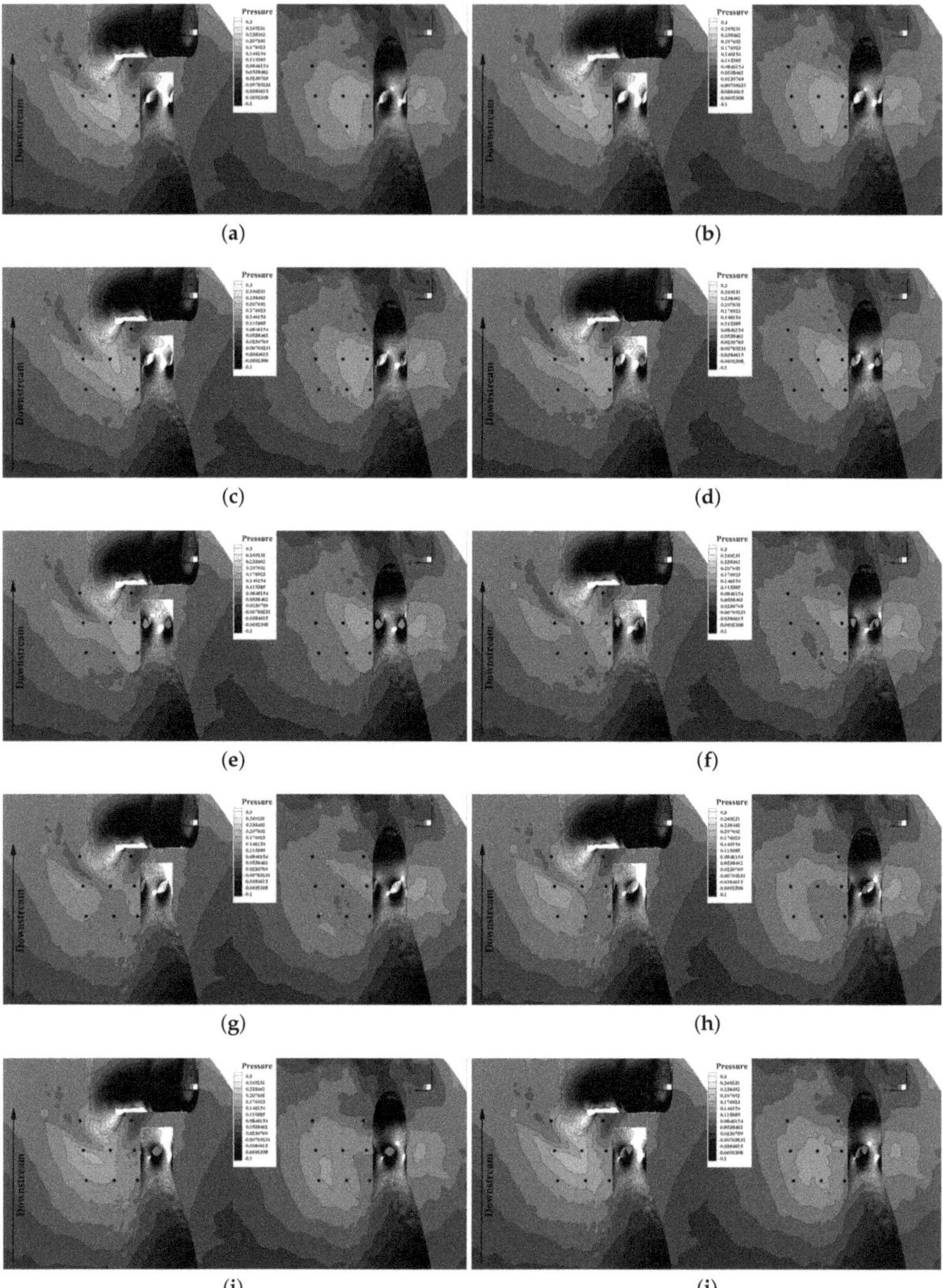

Figure 33. Contour plots of the fluctuating hull pressures (unit: N) at the key blade angle (**a**) $\theta = 0.0°$, (**b**) $\theta = 5.8°$, (**c**) $\theta = 11.6°$, (**d**) $\theta = 17.4°$, (**e**) $\theta = 23.2°$, (**f**) $\theta = 34.8°$, (**g**) $\theta = 46.4°$, (**h**) $\theta = 58.0°$, (**i**) $\theta = 69.6°$, and (**j**) $\theta = 81.2°$ with/without rudder. Note fluid density in RANS is 1.0 kg/m^3 to be directly compared to the BEM results. Black scatters denote the transducer locations.

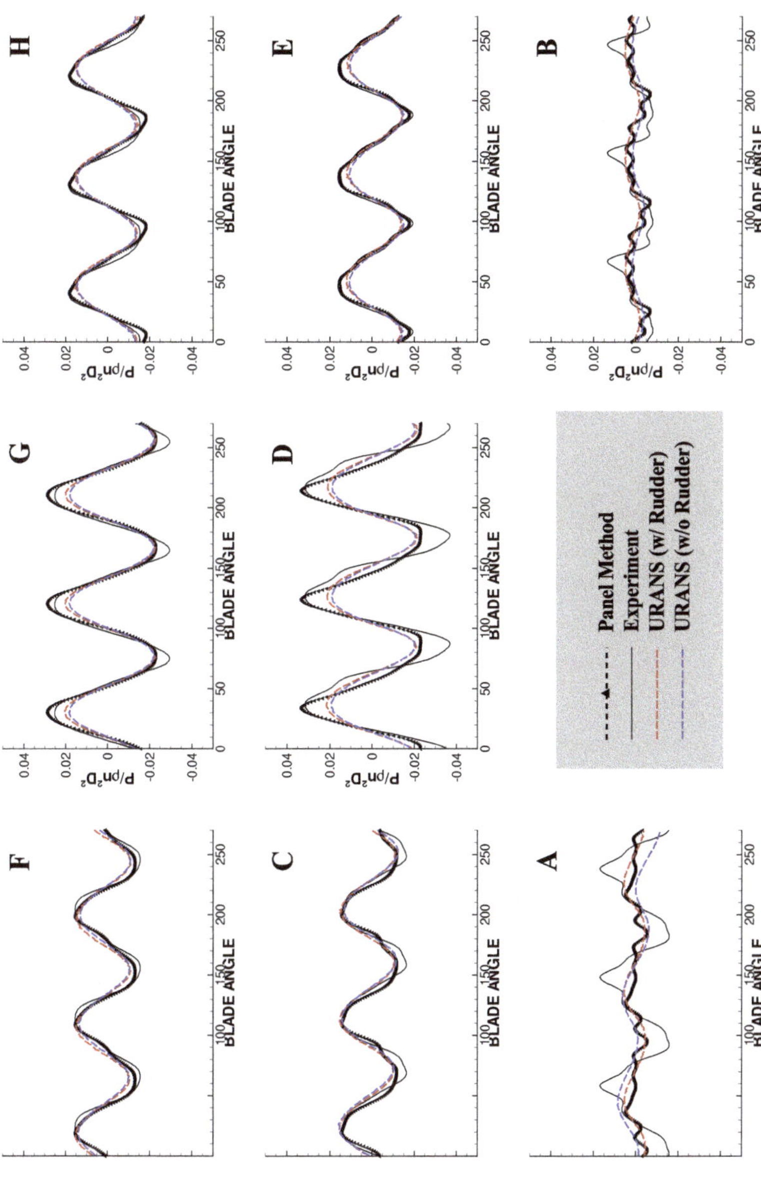

Figure 34. Comparisons of the hull pressures ($P/\rho n^2 D^2$; y axis) in wetted condition between the pressure-BEM, RANS, and the experiments during the three-fourths period of propeller revolution. Results are only shown between 0° to 270° in blade angle (x axis). Locations of the transducers (A–H) are indicated in the figures.

3.3.2. Cavitating Condition: Effects of Tip Vortex Cavitation on the Hull Pressures

Figure 35 shows the hull pressures in cavitating condition with $\sigma_n = 3.507$. In this case, both approaches predict the strong high-frequency components in the pressures harmonics, compared to the wetted condition (Figure 34) at all transducer locations. Transducers F, G, and H in the upstream regions seem to follow the overall fluctuation patterns of the measurements. Both the harmonic amplitudes and high-order harmonics are in reasonable agreement. For other transducers located relatively downstream, BEM predicts somewhat differences in the harmonic amplitudes. The non-axisymmetric sheet cavity, tip vortex cavity, and the wake field compositively provide a strong influence on the hull pressures that the BEM might not be able to follow with general solutions.

Quantitative comparisons are presented in Figure 36 which shows the pressure pulse amplitudes up to the 5th harmonics. Compared to the cavitating conditions with tip vortex cavitation, the fully-wetted results (Figure 37) has weaker high order amplitudes which vanish rapidly as the order increases. On the other hand, the cavitating condition has strong high order pulses on the overall locations. Near the propeller and downstream locations (transducers A–E), the highest order term behaves as the significant encouragement on the pressure frequency due to the close distance from the tip vortex cavitation. Tip vortex cavity, which adjusts its transverse sections in space and time by satisfying the boundary conditions, improves the correlations with the measurements by enabling the present method to detect high-frequency components in the pressure fluctuations. Sheet cavity provides some high frequency impact on the pressure fluctuations but it soon decreases in the higher order.

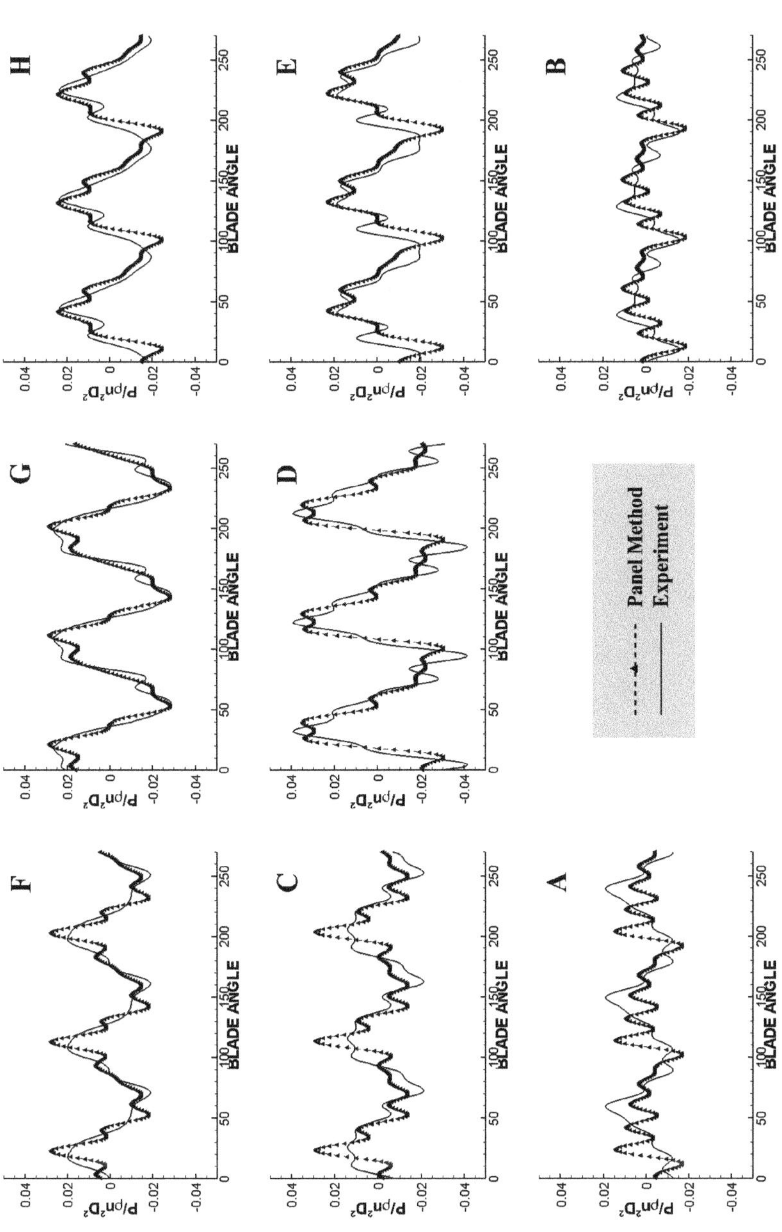

Figure 35. Comparisons of the hull pressures ($P/\rho n^2D^2$; y axis) in cavitating conditions between the pressure-BEM and the experiments during the three-fourths period of propeller revolution. Results are only shown between 0° to 270° in blade angle (x axis). Locations of the transducers (A–H) are indicated in the figures.

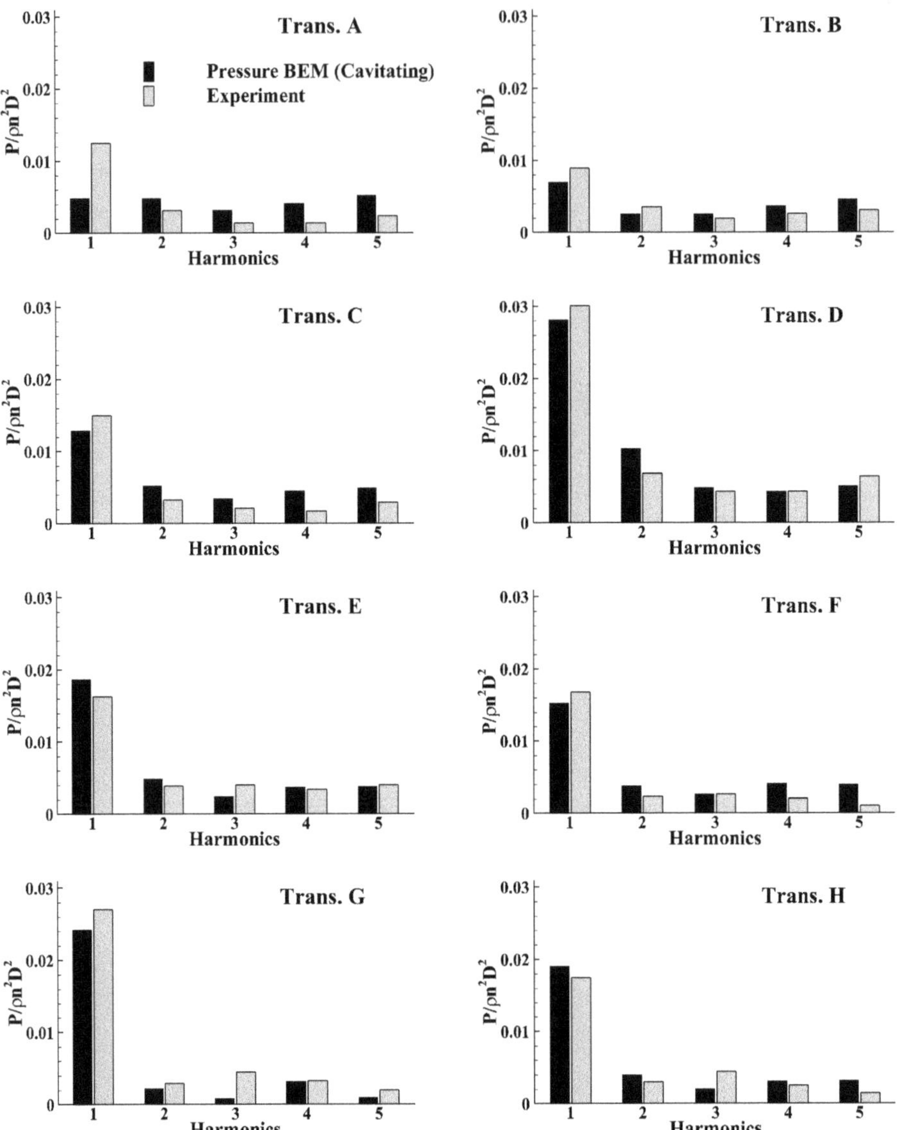

Figure 36. Comparisons of the pressure harmonics predicted by the BEM in cavitating conditions including the tip vortex cavitation model with the measurements [2].

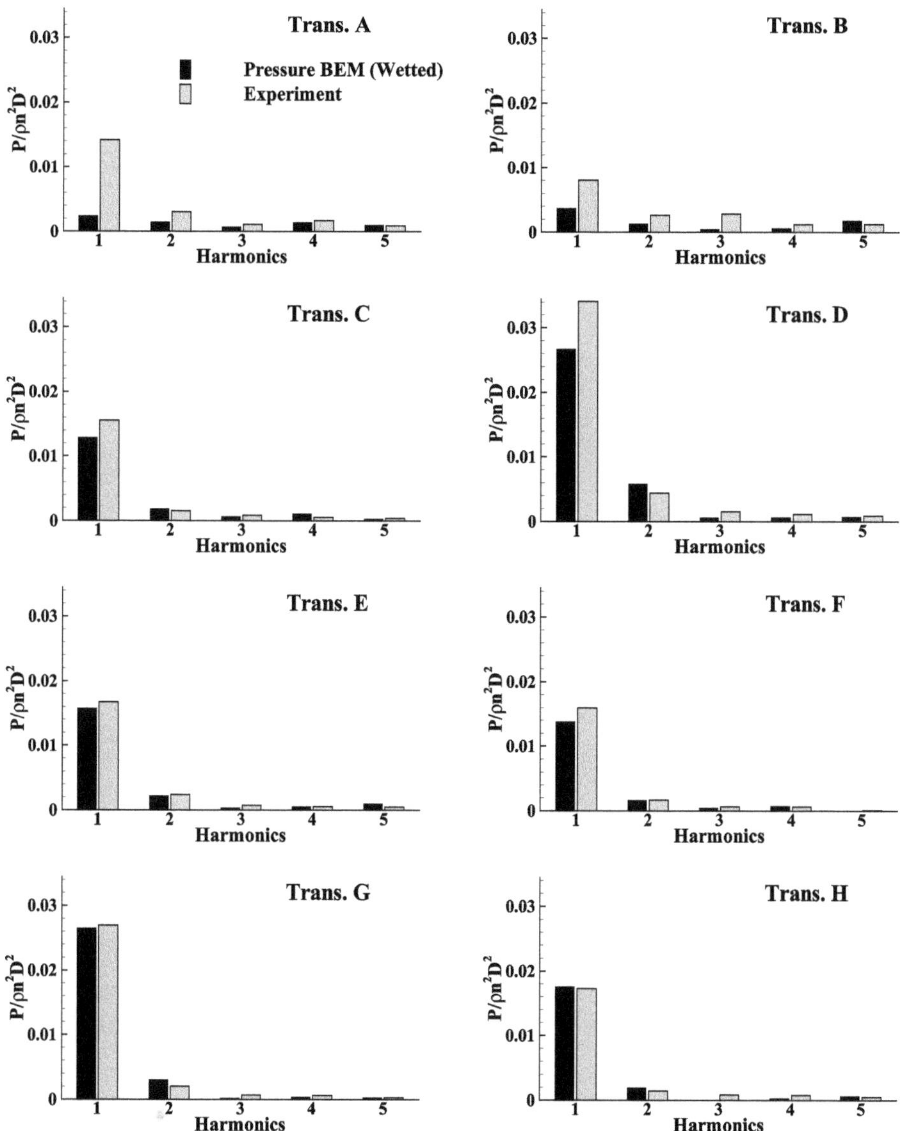

Figure 37. Comparisons of the pressure harmonics predicted by the BEM in fully wetted condition with the measurements [2].

4. Conclusions and Future Work

In this article, several modifications of previously developed methods are introduced to predict the evolution and shape of developed tip vortex cavitation and its effect on the hull pressures. An elaborate algorithm was presented that determines the shape of the tip vortex cavity in time and space, by requiring the pressure over its surface to be equal to vapor pressure. This algorithm is capable of handling excessive twisting of the tip vortex cavity, and sets criteria to prevent the cavity from reaching a minimum radius below which a viscous core might appear, that the current method cannot model. The method allows for tip vortex cavities with transverse sections of general shape, which vary in both space and time.

First, the methods are applied in the case of non-cavitating flow in inclined shaft flow, and the case of a propeller operating behind a hull with a rudder. In both cases, the unsteady wake alignment model is employed. The results from our methods are compared with those measured in an experiment, as well as those predicted from unsteady RANS where the propeller was modeled as a solid boundary. In the case of hull and rudder, the hull pressures were also evaluated from unsteady RANS, and the effect of the rudder on the predicted hull pressures was found to be minor. The agreement among the predictions from our method, RANS, and the measurements was found to be satisfactory.

Last, the method was applied to cases where partial cavitation and developed tip vortex cavitation were both present. Results from the present method are shown in the case of a cavitating propeller inside a tunnel with nonuniform inflow, and in the case of propeller behind a hull with rudder. The cavity patterns on the blade and the trajectories of the developed tip vortex cavities are in reasonable agreement with those observed in the experiments. Furthermore, the predicted hull pressures are in reasonable agreement with those measured, and, in particular, the present method captures higher harmonics in the pressure fluctuations, which are also present in the experiment and have been found in the past to be associated with tip vortex cavity pulsations.

Future work includes studying the blade boundary layer effect on the predicted cavities and the corresponding hull pressures, incorporating the unsteady wake–rudder interaction in the numerical scheme and studying the effect of a finite speed of sound for prediction of far-field underwater noise. Besides, considering the variation of the effective wake along the blade surface and in the wake field is expected to improve the propeller loading in behind condition, and consequently the hull pressures near the propeller tip region.

Author Contributions: Conceptualization, S.K. and S.A.K.; methodology, S.K. and S.A.K.; software, S.K.; validation, S.K. and S.A.K.; formal analysis, S.K. and S.A.K.; investigation, S.K. and S.A.K; data curation, S.K.; writing–original draft preparation, S.K.; writing–review and editing, S.K. and S.A.K.; visualization, S.K.; supervision, S.A.K.; project administration, S.A.K.; funding acquisition, S.A.K.; Overall, S.K. 80% and S.A.K. 20%. Both authors agreed on the overall contributions. All authors have read and agreed to the published version of the manuscript.

Funding: This research was funded by the US Office of Naval Research (Grant Number N00014-14-1-0303 and N00014-18-1-2276; Ki-Han Kim) and by Phase VIII of the "Consortium on Cavitation Performance of High Speed Propulsors".

Acknowledgments: Support for this research was provided by the US Office of Naval Research (Grant Number N00014-14-1-0303 and N00014-18-1-2276; Ki-Han Kim) and by Phase VIII of the "Consortium on Cavitation Performance of High Speed Propulsors". Initial RANS models used in this article were provided by Yiran Su and modified for the present cases by the authors.

Conflicts of Interest: The authors declare no conflict of interest.

References

1. Breslin, J.P.; van Houten, R.J.; Kerwin, J.E.; Johnsson, C.A. Theoretical and Experimental Propeller Induced Hull Pressures Arising from Intermittent Blade Cavitation, Loading, and Thickness. *Trans. Soc. Nav. Archit. Mar. Eng.* **1982**, *99*, 111–151.
2. Hallander, J.; Gustafsson, L. *Aquo Model Test for M/t Olympus*; Technical Report RE40116086-01-00-A; SSPA Sweden AB: Göteborg, Sweden, 2013.
3. Hwang, Y.; Sun, H.; Kinnas, S.A. Prediction of Hull Pressure Fluctuations Induced by Single and Twin Propellers. In Proceedings of the Propellers/Shafting '06 Symposium, The Society of Naval Architects & Marine Engineers (SNAME), Williamsburg, VA, USA, 12–13 September 2006.
4. Bosschers, J. Propeller Tip-Vortex Cavitation and its Broadband Noise. Ph.D. Thesis, University of Twente, Enschede, The Netherlands, 2018.
5. Bosschers, J. A Semi-empirical Prediction Method for Broadband Hull Pressure Fluctuations and Underwater Radiated Noise by Propeller Tip Vortex Cavitation. *J. Mar. Sci. Eng.* **2018**, *6*, 49. [CrossRef]

6. Göttsche, U.; Lampe, T.; Scharf, M.; Abdel-Maksoud, M. Evaluation of Underwater Sound Propagation of a Catamaran with Cavitating Propellers. In Proceedings of the 6th International Symposium on Marine Propulsors (SMP'19), Rome, Italy, 26–30 May 2019.
7. Tani, G.; Viviani, M.; Felli, M.; Lafeber, F.H.; Lloyd, T.; Aktas, B.; Atlar, M.; Seol, H.; Hallander, J.; Sakamoto, N.; et al. Round Robin Test on Radiated Noise of a Cavitating Propeller. In Proceedings of the 6th International Symposium on Marine Propulsors (SMP'19), Rome, Italy, 26–30 May 2019.
8. Gosda, R.; Berger, S.; Abdel-Maksoud, M. Investigation of Reynolds Number Scale Effects on Propeller Tip Vortex Cavitation and Propeller-Induced Hull Pressure Fluctuations. In Proceedings of the 10th International Symposium on Cavitation (CAV2018), Baltimore, MD, USA, 14–16 May 2018.
9. Su, Y.; Kim, S.; Du, W.; Kinnas, S.A.; Grekula, M.; Hallander, J.; Li, Q. Prediction of the Propeller-induced Hull Pressure Fluctuation via a Potential-based Method: Study of the Rudder Effect and the Effect from Different Wake Alignment Methods. In Proceedings of the 5th International Symposium on Marine Propulsors (SMP'17), Espoo, Finland, 12–15 June 2017.
10. Kim, S.; Su, Y.; Kinnas, S.A. Prediction of the Propeller-induced Hull Pressure Fluctuation via a Potential-based Method: Study of the Influence of Cavitation and Different Wake Alignment Schemes. In Proceedings of the 10th International Symposium on Cavitation (CAV2018), Baltimore, MD, USA, 14–16 May 2018.
11. Kim, S.; Du, W.; Kinnas, S.A. *PROPCAV (Version R2019) User's Manual*; Ocean Engineering Report 18-2; Ocean Engineering Group, UT Austin: Austin, TX, USA, 2019.
12. Kim, S.; Kinnas, S.A. Prediction of Unsteady Developed Tip Vortex Cavitation and its Effect on the Induced Hull Pressures. In Proceedings of the 6th International Symposium on Marine Propulsors (SMP'19), Rome, Italy, 26–30 May 2019.
13. Kim, S. An Improved Full Wake Alignment Scheme for the Prediction of Open/ducted Propeller Performance in steady and Unsteady Flow. Master's Thesis, Ocean Engineering Group, The University of Texas at Austin, Austin, TX, USA, 2017.
14. Greely, D.S.; Kerwin, J.E. Numerical Methods for Propeller Design and Analysis in Steady flow. *Trans. Soc. Nav. Archit. Mar. Eng.* **1982**, *90*, 415–453.
15. Kinnas, S.A.; Fan, H.; Tian, Y. A Panel Method with a Full Wake Alignment Model for the Prediction of the Performance of Ducted Propellers. *J. Ship Res.* **2015**, *59*, 246–257. [CrossRef]
16. Su, Y.; Kinnas, S.A. A Time-Accurate BEM/RANS Interactive Method for Predicting Propeller Performance Considering Unsteady Hull/Propeller/Rudder Interaction. In Proceedings of the 32nd Symposium on Naval Hydrodynamics, Hamburg, Germany, 5–10 August 2018.
17. Lee, H.; Kinnas, S.A. Application of BEM in the Prediction of Unsteady Blade Sheet and Developed Tip Vortex Cavitation on Marine Propellers. *J. Ship Res.* **2004**, *48*, 15–30.
18. Su, Y.; Kim, S.; Kinnas, S.A. Prediction of Propeller-induced Hull Pressure Fluctuations via a Potential-based Method: Study of the Effect of Different Wake Alignment Methods and of the Rudder. *J. Mar. Sci. Eng.* **2018**, *6*, 52. [CrossRef]
19. Tian, Y.; Kinnas, S.A. A Wake Model for the Prediction of Propeller Performance at Low Advance Ratio. *Int. J. Rotating Mach.* **2012**, *2012*, 372364. [CrossRef]
20. Kim, S.; Kinnas, S.A.; Du, W. Panel Method for Ducted Propellers with Sharp Trailing Edge Duct with Fully Aligned Wake on Blade and Duct. *J. Mar. Sci. Eng.* **2018**, *6*, 89. [CrossRef]
21. Lee, H. Modeling of Unsteady Wake Alignment and Developed Tip Vortex Cavitation. Ph.D. Thesis, Ocean Engineering Group, The University of Texas at Austin, Austin, TX, USA, 2002.
22. Kinnas, S.A.; Fine, N.E. A Numerical Nonlinear Analysis of the Flow Around 2-D and 3-D Partially Cavitating Hydrofoils. *J. Fluid Mech.* **1993**, *254*, 151–181. [CrossRef]
23. Kinnas, S.A. A Note on the Bernoulli Equation for Propeller Flows: The Effective Pressure. *J. Ship Res.* **2006**, *54*, 355–359.
24. Arndt, R.; Arakeri, V.; Higuchi, H. Some Observations of Tip-vortex Cavitation. *J. Fluid Mech.* **1991**, *229*, 269–289. [CrossRef]
25. Choi, J.; Kinnas, S.A. An Unsteady Three-dimensional Euler Solver Coupled with a Cavitating Propeller Analysis Method. In Proceedings of the 23rd Symposium on Naval Hydrodynamics, Val de Reuil, France, 17–22 September 2000.

26. Boswell, R.; Jessup, S.; Kim, K.; Dahmer, D. *Single Blade Loads on Propellers in Inclined and Axial Flows*; Technical Report DTNSRDC-84/084, David W. Taylor Naval Ship Research and Development Center, Bethesda, MD, USA, 1984.
27. Kerwin, J.E.; Lee, C. Prediction of Steady and Unsteady Marine Propeller Performance by Numerical Lifting-Surface Theory. *Trans. Soc. Nav. Archit. Mar. Eng.* **1978**, *86*, 218–256.
28. Mishima, S.; Kinnnas, S.A.; Egnor, D. *The CAvitating PRopeller EXperiment (CAPREX), Phases I & II*; Technical Report; Department of Ocean Engineering, MIT: Cambridge, MA, USA, 1995.
29. Choi, J.; Kinnas, S.A. A 3-D Euler Solver and Its Application on the Analysis of Cavitating Propellers. In Proceedings of the 25th ATTC American Towing Tank Conference, Iowa City, IA, USA, 24–25 September 1998.
30. He, L.; Kinnas, S.A. Numerical Simulation of Unsteady Propeller/rudder Interaction *Int. J. Nav. Archit. Mar. Eng.* **2017**, *9*, 677–692.
31. Kim, S.; Kinnas, S.A. Prediction of Performance of Tip Loaded Propeller and Its Induced Pressures on the Hull (under review). In Proceedings of the ASME 39th International Conference on Ocean, Offshore and Arctic Engineering (OMAE), Fort Lauderdale, FL, USA, 28 June–3 July 2020.

Article

Cavitation on Model- and Full-Scale Marine Propellers: Steady and Transient Viscous Flow Simulations at Different Reynolds Numbers

Ville Viitanen [1,*], Timo Siikonen [2] and Antonio Sánchez-Caja [1]

1 VTT Technical Research Centre of Finland Ltd, 02150 Espoo, Finland; Antonio.Sanchez@vtt.fi
2 Department of Mechanical Engineering, Aalto University, 02150 Espoo, Finland; timo.siikonen@aalto.fi
* Correspondence: ville.viitanen@vtt.fi

Received: 9 January 2020; Accepted: 5 February 2020; Published: 20 February 2020

Abstract: In this paper, we conducted numerical simulations to investigate single and two-phase flows around marine propellers in open-water conditions at different Reynolds number regimes. The simulations were carried out using a homogeneous compressible two-phase flow model with RANS and hybrid RANS/LES turbulence modeling approaches. Transition was accounted for in the model-scale simulations by employing an LCTM transition model. In model scale, also an anisotropic RANS model was utilized. We investigated two types of marine propellers: a conventional and a tip-loaded one. We compared the results of the simulations to experimental results in terms of global propeller performance and cavitation observations. The propeller cavitation, near-blade flow phenomena, and propeller wake flow characteristics were investigated in model- and full-scale conditions. A grid and time step sensitivity studies were carried out with respect to the propeller performance and cavitation characteristics. The model-scale propeller performance and the cavitation patterns were captured well with the numerical simulations, with little difference between the utilized turbulence models. The global propeller performance and the cavitation patterns were similar between the model- and full-scale simulations. A tendency of increased cavitation extent was observed as the Reynolds number increases. At the same time, greater dissipation of the cavitating tip vortex was noted in the full-scale conditions.

Keywords: marine propeller; CFD; cavitation simulation; turbulence modeling; scale effects

1. Introduction

Marine propeller cavitation can appear in a range of different forms, depending on the operating conditions of the propeller. Typical cavitation types encountered include steady attached sheet cavitation on the blade surface, transient bubble or cloud cavitation on the blade or in the wake, and vortex cavitation at the propeller tip and hub. The propeller cavitation can cause many detrimental effects, and the various types of cavitation usually result in different consequences from the points of view of the propeller material (e.g., erosion), the passengers (e.g., discomfort), or the ship's mission (e.g., identification and detectability). This further amplifies the need of accurate cavitation prognoses via experimental and numerical methodologies.

The propeller performance characteristics can notably vary in non-cavitating conditions at different Reynolds number regimes [1,2]. These aspects have been addressed before [2–6]: for instance, Rijpkema et al. [1] rigorously analyzed two different marine propellers using viscous computational fluid dynamics (CFD), which operated at different Reynolds numbers in non-cavitating conditions. Baltazar et al. [5,6] examined the $\gamma - Re_\theta$ transition model for a wetted propeller performance prediction. Sánchez-Caja et al. [2] investigated the scale effects of a CLT propeller experimentally and numerically. Figure 1 shows an example of paint tests on rough and smooth surfaces representative of full- and model-scale flow regimes. In laminar flow zones, the paint streaks on the blade surface tend to deviate from a circumferential path towards a radial direction (smooth surface) due to the lower shear stress. Conversely, the paths in the streaks corresponding to the rough surface are purely circumferential, which is indicative of turbulent flow. Typically, a distinct curve can be drawn, which separates these two flow zones, as in Figure 1. The location of transition from laminar to turbulent flow, and the extent of the different flow zones, are quite sensitive to the Reynolds number regime in which the propeller operates.

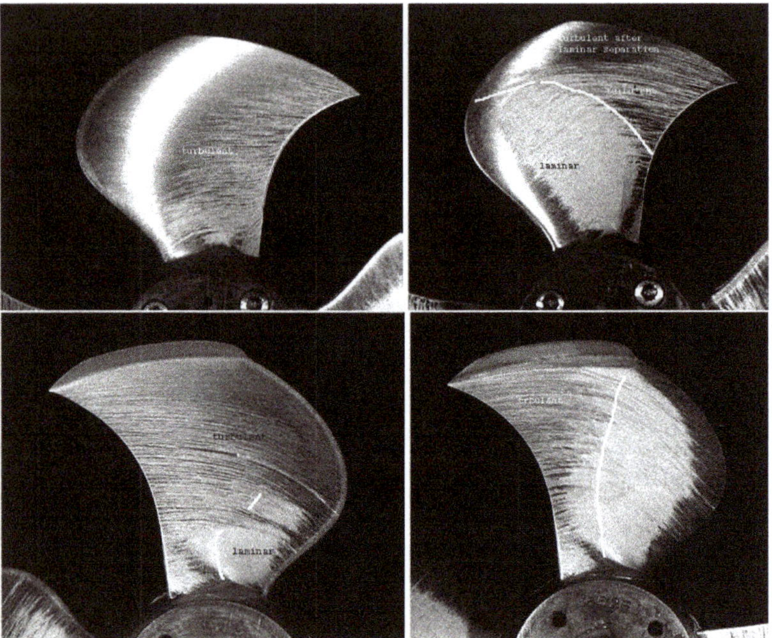

Figure 1. An example of paint test results for a CLT propeller [2]. The left frames show rough surfaces and the right frames smooth surfaces. Reproduced with permission from Juan Gonzalez-Adalid (Sistemar).

Reynolds number effects on the performance of a cavitating propeller or on the various forms of propeller cavitation structures are, however, less often reported [7,8]. The aim of this paper is to investigate numerically the scale effects on propeller cavitation, both in cases that feature steady and unsteady cavitation. Our goal is to gain deeper insight into the differences that may be found when predicting full-scale cavitation performance based on model test results. In the case of conventional, non-depressurized cavitation tanks, the differences also result from the impossibility to attain the cavitation number identity simultaneously at all radial stations. However, conventional facilities can use higher RPMs, guaranteeing larger Reynolds numbers, i.e., turbulent flow conditions over the blades. This may avoid the use of artificial leading edge roughness usually utilized in depressurized tanks due to the low rotational speed that may be required for cavitation scaling. In this study, we simulate viscous cavitating flows in model scale as they occur in conventional tunnels.

We investigate the model- and full-scale propellers in a uniform inflow condition. The propeller performance is predicted numerically utilizing viscous CFD. The range of Reynolds numbers studied covers those usually applied in the propeller open-water and cavitation tunnel model test ($Re = \mathcal{O}(10^5 \ldots 10^6)$) to typical full-scale values ($Re = \mathcal{O}(10^7 \ldots 10^8)$). We investigate two types of marine propellers: a conventional one and a tip-loaded one. Model-scale conditions are studied with four different turbulence modeling approaches: the shear stress transport (SST), both with and without transition modeling; an anisotropic Reynolds stress model; and with delayed-detached eddy simulation (DDES), which is a hybrid RANS/LES model (Reynolds-averaged Navier–Stokes/large eddy simulation). Full-scale conditions are studied with the SST turbulence model and with DDES.

This paper is organized as follows. Section 2 details the utilized viscous flow solver FINFLO, and Section 3 describes the test cases. In Section 4, we validate our numerical approach and assess the sensitivity of the model- and full-scale results to the used grid and time steps. We investigate the cavitation at different Reynolds number regimes in Section 5, and the near-blade flow including skin-friction and pressure distributions in Section 6. Finally, conclusions are drawn in Section 7.

2. Flow Solution

2.1. Governing Equations

The flow model applied is based on a homogeneous mixture flow assumption, and a compressible form of the Navier–Stokes equations is solved [9,10]. The continuity and momentum equations are solved for the mixture, and the energy equations are included to predict the correct acoustic signal speeds. The governing equations are

$$\begin{aligned} \frac{\partial \alpha_k \rho_k}{\partial t} + \nabla \cdot \alpha_k \rho_k \mathbf{V} &= \Gamma_k, \\ \frac{\partial \rho \mathbf{V}}{\partial t} + \nabla \cdot \rho \mathbf{V}\mathbf{V} + \nabla p &= \nabla \cdot \boldsymbol{\tau}_{ij} + \rho \mathbf{g}, \end{aligned} \tag{1}$$

where p is the pressure, $\mathbf{V}$ the absolute velocity in a global non-rotating coordinate system, and $\boldsymbol{\tau}_{ij}$ the stress tensor, α_k a void (volume) fraction of phase k, ρ_k the density, t the time, Γ_k the mass-transfer term, and $\mathbf{g}$ the gravity vector. The void fraction is defined as $\alpha_k = \mathcal{V}_k/\mathcal{V}$, where $\mathcal{V}_k$ denotes the volume occupied by phase k of the total volume, $\mathcal{V}$. For the mass transfer, $\sum_k \Gamma_k = 0$ holds, and consequently only a single mass-transfer term is needed.

Although the phase temperatures do not play a significant role in cavitation, the energy equations are always solved in the present method. The aim is to apply a compressible form of the equations. To predict the correct acoustic signal speeds, a complete model is needed. The phase temperatures T_g and T_l also have some influence on the solution via the material properties that are calculated as functions of the pressure and phase temperatures. The calculation of the material properties is described in [11]. The sound speed c for a two-phase mixture is defined as

$$\frac{1}{\rho c^2} = \frac{\alpha}{\rho_g c_g^2} + \frac{1-\alpha}{\rho_l c_l^2} \quad \text{and} \quad \frac{1}{c_k^2} = \frac{\partial \rho_k}{\partial p} + \frac{1}{\rho_k}\frac{\partial \rho_k}{\partial h_k}. \tag{2}$$

In the expressions above, the indices g and l refer to gas and liquid phases, respectively, and h_k denotes the enthalpy of phase k.

The energy equations for phase $k = g$ or l are written as

$$\begin{aligned} &\frac{\partial \alpha_k \rho_k (e_k + \frac{V^2}{2})}{\partial t} + \nabla \cdot \alpha_k \rho_k (e_k + \frac{V^2}{2}) \mathbf{V} = \\ -\nabla \cdot \alpha_k \mathbf{q}_k + \nabla \cdot \alpha_k \boldsymbol{\tau}_{ij} \cdot \mathbf{V} + q_{ik} &+ \Gamma_k (h_{k\mathrm{sat}} + \frac{V^2}{2}) + \alpha_k \rho_k \mathbf{g} \cdot \mathbf{V}. \end{aligned} \tag{3}$$

Here, e_k is the specific internal energy, $\mathbf{q}_k$ the heat flux, q_{ik} interfacial heat transfer from the interface to phase k, and $h_{k\mathrm{sat}}$ saturation enthalpy. Since $\sum_k \Gamma_k = 0$, by adding the energy equations together, the following relationship is obtained between the interfacial heat and mass transfer,

$$\Gamma_g = -\frac{q_{ig} + q_{il}}{h_{g\mathrm{sat}} - h_{l\mathrm{sat}}} \quad \text{and} \quad q_{ik} = h'_{ik}(T_{\mathrm{sat}} - T_k). \tag{4}$$

Above, h'_{ik} is a heat transfer coefficient between the phase k and the interface. The interfacial heat transfer coefficients are based on the mass transfer, as shown in Section 2.3.

As compared to the single-phase flow, the momentum and total continuity equations in the homogeneous model do not change, except for the material properties like density and viscosity, which are calculated as

$$\rho = \sum_k \alpha_k \rho_k \quad \text{and} \quad \mu = \sum_k \alpha_k \mu_k, \tag{5}$$

where μ is the dynamic viscosity. The turbulence effects are currently handled using single-phase models for the mixture.

The finite volume method is utilized to discretize the governing equations. The viscous fluxes as well as the pressure terms are centrally differenced. For the convective part, the variables on the cell surfaces are evaluated using a third-order upwind-biased monotonic upstream-centered scheme for conservation laws (MUSCL) interpolation. A compressive flux limiter has been shown to improve the predicted tip vortex cavitation pattern [10], and such is applied for the void fraction in the present study. For the time derivatives, a second-order three-level fully implicit method is used.

2.2. Turbulence Modelling

Nominally, a Reynolds-averaged form of the Navier–Stokes equations is used, and the DDES approach [12] that combines RANS and LES is also applied in the same form. In the present study, the calculations are performed up to the wall both in the model- and full-scale simulations, avoiding the use of wall functions. The height of the first cell was adjusted such that the non-dimensional wall distance $y^+ = \rho u_\tau y/\mu \lesssim 1$ for the first cell, with $u_\tau = \sqrt{\tau_w/\rho}$ being the friction velocity, τ_w the wall shear stress, and y is the normal distance from the solid surface to the center point of the cell next to the surface.

The base turbulence closure applied in the present calculations is the SST $k-\omega$ model [13]. That is a zonal model, referring to the formulation where the $k-\omega$ equations are solved only inside the boundary layer, and the standard $k-\varepsilon$ equations, transformed to the ω-formulation, are solved away from the walls. The effect of transition is investigated using the $\gamma - Re_\theta$ approach in model scale. In the applied transition model (also known as a Local Correlation-Based Transition Model LCTM) two extra field equations are inserted for turbulence modeling [14–16]. The purpose of the transport equation for turbulence intermittency γ, is to produce variable γ, which is coupled with the SST turbulence model to trigger transition in the boundary layer:

$$\frac{\partial(\rho\gamma)}{\partial t} + \frac{\partial(\rho u_j \gamma)}{\partial x_j} = P_\gamma + \frac{\partial}{\partial x_j}\left[\left(\mu + \frac{\mu_t}{\sigma_f}\right)\frac{\partial \gamma}{\partial x_j}\right]. \tag{6}$$

Here, P_γ is the contribution from all of the source and destruction terms and σ_f is a Schmidt number. The transport equation for the *local* transition onset momentum thickness Reynolds number $\widetilde{Re}_{\theta t}$ captures the non-local effects of the free-stream turbulence intensity into the boundary layer:

$$\frac{\partial(\rho \widetilde{Re}_{\theta t})}{\partial t} + \frac{\partial(\rho u_j \widetilde{Re}_{\theta t})}{\partial x_j} = P_{\theta t} + \frac{\partial}{\partial x_j}\left[\sigma_{\theta t}(\mu + \mu_t)\frac{\partial \widetilde{Re}_{\theta t}}{\partial x_j}\right]. \tag{7}$$

The underlying SST $k-\omega$ model is isotropic, i.e., it predicts the Reynolds stress tensor according to the Boussinesq approximation. In other words, no individual modeling is employed for each of the normal turbulent stresses; only their sum, $k = \overline{u'_i u'_i}/2$, is modeled. To account for turbulence anisotropy, the EARSM can be utilized. The idea in algebraic Reynolds stress models is to allow a nonlinear relationship between the rate of mean strain and the Reynolds stresses, but without the need of solving additional partial differential equations for each of the six Reynolds stresses. The EARSM relies on the two-equation formulation, but with the Reynolds stress anisotropy tensor evaluated from a linear pressure-strain model, and it forms a physically well-founded strategy among two-equation models for improved prediction of flows involving complex features, such as streamline curvature effects and system rotation [17,18].

In this study, DDES (Delayed DES) is also based on the SST $k-\omega$-model [19]. DDES is a slightly modified version of the detached-eddy simulation (DES). DES and DDES reduce to a RANS model in regions where the largest turbulent fluctuations are of a smaller size to the local grid spacing. Both are hybrid RANS/LES models, and function as an LES subgrid-scale model in regions where the local turbulent phenomena are of greater size than the local grid spacing [20]. A time-accurate solution is made to resolve turbulent fluctuations. In DES, the equation for the turbulent kinetic energy (k) can be written with a modified dissipation term as [21]:

$$\rho \frac{Dk}{Dt} = P - \frac{\rho k^{3/2}}{l_{DES}} + D, \tag{8}$$

where P is the production of turbulence, l_{DES} is the length scale, and D is the diffusion term. The DES length scale is computed as the minimum of the RANS length scale, $l_{RANS} = \sqrt{k}/\beta^* \omega$, and the local resolution Δ. Here, $\beta^* = 0.09$ was a model constant. The parameter Δ is evaluated as the minimum of the local wall distance, and the grid resolution $\max(\Delta x_i)$, where Δx_i denotes the thickness of the cell in different index directions. The DES length scale is then

$$l_{DES} = \min(C_{DES}\Delta, l_{RANS}), \tag{9}$$

and the coefficient C_{DES} is computed from

$$C_{DES} = (1 - F_1)C_{DES}^{k-\varepsilon} + F_1 C_{DES}^{k-\omega}, \tag{10}$$

where the constants are $C_{DES}^{k-\varepsilon} = 0.61$, $C_{DES}^{k-\omega} = 0.78$, and F_1 is Menter's blending function [19]. Furthermore, when utilizing DDES, the length scale is replaced by the expression [12]

$$l_{DDES} = l_{RANS} - F_1 \max(0, l_{RANS} - C_{DES}\Delta). \tag{11}$$

Here, $F_1 \to 1$ outside the boundary layer, and the length scale becomes $l_{DDES} = C_{DES}\Delta$ if the grid spacing permits. The DDES variant of DES aims to improve the accuracy compared to Equation (9), which has, in some instances, been observed to cause grid-induced separation.

2.3. Mass and Energy Transfer

A number of mass-transfer models have been suggested for cavitation [22]. Usually, the mass-transfer rate is proportional to a pressure difference from a saturated state or to a square root of that. In this study, the mass-transfer model is similar to that of Choi and Merkle [23]:

$$\Gamma_l = \frac{\rho_l \alpha_l \min\left[0, p - p_{\text{sat}}\right]}{\frac{1}{2}\rho_\infty V_\infty^2 (L_{\text{cav}}/V_{\text{cav}})\tau_l} + \frac{\rho_g \alpha_g \max\left[0, p - p_{\text{sat}}\right]}{\frac{1}{2}\rho_\infty V_\infty^2 (L_{\text{cav}}/V_{\text{cav}})\tau_g}, \tag{12}$$

where p_{sat} is the saturation pressure, ρ_∞ the reference (inlet) density, and V_∞ the corresponding velocity. The evaporation time constants were made non-dimensional using the reference length L_{cav} and the velocity related to cavitation (V_{cav}). In some cases, such as on a propeller blade, the cavitation length and velocity differ from the reference length L_{ref} and the reference velocity (V_∞). The time constants correspond to the parameters of the original model as $\tau_l = 1/C_{\text{dest}}$ and $\tau_g = 1/C_{\text{prod}}$. The empirical parameters of the cavitation model are calibrated in [24].

In the present method, the saturation pressure was based on the free-stream temperature, and the gas phase was assumed to be saturated (i.e., $T_g = T_{\text{sat}}$). Liquid temperature varies less because of the mass and energy transfer. As the gas temperature was forced to be $T_g = T_{\text{sat}}$, $q_{ig} = 0$. From Equation (4), the interfacial heat transfer can be solved for the liquid phase

$$q_{il} = -(h_{g\text{sat}} - h_{l\text{sat}})\Gamma_g - q_{ig} = (h_{g\text{sat}} - h_{l\text{sat}})\Gamma_l. \tag{13}$$

Using Equations (12) and (13), the interfacial transfer terms in the continuity and energy equations can be solved.

To decrease the oscillations in the solution owing to the rapid changes in the mass transfer, the mass-transfer rate was under-relaxed between the iteration cycles as $\Gamma_l^{n+1} = \alpha_\Gamma \Gamma_l^* + (1-\alpha_\Gamma)\Gamma_l^n$, where $\alpha_\Gamma = 0.5$ is an under-relaxation factor, n refers to the iteration cycle, and Γ_l^* is calculated from Equation (12). For small values $|\Gamma_l^n| < 0.1\,(\text{kg/m}^3\text{s})$, under-relaxation was not applied. The under-relaxation factor and the limit are quite arbitrary, although tested by numerous simulations.

2.4. Solution Algorithm

The solution method is a segregated pressure-based algorithm where the momentum equations are solved first, and then a pressure-velocity correction is made. The basic idea in the solution of all equations is that the mass balance is not forced at every iteration cycle, but rather the effect of the mass error is subtracted from the linearized conservation equations. A pressure correction equation was derived from the continuity equation, Equation (1), linked with the linearized momentum equation. The velocity-pressure coupling is based on the corresponding algorithm for a single-phase flow [9], and is described in more detail in [10,25].

Quasi-steady simulations exploit the fact that the governing equations can yield a steady-state solution, when the equations are expressed in the coordinate system that is rotating with the propeller. Absolute velocities are used in the solution, and the rotational movement of the propeller is accounted for in the convection velocity and as source terms in the y- and z-momentum equations as the propeller is rotating around the x-axis. The equations are iterated until the global force coefficients and the L_2 norms of the main variables have stabilized at a sufficiently low level. In transient simulations, the governing equations are integrated in physical time, and the whole computational domain is rotated at the propeller rate of revolution. A steady-state solution is sought within each physical time step, and the equations are iterated until the L_2 norms of the main variables have decreased by a sufficient amount, i.e., at least 2–3 orders of magnitude. Approximately one hundred inner iterations were required within each physical time step. A physical time step corresponding to 0.5 degrees of the propeller rotation was used in all transient simulations.

3. Test Cases

Two propellers are considered in this study: the Potsdam propeller test case (PPTC) propeller [26] and a tip loaded propeller (TLP) propeller. The former is a more conventional five-bladed propeller with moderate skew, whereas the latter is a pressure-side oriented tip blade propeller with enhanced efficiency. Photographs of the propellers are shown in Figure 2. The principal data of the studied propellers is given in Table 1 for the PPTC and in Table 2 for the TLP.

The advance coefficient and cavitation number are defined as

$$J = \frac{V_A}{nD} \quad \text{and} \quad \sigma_n = \frac{p - p_{\text{sat}}}{\frac{1}{2}\rho(nD)^2}, \tag{14}$$

respectively, where V_A is the propeller speed of advance, n the propeller rate of revolutions, D the propeller diameter, p is the pressure, p_{sat} the saturation pressure, and ρ the fluid density. The thrust and torque of the propeller are nondimensionalized as

$$K_T = \frac{T}{\rho n^2 D^4} \quad \text{and} \quad K_Q = \frac{Q}{\rho n^2 D^5}, \tag{15}$$

respectively, where T denotes the thrust and Q the torque of the propeller. The open-water efficiency of the propeller is $\eta_0 = J/(2\pi)\, K_T/K_Q$. The Reynolds number is defined as

$$Re = \frac{\rho c_{0.7}\sqrt{V_A^2 + (0.7\omega D/2)^2}}{\mu}, \tag{16}$$

where $\omega = 2\pi n$, and $c_{0.7}$ denotes the chord length at $r/R = 0.7$ radius.

Table 1. Main geometric parameters of the Potsdam propeller test case (PPTC) propeller [26].

Diameter [m]	0.250
Pitch ratio at $r/R = 0.7$	1.635
Chord at $r/R = 0.7$	0.417
EAR	0.779
Skew [°]	18.837
Hub ratio	0.300
Number of blades	5
Rotation	Right-handed

Table 2. Main geometric parameters of the tip loaded propeller (TLP) propeller.

Diameter [m]	0.250
Pitch ratio at $r/R = 0.7$	1.465
EAR	0.594
Number of blades	5
Rotation	Right-handed

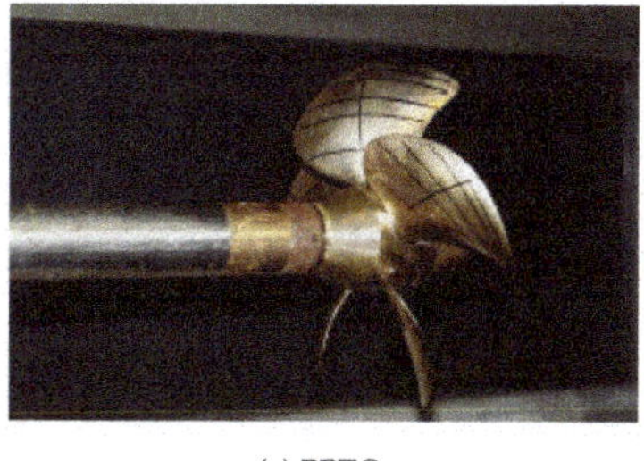

(**a**) PPTC.

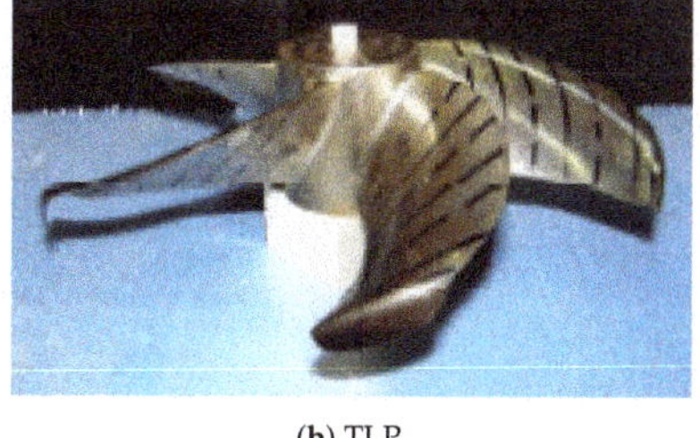

(**b**) TLP.

Figure 2. Photographs of the investigated propellers. The black circles on the blades show the reference radii. Photograph of PPTC taken from Ref. [26] and reproduced with permission from Lars Lübke (SVA). Photograph of TLP reproduced with permission from Michael Brown (ONR).

The investigated conditions for the cavitating cases are listed in Table 3 for each propeller. We study the cavitating conditions for the PPTC at five different Reynolds numbers, and for the TLP at two different Reynolds numbers. Two different cases are studied with the PPTC using different J and σ_n, given in Table 3. The Case 1 yields mostly steady vortex and sheet cavitation patterns, whereas the Case 2 exhibits unsteady cavitation phenomena on the propeller blades and in the slipstream. The model-scale conditions were chosen to match the experimental conditions (for PPTC Case 1, this corresponds to Re_2; the Re_1 was chosen to assess also a lower Reynolds number in model scale for this operating point), and the full-scale conditions were selected to ensure a sufficiently high Reynolds number. The reference length and velocity scales related to the Merkle cavitation model, discussed in Section 2.3, were here the chord and the speed at $r/R = 0.7$ radius. We used the same values for the empirical parameters in the cavitation model in model- and full-scale simulations, and they were $C_{\text{dest}} = C_{\text{prod}} = 369$.

Table 3. Test conditions for the cavitating cases.

Quantity	PPTC, Case 1	PPTC, Case 2	TLP
J	1.019	1.269	0.729
σ_n	2.024	1424	4.092
K_T	0.374	0.206	0.493
n [rps]	25.0	25.0	15.0
Re, model scale	$Re_1 = 5.0 \times 10^5$, $Re_2 = 1.3 \times 10^6$	1.4×10^6	9.7×10^5
Re, full scale	2.9×10^7	3.0×10^7	5.7×10^7

The Case 1 for the PPTC is studied in model scale using the SST, the transition model (TM) and the EARSM. This condition is studied in full scale using SST. The Case 2 for the PPTC is studied employing URANS (unsteady RANS) with transient SST simulations and DDES in model scale and full scale. The TLP is studied in model scale using the SST and the TM. This condition is studied in full scale using SST.

3.1. Computational Setup

Two grids were constructed for both propellers: a model-scale grid and a full-scale grid. The model-scale grids were generated to simulate the experimental set-ups. The computational grids for the full scale simulations were scaled from the model scale grids, and additional cells were clustered close to solid surfaces to ensure that $y^+ \approx 1$ for the simulations. The scaling factor for PPTC was 12, and for the TLP this was 26.

The computational domain is shown in Figure 3 for the PPTC propeller. Corresponding design for the computational domain was used for TLP propeller. The calculations were performed with three grid densities. On the coarse grid level, every second point in all directions is removed compared to the finer level grid. A solution on the coarse grid is used as an initial guess for the computations performed on the next finer grid level. The fine computational grid used for the PPTC propeller consists of ~6.5 million cells in 36 grid blocks, the medium grid consists of 0.8 million cells, and the coarse grid of 0.1 million cells. The fine grid for the TLP consists of roughly 10 million cells in 45 grid blocks, the medium grid consists of 1.3 million cells, and the coarse grid of 0.2 million cells. The different grids on the suction sides of the propeller blades are shown in Figure 4. The surface grid on the pressure sides of the blades is similar. A similar grid for the PPTC was used before [10,25], where the numerical simulations with the hybrid RANS/LES model showed very good agreement also with respect to the local flow LDV (laser Doppler velocimetry) measurements in the propeller wake. The grid density in the wake region was further increased for this study. Due to the symmetric nature of the problem of a propeller operating in uniform inflow, only one blade is modeled. The blades, hub, and shaft are modeled as no-slip rotational surfaces. A velocity boundary condition is applied at the inlet and a pressure boundary condition is applied at the outlet. A slip boundary condition is applied at the simplified cavitation tunnel walls.

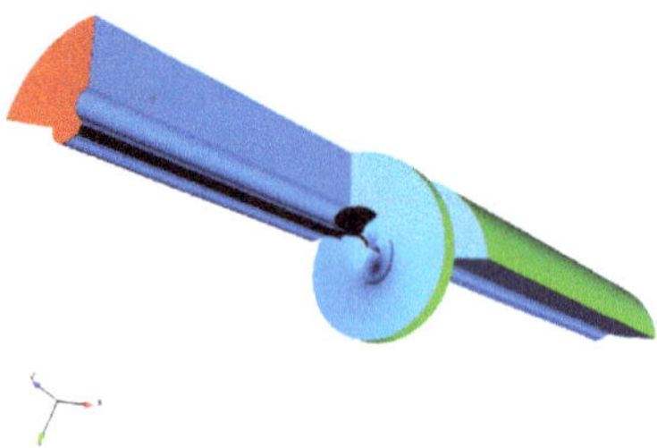

Figure 3. Perspective view of the computational domain for the PPTC. The propeller, hub, and shaft are colored with black. The inlet face is colored with red, the cyclic boundaries with blue, and the simplified tunnel wall with green.

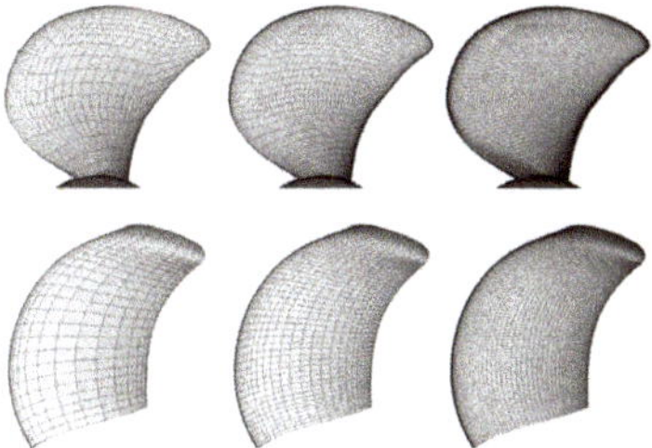

Figure 4. Grid density on the suction sides of the PPTC (top row) and TLP (bottom row) propeller blades. Coarse (left), medium (middle), and fine (right) model-scale grids shown.

The whole computational domains are considered as rotating with the given rate of rotation. The inflow velocity is set based on the advance numbers of the propeller, and the background pressure level is set based on the cavitation number. Low turbulent intensity (1%) and eddy viscosity ($\mu_t/\mu = 10$) ratio were used as initial conditions in all calculations. We note here that the sensitivity of the transition model to the initial turbulence field quantities in the model-scale cavitating propeller performance predictions needs be assessed in the future. The inlet is located roughly five propeller diameters upstream of the propeller, and the outlet is located ten diameters downstream of the propeller. For the PPTC propeller, the radius of the whole computational domain is 0.3385 m in model scale, which corresponds to the cross-sectional area of the experimental cavitation tunnel. The radius of 0.7 m in model scale was used for the computational domain for the TLP, which was larger in cross-sectional area than in the experimental tank. Typically, a steady-state case requires around one day of computational time on fine grid resolution with approximately 20 CPUs, whereas an unsteady case requires approximately 3 days with the similar machine set-up to reach one full revolution.

4. Validation of the Numerical Method

In this section, we perform a validation study of our numerical simulations in terms of the predicted global forces and cavitation patterns. Also, a grid and time step sensitivity study is carried out. We consider both the model- and full-scale conditions in the validation. Section 4.1 concerns the global performance characteristics of the propellers, Section 4.2 the cavitation observation, and Section 4.3 shows a grid and time step sensitivity analysis.

4.1. Global Forces

The open-water characteristics in non-cavitating conditions for the propellers are shown in Figure 5. We produced the open-water curves for the PPTC from $J = 0.5 \ldots 1.6$ and for the TLP from $J = 0.5 \ldots 1.2$ with the SST $k - \omega$ turbulence model in model-scale (MS) conditions. Selected points are also simulated with the EARSM and $\gamma - Re_\theta$ models, as well as in full-scale (FS) conditions. The SST and EARSM simulations in model scale capture the propeller open-water performance with good accuracy for both propellers. EARSM predicts slightly lower torque for the TLP propeller than the SST model. The transition simulations predict 2–3% higher thrust coefficient than the SST in model scale for the propellers, at the investigated points. The transition simulations, as well the full-scale simulations, yield a slightly lower torque coefficient, and consequently, a little higher efficiency for the propeller. The predicted thrust coefficients and their comparison with the experimental values are given in Table 4. Greater differences with respect to the experiments are seen in the unsteady case, where it is also noteworthy that the model- and full-scale simulations both gave the same thrust coefficient.

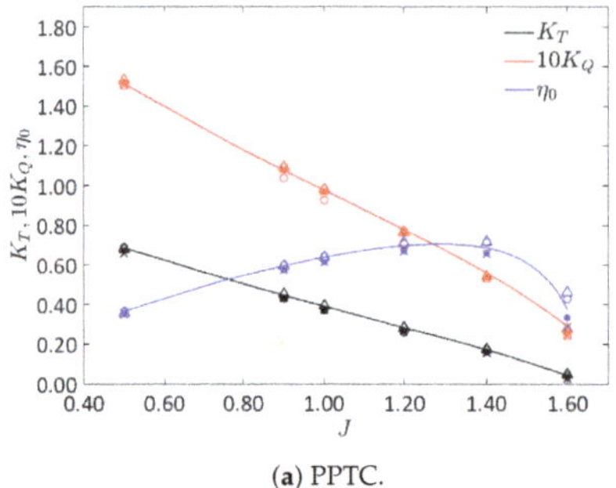

(**a**) PPTC.

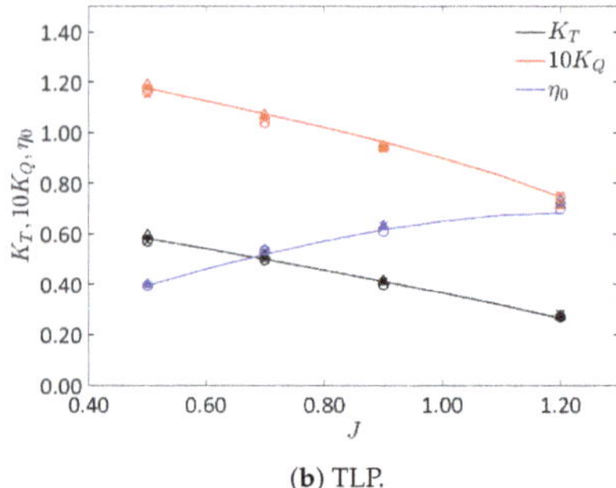

(**b**) TLP.

Figure 5. Open-water characteristics for the propellers. Solid lines: experimental results; • symbols denote corresponding $k-\omega$ model-scale simulations; × symbols denote EARSM model-scale simulations; △ symbols denote $\gamma - Re_\theta$ model-scale simulations; ○ symbols denote $k-\omega$ full-scale simulations.

The thrust coefficients in the cavitating conditions are compared to the model-scale experiments in Table 4. All steady-state RANS simulations yield results within 4% of the experimental values. A greater difference is seen at the PPTC Case 1 conditions for the lowest investigated Re, where both the SST and TM underpredict the propeller thrust. The SST and TM give similar results for the TLP. The results at full-scale Reynolds numbers tend to have the least deviation with respect to the experiments. For the transient simulations (PPTC Case 2), the model-scale DDES gives a slightly lower and the full-scale DDES a slightly higher thrust coefficient. The URANS simulations yield a greater difference when compared to the experiments. The SST predicts a smaller thrust value than the DDES in model scale, and a higher in full scale. The TM significantly underpredicts the propeller thrust. Although not shown here, we note that the TM gave a very steady time history of the propeller global forces, with next to no fluctuations in the K_T and K_Q values. The simulations with the other turbulence modeling approaches (both URANS SST and the DDES in model- and full-scale), then again, predicted a more unsteady cavitation on the blades and consequently also highly oscillating global force time histories. See also Sections 4.2 and 4.3.

Table 4. Predicted thrust coefficients in cavitating conditions.

Case	Thrust Coefficient	Difference to Experiments
PPTC, Case 1, $Re_1 = 5.0 \times 10^5$, SST	0.360	−4%
PPTC, Case 1, $Re_1 = 5.0 \times 10^5$, TM	0.361	−4%
PPTC, Case 1, $Re_2 = 1.3 \times 10^6$, SST	0.369	−1%
PPTC, Case 1, $Re_2 = 1.3 \times 10^6$, TM	0.364	−3%
PPTC, Case 1, $Re = 2.9 \times 10^7$, SST	0.373	0%
PPTC, Case 2, $Re = 1.4 \times 10^6$, SST	0.195	−5%
PPTC, Case 2, $Re = 3.0 \times 10^6$, TM	0.183	−12%
PPTC, Case 2, $Re = 1.4 \times 10^6$, DDES	0.204	−1%
PPTC, Case 2, $Re = 3.0 \times 10^7$, SST	0.212	3%
PPTC, Case 2, $Re = 3.0 \times 10^7$, DDES	0.208	1%
TLP, $Re = 9.7 \times 10^5$, SST	0.503	2%
TLP, $Re = 9.7 \times 10^5$, TM	0.505	2%
TLP, $Re = 5.7 \times 10^7$, SST	0.500	1%

4.2. Cavitation Observation

The model-scale cavitation patterns of the propellers are compared to the experimental observations. We compare both the near-blade cavitation and the overall cavitation patterns also in the wake. The cavitation extents based on the experiments are compared to the simulations with the different turbulence modeling approaches. The cavitation is visualized in the numerical results by the iso-surface of the void fraction value 0.1. In this section, all model-scale results for PPTC Case 1 conditions are those with the higher Reynolds number, that is, $Re_2 = 1.3 \times 10^6$.

The model-scale cavitation patterns of the PPTC Case 1 are compared in Figures 6 and 7. The former figure shows the cavitation type near the blades, and the latter figure shows the cavitation structures also in the propeller slipstream. The experiments are compared to the results obtained with the different turbulence models. The propeller has strong tip vortex and hub vortex cavitation, which is visible in the experiments and in the simulations. The mean shape and extent of the root cavitation, as well as the tip and hub vortex cavitation, are captured well. The tip and hub vortex cavitation extending far behind the propeller are captured exceptionally well, as shown in Figure 7. Also, the modal shapes of the cavitating tip vortex are qualitatively well captured. Streak cavitation was observed in the experiments at several radial locations near the leading edge of the blade, whereas the simulation predicts sheet cavitation at the leading edge. The global cavitation is predicted similarly with each of the utilized turbulence models, although the transition model yields a bit narrower hub cavitating hub vortex. Little difference is seen between the utilized turbulence modeling approaches for the PPTC Case 1 conditions.

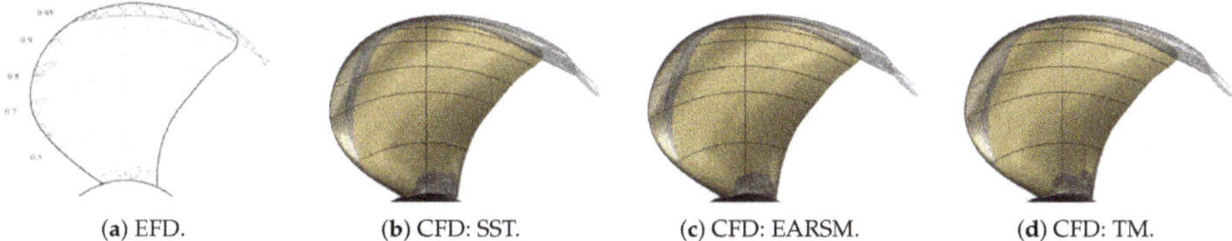

(**a**) EFD. (**b**) CFD: SST. (**c**) CFD: EARSM. (**d**) CFD: TM.

Figure 6. Comparison of model-scale near-blade cavitation observation for the PPTC Case 1. The EFD result is given in [27], and reproduced with permission from Lars Lübke (SVA).

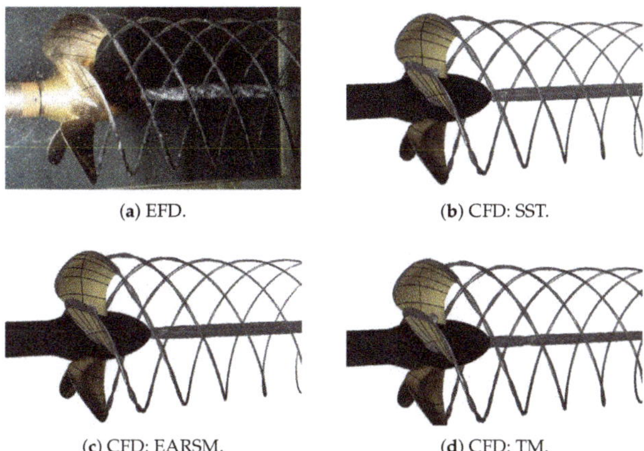

(**a**) EFD. (**b**) CFD: SST.

(**c**) CFD: EARSM. (**d**) CFD: TM.

Figure 7. Comparison of model-scale cavitation patterns for the PPTC Case 1. The photograph from experiments is given in [26], and reproduced with permission from Lars Lübke (SVA).

The model-scale cavitation patterns of the PPTC Case 2 are compared in Figures 8 and 9, the former figure showing the cavitation type near the blades, and the latter the cavitation structures in the propeller slipstream. Also here, the experiments are compared to the results obtained with the different turbulence models. The propeller has a finer tip vortex cavitation and stronger root cavitation. The root cavitation extends past the blade trailing edge and sheds unsteady vapor structures in the wake. Bubble cavitation on the blade is observed in the experiments, covering roughly the entire blade between the root cavitation zone and the tip. The SST simulations and the DDES predict larger root cavitation extent than is visible in the experiments; additionally, the bubble cavitation on the blade appears as a fine sheet cavity. The TM shows a quite extensive root cavitation and a small cavity spot on the blade. The tip vortex cavitation incepts at roughly the same location in the experiments and in the simulations in Case 2. Unsteady shedding of the root cavitation occurs in the experiments in Case 2, subsequently transforming into cloudy cavitation in the propeller slipstream. The SST simulations and the DDES also predict unsteady shedding of cavitation structures from the root, and these attain a slightly more visible and organized form (Figure 9). TM simulations for PPTC Case 2 fail to predict the slipstream cavitation; the transition model also predicted a notably lower K_T for this condition (cf. Table 4). The propeller also experiences pressure side cavitation at this operating condition, which is further assessed in Section 5.

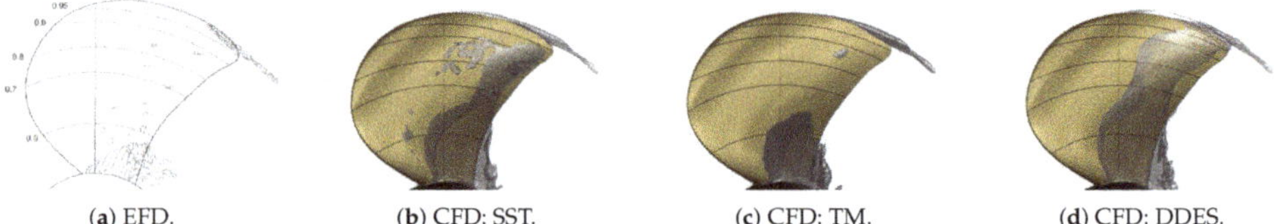

(**a**) EFD. (**b**) CFD: SST. (**c**) CFD: TM. (**d**) CFD: DDES.

Figure 8. Comparison of model-scale near-blade cavitation observation for the PPTC Case 2. The EFD result is given in [26], and reproduced with permission from Lars Lübke (SVA).

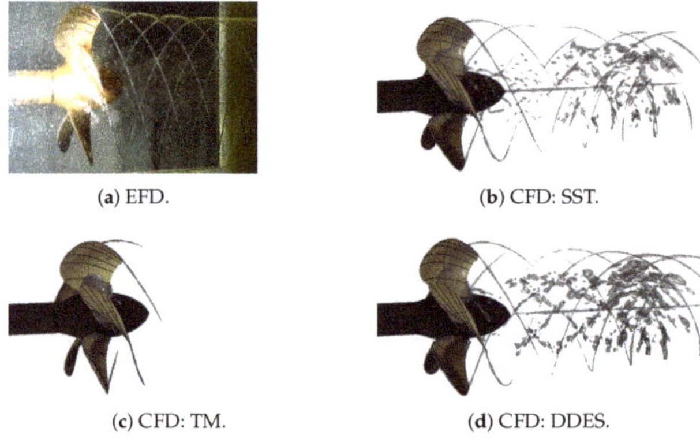

(**a**) EFD. (**b**) CFD: SST.

(**c**) CFD: TM. (**d**) CFD: DDES.

Figure 9. Comparison of model-scale cavitation patterns for the PPTC Case 2. The photograph from experiments is given in [26], and reproduced with permission from Lars Lübke (SVA).

The model-scale cavitation extent of the TLP propeller is compared in Figure 10 for cavitation patterns near the blades, and in Figure 11 for overall cavitation extent of the propeller. We show the experimental results together with those obtained with the different turbulence models. The propeller has sheet cavitation upward of $0.9R$ toward the tip which incepts at the leading edge. The simulations predict this, although the sheet starts at a slightly lower radius than in the experiments. The sheet cavitation covers almost the entire chord of the blade, and turns cloudy close to the trailing edge. The root cavitation is observed neither in the experiments nor the simulations. The upper part of the sheet cavity transforms into a cavitating tip vortex as it departs the blade. The cavitating tip vortex is relatively fine and visible both in the experiments as well as in the simulations. As we see in Figure 11, the cavitating tip vortex extends far in the propeller wake, and this behavior is also captured well with the numerical simulations with all utilized turbulence models. Also, the secondary fine tip vortex cavitation due to the end plate is captured in the CFD simulations, although it quickly diminishes after departing the blade. Again, the global cavitation is predicted similarly with each of the utilized turbulence models.

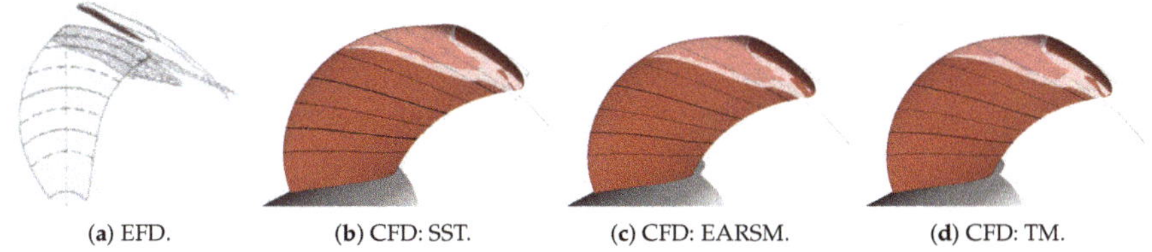

(a) EFD. (b) CFD: SST. (c) CFD: EARSM. (d) CFD: TM.

Figure 10. Comparison of near-blade cavitation patterns for the TLP.

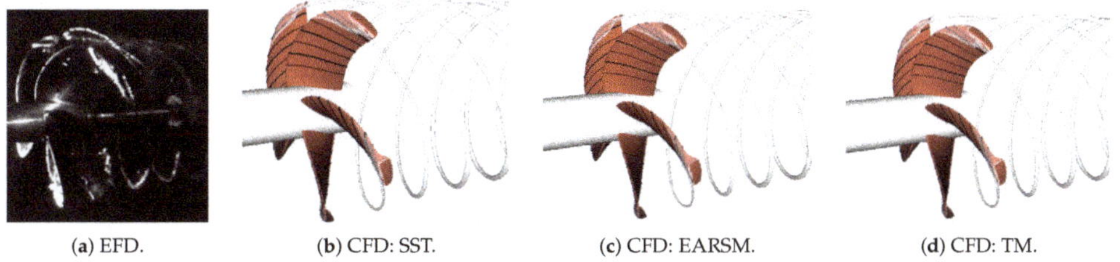

(a) EFD. (b) CFD: SST. (c) CFD: EARSM. (d) CFD: TM.

Figure 11. Comparison of model-scale cavitation patterns for the TLP.

4.3. A Grid and Time Step Sensitivity Study in Model- and Full-Scale Conditions

Next, we carry out a grid sensitivity investigation for the PPTC Case 1. We consider the model- and full-scale simulations at three different numerical grid densities. The propeller performance characteristics in model- and full-scale are given in Table 5. The simulated performance characteristics are compared to the model test results.

Table 5. Predicted thrust and torque coefficients with the different grid densities for PPTC Case 1. The percentage difference to the model-scale values from the experiments is given in parentheses.

Case	Thrust Coefficient, K_T	Torque Coefficient, $10K_Q$	Open-water Efficiency, η_0
Experiments (model scale)	0.374	0.970	0.625
Coarse grid, model scale	0.343 (−9%)	0.874 (−11%)	0.637 (3%)
Medium grid, model scale	0.352 (−6%)	0.907 (−7%)	0.630 (1%)
Fine grid, model scale	0.369 (−1%)	0.950 (−2%)	0.629 (1%)
Coarse grid, full scale	0.338 (−11%)	0.833 (−16%)	0.658 (5%)
Medium grid, full scale	0.358 (−4%)	0.889 (−9%)	0.654 (4%)
Fine grid, full scale	0.373 (−0%)	0.940 (−3%)	0.644 (3%)

We note that the finest grids produce results that are within few per cents of the experimental values. The full-scale simulations yield a bit higher thrust coefficient, a slightly lower torque coefficient, and consequently a higher open-water efficiency. The propeller efficiency, again, does not show as great dependency on the utilized grid density in either model- or full-scale as the global forces do. Both the model- and full-scale performance characteristics converge monotonically toward the experimental values.

Figure 12 displays the near-blade cavitation results, and Figure 13 shows the overall cavitation extent obtained with different numerical grid densities. Furthermore, we compare the pressure in the wake in Figure 14 with the three grids. We evaluated the pressure coefficient, $C_p = 2(p - p_\infty)/(\rho_\infty n^2 D^2)$, in the wake at the track of the tip vortex.

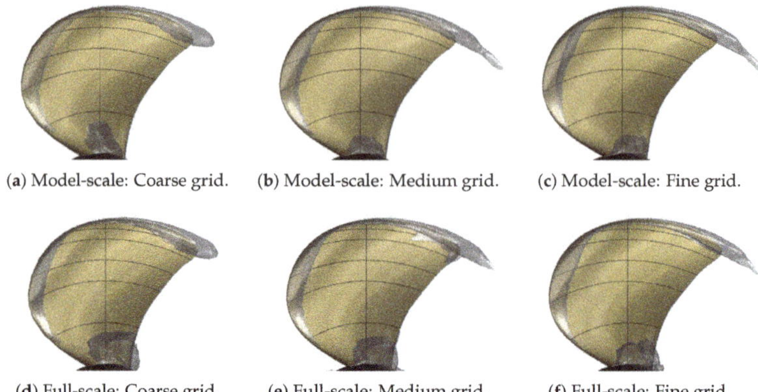

(**a**) Model-scale: Coarse grid. (**b**) Model-scale: Medium grid. (**c**) Model-scale: Fine grid.

(**d**) Full-scale: Coarse grid. (**e**) Full-scale: Medium grid. (**f**) Full-scale: Fine grid.

Figure 12. Comparison of near-blade cavitation results with different grid densities. PPTC Case 1 conditions, results with the SST $k-\omega$ model.

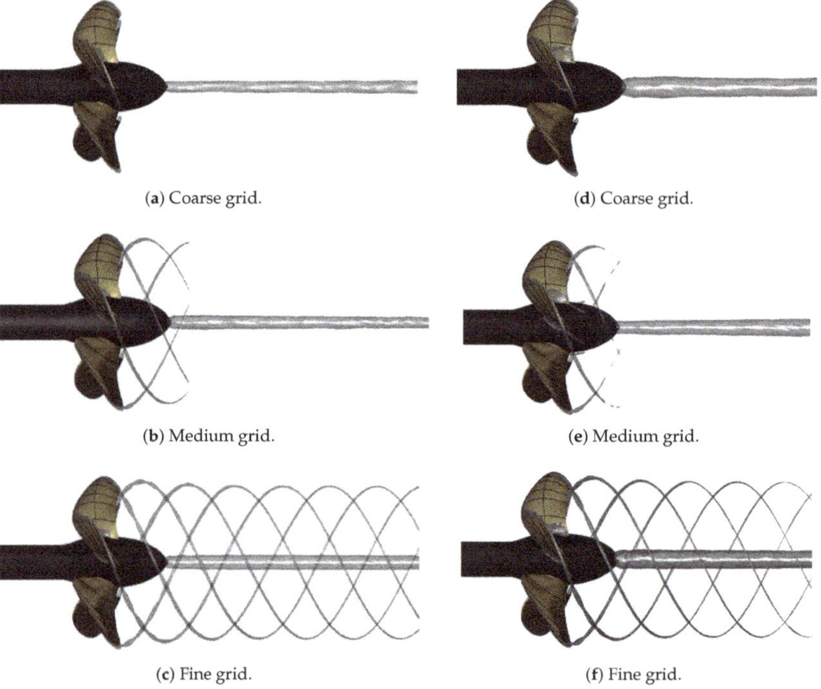

(**a**) Coarse grid. (**d**) Coarse grid.

(**b**) Medium grid. (**e**) Medium grid.

(**c**) Fine grid. (**f**) Fine grid.

Figure 13. Comparison of cavitation patterns for the PPTC propeller with different grid densities. On the left: model-scale simulations. On the right: full-scale simulations.

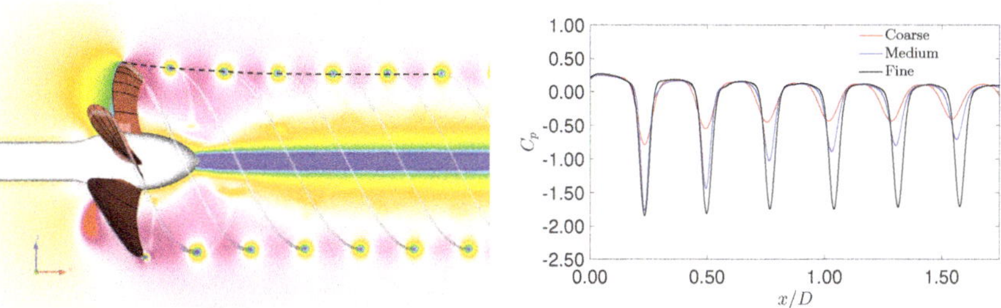

Figure 14. Pressure coefficient in the track of the tip vortex with different grid densities for the PPTC Case 1 conditions. Model-scale SST simulations. The figure on the left shows the fine grid SST simulations and the track as dashed line along which the C_p is evaluated.

The cavitation for PPTC Case 1 near the propeller blades is similar to the different utilized grids. All grids predict sheet cavitation at the leading edge which then transforms to the tip vortex cavitation. The root cavitation generally grows in extent as the grid is refined, but otherwise the near-blade cavitation is mostly independent of the utilized grid densities. We observe greater differences in the results between the different grids in the propeller wake. The coarse grid simulations yield a very limited vortex cavitation, as the tip vortex cavitation seizes just after the blade trailing edge. We do not, however, observe such a clear influence of the grid density on the hub vortex cavitation. The medium grid simulations give the hub vortex cavitation that match better the fine grid results, and the experiments, but the tip vortex cavitation extends less than one diameter in the axial direction. The modal shapes are also barely distinguishable in the medium grid solution. Figure 14 shows the decay in the pressure minima in model scale is notably greater for the coarse and medium grids as the axial distance from the propeller increases. The pressure coefficient in the first vortex core for the coarse grid is significantly less than the vapor pressure, and therefore the tip vortex cavitation terminates just after the blades. For the medium grid, the first tip vortex core on the curve maintains a low C_p level, and we observed a cavitating vortex in Figure 13b. Beyond this, the pressure coefficient decreases and the tip vortex cavitation disappears.

The skin-friction and pressure coefficients for the different grid densities are compared in Figures 15 and 16 for model- and full-scale conditions for the PPTC, respectively. We defined the skin-friction coefficient as $C_f = 2|\tau_{ij}|/(\rho_\infty n^2 D^2)$. In these figures, the coefficients are plotted on both sides of the blades, extracted from the radius $r/R = 0.7$. Furthermore, the skin-friction and pressure coefficients are visualized on the blades in Figures 17 and 18 for the model- and full-scale conditions, together with surface-flow LIC (line integral convolution). The skin-friction coefficient shows greater sensitivity to the utilized grid density, especially on the suction sides of the blades. This is expected, as the wall y^+ values deteriorate as the grid resolution decreases. The pressure coefficient, however, exhibits much less variation between the different grid densities both in model- and full-scale simulations. The pressure coefficients at the $r/R = 0.7$ practically overlap each other in Figures 15 and 16 for the most part of the blade. The increase in the pressure after the leading edge (LE) sheet cavity closure is not visible in the coarse grid simulations, whereas the medium and fine grid simulations show clearly elevated C_p. The coarse grid C_p is very smooth in both model- and full-scale conditions, as we see also in Figures 15 and 16. The fine grid C_p shows quite sharp a jump at the cavity closure line in model- and full-scale conditions. Significantly lower skin-friction values are seen underneath the sheet cavitation near the LE. The C_f shows also quite sharp an increase after the cavity closure in the fine grid simulations, being smoother for the coarser grid solutions. In addition, the surface flow underneath the cavity deviates drastically from the typical (turbulent) circular paths in the model-scale medium and fine grid solutions. In model-scale conditions, the surface flows show steeper turns back toward the LE than in full scale. The coarse grid solutions fails to show similar phenomena. Interestingly, judging by the surface flow visualization, the medium grid solution in model scale predicts the cavity closure line a bit farther from the LE than the fine grid solution. The opposite is noted in full-scale, where the fine grid solution now predicts the closure line farther from the LE than the medium grid solution (we note that this behavior was not so clear in the blade C_p or C_f distributions, Figures 15 and 16). In full scale, the surface flow underneath the sheet cavity is oriented differently between the medium and fine grid solutions, as it turns upward toward the tip closer to the LE with the medium grid.

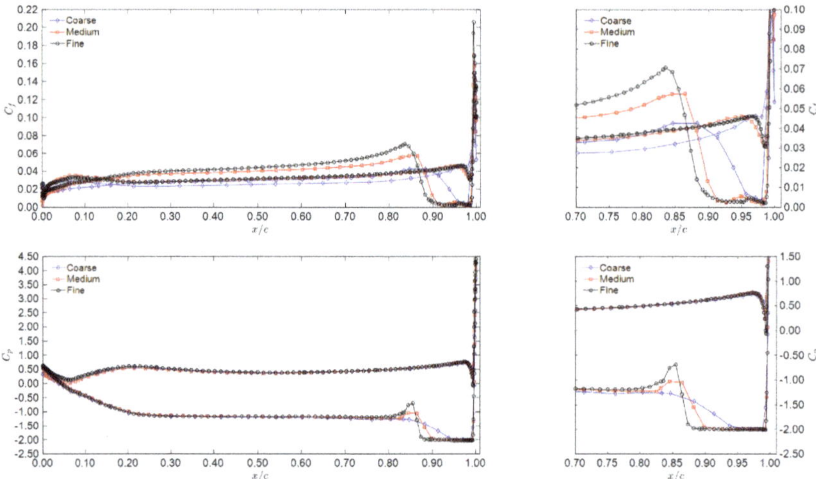

Figure 15. Comparison of skin-friction and pressure coefficient distributions at the radius $r/R = 0.7$ on the blades. PPTC Case 1 conditions in model scale. The results are from the three grid densities. The TE is at $x/c = 0$ and the LE at $x/c = 1$.

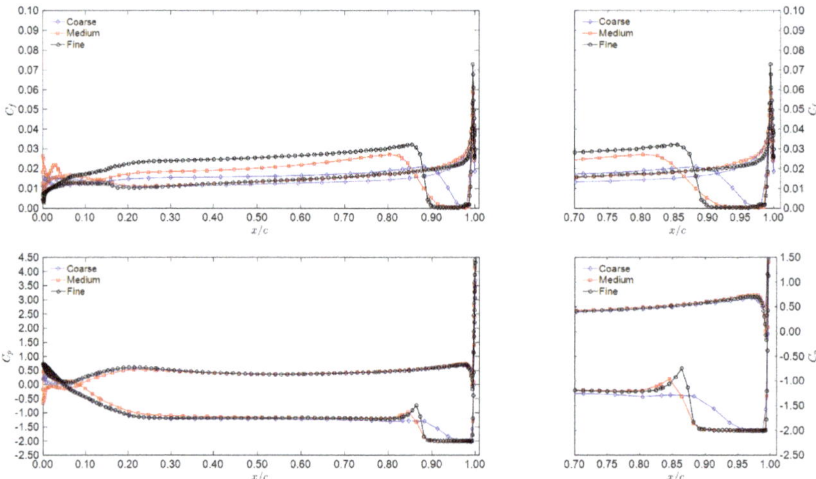

Figure 16. Comparison of skin-friction and pressure coefficient distributions at the radius $r/R = 0.7$ on the blades. PPTC Case 1 conditions in full scale. The results are from the three grid densities. The TE is at $x/c = 0$ and the LE at $x/c = 1$.

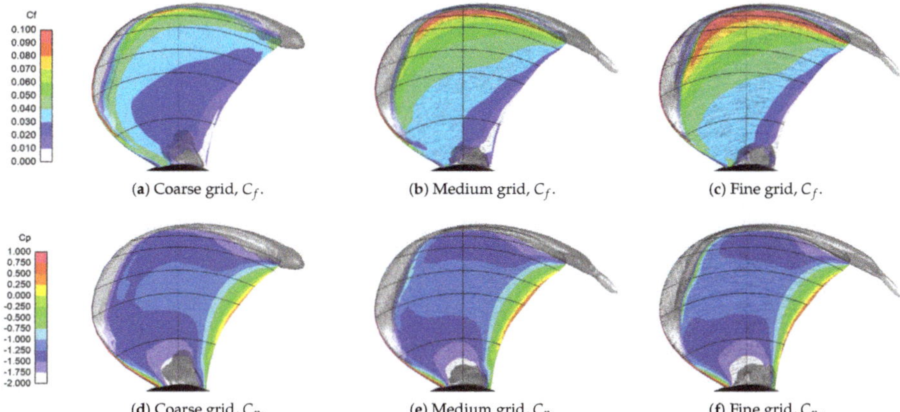

(**a**) Coarse grid, C_f. (**b**) Medium grid, C_f. (**c**) Fine grid, C_f.

(**d**) Coarse grid, C_p. (**e**) Medium grid, C_p. (**f**) Fine grid, C_p.

Figure 17. Comparison of near-blade cavitation results with different grid densities. Top row: skin-friction coefficient. Bottom row: pressure coefficient. Model-scale simulations for PPTC Case 1.

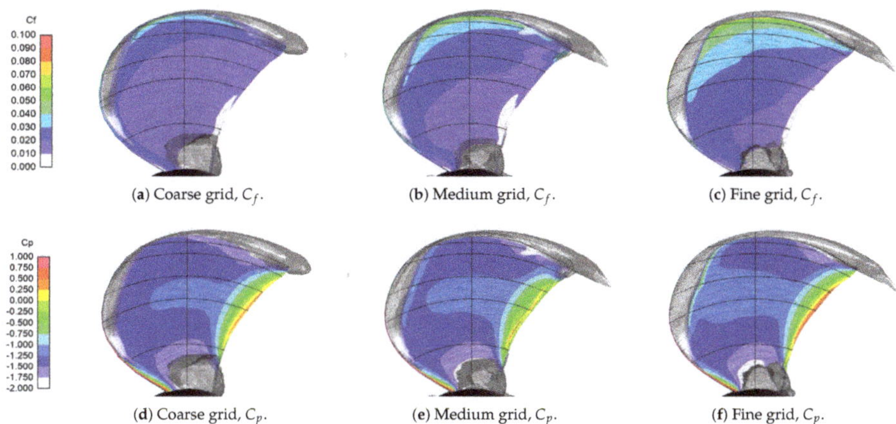

(**a**) Coarse grid, C_f. (**b**) Medium grid, C_f. (**c**) Fine grid, C_f.

(**d**) Coarse grid, C_p. (**e**) Medium grid, C_p. (**f**) Fine grid, C_p.

Figure 18. Comparison of near-blade cavitation results with different grid densities. Top row: skin-friction coefficient. Bottom row: pressure coefficient. Full-scale simulations for PPTC Case 1.

Next, we investigate the sensitivity of the simulations with respect to the time-step. The DDES of the unsteady cavitation (PPTC Case 2) is considered both in model and full scale. We conduct the analyses using the medium grids. In the analyses, we use the time-steps that correspond to 0.5, 1.0, and 2.0 degrees of propeller rotation. Additionally, we compare the DDES with URANS simulations using the SST $k - \omega$ in both model- and full-scale conditions, where the time-step corresponding to 0.50 degrees of propeller rotation is used. Note that the time step is relative large for the DDES simulation. In all of the results depicted here, we first simulated the case with the URANS SST for one propeller revolution with $\Delta t \mathrel{\widehat{=}} 0.50^\circ$, after which we started the DDES and switched to the other time steps. The time histories of the thrust and torque coefficients are shown in Figures 19 and 20. The quantities ΔK_T and ΔK_Q denote the ratio of the computed value and the corresponding measured model-scale value, that is, $\Delta K_T = K_{T,CFD}/K_{T,EFD}$, for instance. The thrust coefficients in the frequency domain are shown in Figure 21. The average value of the model-scale K_T with DDES does not show great dependency on the time step. The URANS SST gives a lower thrust coefficient than the DDES in model scale. The average K_Q shows more variation with respect to the utilized time step. The average value of the full-scale torque coefficient seems to reduce as the time is reduced. More frequency content in the global forces is seen as the time increment is reduced. Especially in full-scale conditions, the difference between the smallest time step to the other simulations is relatively large, and the URANS SST yield rather similar results to the DDES with twice the time-step. The DDES with $\Delta t \mathrel{\widehat{=}} 2.00°$ does not yield comparable frequency content in the time histories.

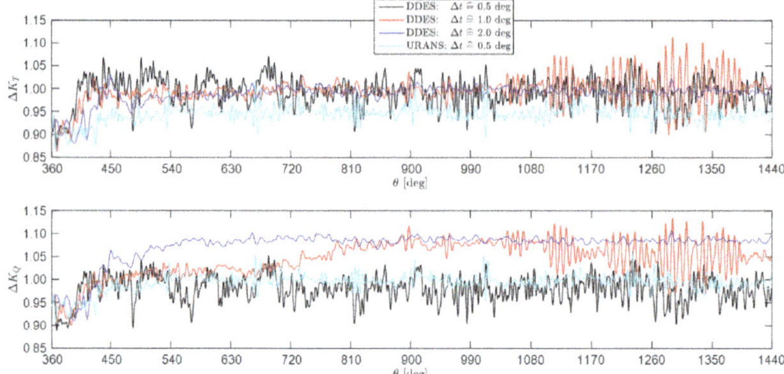

Figure 19. Time histories of the propeller thrust coefficient for PPTC Case 2. Model-scale DDES and URANS simulations with different time-steps.

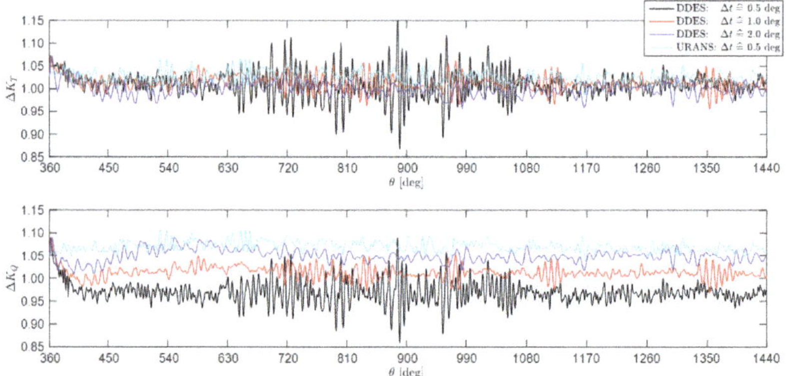

Figure 20. Time histories of the propeller thrust coefficient for PPTC Case 2. Full-scale DDES and URANS simulations with different time-steps.

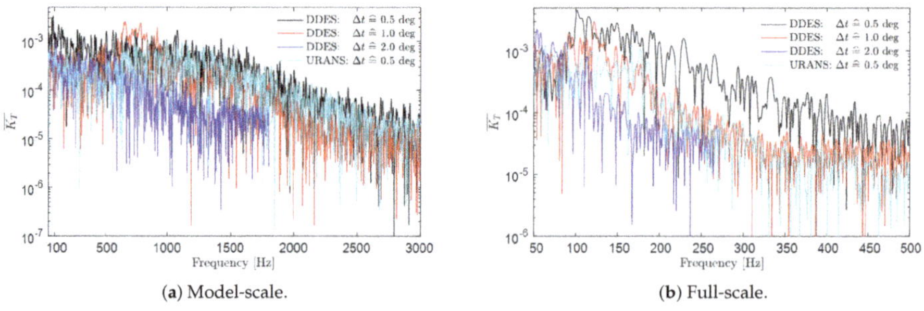

(**a**) Model-scale. (**b**) Full-scale.

Figure 21. Propeller thrust coefficient in the frequency domain for PPTC Case 2.

5. Model- and Full-Scale Propeller Cavitation

5.1. Cavitation Patterns for PPTC Propeller, Case 1 Conditions

We compare the cavitation patterns in model- and full-scale for the PPTC Case 1. Cavitation close to the blades is visualized in Figure 22, and the overall cavitation patterns are shown in Figure 23. The near-blade cavitation patterns, that is, the root cavitation as well as the LE sheet cavitation transforming into tip vortex cavitation, are very similar between the different simulation conditions. We observe little difference in the model-scale simulations using the SST or TM at the investigated Reynolds numbers. The root cavitation area appears larger in the full-scale conditions, extending past the trailing edge of the blades. The model-scale simulations at Re_1 yield a slightly narrower tip vortex cavitation, especially farther downstream from the propeller. The tip and hub vortex shape and extent were well reproduced also in full-scale conditions, although the extent of the tip vortex cavitation is slightly less in full scale.

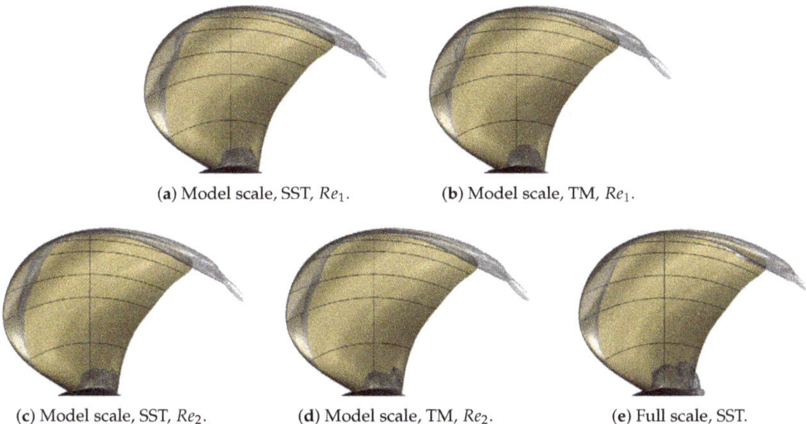

(**a**) Model scale, SST, Re_1. (**b**) Model scale, TM, Re_1.

(**c**) Model scale, SST, Re_2. (**d**) Model scale, TM, Re_2. (**e**) Full scale, SST.

Figure 22. Comparison of the predicted near-blade cavitation observation for the PPTC Case 1 in model- and full-scale conditions.

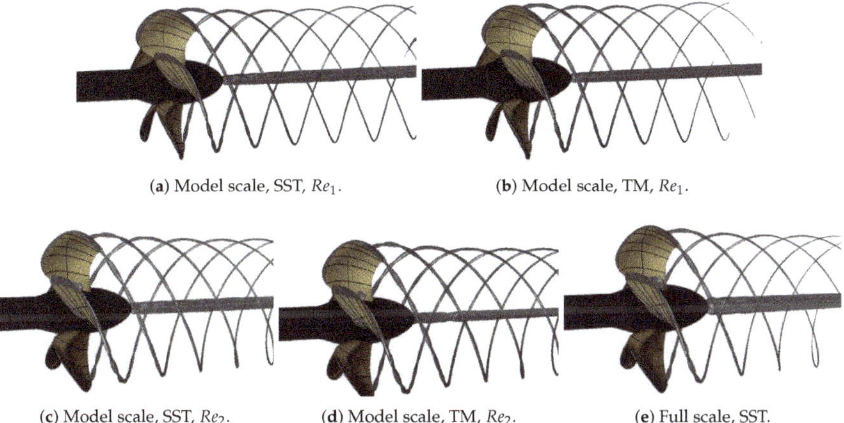

(**a**) Model scale, SST, Re_1. (**b**) Model scale, TM, Re_1.

(**c**) Model scale, SST, Re_2. (**d**) Model scale, TM, Re_2. (**e**) Full scale, SST.

Figure 23. Comparison of the predicted cavitation observation for the PPTC Case 1 in model- and full-scale conditions.

Next, we investigate the rate of evaporation within the cavity volume in for the PPTC propeller in Case 1 conditions. Figure 24 shows the results on the blade, and Figure 25 with a closer look near the LE. Here, the rate of evaporation is visualized at various cuts in the cavity volume in the axial direction. Also the surface flow on the blades are shown. The boundary layer flow is also discussed in Section 6. The figure depicts both the model- and full-scale results for the propeller, and the model-scale results are those obtained with the SST and the transition models. The rate of evaporation in model scale is strong at the core of the tip vortex, whereas condensation occurs at the outer edge of the cavitating vortex. Similar trends are also present in full-scale, although the rate is less. A more complex interplay of the evaporation and condensation takes place on the sheet cavities close to the blades: the LE and root areas. The sheet cavity closure line near the leading edge, and corresponding transition between the evaporation and condensation, shifts toward a greater chord length from the leading edge in full-scale conditions. This is quite clearly evidenced also by the orientation of the blade surface flow, where the streamlines on the surface are straighter in full scale until the location of the re-entrant jet. The flow within the sheet cavitation near the leading edge is more complex in model scale.

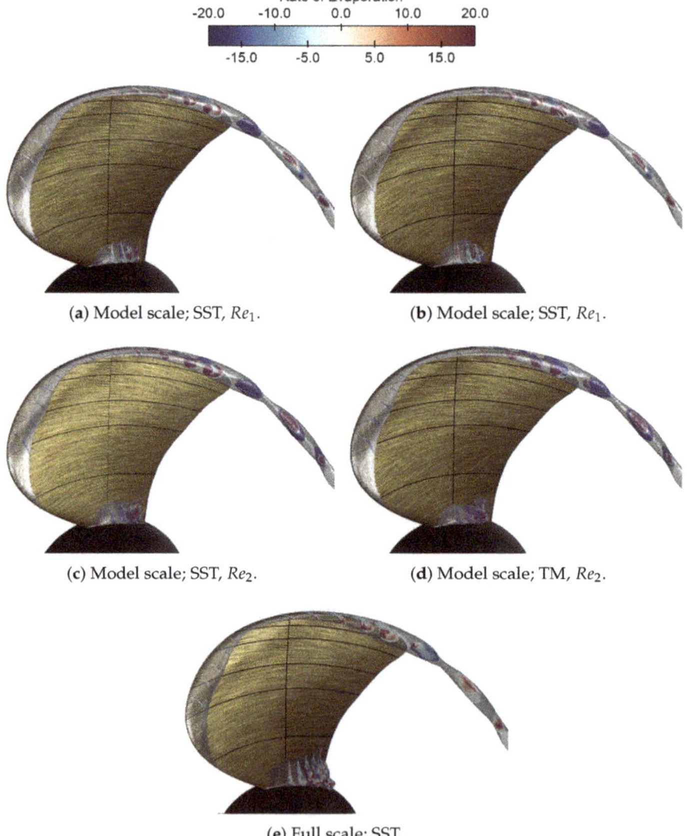

(**a**) Model scale; SST, Re_1.

(**b**) Model scale; SST, Re_1.

(**c**) Model scale; SST, Re_2.

(**d**) Model scale; TM, Re_2.

(**e**) Full scale; SST.

Figure 24. Rate of evaporation [kg/m^{3}s] within the cavity volume in model and full scale for the PPTC Case 1.

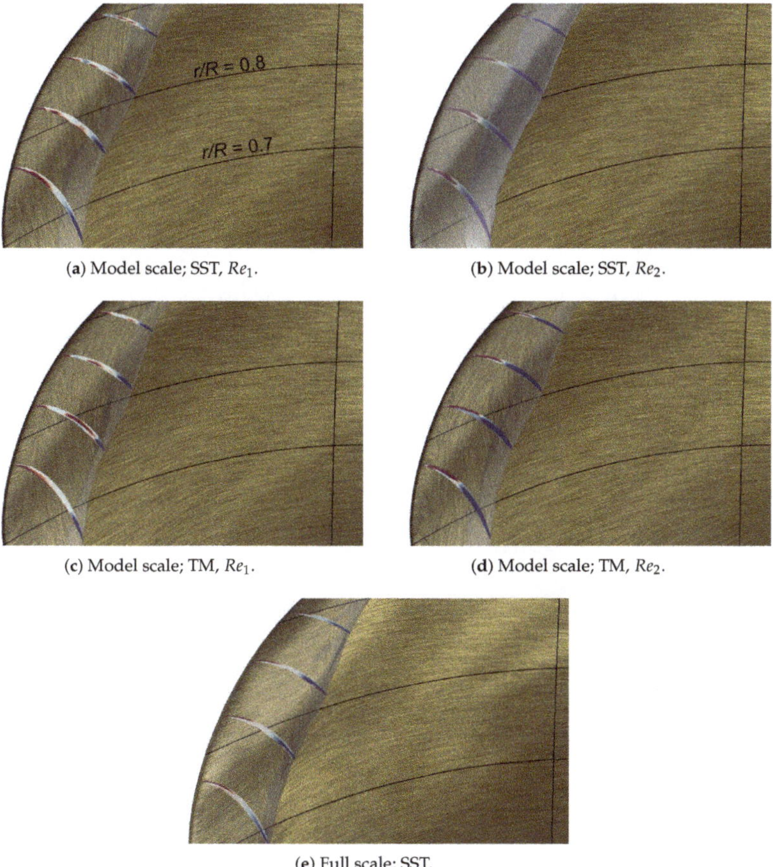

(**a**) Model scale; SST, Re_1.

(**b**) Model scale; SST, Re_2.

(**c**) Model scale; TM, Re_1.

(**d**) Model scale; TM, Re_2.

(**e**) Full scale; SST.

Figure 25. Rate of evaporation [kg/m^{3}s] within the cavity volume in model and full scale for the PPTC Case 1. Closer view near the LE. The color scale is the same as in Figure 24.

We note some differences in the sheet cavitation structure in the model-scale conditions, depending on the *Re* and the utilized turbulence model. The surface flow is directed almost straight upwards toward the tip with SST at both investigated Reynolds numbers, and more waviness is seen at lower *Re*. The chordwise location of the line dividing the regions of transition from the circular to upward surface flow just after the leading edge shifts closer to the LE at the higher *Re* (similar shift is seen with TM). At model-scale Re_2, the SST and TM show clear twofold behavior: strong evaporation close to the LE and condensation after approximately half of the length of the cavity. We observe that condensation occurs within a fine layer just close to the blade, under the region of the strong evaporation. At model-scale Re_1, we see a more mixed behavior with two distinct regions of evaporation and condensation.

5.2. Cavitation Patterns for PPTC Propeller, Case 2 Conditions

Next, in Figure 26 we compare the cavitation patterns close to the blades and in the slipstream of the propeller in model- and full scale for the PPTC Case 2. Furthermore, the cavitation behind the propellers and on the pressure sides of the blades is shown in Figure 27. The near-blade cavitation pattern is again similar in model- and full-scale conditions. The fine sheet cavity covers a large part of the propeller, as it extends from the root to the tip. The inception of the tip vortex cavitation appears similar between the simulations. Unsteady shedding of the root cavitation occurs in both simulations, subsequently transforming into cloudy cavitation in the propeller wake. The full-scale DDES predict clearer cavitation structures just behind the blades, whereas the model-scale DDES shows more extensive cavitation farther downstream of the propeller. We continue with the analyses of the unsteady cavitation shedding in Figure 28.

In Figure 27, we note that the propellers exhibit a sheet cavitation on the pressure side, originating from the root and extending up until the tip of the blade. As it approaches the tip, the pressure side sheet cavitation turns very fine, and ultimately joins the suction side at the propeller tip to form the tip vortex cavitation. This is also visible in Figure 26a,b. Although not shown here, we note that this extension of the fine pressure side sheet cavitation was not so clearly visible in the results obtained with the RANS models utilized. The propellers also show root cavitation on the pressure sides of the blades. A range of cavitation structures of varying sizes appear in the propeller wake, as we also noted earlier in Figures 9 and 26. Again, we observe clearer these cavitation structures just behind of the full-scale propeller, whereas in model-scale conditions, they appear farther downstream.

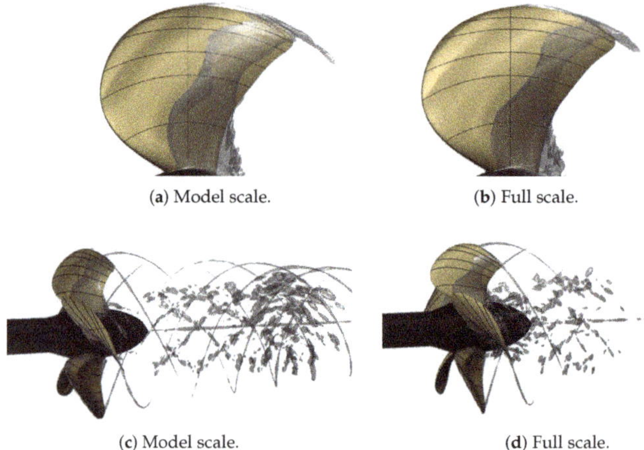

(**a**) Model scale. (**b**) Full scale.

(**c**) Model scale. (**d**) Full scale.

Figure 26. Comparison of the predicted cavitation patterns for the PPTC Case 2 in model- and full-scale conditions. Both results are from DDES.

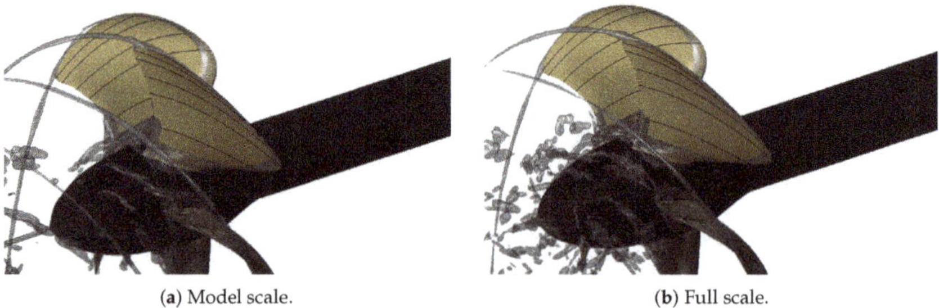

(**a**) Model scale. (**b**) Full scale.

Figure 27. Comparison of the predicted cavitation patterns behind the propeller and on the pressure side of the blades for the PPTC Case 2 in model- and full-scale conditions. Both results are from DDES.

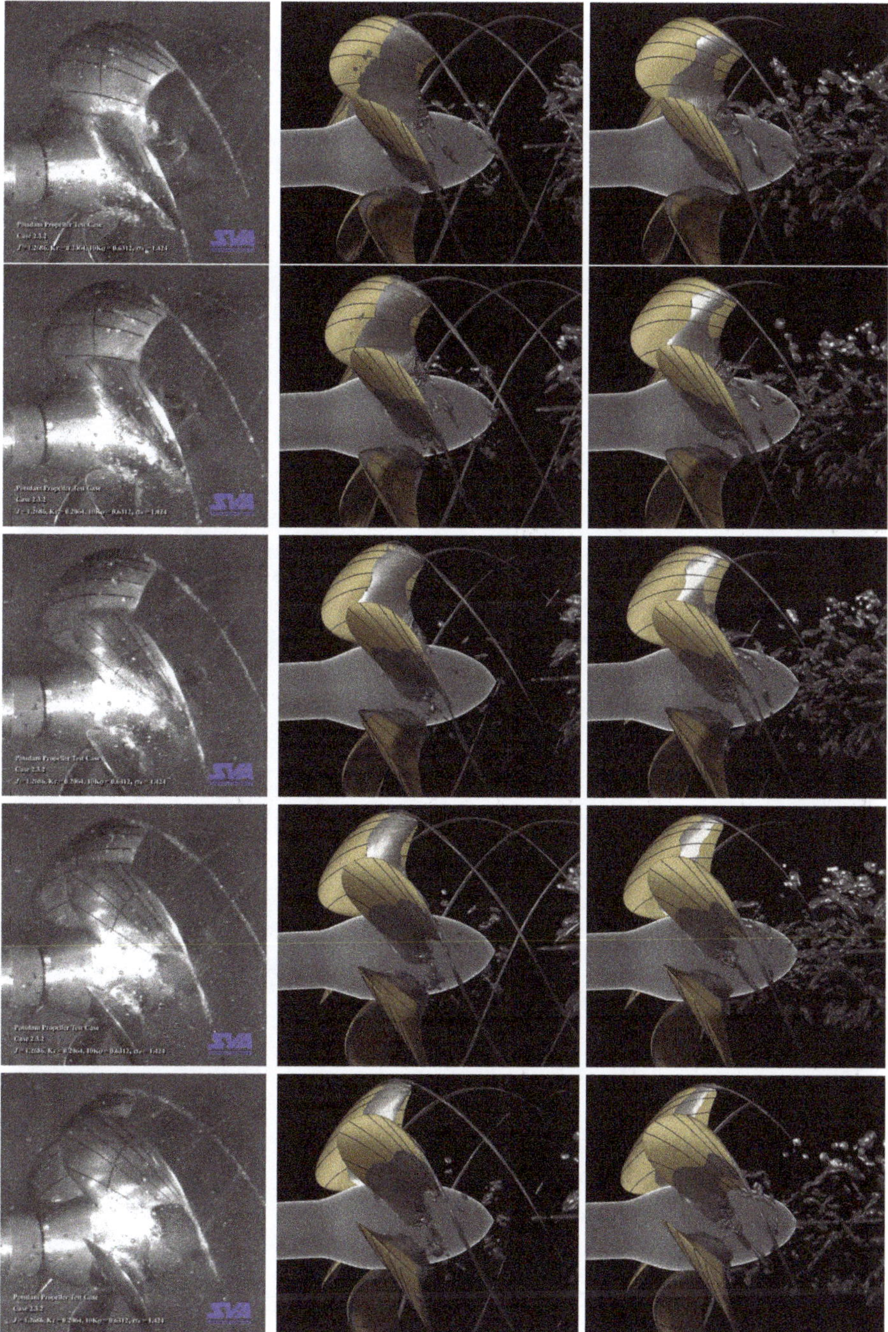

Figure 28. Cavitation growth and detachment in model- and full-scale PPTC Case 2 conditions. Left column: Experiments (model scale), reproduced with permission from Lars Lübke (SVA). Middle column: DDES Model scale. Right column: DDES Full scale. The time increment between each frame corresponds to 10 degrees of propeller rotation.

To visualize the model- and full-scale cavitation dynamics for the PPTC Case 2 conditions, we show a series of snapshots of the DDES for this case in Figure 28. In the figure, we view the simulated cavitation growth and detachment in model- and full-scale conditions together with the still frames from the video recording of the cavitation tunnel experiments (https://player.vimeo.com/video/159048901). We show five frames in the figure, and the time increment between each frame corresponds to 10 degrees of propeller rotation, corresponding to $\Delta t = 1.389$ ms in model scale and 9.259 ms in full scale. The propeller has root cavitation both on the suction and pressure sides of the blades. From these, unsteady vapor structures are shed to the wake. Particularly close to the hub surface, these appear as relatively long filaments. Comparing the model- and full-scale conditions, the length of these filaments is greater in full scale. The full-scale DDES show more extensive cavitation structures just behind the blades. Additionally, the vapor structures in the "cloud cavitation" region in propeller wake were more dynamic in the full-scale simulations, that is, they exhibited a relatively higher frequency "bursting" behavior. These structures take place also closer to the propeller in the full-scale conditions, cf. Figures 9 and 26. We observed a similar behavior when comparing the unsteady cavitation characteristics in model- and full-scale SST simulations, although these are not shown here. The tip vortex cavitation has some instability in the DDES: its extent and shape featured rapid changes occasionally in the simulations. This behavior was likely caused by the bursting of the slipstream "cloudy" cavitation, and the subsequent pressure waves caused by the collapse. In the experiments, some instability in the tip vortex cavitation is also seen, although similar appearance/disappearance was not observed. The shape or the extent of the fine sheet cavitation on the blades, upward of approximately $r/R \approx 0.5$, does not appear very transient in either simulation; occasional appearance of bubbly cavitation is visible in the experiments.

5.3. Cavitation Patterns and Evaporation Rate for TLP Propeller

We compare the cavitation patterns of the TLP propeller in the model- and full-scale conditions. Figure 29 shows both the cavitation close to the blades and the overall cavitation patterns. The results obtained with the SST model are shown. The cavitation patterns on the blades mostly resemble each other in model- and full-scale conditions. The model-scale cavitation detaches from the blade before the trailing edge, and the sheet cavitation covers the entire blade surface in full scale. In the full-scale conditions, this cavity detachment is different, as it remains attached to the blade until the trailing edge. The sheet cavitation transforms into vortex cavitation before the tip of the blade in both cases. Cavitation of the secondary tip vortex due to the endplate is visible in both simulations, although the very fine vortex cavitation does not extend very far in either case. The tip vortex shape and extent in the full-scale conditions is similar to the model-scale observations, although the extent of the tip vortex cavitation is less in full scale.

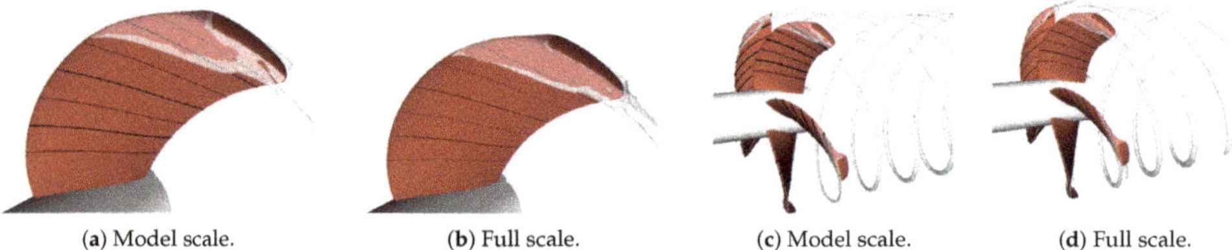

(**a**) Model scale. (**b**) Full scale. (**c**) Model scale. (**d**) Full scale.

Figure 29. Comparison of the predicted cavitation patterns for the TLP in model- and full-scale conditions. Both results are from SST simulations.

We study the rate of evaporation within the cavity volume for the TLP propeller in Figure 30. Similarly to the above, the rate of evaporation is visualized at various cuts in the cavity volume in the axial direction. Also the surface-restricted streamlines on the blades are shown. The figure shows both the model- and full-scale results for the propeller, and the model-scale results are those obtained with the SST and the transition models. The boundary layer flow is further discussed in Section 6. The rate of evaporation in model scale is stronger for the TLP at the core of the tip vortex, compared to the PPTC case. The TLP propeller has a little lower J value, and the cavitating tip vortex is finer. Condensation occurs at the outer edge of the cavitating vortex. These trends are also present in full scale, and again the rate of evaporation is less. The sheet cavity on the blade surface is quite thick. Evaporation seems to dominate closer to the leading edge, whereas more condensation is seen closer to the mid-chord position. The surface-restricted streamlines are straighter in full scale, also underneath the sheet cavity. The flow within the sheet cavitation is more complex in model scale, which is visible in both the SST and transition simulations. The model-scale simulations with the SST and transition model yield relatively similar results.

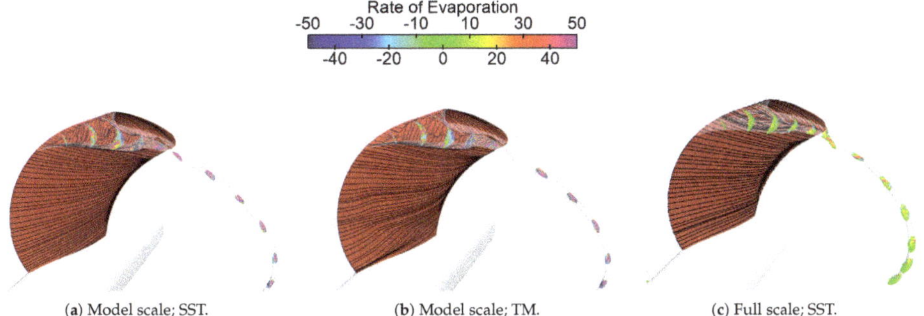

Figure 30. Rate of evaporation [kg/m^{3}s] within the cavity volume in model and full scale for the TLP propeller. Perspective views near one blade. Cavitation is shown as the grey transparent surface, and the surface-restricted streamlines on the blades as the black curves.

6. Near-Blade Flow: Skin-Friction and Pressure Distributions

6.1. Near-Blade Flow Analysis for PPTC

We show a comparison of the skin-friction and pressure coefficients in Figures 31 and 32 in model- and full-scale conditions. The distributions are compared on the radius $r/R = 0.7$ for PPTC Case 1. The vertical scale is limited in these graphs to better visualize the results, and the peak skin-friction and pressure coefficient values at the LE are not shown. The highest values were produced in the SST and TM simulations with Re_1, while the lowest in the full-scale conditions. The C_p values at the leading edge are nearly identical in all simulations.

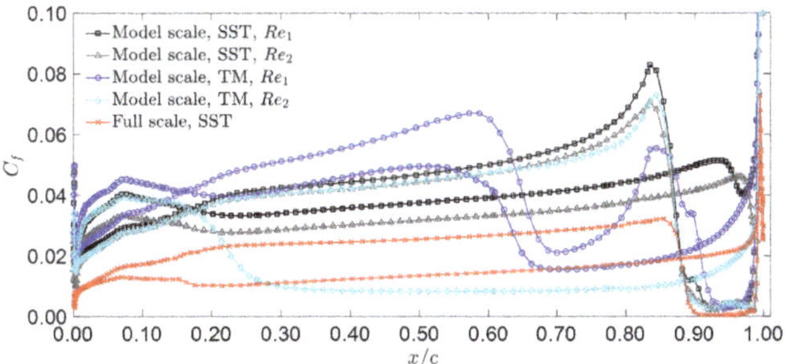

Figure 31. Comparison of skin-friction coefficients at the radius $r/R = 0.7$ on the blades. PPTC Case 1 conditions in model- and full-scale.

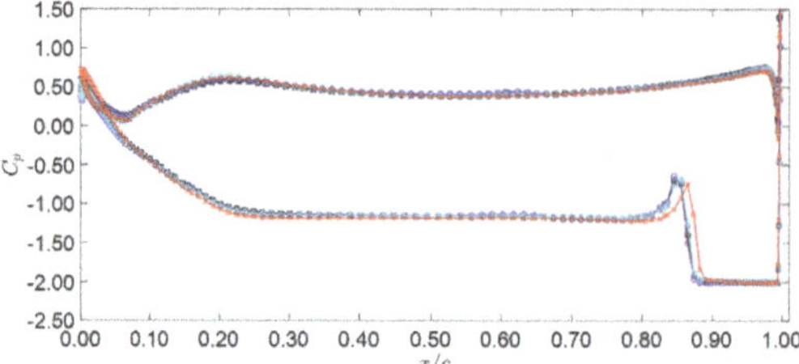

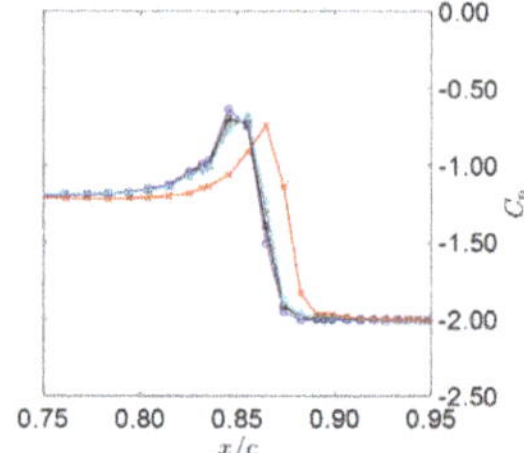

Figure 32. Comparison of pressure coefficient at the radius $r/R = 0.7$ on the blades. PPTC Case 1 conditions in model- and full-scale. The legend is the same as in Figure 31.

Figure 33 shows the skin-friction and pressure coefficients on the blades in model- and full-scale conditions. For reference, the same radii as before are shown as the black circles drawn on the blades. The boundary layer flow was mostly circumferentially directed along the blade for the model- and full-scale SST simulations. Flow separation takes place at the root cavitation close to the TE in all simulations. The SST model predicts fully turbulent flow on the wetted parts in model scale, and little difference in the surface flow is seen between the two model-scale SST simulations. A clear difference in the surface streamlines for the PPTC in Figures 24, 25 and 33 is seen in the areas covered by the sheet cavitation. The model scale simulations using the SST and TM show a main flow direction under the sheet cavity before the cavity closure line that is more wavy and radially directed, whereas the surface streamlines are more organized and circumferentially oriented in full scale. The TM at higher Re predicts a laminar flow region below $r/R \approx 0.6$, and the TM with lower Re shows laminar flow zones extending slightly above $r/R \approx 0.6$, but transitioning to turbulent flow before the mid-chord. The effect of cavitation on the surface restricted streamlines is significant. The re-entrant jets in both model- and full-scale simulations are directed towards the cavitating tip vortex at the closure line of the sheet cavitation as observed before in model-scale conditions [10,24].

Generally, the skin-friction coefficient drops to very small values in the areas underneath the cavitation and attains greater values in the wetted parts of the blades. The full-scale skin-friction coefficient values are mostly below those seen in model-scale. We observe a peak in the C_f immediately after the cavity closure in all model-scale simulations; this is not visible in full-scale conditions. The skin-friction coefficient under the LE sheet cavity appears as more uniform in the model-scale SST and TM simulations with the higher Re_2, when compared to the results with the lower Reynolds number. The C_f exhibits some deviations between the model-scale SST simulations, obtaining greater values at smaller radial locations with the higher Reynolds number. At $r/R = 0.7$, the simulations with higher Re produces greater C_f magnitude. The TM simulations with the lower Re predicts laminar flow regions below $r/R = 0.7$, visible in the surface flow orientation and as a lower C_f. The TM simulations with Re_2 predicts a laminar flow zone roughly below $r/R \approx 0.6$, extending almost the entire chord length. Exceptions are seen in the root area, where flow separation takes place due to the cavitation. The skin-friction distributions between the TM simulations at the investigated Re are quite different.

The TM at Re_1 shows a peculiar chordwise development of the C_f: small values under the LE sheet cavitation rebound to higher values after the closure line (corresponding to the other simulations), again attaining small values in the laminar flow zone, after which flow transitions to a turbulent region, the skin-friction coefficient increases and the surface flow changes to a more circumferential orientation. A similar development is seen on the pressure side, but without the low C_f due to lack of cavitation there: laminar flow region extends to $x/c \approx 0.65$, after which the skin-friction starts to increase, and the surface flow (not shown here) transforms from nearly radial to more circumferential orientation. The TM at the higher Reynolds number predicts a skin-friction distribution in line with the model-scale SST simulations above approximately $r/R = 0.6$ on the suction side, albeit with slightly smaller values for the C_f. On the blade pressure side, however, the TM simulations at the higher Reynolds number show strange development. After the LE peak in the C_f, the model predicts laminar flow with a nearly constant skin-friction coefficient. The coefficient begins to increase only once we have reached a chordwise position of $x/c \approx 0.3$. Inspection of the surface flow on the pressure side revealed similar observations. The reason for this is not clear, and further studies should be carried out in the future.

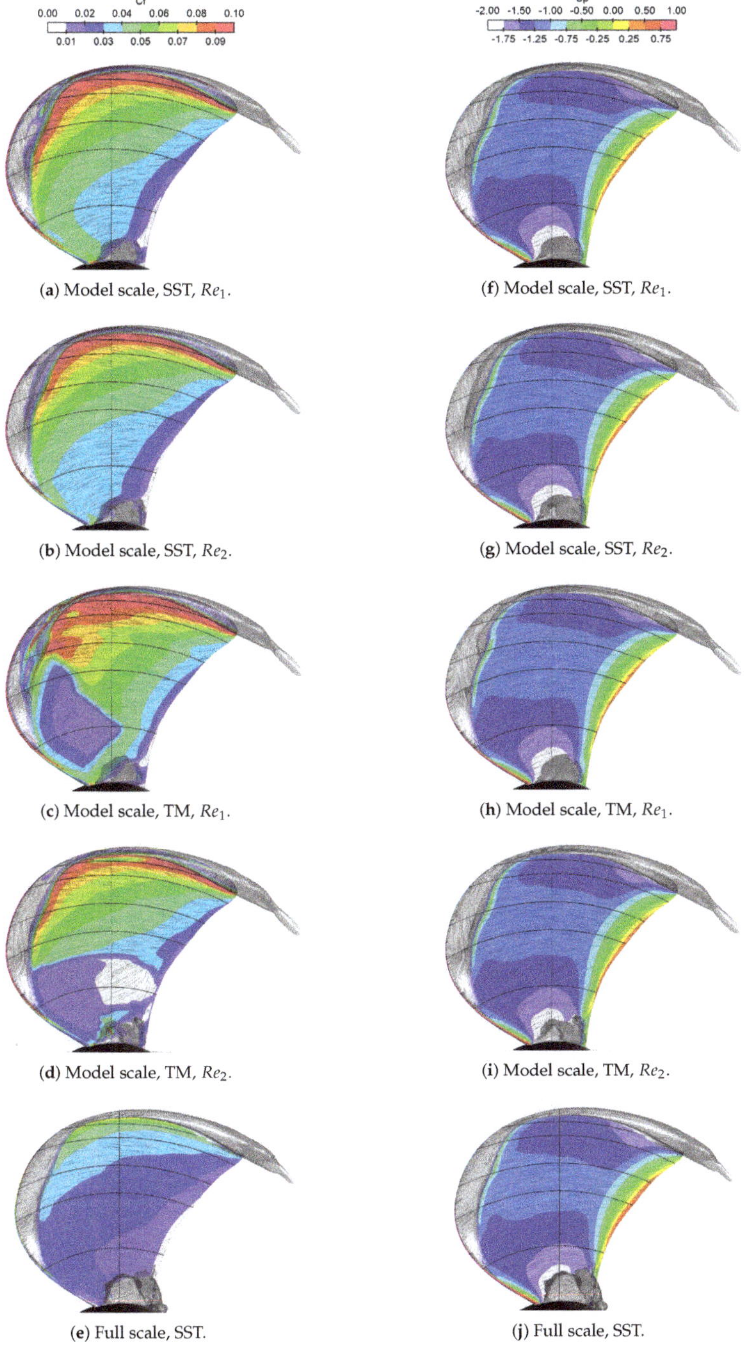

(**a**) Model scale, SST, Re_1.

(**f**) Model scale, SST, Re_1.

(**b**) Model scale, SST, Re_2.

(**g**) Model scale, SST, Re_2.

(**c**) Model scale, TM, Re_1.

(**h**) Model scale, TM, Re_1.

(**d**) Model scale, TM, Re_2.

(**i**) Model scale, TM, Re_2.

(**e**) Full scale, SST.

(**j**) Full scale, SST.

Figure 33. Comparison of near-blade skin-friction and pressure coefficients for PPTC Case 1. Left column: C_f. Right column: C_p.

The overall blade pressure distributions are very similar in the model-scale SST and TM simulations, and little sensitivity is seen with respect to the Reynolds number. The low pressure region at the middle of the blade is slightly narrower in full-scale conditions (Figure 33j). Also, a higher surface pressure in the tip area, below the cavitation zone, is seen for the full-scale conditions. The root cavitation extent is larger in full scale, and the low pressure region near the root covers a larger area in full scale as well. The pressure coefficient distributions at the $r/R = 0.7$ radius, shown in Figure 32, are very similar between all simulations, except for the TE region and the location of the cavity closure. The surface pressure is limited to the vapor pressure between the LE and $x/c \approx 0.85$, after which it increases rapidly. This increase takes place sooner and a bit sharper in full-scale conditions, whereas the model scale result appears as quite insensitive of the Reynolds number or the used turbulence models. The strength of the pressure increase, peaking at a value bit below -0.5, is similar between all simulations.

6.2. Near-Blade Flow Analysis for TLP

The skin-friction and pressure distributions on the blades of the TLP are shown in Figures 34 and 35, where we compare the model- and full-scale conditions. For reference, the same radii as before are shown as the black circles drawn on the blades.

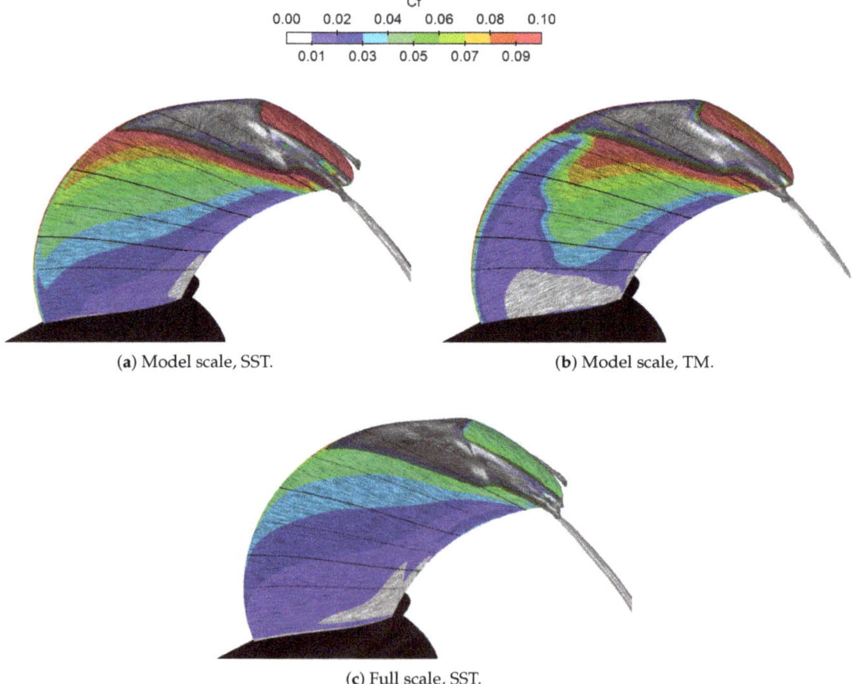

(**a**) Model scale, SST.

(**b**) Model scale, TM.

(**c**) Full scale, SST.

Figure 34. Comparison of near-blade skin-friction for TLP.

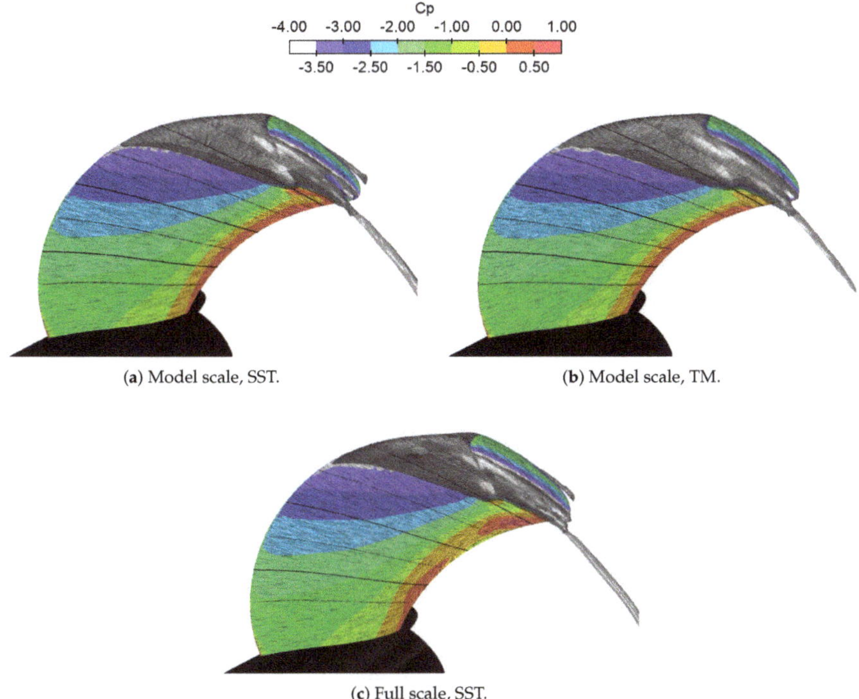

(**a**) Model scale, SST.

(**b**) Model scale, TM.

(**c**) Full scale, SST.

Figure 35. Comparison of blade pressure coefficient for TLP.

The boundary layer flow for TLP is mostly circumferentially directed along the blade for the model- and full-scale simulations. The surface flow turns upward before the TE in full-scale conditions, which appears also in the C_f distribution as lower values. Otherwise, the shapes of the C_f distributions between the two SST simulations are quite alike, albeit the full-scale conditions yield less chordwise variation and overall lower values for the skin-friction coefficient. The SST model predicts fully turbulent flow on the wetted parts in model scale, and little difference in the surface flow is seen between the SST simulations. The TM predicts laminar flow regions near the LE and close to the root below $r/R = 0.4$; this is also visible as the zones of smaller values in the skin-friction distributions when compared to the model-scale SST. The laminar flow regions predicted by the TM do not have a visible influence on the C_p. The TM predicts a little lower skin-friction coefficient on the end plate, although we do not see much differences in the surface flow between the SST and TM in model scale there. A clear difference in the surface streamlines is seen in the areas covered by the sheet cavitation, especially in model-scale conditions. The model-scale simulations show a significantly more wavy and partly radially directed main flow direction under the sheet cavitation. Small spots of flow separation appear, evidenced also as quite complex phase change phenomena (cf. Figure 30). The flow underneath the sheet cavity is more regular and circumferentially oriented in full-scale than in the model-scale simulations. The sheet cavitation turns narrower but remains attached until the TE, which is also visible as the region of lower C_f values that extend to the trailing edge. We do not, however, see a similar effect in the C_p, which remains low on the blade surface throughout the chord. The skin-friction coefficient underneath the cavitation attains lower values compared to the wetted parts of the blade. The blade pressure distributions are mainly very similar between the model- and full-scale simulations. The different surface flow behavior underneath the sheet cavitation has little effect in the blade pressure distributions which are limited by the vapor pressure. The pressure is slightly smaller in full scale near the detachment of the sheet cavity, and then grows to larger values near the trailing edge.

7. Discussion and Conclusions

We have numerically investigated single- and two-phase flows around marine propellers in open-water conditions at different Reynolds number regimes. Two types of marine propellers were considered—a conventional and a tip-loaded one—and both steady and unsteady propeller cavitation were investigated. We carried out the simulations using a homogeneous compressible two-phase flow model with RANS and hybrid RANS/LES turbulence modeling approaches. Transition was accounted for in the model-scale simulations by employing a transition model. The sensitivity of the results to the utilized numerical grid and time step size was studied. We conducted the sensitivity studies of the cavitation simulations with respect to the propeller global performance characteristics and cavitation patterns, as well as to the blade loading in terms of skin-friction and pressure distributions. These investigations were done both in model- and full-scale conditions.

The model-scale propeller performance and cavitation patterns are captured well in the numerical simulations, with generally little difference between the utilized turbulence models in the model-scale conditions. The global performance characteristics agreed favorably with the experimental values, the deviations being no larger than few per cent on the finest numerical grids. In addition, the unsteady hybrid RANS/LES simulations predicted the average global performance that matched closely with the experiments. The model-scale URANS performed scantly worse. Particularly, the TM predicted a large difference with respect to the experiments in the unsteady case: the low-pressure region on the blades was not sufficient for the inception of cavitation there.

The different cavitation types for both propellers were mostly replicated with the numerical simulations. Especially, the tip and hub vortex cavitation shapes and extents were captured well, in both the steady and unsteady cases. Also, the modal shapes of the cavitating tip vortex were present in the simulation results. We noted an exceptional deviation to the experiments with TM in the investigated unsteady case, however, which merits a further assessment.

The full-scale performance characteristics were also compared to the model-scale values: the simulations indicated a slightly increased thrust and a lower torque coefficient in full-scale conditions, both for wetted and cavitating cases. Considering the cavitation observation, we noted corresponding cavitation types between the model- and full-scale conditions. We observed a tendency of increased cavitation extent as the Reynolds number increases. The tip vortex cavitation in quasi-steady simulations was thicker especially near the blades in the full-scale conditions. The hub vortex for the PPTC propeller was clearly thicker in the full-scale simulation. A similar trend revealed when we compared the model-scale results for the PPTC at the different Reynolds numbers. The root cavitation was also more extensive in full scale. The same trend was observed in the tip vortex cavitation for the TLP propeller: the thickness of vortex cavitation was larger in full scale than in model scale. We observed that the phase change phenomena obtained a slower rate in full-scale conditions. At the same time, dissipation of the cavitating tip vortices was greater in the full-scale simulations. In the case of unsteady computations, the tip vortex cavitation extent was also diminished in full scale. The "cloudy" cavitation in the propeller wake was, then again, more extensive in full scale also appearing closer to the propeller. The simulated cavitation detachment was more comprehensive in full scale. The vapor structures in the "cloud cavitation" region in propeller wake were more dynamic in the full-scale simulations exhibiting a relatively higher frequency "bursting" behavior.

The surface flow on the blades differed clearly between the model- and full-scale simulations in the areas covered by the sheet cavitation for both investigated propellers. The near-blade flow underneath the sheet cavity was more regular and circumferentially oriented in full-scale than in the model-scale conditions. The model-scale simulations show a more wavy and radially directed surface flow direction under the sheet cavitation.

The skin-friction coefficients were lower in the areas covered by the sheet cavitation in the case of steady cavitation simulations. The blade surface pressure is limited by the vapor pressure under the cavities. As expected, the skin-friction coefficient showed a significant grid resolution and Reynolds number dependency. The blade pressure coefficient distribution, then again, did not show as strong sensitivity. The finer grids predicted quite sharp an increase in the chordwise C_p after the LE sheet cavity closure, whereas we noted a lot smoother growth with the coarse grids. The overall blade pressure distributions are very similar in the model-scale SST and TM simulations, and little sensitivity is seen with respect to the Reynolds number or the presence of laminar flow regions. The increase in C_p at the LE sheet cavity closure appears sharper and takes place closer to the LE in full-scale conditions.

The unsteady cavitation dynamics in the propeller wake for PPTC Case 2 is extremely complex. The simulations yield the void fractions greater than visible in the cavitation tank test recordings. Predictions of the "cloudy" or bubbly cavitation flow appearances with finite grid densities is very difficult using homogeneous mixture models. In this sense, better results could be achieved with more holistic modeling approaches that are able to better account for the phase change physics, such as inhomogeneous or multi-scale flow models. An extended model of the former type is under development by the authors, and we have obtained promising initial results of propeller cavitation using such an approach.

Author Contributions: Conceptualization, V.V., T.S. and A.S.-C.; methodology, V.V.; software, V.V. and T.S.; validation, V.V., T.S. and A.S.-C.; formal analysis, V.V.; investigation, V.V.; resources, V.V. and A.S.-C.; data curation, V.V. and T.S.; writing–original draft preparation, V.V.; writing–review and editing, T.S. and A.S.-C.; visualization, V.V.; supervision, T.S. and A.S.-C.; project administration, A.S.-C.; funding acquisition, A.S.-C. All authors have read and agreed to the published version of the manuscript.

Funding: This research received no external funding.

Acknowledgments: The authors are very grateful to Michael Brown from NSWCCD for providing the experimental results of the TLP propeller in the forms of open-water data and cavitation tunnel results.

Conflicts of Interest: The authors declare no conflict of interest.

Abbreviations

The following abbreviations are used in this manuscript:

CFD	Computational fluid dynamics
CLT	Contracted and loaded tip
CPU	Centra processing unit
DES	Detached eddy simulation
DDES	Delayed detached eddy simulation
EARSM	Explicit algebraic Reynolds stress model
EFD	Experimental fluid dynamics
LDV	Laser Doppler velocimetry
LE	Leading edge
LES	Large eddy simulation
MUSCL	Monotonic upstream-centred scheme for conservation laws
PPTC	Potsdam propeller test case
RANS	Reynolds averaged Navier–Stokes
SST	Shear stress transport
TE	Trailing edge
TLP	Tip loaded propeller
TM	Transition model

References

1. Rijpkema, D.; Baltazar, J.; Falcão de Campos, J. Viscous flow simulations of propellers in different Reynolds number regimes. In Proceedings of the Fourth International Symposium on Marine Propulsors, Austin, TX, USA, 31 May–June 4 2015.

2. Sánchez-Caja, A.; González-Adalid, J.; Pérez-Sobrino, M.; Sipilä, T. Scale effects on tip loaded propeller performance using a RANSE solver. *Ocean Eng.* **2014**, *88*, 607–617. [CrossRef]
3. Dong, X.Q.; Li, W.; Yang, C.J.; Noblesse, F. RANSE-based Simulation and Analysis of Scale Effects on Open-Water Performance of the PPTC-II Benchmark Propeller. *J. Ocean Eng. Sci.* **2018**, *3*, 186–204. [CrossRef]
4. Müller, S.B.; Abdel-Maksoud, M.; Hilbert, G. Scale effects on propellers for large container vessels. In Proceedings of the First Internatioal Symposium on Marine Propulsors, Trondheim, Norway, 22–24 June 2009.
5. Baltazar, J.; Rijpkema, D.; Falcão de Campos, J. On the use of the $\gamma - Re_\theta$ transition model for the prediction of the propeller performance at model-scale. *Ocean Eng.* **2018**, *170*, 6–19. [CrossRef]
6. Baltazar, J.; Rijpkema, D.; Falcão de Campos, J. Prediction of the Propeller Performance at Different Reynolds Number Regimes with RANS. In Proceedings of the 6th International Symposium on Marine Propulsors, SMP'19, Rome, Italy, 26–30 May 2019.
7. Amromin, E. Estimations of scale effects on blade cavitation. *J. Phys. Conf. Ser. IOP Publ.* **2015**, *656*. [CrossRef]
8. Viitanen, V.; Siikonen, T.; Sánchez-Caja, A. Numerical Viscous Flow Simulations of Cavitating Propeller Flows at Different Reynolds Numbers. In Proceedings of the 6th International Symposium on Marine Propulsors, SMP'19, Rome, Italy, 26–30 May 2019.
9. Miettinen, A.; Siikonen, T. Application of pressure- and density-based methods for different flow speeds. *Int. J. Numer. Methods Fluids* **2015**, *79*, 243–267. doi:10.1002/fld.4051. [CrossRef]
10. Viitanen, V.M.; Siikonen, T. Numerical simulation of cavitating marine propeller flows. In Proceedings of the 9th National Conference on Computational Mechanics (MekIT'17), Trondheim, Norway, 11–12 May 2017; International Center for Numerical Methods in Engineering (CIMNE): Trondheim, Norway, 2017; pp. 385–409. ISBN 978-84-947311-1-2.
11. Miettinen, A. *Simple Polynomial Fittings for Steam, CFD/THERMO-55-2007*; Report 55; Laboratory of Applied Thermodynamics, Aalto University: Espoo, Finland, 2007.
12. Spalart, P.R.; Deck, S.; Shur, M.; Squires, K.; Strelets, M.K.; Travin, A. A new version of detached-eddy simulation, resistant to ambiguous grid densities. *Theor. Comput. Fluid Dyn.* **2006**, *20*, 181–195. [CrossRef]
13. Menter, F.R. Two-equation eddy-viscosity turbulence models for engineering applications. *AIAA J.* **1994**, *32*, 1598–1605. [CrossRef]
14. Langtry, R.B. A Correlation-Based Transition Model Using Local Variables for Unstructured Parallelized CFD Codes. Ph.D. thesis, University of Stuttgart, Stuttgart, Germany, 2006.
15. Menter, F.; Langtry, R.; Likki, S.; Suzen, Y.; Huang, P.; Völker, S. A correlation-based transition model using local variables Part I: model formulation. *J. Turbomach.* **2006**, *128*, 413–422. [CrossRef]
16. Langtry, R.; Menter, F.; Likki, S.; Suzen, Y.; Huang, P.; Völker, S. A correlation-based transition model using local variables Part II: Test cases and industrial applications. *J. Turbomach.* **2006**, *128*, 423–434. [CrossRef]
17. Hellsten, A. Curvature corrections for algebraic Reynolds stress modeling: a discussion. *AIAA J.* **2002**, *40*, 1909–1911. [CrossRef]
18. Hellsten, A.; Wallin, S.; Laine, S. Scrutinizing curvature corrections for algebraic Reynolds stress models. In Proceedings of the 32nd AIAA Fluid Dynamics Conference, AIAA, St. Louis, MO, USA, 24–26 June 2002; AIAA Paper 2002–2963.
19. Menter, F. Influence of freestream values on $k - \omega$ turbulence model predictions. *AIAA J.* **1992**, *30*, 1657–1659. [CrossRef]
20. Shur, M.; Spalart, P.; Strelets, M.; Travin, A. Detached-Eddy Simulation of an Airfoil at High Angle of Attack. In Proceedings of the 4th International Symposium on Engineering Turbulence Modelling and Experiments, Ajaccio, Corsica, France, 24–26 May 1999; pp. 669–678.
21. Strelets, M. Detached eddy simulation of massively separated flows. In Proceedings of the 39th AIAA Aerospace Sciences Meeting and Exhibit, Reno, NV, USA, 8–11 January 2001; AIAA Paper 2001-0879.
22. Frikha, S.; Coutier-Delgosha, O.; Astolfi, J.A. Influence of the cavitation model on the simulation of cloud cavitation on 2D foil section. *Int. J. Rotat. Mach.* **2009**, *2008*. [CrossRef]
23. Choi, Y.H.; Merkle, C.L. The application of preconditioning in viscous flows. *J. Comput. Phys.* **1993**, *105*, 207–230. [CrossRef]
24. Sipilä, T. RANS Analyses of Cavitating Propeller Flows. Licentiate Thesis, School of Engineering, Aalto University, Espoo, Finland, 2012.

25. Viitanen, V.M.; Hynninen, A.; Sipilä, T.; Siikonen, T. DDES of Wetted and Cavitating Marine Propeller for CHA Underwater Noise Assessment. *J. Marine Sci. Eng.* **2018**, *6*, 56. [CrossRef]
26. Heinke, H.J. *Potsdam Propeller Test Case (PPTC). Cavitation Tests with the Model Propeller VP1304;* SVA Potsdam Model Basin Report No.3753; Schiffbau-Versuchsanstalt Potsdam GmbH: Potsdam, Germany, 2011.
27. Viitanen, V.; Hynninen, A.; Lübke, L.; Klose, R.; Tanttari, J.; Sipilä, T.; Siikonen, T. CFD and CHA simulation of underwater noise induced by a marine propeller in two-phase flows. In Proceedings of the Fifth International Symposium on Marine Propulsors (smp'17), Espoo, Finland, 12–15 June 2017.

Journal of
Marine Science
and Engineering

Article

Suppression of Tip Vortex Cavitation Noise of Propellers using PressurePores^{TM} Technology

Batuhan Aktas [1], Naz Yilmaz [2,*], Mehmet Atlar [1], Noriyuki Sasaki [1], Patrick Fitzsimmons [1] and David Taylor [3]

1 Department of Naval Architecture, Ocean and Marine Engineering, University of Strathclyde, Glasgow G1 1XQ, UK; batuhan.aktas@strath.ac.uk (B.A.); mehmet.atlar@strath.ac.uk (M.A.); noriyuki.sasaki@strath.ac.uk (N.S.); patrick.fitzsimmons@strath.ac.uk (P.F.)

2 Naval Architecture and Marine Engineering Department, Maritime Faculty, Bursa Technical University, 152 Evler Mahallesi, Eğitim Caddesi, 16330 Bursa, Turkey

3 Oscar Propulsion Ltd, West Sussex RH12 3EH, UK; david.taylor@oscarpropulsion.co.uk

* Correspondence: naz.yilmaz@btu.edu.tr

Received: 21 January 2020; Accepted: 6 February 2020; Published: 1 March 2020

Abstract: This study aims to demonstrate the merits of pressure-relieving holes at the tip region of propellers, which is introduced as "PressurePores^{TM}" technology as a retrofit on marine propellers to mitigate tip vortex cavitation noise for a quieter propeller. Shipping noise originates from various sources on board a vessel, amongst which the propeller cavitation is considered to dominate the overall radiated noise spectrum above the inception threshold. Thus, by strategically introducing pressure-relieving holes to modify the presence of cavitation, a reduction in the overall cavitation volume can be achieved. This mitigation technique could consequently result in a reduction of the radiated noise levels while maintaining the design efficiency as much as possible or with the least compromise. The strategic implementation of the holes was mainly aimed to reduce the tip vortex cavitation as this is one of the major contributors to the underwater noise emissions of a ship. In this paper, the details and results of a complementary numerical and experimental investigation is presented to further develop this mitigation concept for underwater radiated noise (URN) and to validate its effectiveness at model scale using a research vessel propeller. An overall finding from this study indicated that a significant reduction in cavitation noise could be achieved (up to 17 dB) at design speed with a favourable strategic arrangement of the pressure pores. Such a reduction was particularly evident in the frequency regions of utmost importance for marine fauna while the propeller lost only 2% of its efficiency.

Keywords: PressurePores^{TM}; pressure relief holes; underwater radiated noise (URN); cavitation noise mitigation; experimental hydrodynamics; computational fluid dynamics (CFD)

1. Introduction

The technological developments over the last half-century have revolutionised the world that we live in. One of the main driving factors for such swift advancement is the globalisation of the world. Commercial shipping has contributed the globalisation by providing the most efficient means of transportation of bulk materials. With the ever-increasing world population, the volume of commercial shipping has been experiencing an increasing trend over the last five decades. Unfortunately, this has also resulted in the elevation of emissions produced by the maritime industry [1].

One of the most adverse by-products of commercial shipping has been underwater radiated noise (URN) emission [2]. The extraordinary expansion of the world fleet has resulted in increased levels of ambient noise in the world's seas, especially in the low-frequency domain [3]. Unfortunately, this domain is also utilised by marine mammals for their various fundamental living activities. Thus, exposing them to such an abrupt change in ambient noise levels may disorient them or disrupt their communication signals, leading to behavioural changes of these mammals and hence local extinction [4,5].

Within the framework described above, the recently conducted PressurePores^{TM} Technology development project (Patent Application Number PCT/GB2016/051129) aimed to explore the merits of implementing pressure-relieving holes (PressurePores) on marine propellers to mitigate the cavitation induced noise for a more silent propeller. This paper presents a review and results of the experimental and computational study conducted to develop this technology.

Before the pressure-relieving holes implementation, different methods in the literature were reviewed for the mitigation of cavitation and resulting noise. Firstly, the studies of blade geometry modification were investigated. According to [6], the main source of the noise, the pressure fluctuations, can be reduced with propeller geometry modifications. For example, larger skew angles can affect the cavitation dynamics reducing pressure fluctuations, noise and vibration [7–9]. Another solution for cavitating noise reduction is by increasing the number of blades which can also reduce the unsteady force on each propeller blade [10]. By improving the finishing of the blade surfaces, modifying the trailing edge [11], changing the blade area, or optimisation of blade pitch distribution might also be further solutions for the mitigation of cavitation noise.

In this study, a literature review was conducted for the pressure-relieving holes method as a solution to cavitating noise mitigation. This review revealed that in the late 1990s, the Indian Institute of Technology in Bombay conducted research involving cavitation noise on marine propellers [12]. In their research study, Sharma et al [12] tried to delay the onset of the tip vortex cavitation and to reduce noise emissions without influencing the propeller performance adversely. They achieved a noise reduction by modifying model propellers by drilling 300 holes of 0.3 mm diameter in each blade. The holes were drilled at the tip and the leading edge areas of the blades. Sharma et al.'s tests indicated that the dominant cavitation type at inception was the tip vortex cavitation under any testing conditions. The modifications did not demonstrate any measurable influence on the performance characteristics of any of the propellers tested, but it had a great influence on the development of the Tip Vortex Cavitation (TVC).

The resulting acoustic benefit obtained in the Sharma et al. study [12] was a great improvement by a substantial attenuation of the low-frequency spectral peaks, as shown in Figure 1. While the test results with the original (not modified) propellers showed a consistent rise of spectrum levels throughout the frequency range, as the advance coefficients were reduced, this was not the case for the modified propellers with reduced spectral peaks. Also, the advance coefficients had a weak effect on the noise levels which was attributed to the consequences of the modification where the tips were unloaded. Furthermore, the suction peak in the leading edge was also minimised while the TVC strength was decreased due to the increase in the angle of incidence.

Figure 1, which was taken from [12], also presents a comparison of the noise characteristics for the original and the two modified propellers, A and B, at the advance coefficient of J = 0.38. In such a low J value, the improvement was more significant. Particularly for low frequencies, between 1 and 2 kHz, a reduction of about 15 dB was observed in the noise levels of both propellers. Sharma et al. concluded that "the modifications carried out had no measurable influence on the performance characteristics of the basic propellers". However, they achieved a delay in the onset of the cavitation and significant noise reductions. One interesting point to note in Sharma et al.'s work was that all the propellers were tested in uniform flow conditions. This inherently disregarded the effect of the ship hull (wake) on the propeller flow, which is one of the most significant contributors to cavitation dynamics and hence induced radiated noise.

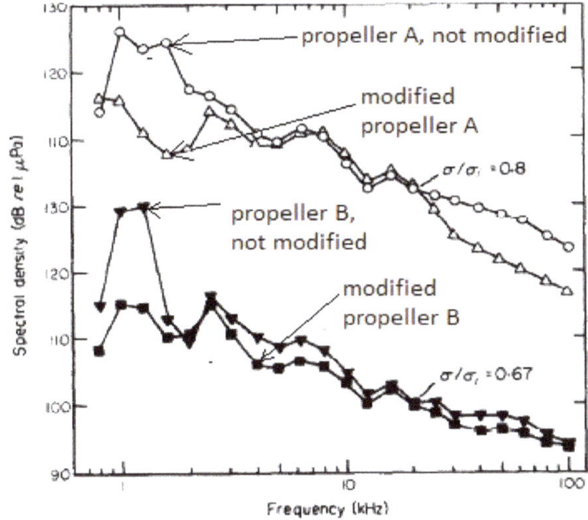

Figure 1. Influence of blade modification on cavitation noise for J = 0.38. Reproduced from [12] with permission from Elsevier, 1990.

To explore the merits of the pressure relief holes, a pilot experimental study was conducted in The Emerson Cavitation Tunnel (ECT) of Newcastle University as part of an MSc thesis by Xydis [13]. The model scale propeller used for this study (Figure 2) was an existing propeller model with a diameter of 0.35 m, which was based on the as-fitted propeller of a 95,000 tonnes tanker with four blades and an expanded blade area ratio (BAR) of 0.524. There were two further replicas of this model propeller, which had previously been used for coating research; all were made of aluminium. In the pilot study, the blue coloured, anodized model (without drilled holes) was used to establish a base line (or reference) performance measurement. The other two models were modified with pressure-relieving holes on their blades and tested for comparison with the reference propeller. To see the effect of the holes on the two observed types of cavitation (sheet and tip vortex), the 2nd model was modified with a number of small holes drilled near the blade tip region above 90% of the propeller radius, while the 3rd model propeller had tip holes and mid-span holes above 60% of the blade radius, respectively. The three models used are shown in Figure 2 and called "Base", "Tip modified" and "Sheet modified" related to the intact propeller (no holes), their intended effect on the tip vortex and sheet cavitation, respectively. To simulate more realistic operational conditions, these propellers were tested behind a wake using 2D wake screen.

Figure 2. From left to right: "Base" propeller; "Tip" (region) modified propeller; and mid and tip (region) modified, "Sheet" propeller. Reproduced from [13] with permission from Newcastle University.

A summary of the comparative propeller performance measurements in terms of the thrust and torque curves of the three model propellers is shown in Figure 3. A large number of holes spread over the mid and tip region of the "Sheet" propeller blades gave a significant reduction in the thrust, torque and efficiency. The more conservative number of pressure relief holes concentrated around the tip region for the "Tip" modified propeller did not produce such a significant impact on the thrust and torque compared to the "base" propeller as also shown in Figure 3. These experimental results, moreover, are in very close agreement with the established CFD models (omitted to shorten the length of the paper) which shows up to 13.8% cavitation volume reduction and 0.5% loss of efficiency.

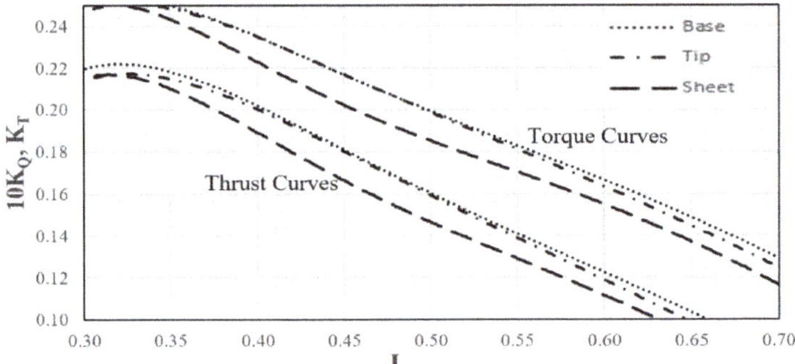

Figure 3. Comparative non-dimensional thrust and torque coefficients in open water and cavitation condition. Reproduced from [13] with permission from Newcastle University.

The noise measurements shown in Figures 4 and 5 correspond to typical operating conditions of this vessel. They reveal up to a 10 dB reduction in the sound pressure levels (SPL) for the mid-frequency region (300 Hz to 2 kHz) as well as in the high-frequency region (10 to 20 kHz) for the tip modified propeller and for advance coefficients of J = 0.55 and J = 0.5, respectively.

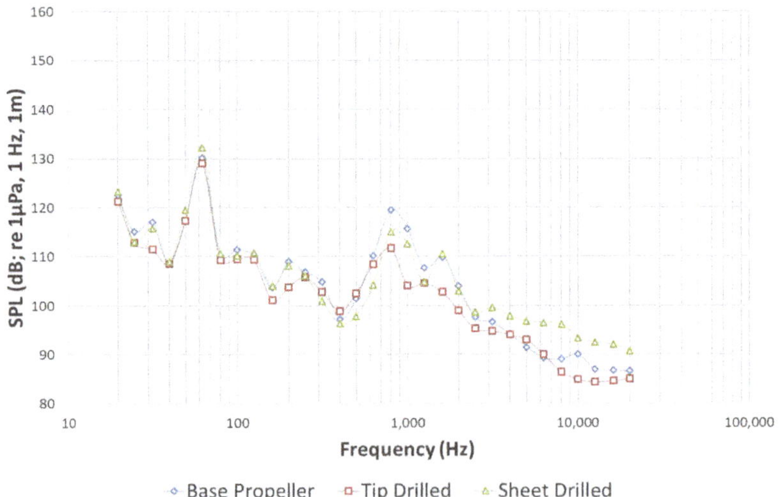

Figure 4. Comparative sound pressure levels of three model propellers J = 0.55 and cavitation condition. Reproduced from [13] with permission from Newcastle University.

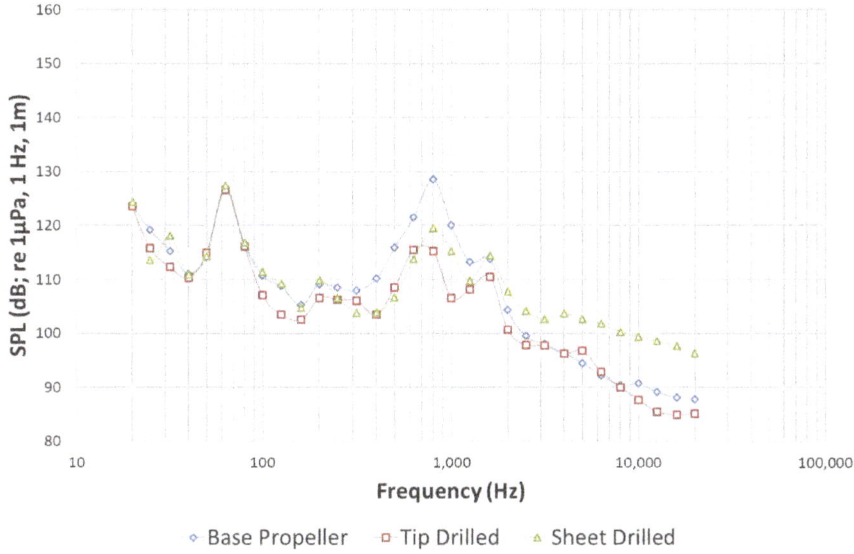

Figure 5. Comparative sound pressure levels of three model propellers tested for J = 0.50 and cavitation condition. Reproduced from [13] with permission from Newcastle University.

While the study demonstrated some encouraging signs of the radiated noise reduction, the level of the reduction in cavitation extent to support this mitigation technique needed more sophisticated and detailed observations. Inspired by this MSc study, and based on the model propellers tested in the same study [13], a comprehensive Computational Fluid Dynamics (CFD) investigation was conducted by Aktas et al. [14] to demonstrate the effectiveness of this mitigation method using the propeller of Newcastle University's Research Vessel, *The Princess Royal*. This mitigation method was later patented as the PressurePores^{TM} Technology by the sponsoring company. Based on the results of this CFD investigation, the best performing cases with the strategically selected PressurePores^{TM} were applied on the Princess Royal's propeller model to be tested at a towing tank for efficiency measurements. Cavitation tunnel tests were performed for the cavitation characteristics and noise measurements and to compare with the CFD investigations.

The details and results of the above mentioned computational and experimental investigations on the PressurePores^{TM} technology are presented in the remaining sections of this paper. Therefore, Section 2 describes the model propeller of research vessel *The Princess Royal* together with the experimental set-up and test conditions for both the CFD simulations and the cavitation tunnel tests which were conducted in the University of Genova Cavitation Tunnel (UNIGE). Section 3 presents the prototype testing including cavitation tunnel test observations in UNIGE and radiated noise measurements together with propeller performance tests conducted in the CTO towing tank of Gdansk. In Section 4, the details and results of the CFD model of the cavitating propeller are presented. Finally, Section 5 presents the main conclusions obtained from the oveall study.

2. Experimental Investigations

The Experimental Fluid Dynamics (EFD) approach adopted in this study used a propeller model modified for two versions of the PressurePores^{TM} technology, a cavitation tunnel and a towing tank as described in the following including the experimental test matrix.

2.1. Propeller Model; The Princess Royal Propeller

The propeller model used for both tests represented the port side propeller of *The Princess Royal* [15] with a scale ratio of 3.41, giving a 220 mm model propeller diameter. The reason was selecting this propeller as the test case two folds: firstly, this vessel has become a benchmark vessel worldwide for URN and cavitation investigations; secondly, the Authors had extensive information and access to this vessel for future use in validation studies at full-scale as part of the PressurePore^{TM} technology project.

The propeller model was manufactured with high accuracy in consideration of cavitation testing as shown by the deviation contour plot given in Figure 6. The principal dimensions of the full-scale propeller are given in Table 1.

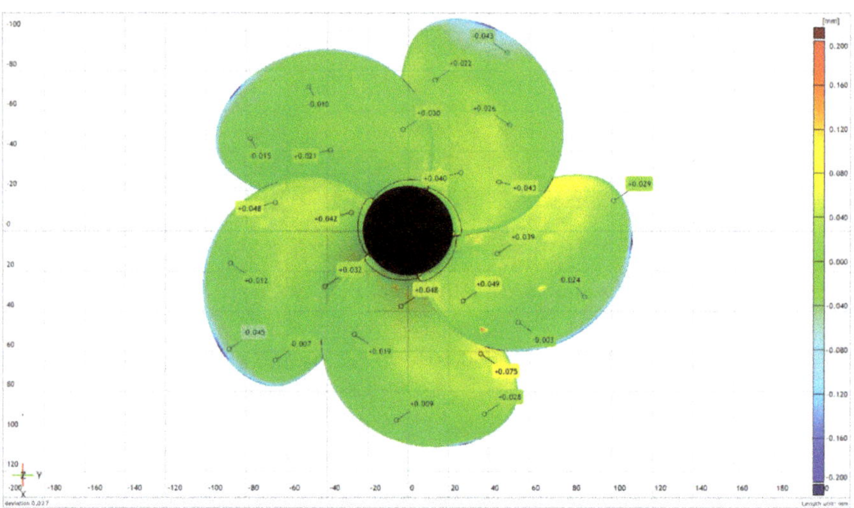

Figure 6. Manufacturing accuracy Image of the Princess Royal Propeller.

Table 1. Propeller main characteristics and particulars.

Parameters	Value	Unit
Number of blades	5	–
Full scale propeller diameter	0.75	(m)
Model scale propeller diameter	0.22	(m)
Pitch ratio	0.8475	–
Blade area ratio	1.057	–
Rake	0	(deg)

2.2. Application of the PressurePores^{TM} Technology

The application of the PressurePore^{TM} technology to this benchmark test propeller used the knowledge and experience gained through CFD studies conducted both with this propeller, as reported in Section 3, and another the preliminary test case propeller for a 95,000 tonnes merchant tanker as reviewed in Section 1.

In establishing the PressurePore^{TM} technology, different variations of the number of holes, hole size, hole axis directions (i.e., normal to the blade and shaft axis) were investigated by referring to the results of the Sharma et al. [12], and the recent study by Xydis [13] using CFD. An adaptive mesh refinement technique for evaluating the impacts of these variations on the propeller performance (K_T, K_Q and η_0) in cavitating flow conditions was used. In these investigations, the pore configurations applied were simulated initially by using a non-optimised TVC model but later using the advanced adaptive mesh refinement technique of Yilmaz et al. [16] as described and discussed in Section 4.

From the CFD investigations with different pore configurations, two were selected and adopted on *the Princess Royal* model propeller which was tested at the CTO towing tank for accurate prediction of the propeller open water performance parameters. These tests were followed by further experiments conducted in the UNIGE cavitation tunnel for the cavitation observation and underwater noise measurements. The two-pore configurations applied are shown in Figures 7 and 8, as "*Modified Propeller*" and "*Modified Propeller-2*", respectively. As a result of the CFD investigation and considering the practicality of the pores implementation the pore diameter was selected as 1 mm. The *Modified propeller* had 33 pores while the *Modified-2 propeller* had 17 pores.

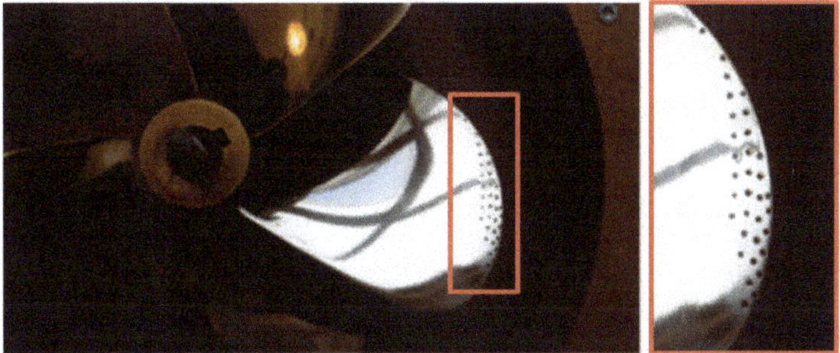

Figure 7. Princess Royal "Modified Propeller".

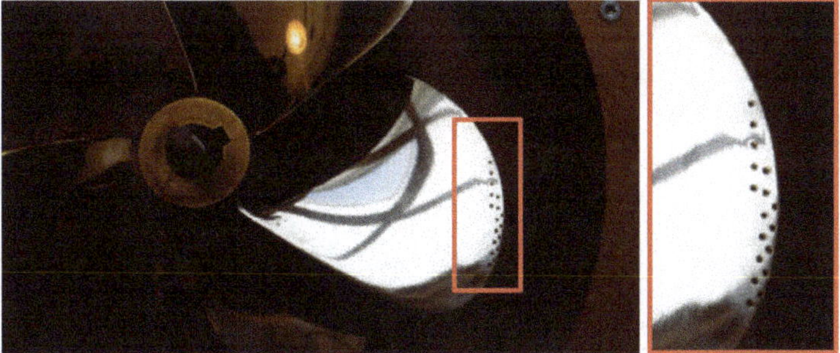

Figure 8. Princess Royal "Modified Propeller-2".

Comprehensive numerical investigations were carried out to determine optimum PressurePore geometry and distribution. These investigations involved varying number, size, shape and finish of the pressure-relieving holes as well as their strategic arrangement (distribution) over the blades. The focus of the investigations was to reduce the cavitation volume, mainly due to TVC while maintaining propeller efficiency.

2.3. Test Facilities

Tests were conducted in the medium-size cavitation Tunnel of the University of Genoa (UNIGE) and in the large towing tank of the Centrum Techniki Okrętowej S.A. (CTO) Model Basin in Gdansk.

The UNIGE tunnel is a Kempf & Remmers (K&R) closed water circuit tunnel, schematically represented in Figure 9. The tunnel has a square testing section of 0.57 m × 0.57 m, having a total testing section length of 2 m. The nozzle contraction ratio is 4.6:1, allowing a maximum tunnel flow speed of 8.5 m/s in the test section. The tunnel is equipped with a K&R H39 dynamometer, which measures the propeller thrust, torque and rate of revolution.

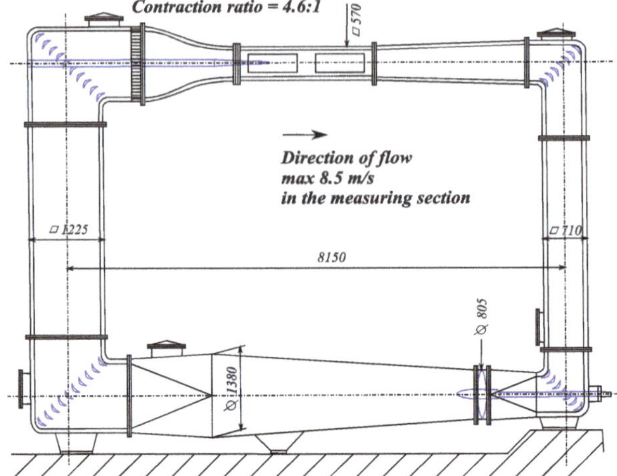

Figure 9. University of Geneva (UNIGE) cavitation tunnel.

The CTO towing tank is approximately 270 m × 12 m × 6 m in length, breadth and depth, respectively and is fitted with a towing carriage having a maximum speed of 12 m/s. The performance of the propeller model before and after the application of the PressurePores^{TM} technology was measured using a standard open water dynamometer, as shown in Figure 10.

Figure 10. CTO Towing tank open water test set-up.

2.4. Test Setup and Test Matrix

The test set-up used for the cavitation tunnel is shown in Figure 11. To simulate realistic operational conditions, the tunnel tests were carried out behind a simulated (nominal) wake field which was produced based on the wake survey conducted with *The Princess Royal* model at the Ata Nutku Towing tank of Istanbul Technical University [17]. For this purpose, a wire mesh wake screen was constructed upstream of the propeller and was verified using a 2D Laser Doppler Velocimetry (LDV) device. The cavitation tunnel setup is schematically presented in 11.

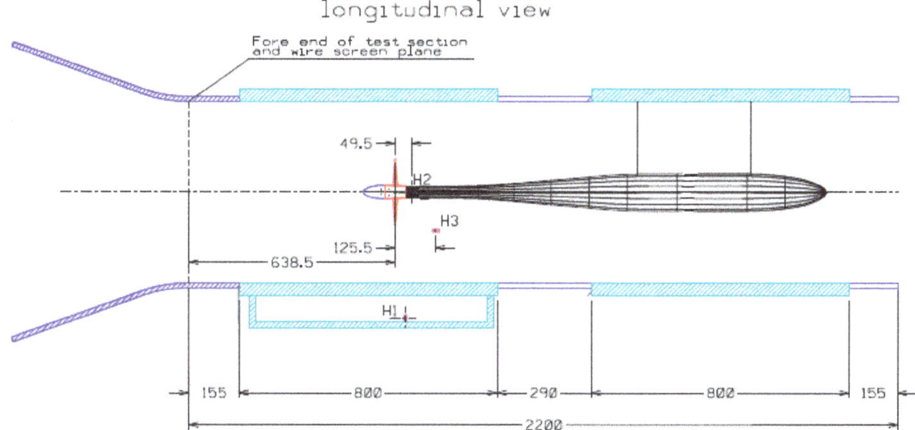

Figure 11. Cavitation tunnel setup, longitudinal view.

The comparative velocity distributions of the simulated wake in the UNIGE tunnel and the nominal wake measured in the ITU towing tank are shown in Figure 12. A part of the simulated wake data is missing due to the shaft blocking the LDV paths since the measurements could only be made from the starboard side of the test section.

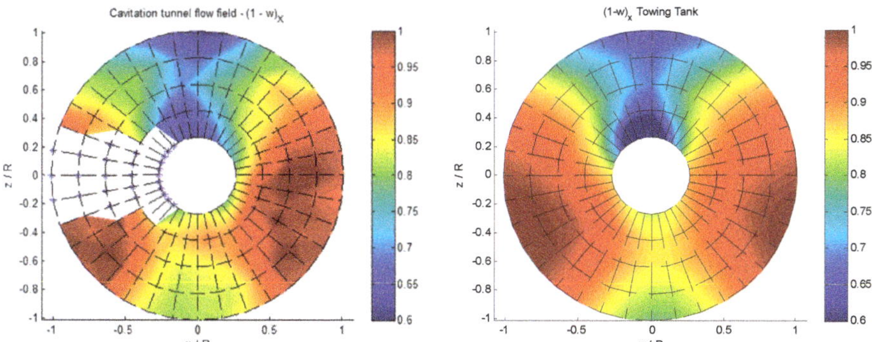

Figure 12. Nominal wake field: Simulated in cavitation tunnel (**Left**); Measured at towing tank (**Right**).

Based on typical in-service operational conditions of *The Princess Royal*, which correspond to 10.5 kn and 15.1 kn vessel speeds, the cavitation tunnel test matrix is as shown in Table 2.

Table 2. Full-scale operational conditions during sea trials.

Condition	Engine (RPM)	Shaft (rps)	STW (kn)	K_T	$10K_Q$	σ_N (nD)
V1	1500	14.3	10.5	0.211	0.323	1.91
V2	2000	19.0	15.1	0.188	0.318	1.07

In Table 2, STW represents the vessel speed through the water. K_T and K_Q are the standard thrust and torque coefficients, respectively, while the cavitation number is defined based on the propeller shaft speed using Equation (1):

$$\sigma_n = \frac{P_a + \rho g h_s - P_v}{0.5\rho(nD)^2} \quad (1)$$

where P_a is the atmospheric pressure, g is the gravitational acceleration, ρ is the density of water, h_s is the shaft immersion of the propeller, P_v is the vapour pressure, n is the propeller shaft speed in rps, and finally D is the diameter of the propeller. Table 3 shows the non-dimensional parameters used for some of the performance and operational characteristics of the propeller.

Table 3. Non-dimensional performance and operational parameters for propellers.

Performance Characteristics	Symbol	Formula
Thrust coefficient	K_T	$\frac{T}{\rho n^2 D^4}$
Torque coefficient	K_Q	$\frac{Q}{\rho n^2 D^5}$
Advance coefficient	J	$\frac{V_a}{nD}$
Efficiency	η_0	$\frac{J \times K_T}{2\pi \times K_Q}$

where T is the thrust, V_a is the advance velocity, Q is the torque and η_0 is the propeller efficiency.

The model scale test conditions were specified according to the thrust coefficient identity. As shown in Table 2, while Condition V2 corresponded to the actual service speed (about 15 knots) of the research vessel, Condition V1 corresponded to the 10.5 kn speed condition.

The cavitation tunnel tests were completed in three stages: the first stage involved the tests with the original propeller model (*Intact propeller*), the second stage involves the propeller model with 33-1 mm pores on each blade (*Modified propeller*); and the third and final stage, with 17-1 mm pores on each blade (*Modified-2 propeller*) which was achieved by closing a half of the pores on the Modified propeller with an epoxy material and smoothing them with care.

During the tests, the water quality was assessed based on the dissolved oxygen content of the tunnel which was monitored by using the ABB dissolved oxygen sensor, model 8012/170, coupled with an ABB analyser model AX400.

3. Prototype Testing

3.1. Cavitation Observations

To be able to make qualitative comparisons between the cavitation experienced by the intact and modified propeller cases, cavitation observations were carried out at the UNIGE Cavitation tunnel. For this purpose, a mobile stroboscopic system was utilised to visualize and record the cavitation phenomenon on and off the propeller blades. The cavitation recordings were made with three Allied Vision Tech Marlin F145B2 Firewire Cameras, with a resolution of 1392 × 1040 pixels and a frame rate up to 10 fps.

Remarks on the cavitation observation for the intact propeller, Modified Propeller and Modified Propeller-2 cases for Condition V1 and V2 are provided in Table 4. Some sample images are also shown in Figures 13 and 14 for the three propellers alongside Case V1 and V2 in this respective order. From the images and the remarks in Table 4, it is evident that with the introduction of PressurePores^{TM}, tip vortex cavitation experienced by the intact propeller was disrupted. For the Modified propeller case, in both conditions, the tip vortex cavitation almost disappeared. For the Modified Propeller-2 case, the PressurePores^{TM} changed the nature of the steady, solid tip vortex cavitation to a cloudier line with less strength and reduced core diameter.

Table 4. Cavitation observation for the intact propeller, Modified Propeller and Modified Propeller-2 cases for Condition V1 and V2.

Condition	K_T	σ_N	*"Intact Propeller"* Observations	*"Modified Propeller"* Observations	*"Modified Propeller-2"* Observations
V1	0.211	1.91	TVC everywhere, starting from blade L.E.; S.S. sheet cavitation at 0°, from 0.8R to the tip, for 15% of the chord at 0.8R, 100% at 0.97R; S.S. sheet cavitation at 180°, from 0.85R to the tip, for 10% of the chord at 0.85R.	Pores cavitation everywhere; TVC at 0° and 180°, only cloudy vortex at other positions; S.S. sheet cavitation at 0°–45° from 0.8R for 10% of the chord, merging with holes cavitation at outer radii; S.S. sheet cavitation at 180°, from 0.85R for 5% of the chord, merging with holes cavitation at outer radii.	Pores cavitation everywhere; TVC everywhere, at 90° and 270° the cavitating core is at inception; S.S. sheet cavitation at 0°, from 0.8R, for 15% of the chord, at 180°, from 0.85R for 10% of the chord.
V2	0.188	1.07	TVC everywhere, starting from blade L.E.; double vortex at 0°–60°; S.S. sheet cavitation at 0°, from 0.8R to the tip, for 50% of the chord at 0.8R, 100% at 0.85R; S.S. sheet cavitation at 90° and 270°, from 0.9R for 10% of the chord; S.S. sheet cavitation at 180°, from 0.83R to the tip, for 50% of the chord at 0.83R, 100% of the chord at 0.92R	Pores cavitation everywhere; TVC everywhere, with double vortex at 0°–60°; S.S. sheet cavitation at 0°–45° from 0.8R for 30% of the chord, merging with holes cavitation at outer radii; S.S. sheet cavitation at 180°, from 0.83R for 20% of the chord, merging with holes cavitation at outer radii.	Pores cavitation everywhere; TVC everywhere, the cavitating core is now well developed but still presents unstable behaviour; double vortex at 0°; S.S. sheet cavitation at 0° from 0.8R for 40% of the chord, at 180° from 0.8R for 30% of the chord.

Figure 13. Intact vs. Modified propeller and Modified Propeller-2 in condition V1 and viewed from starboard.

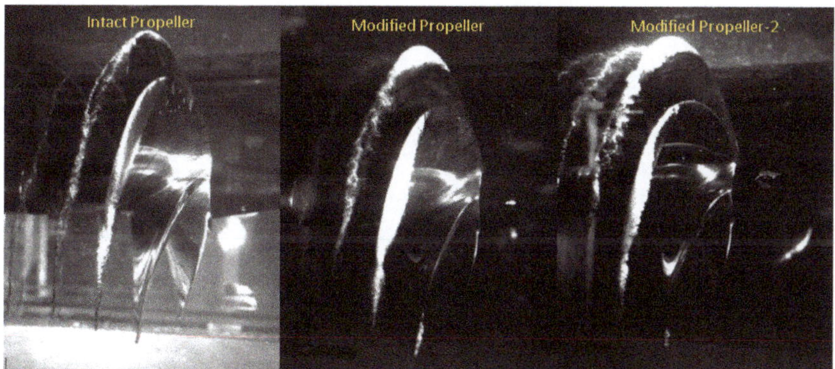

Figure 14. Intact vs. Modified propeller and Modified Propeller-2 in condition V2 and viewed from starboard.

3.2. Radiated Noise Measurements

In Section 2, Figure 11 showed a schematic of the setup adopted during these tests, including the positions of three hydrophones. As shown in Figure 11, two hydrophones were mounted on fins downstream of the propeller: one on the port side at the same vertical position as the propeller shaft (H2); the other (H3) on the starboard at a lower vertical position. The third hydrophone (H1) was mounted in an external plexiglass tank filled with water and mounted on the bottom window of the test section. The measurements from H3 were used for the noise results presented throughout this paper.

Moreover, the noise tests were also repeated at least three times. For the post-processing of the noise measured, the ITTC [18] guidelines for model scale noise measurements were followed.

The average Power Spectral Density, G(f) in Pa^2/Hz, was computed from each sound pressure signal p(t) using Welch's method of averaging modified spectrograms. The Sound Pressure Power Spectral Density Level, L_p, is then represented by Equation (2):

$$L_p(f) = 10\log_{10}\left(\frac{G(f)}{p_{ref}^2}\right)(\text{dB re } 1\ \mu\text{Pa}^2/\text{Hz}) \tag{2}$$

where $p_{ref} = 1$ μPa.

The background noise of the facility and setup was measured by replacing the propeller with a dummy hub while applying the same conditions of the shaft revolutions, flow speed and vacuum. Only one series of the background noise measurements were carried out since the tunnel operational conditions do not vary significantly when changing from the intact to the modified propeller cases.

Based on the comparison of the total noise measured and background noise, the net sound pressure levels of the propeller were analysed with the following procedure in [19]:

1. Signal to noise ratio higher than 10 dB: No correction made
2. Signal to noise ratio higher than 3 dB but lower than 10dB:

$$\text{L}_{\text{PN}} = 10\text{Log}_{10}\left[10^{(\text{L}_{\text{Ptot}}/10)} - 10^{(\text{L}_{\text{Pbg}}/10)}\right] \tag{3}$$

3. Signal to noise ratio lower than 3 dB:

$$\text{Results disregarded} \tag{4}$$

The net sound pressure levels may be scaled to a reference distance of 1-m exploiting measured transfer functions, or simply according to Equation (5):

$$\text{L}_{\text{PN@1m}} = \text{L}_{\text{PN}} + 20\text{Log}_{10}(\text{r}) \tag{5}$$

where r is the distance between propeller (acoustical centre) and sensor.

The latter formulation has been used in the present work. The acoustical centre of the propeller was defined with respect to the centre of the propeller disk.

Based upon the above-described post-processing, Figures 15–18 shows the measured noise levels for the *intact propeller*, *Modified propeller* and *Modified Propeller-2* in the narrow and Third-octave band for condition V1 and V2. In both cases, significant reductions in the radiated noise levels can be observed over a frequency range from 200 Hz to 1 kHz. For the service speed condition V2, the reductions are consistent almost throughout the entire frequency range tested. Whereas for condition V1, the pores caused some increase in the URN in the high-frequency region.

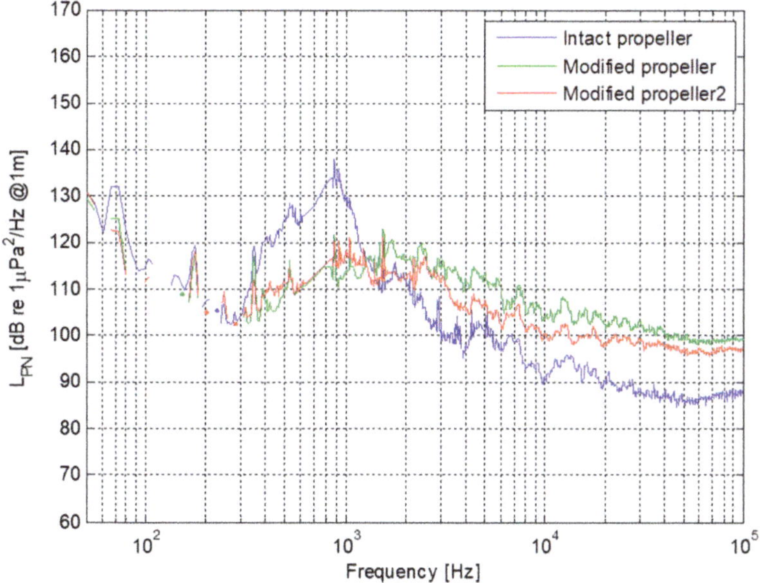

Figure 15. 2 Net noise levels: Intact, Modified propeller and Modified Propeller-2 in the 1 m (narrowband), condition V1, hydrophone H3.

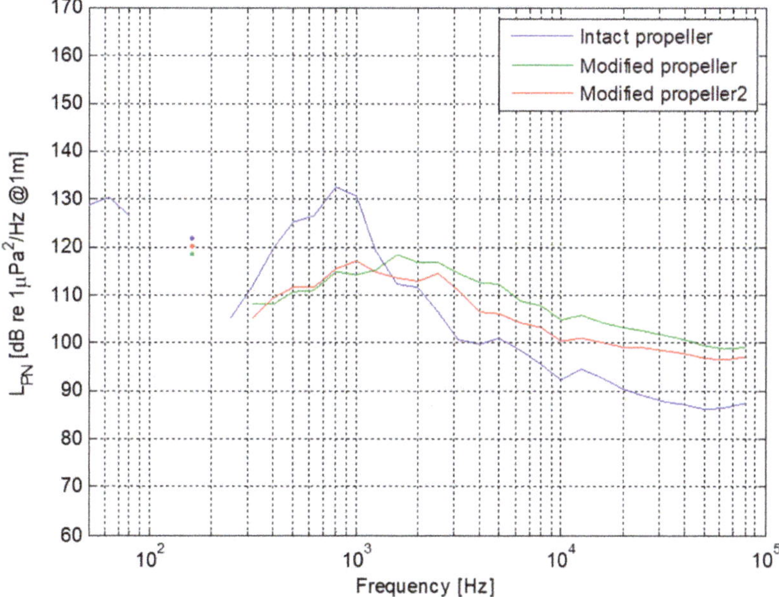

Figure 16. Comparison between Intact, Modified Propeller and Modified Propeller-2 net noise levels at 1 m (one third octave band), condition V1, hydrophone H3.

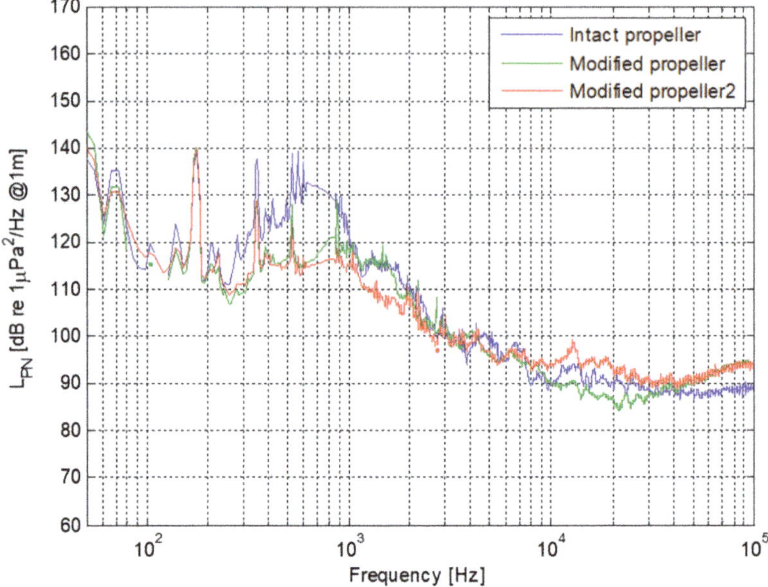

Figure 17. Comparison between Intact, Modified propeller and Modified Propeller-2 net noise levels at 1 m (narrowband), condition V2, hydrophone H3.

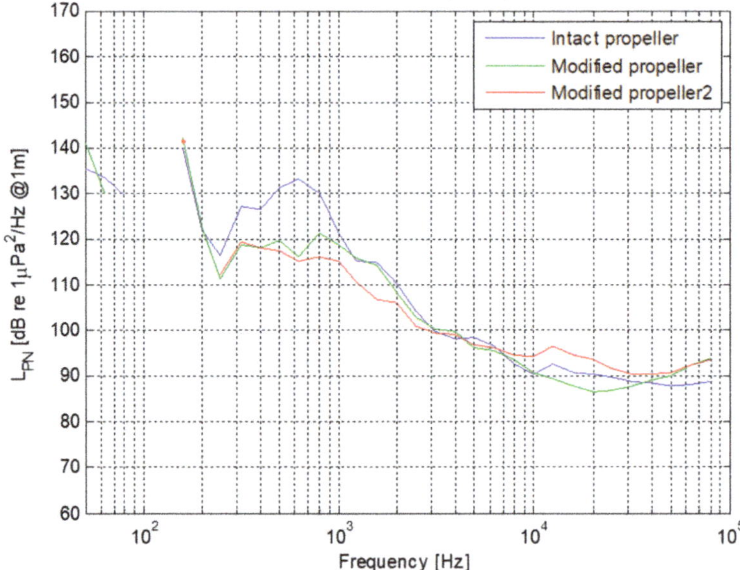

Figure 18. Comparison between Intact, Modified Propeller and Modified Propeller-2 net noise levels at 1 m (one third octave band), condition V2, hydrophone H3.

Figure 19 presents the net difference between the noise levels of the intact propeller and both modified propellers for Condition V2 as measured by hydrophone H3 to demonstrate the effectiveness of the PressurePore^{TM} technology. As shown in Figure 19, a maximum of 17 dB reduction is possible by using this technology at the critical low-frequency region of the URN spectrum.

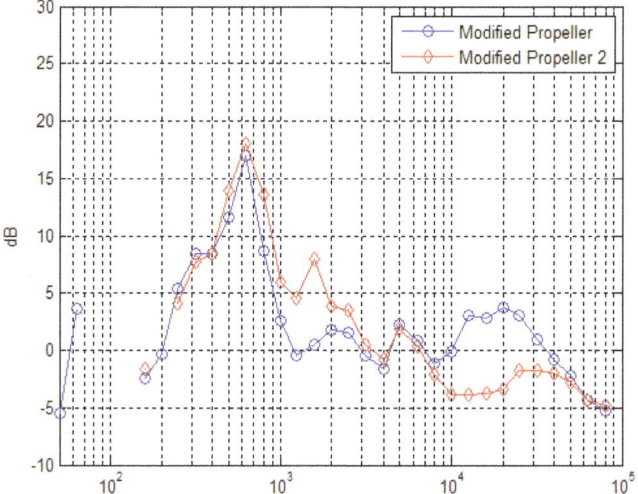

Figure 19. Noise reduction with application of pressure relief holes in Third Octave band for condition V2, measured at hydrophone H3.

3.3. Propeller Performance Tests

This section presents details and results of the propeller open water tests conducted at the CTO towing tank. The purpose of these tests was to determine the propeller performance characteristics (thrust, torque and efficiency) before and after applying the PressurePores^{TM} technology.

During these tests, the rate of the propeller shaft revolutions was selected for Reynolds Numbers above the critical threshold of 500,000. To confirm the typical convergence of the measurements, for the single advance ratio of J = 0.6, the tests were repeated for three additional values of the Reynolds number. Figure 20 shows the resulting test data analysed for thrust, torque and efficiency coefficients as fitted by 4th-degree polynomials for the three propeller test cases.

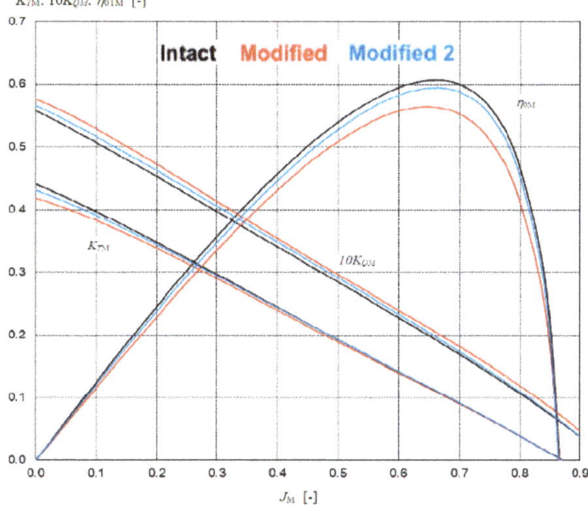

Figure 20. Open water characteristics of *The Princess Royal* model propeller (*intact*) before and after the application of PressurePores^{TM} (*Modified propeller* and *Modified Propeller-2*).

The principal operating condition of *The Princess Royal* propeller is very close to Advance Coefficient J = 0.5. As shown in Figure 20, the open water tests indicated that there is a 2% loss of thrust and 4% gain in torque which consequently results in a propeller efficiency loss of 5.7% for the Modified Propeller compared to the intact propeller. For the Modified Propeller-2 case, with only half of the pores of the Modified propeller, the loss in thrust was about 0.1% gain in torque was 2.2% giving an efficiency loss of 2.3%.

4. Numerical Investigations (CFD Approach)

The effect of the drilled holes on the propeller performance and cavitation dynamics were simulated using CFD (Star CCM+) on the Base (intact) and Modified propeller. This allowed the expected cavitation volume reduction to be estimated at the modification stage of the propeller and selecting the favourable pressure pores arrangement using the new meshing refinement approach, MARCS, as explained in detail in the following.

4.1. Mesh Adaption Refinement Approach for Cavitation Simulations (MARCS)

An advanced mesh refinement technique for capturing tip vortex cavitation in a propeller slipstream was proposed [20], where preliminary results were presented for a limited range of tip vortex extensions for two benchmark propeller models (PPTC and INSEAN E779A). The method was further developed by using the INSEAN E779A propeller, allowing a greater extension of the TVC in the propeller slipstream to be achieved [16]. This mesh refinement approach (MARCS) was further applied on *The Princess Royal* propeller model, capturing an even greater extension of the tip cavitating tip vortex development in the propeller slipstream [21].

In MARCS, the adaptive mesh refinement was created only in the region where the tip vortex cavitation may occur. First, the upper limit of absolute pressure in the solution was determined by creating a threshold region in Star CCM+ (Figure 21, Left). In such cavitation simulations, the volume fraction of vapour shows the volume where the absolute pressure is below the saturation pressure of water, thus identifying the cavitating volume. A threshold region was created by increasing the saturation pressure from a default saturation pressure, 3169 [Pa] to a higher value, 17,000 [Pa] thus generating, the pink region shown in Figure 21, Left. This artifice provided an indication of the volumetric trajectory on which to generate a fine mesh (Figure 21 Right) for accurately capturing the pressure-drop correctly and tracking the cavity bubbles in the propeller slipstream.

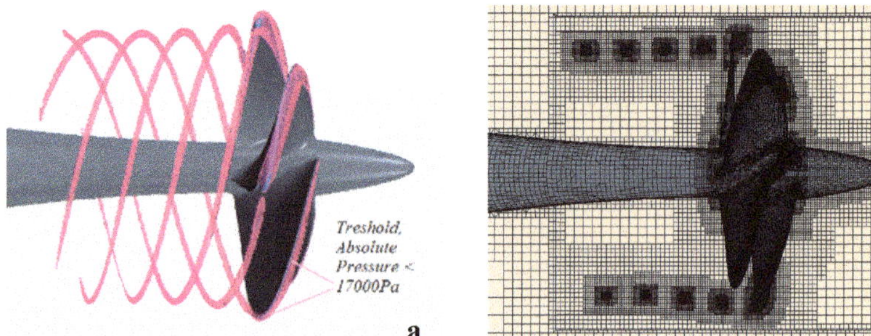

Figure 21. Tip vortex cavitation refinement (MARCS); (**a**) Left; Treshold Region, (**b**) Right; Generated Mesh.

The numerical mesh is an unstructured trimmed grid, and basic prismatic cells are applied near to the blade surface for resolving the boundary layer (trying to keep y+ close to 1). MARCS was also used for local mesh refinement to be able to simulate TVC and evaluate the benefit of using the drilled holes in the presence of more realistic TVC extents. Additionally, for the tip drilled propeller, different mesh arrangements were generated around the drilled holes. Smaller mesh size for the pressure pore surfaces was used to capture the cylindrical hole shape properly as shown in Figure 22.

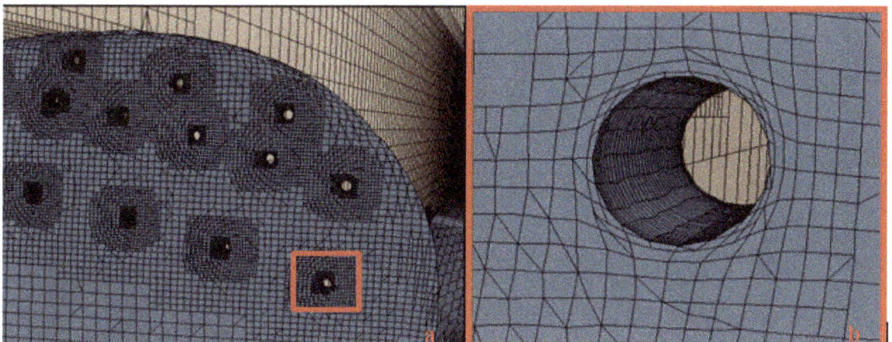

Figure 22. Local drilled hole refinements for capturing accurate flow through the holes (**a**) Left; Propeller tip, (**b**) Right; Zoom to the hole.

For evaluating the quality of the numerical results, an uncertainty analysis was conducted for the essential grid and time step sensitivity on the numerical solutions. For these cases, it was concluded that the most dominant errors were due to discretisation errors instead of iterative errors which were neglected accordingly. Uncertainty calculations followed the approach of Stern et al. [22]; this required at least 3 different cases (such as computational grids, time step values etc.) and estimated the uncertainty values for the calculations. This analysis method had been previously applied for non-cavitating and cavitating propellers, including the intact model of *The Princess Royal* [16,21].

4.2. Numerical Model

The commercial CFD software, STAR-CCM+ finite volume stress solver, was used in the present simulations to solve the governing equations (such as continuity and momentum) [23].

In CFD procedures, fluid flows are simulated using various methodologies depending on the nature of the flow problem and the availability of computational resources. Numerical methods can be broadly categorized as Reynolds Averaged Navier-Stokes (RANS), Detached Eddy Simulation (DES), Large Eddy Simulation (LES) and Direct Numerical Simulation (DNS). While RANS solvers are widely used for open water simulations to predict propeller performance coefficients, scale-resolving simulations such as DES and LES models are commonly required for calculating turbulent cavitating flows.

With the selection of the LES turbulence model for the tip vortex cavitation simulations of the base (*intact)* propeller model and *modified* model, different time step values were tried based on the time steps recommended by ITTC and others in the open literature. Hence the time step was calculated such that the propeller rotates between 0.5 and 2 degrees per time step [24]. Finally, a time step value of $\Delta t = 5 \times 10^{-5}$ s, corresponding to 0.59 degree of propeller rotation, was used for the cavitation simulations.

For cavitation modelling, while a Volume of Fluid (VOF) model was used in describing the multiphase flow, the present modelling used the Schnerr-Sauer model which is based on the Rayleigh-Plesset Equation (see STAR-CCM+ [23]).

The numerical domain for this study consisted of a static domain representing the cavitation tunnel and a rotational domain around the propeller employing a sliding mesh approach. The domain boundaries are defined as velocity inlet and pressure outlet. The tunnel wall, as well as the propeller, were defined as the wall-type of boundary conditions. The rotating domain passes through the gap between shaft and hub.

4.3. Results

In the following section, the results of the CFD investigations of *The Princess Royal* model propeller (intact and modified) in cavitating conditions are presented and discussed using cavitation patterns, including TVC extension. The cavitation pattern images are also compared with the EFD results of the cavitation observations for validation purposes.

Cavitation Pattern including TVC

Cavitation patterns derived using MARCS (Section 4.1) are shown in Figure 23 for the *intact* propeller (in top row figures) and *Modified* propeller (in bottom row figures), thus allowing the cavitation volume reduction to be estimated by CFD in selecting the favourable pressure pores arrangement.

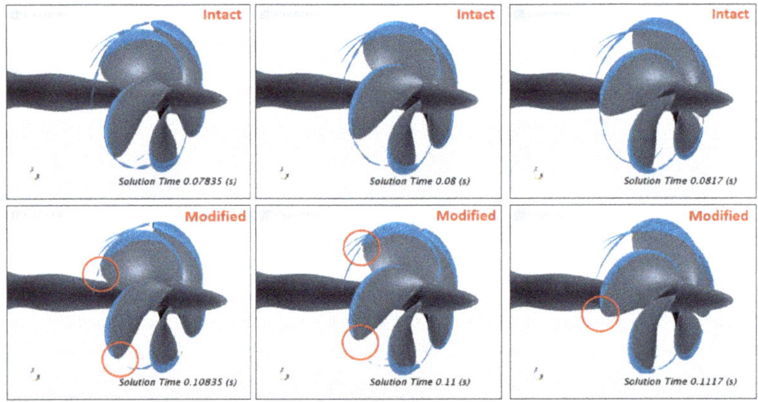

Figure 23. Pattern comparisons between intact and modified propellers for different blade positions. (**Top row**: Intact propeller, **Bottom row**: Modified propeller).

Figures 24 and 25 also compare the cavitation pattern images obtained through the EFD observations in the UNIGE cavitation tunnel and CFD computations for the intact and modified propellers. Good agreement was found for the visual cavitation dynamics, TVC size and extent. The expected cavitation volume reduction was also confirmed in both EFD and CFD results.

Furthermore, while Table 5 demonstrates K_T comparison between CFD calculations and EFD results for the intact propeller, Table 6 compares K_T and cavitation volume values that have been obtained from CFD computations for the intact and modified propeller.

Table 5. Comparison of K_T values for intact propeller between the EFD and computational fluid dynamics (CFD) results.

	Intact Propeller	
	EFD	CFD
K_T	0.211	0.225
$\Delta\%$ K_T	-	6.6%

Table 6. K_T and cavity volume comparison between Intact and Modified Propeller (CFD).

	CFD	
	Intact	Modified
Cavitation volume (m^3)	8.11×10^{-6}	6.47×10^{-6}
K_T	0.225	0.222
$\Delta\%$ K_T	–	−1.3%
$\Delta\%$ cavitation volume	–	−20.1%

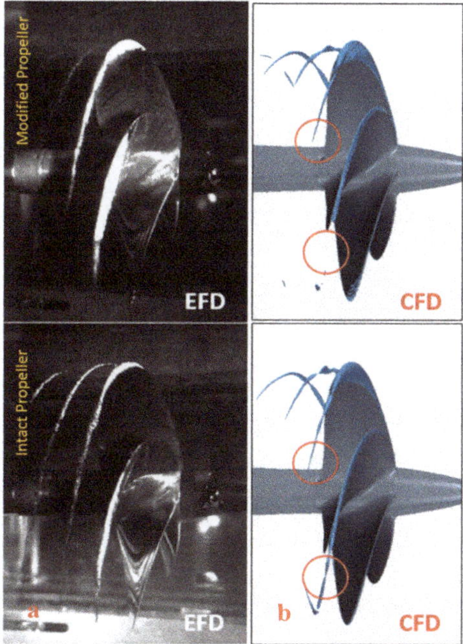

Figure 24. Comparison between (**a**) EFD (Left) and (**b**) CFD (Right) results for intact and modified Propellers.

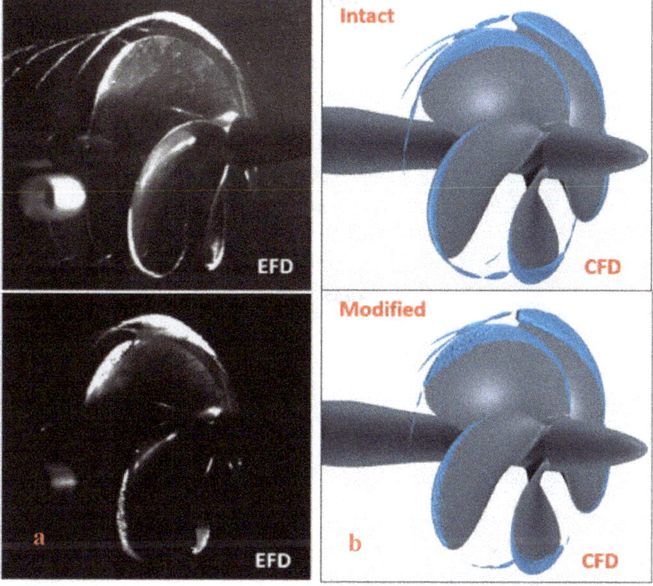

Figure 25. Comparison between (**a**) EFD (**Left**) and (**b**) CFD (**Right**) for intact and modified propellers.

The EFD and CFD results confirmed that the application of the Pressure Pores in the tip region of the propellers results in cavitation volume reduction. This can be observed through the decreased TVC appearance particularly in Figure 24. Although the Authors have not yet developed a CFD module for predicting cavitating noise levels within the scope of this study, the EFD measured reductions in noise from the drilled propeller blades are consistent with the cavitation volume reductions (up to 20%) estimated in CFD simulations (Table 6). This CFD modelling approach, based on MARCS, was used extensively in the selection process for the most favourable Pressure Pores arrangements as described in more detail in [14].

5. Conclusions

A complementary combination of experimental and numerical investigations was conducted to develop and explore the benefits of the PressurePores™ concept to mitigate the URN levels of a cavitating marine propeller. Following a pilot study in a cavitation tunnel, which showed promising results, the extensive CFD simulations were conducted for further development of this mitigation technique and establish its strategic application.

To accurately simulate the effects of PressurePores™ on the TVC of a propeller, a recently developed advanced adaptive meshing technique (MARCS) was coupled with a commercial CFD code to capture cavitating propeller flow properties. With the understanding achieved through the CFD simulations, the PressurePores™ technology was applied on a prototype propeller and then validated by using model tests conducted in the University of Genoa cavitation tunnel and the CTO towing tank for the cavitation characteristics, noise and efficiency.

The test results conducted with the model propeller of a research vessel and two different combinations of the PressurePores™ technology revealed that significant reductions in the measured propeller noise levels can be achieved.

Comparative test results for the "Modified Propeller-2" test case indicated a noise reduction as high as 17 dB compared to the unmodified propeller. This was achieved particularly in the frequency region which is of utmost importance for some marine mammals. For this configuration, towing tank results showed about a 2% loss in the propeller efficiency.

The test results for the first "Modified Propeller" showed further superior underwater noise reduction in the same high-frequency range but with a higher loss (of about 5.7%) in propeller efficiency.

The study shows that a recently developed advanced CFD modelling application can simulate sheet and tip vortex cavitation characteristics for intact blades and this with pressure relief holes. The CFD studies showed up to 13.8% cavitation volume reduction for the pressure pores applied on the earlier mentioned commercial tanker propeller used in [13], while the corresponding cavitation tunnel tests showed significant noise emission reductions (up to 10 dB) with only 0.5% loss of efficiency. This suggests that PressurePores™ technology may be a useful and attractive noise mitigation technique for the retrofit of noisy marine propulsors within the framework of increasing scrutiny on the marine noise pollution from commercial shipping.

Author Contributions: Author contributions for various aspects of the manuscript is as following: Conceptualization, M.A., B.A. and D.T.; methodology, B.A, M.A. and N.Y.; software, N.Y., B.A. and N.S.; validation, N.Y., B.A. and P.F.; investigation, N.Y., B.A. and M.A; resources, M.A. and D.T.; data curation, B.A., N.Y.; writing—original draft preparation, B.A., N.Y., M.A., P.F. and N.S.; writing—review and editing, B.A., N.Y., M.A.; supervision, M.A., N.S., P.F. and D.T.; project administration, M.A.; funding acquisition, D.T. All authors have read and agreed to the published version of the manuscript.

Funding: This research was funded by OSCAR Propulsion Ltd.

Acknowledgments: The access provided to High-Performance Computing for the West of Scotland (Archie-West) through EPSRC grant no. EP/K000586/1 is gratefully acknowledged.

Conflicts of Interest: The authors declare no conflicts of interest.

References

1. International Maritime Organization. Noise from commercial shipping and its adverse impacts, on marine life. *Mar. Environ. Prot. Comm. Int. Marit. Organ. (MEPC)* **2013**, *66*, 17.
2. Ross, D. *Mechanics of Underwater Noise*; Peninsula Publishing: CA, USA, 1976.

3. Frisk, G.V. Noiseonomics: The relationship between ambient noise levels in the sea and global economic trends. *Sci. Rep.* **2012**, *2*, 2–5. [CrossRef]
4. Richardson, W.J.; Greene, C.R., Jr.; Malme, C.I.; Thomson, D.H. Marine mammals and noise. *J. Exp. Mar. Biol. Ecol.* **2013**, *210*, 161–163.
5. White, P.; Pace, F. The Impact of Underwater Ship Noise on Marine Mammals. In Proceedings of the 1st IMarEST Ship Noise and Vibration Conference, London, UK, 27–28 January 2010.
6. Chekab, M.A.F.; Ghadimi, P.; Djeddi, S.R.; Soroushan, M. Investigation of Different Methods of Noise Reduction for Submerged Marine Propellers and their Classification. *Am. J. Mech. Eng.* **2013**, *1*, 34–42. [CrossRef]
7. Mautner, T.S. *A Propeller Skew Optimization Method*; Naval Ocean Systems Center: San Diego, CA, USA, 1987.
8. Mosaad, M.A.; Mosleh, M.; El-Kilani, H.; Yehia, W. Propeller Design for Minimum Induced Vibrations. *Port Said Eng. Res. J.* **2011**, *10*, 2.
9. Ji, B.; Luo, X.; Wu, Y. Unsteady cavitation characteristics and alleviation of pressure fluctuations around marine propellers with different skew angles. *J. Mech. Sci. Technol.* **2014**, *28*, 1339–1348. [CrossRef]
10. Andersen, P.; Kappel, J.; Spangenberg, E. Aspects of Propeller Developments for a Submarine. In *The First International Symposium on Marine Propulsors: Smp'09*; Norwegian Marine Technology Research Institute (MARINTEK): Trondheim, Norway, 2009; pp. 551–561.
11. Hyung-Sik, P.; Su-Hyun, C.; Nho-Seong, K. Identification of Propeller Singing Phenomenon through Vibration Analysis of Propeller Blade. In Proceedings of the 15th International Offshore and Polar Engineering Conference, (The International Society of Offshore and Polar Engineers), Seoul, Korea, 19–24 June 2005.
12. Sharma, S.D.; Mani, K.; Arakeri, V.H. Cavitation noise studies on marine propellers. *J. Sound Vib.* **1990**, *138*, 255–283. [CrossRef]
13. Xydis, K. Investigation into Pressure Relieving Holes on Propeller Blades to Mitigate Cavitation and Noise. Master's Thesis, Newcastle University, Newcastle, UK, 2015.
14. Aktas, B.; Yilmaz, N.; Atlar, M. *Pressure-Relieving Holes to Mitigate Propeller Cavitation and Underwater Radiated Noise*; Department of Naval Architecture Ocean and Marine Engineering; Strathclyde University: Glasgow, UK, 2018.
15. Atlar, M.; Aktas, B.; Sampson, R.; Seo, K.C.; Viola, I.M.; Fitzsimmons, P.; Fetherstonhaug, C. A Multi-Purpose Marine Science & Technology Research Vessel for Full-Scale Observations and Measurements. In Proceedings of the International Conference on Advanced Model Measurement Technologies for the Marine Industry, Gdansk, Poland, 17–18 September 2013.
16. Yilmaz, N.; Atlar, M.; Khorasanchi, M. An Improved Mesh Adaption and Refinement Approach to Cavitation Simulation (MARCS) of Propellers. *J. Ocean Eng.* **2019**, *171*, 139–150. [CrossRef]
17. Korkut, E.; Takinaci, A.C. *18M Research Vessel Wake Measurements*; Istanbul Technical University Faculty of Naval Architecture and Ocean Engineering: İstanbul, Turkey, 2013; p. 6.
18. ITTC. Model-scale Propeller Cavitation Noise Measurements. Recommended Procedures and Guidelines 7.5-02-01-05. Available online: https://ittc.info/media/4052/75-02-01-05.pdf (accessed on 19 February 2020).
19. ANSI. Quantities and Procedures for Description and Measurement of Underwater Sound from Ships – Part 1: General Requirements. American National Standard ANSI/ASAS. 2009. Available online: https://www.iso.org/obp/ui/#iso:std:iso:pas:17208:-1:ed-1:v1:en (accessed on 29 February 2020).
20. Yilmaz, N.; Khorasanchi, M.; Atlar, M. An Investigation into Computational Modelling of Cavitation in a Propeller's Slipstream. In Proceedings of the Fifth International Symposium on Marine Propulsion SMP'17, Espoo, Finland, 12–5 June 2017.
21. Yilmaz, N. Investigation of Cavitation Influence on Propeller-Rudder-Hull Interaction. Ph.D. Thesis, University of Strathclyde, Glasgow, Scotland, 2019.
22. Stern, F.; Wilson, R.V.; Coleman, H.W.; Paterson, E.R. Comprehensive Approach to Verification and Validation of CFD Simulations-Part1: Methodology and Procedures. *J. Fluids Eng.* **2001**, *123*, 793–802. [CrossRef]
23. Siemens. STAR-CCM+ User Guide. 2018. Available online: https://www.plm.automation.siemens.com/media/global/en/Siemens%20PLM%20Simcenter%20STAR-CCM_tcm27-62845.pdf (accessed on 29 February 2020).

24. The International Towing Tank Conference. *ITTC Recommended Procedures and Guidelines, Practical Guidelines for Ship-Propulsion CFD, 7.5-03-03-01*; The International Towing Tank Conference: Bournemouth, UK, 2014.

Journal of
Marine Science and Engineering

Article

Surprising Behaviour of the Wageningen B-Screw Series Polynomials

Stephan Helma

Stone Marine Propulsion (SMP) Ltd., SMM Business Park, Dock Road CH41 1DT, UK; sh@smpropulsion.com

Received: 9 December 2020; Accepted: 2 February 2020; Published: 18 March 2020

Abstract: Undoubtedly, the Wageningen B-screw Series is the most widely used systematic propeller series. It is very popular to preselect propeller dimensions during the preliminary design stage before performing a more thorough optimisation, but in the smaller end of the market it is often used to merely select the final propeller. Over time, the originally measured data sets were faired and scaled to a uniform Reynolds number of $2 \cdot 10^6$ to increase the reliability of the series. With the advent of the computer, polynomials for the thrust and torque values were calculated based on the available data sets. The measured data are typically presented in the well-known open-water curves of thrust and torque coefficients K_T and K_Q versus the advance coefficient J. Changing the presentation from these diagrams to efficiency maps reveals some unsuspected and surprising behaviours, such as multiple extrema when optimising for efficiency or even no optimum at all for certain conditions, where an optimum could be expected. These artefacts get more pronounced at higher pitch to diameter ratios and low blade numbers. The present work builds upon the paper presented by the author at the AMT'17 and smp'19 conferences and now includes the extended efficiency maps, as suggested by Danckwardt, for all propellers of the Wageningen B-screw Series.

Keywords: propeller; Wageningen B-screw Series; open-water characteristics; propeller efficiency map; Danckwardt diagram; optimum propeller

---

1. Introduction

1.1. The Wageningen B-Screw Series and Its Polynomial Representation

The Wageningen B-screw Series dates back to 1936 [1], when the first results were published. In the following years, the series was systematically expanded to include more than 120 single propellers. The measured data were presented inter alia in open-water diagrams showing the dimensionless thrust and torque coefficients, K_T and K_Q, and the open-water efficiency η_o as functions of the also dimensionless advance coefficient J:

$$J = \frac{v_a}{nD}, \tag{1}$$

$$K_T = \frac{T}{\rho n^2 D^4}, \tag{2}$$

$$K_Q = \frac{Q}{\rho n^2 D^5} = \frac{P_P}{2\pi\rho n^3 D^5}, \text{ and} \tag{3}$$

$$\eta_o = \frac{J}{2\pi} \cdot \frac{K_T}{K_Q}, \tag{4}$$

where T = measured thrust; Q = measured torque; P_P = propeller power; ρ = water density; n = shaft speed (in s^{-1}); D = propeller diameter; and v_a = speed of advance. If not stated otherwise, all values are in SI base units.

All tested propeller models had a diameter of 240 mm and, hence, different section chord lengths due to the variable blade area ratio and number of propeller blades. The propellers were tested in varying model basins of MARIN using a diverse rate of revolutions resulting in considerably different Reynolds numbers for each propeller in the whole series. Oosterveld and van Oossanen engaged in the formidable tasks of scaling all available open-water data sets to a uniform Reynolds number of $2 \cdot 10^6$ (based on chord length and section advance speed) and calculating polynomials for the thrust and torque coefficients by multiple regressions analysis [2]. With the help of these polynomials, it is possible to calculate these coefficients as functions of the advance coefficient J, the pitch to diameter ratio P/D, the expanded blade area ratio A_e/A_0, and the number of blades Z:

$$K_T = \sum C_{s,t,u,v} \cdot J^s \cdot (P/D)^t \cdot (A_e/A_0)^u \cdot Z^v \text{ and} \tag{5}$$

$$K_Q = \sum C_{s,t,u,v} \cdot J^s \cdot (P/D)^t \cdot (A_e/A_0)^u \cdot Z^v, \tag{6}$$

where $C_{s,t,u,v}$ = coefficient; and s, t, u, and v = whole-number exponents.

These polynomials are nowadays widely used in either selecting the optimum propeller or as a basis for further refinements. It is therefore of utmost significance that these polynomials are consistent and accurate.

1.2. Time Line of Experimental Results

As shown by Helma for the B4-70 propeller, the open-water characteristics have changed substantially over time [3], see Figure 1. The effect on the widely used B_p–δ diagram was also outlined in the same work, see Figure 2.

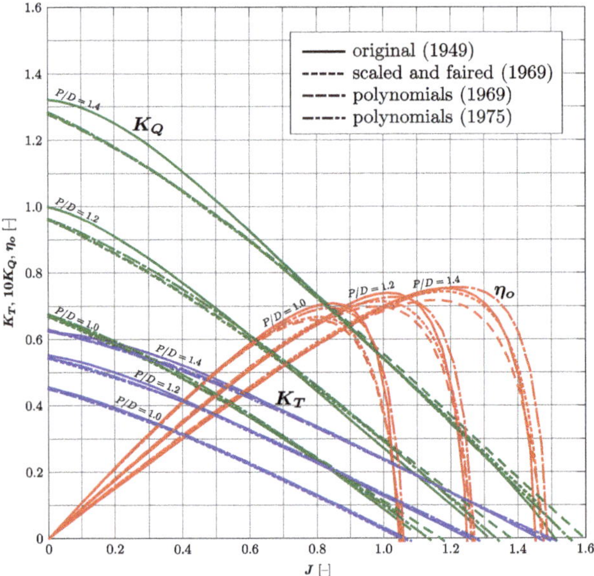

Figure 1. Open-water diagrams originally published by van Lammeren and van Aken in 1949 [4], scaled and faired, the fitted polynomials (both by van Lammeren et al. in 1969 [5]), and the most recent polynomials by Oosterveld and van Oossanen (1975) [2]) of propeller B4-70. The Reynolds number is $2.72 \cdot 10^5$ for the original data, otherwise $2 \cdot 10^6$. Reproduced from [3] with permission from AMT'17, 2017.

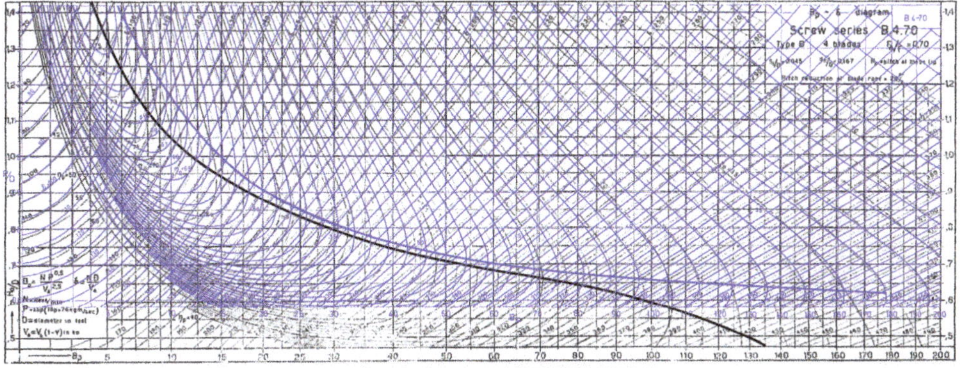

Figure 2. B_P–δ diagram of propeller B4-70 based on the originally published values by van Lammeren and van Aken in 1949 ([original black print, Reynolds number = $2.72 \cdot 10^5$) and on the scaled and faired open-water characteristics by van Lammeren et al. in 1969 (blue, Reynolds number = $2 \cdot 10^6$). Note the different lines for $\eta_{o,max}$ (in bold). $B_p = N \cdot \sqrt{P_P}/v_a^{2.5}$ and $\delta = N \cdot D/v_a$, where N = shaft speed (in min^{-1}); P_P = delivered power (in hp); v_a = advance speed (in kn); and D = propeller diameter (in feet). Reproduced from [4,5], with permissions from SWZ | Maritime and Society of Naval Architects and Marine Engineers, 1949 and 1969.

2. Efficiency Maps

2.1. Basic Efficiency Maps

It is assumed, that the presentation in the form of the standard open-water diagrams for propellers are known to the reader. To recap, these diagrams show three families of curves with the thrust and torque coefficients and the efficiency as functions of the advance coefficient—$K_T(J)$, $K_Q(J)$, and $\eta_o(J)$—for a set of constant P/D-values for each propeller of the Wageningen B-screw Series.

Many authors suggested a different way of representing the nondimensional open-water data (see Appendix A.2); we will call them efficiency maps, as proposed by many of these authors. There are two efficiency maps: the K_Q–J and the K_T–J map. On the K_Q–J efficiency map, we draw the family of $K_Q(J)$ curves for our set of constant P/D-values as for the conventional open-water diagram (thin red lines, please refer to Figure 3a). Instead of adding the family of efficiency curves for constant P/D-values, we add contour lines for the efficiency: $\eta_o\left(J, K_Q(J)\right) = \text{const}$ for a set of selected η_o-values (thin black lines). We can also draw the contour lines for the thrust coefficient: $K_T\left(J, K_Q(J)\right) = \text{const}$ (dashed red lines). So far, this gives us a diagram with exactly the same information content as for the conventional open-water diagram, just displayed in a different way.

We can now draw two lines of maximum propeller efficiency into this basis efficiency map: "$\eta_{o,max}$ for $J = \text{const}$" (bold black line) and "$\eta_{o,max}$ for $P/D = \text{const}$" (bold red line). (Appendix A.1 describes their derivation.) In the open-water diagram, the curves corresponding to the "$\eta_{o,max}$ for $J = \text{const}$" and "$\eta_{o,max}$ for $P/D = \text{const}$" lines would be the envelope to and the line connecting the maxima of all $\eta_o(J)$ curves, respectively.

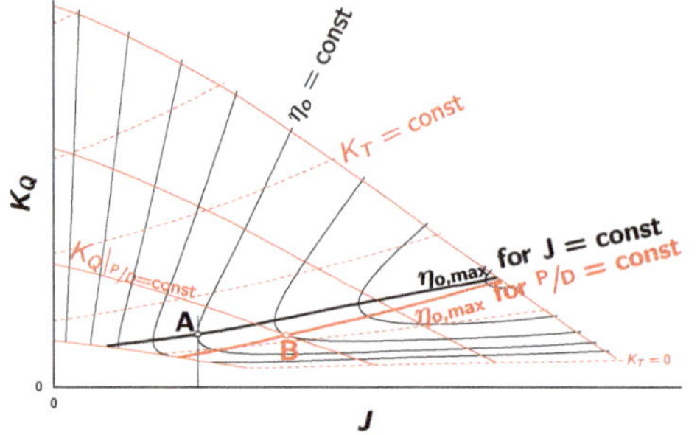

(**a**) Basic efficiency map showing K_Q (thin red lines), K_T = const (dotted red lines), η_o = const (thin black lines), and the construction of "$\eta_{o,max}$ for J = const" (dotted black line and point A, bold black line) and "$\eta_{o,max}$ for $^P/_D$ = const" (point B, bold red line).

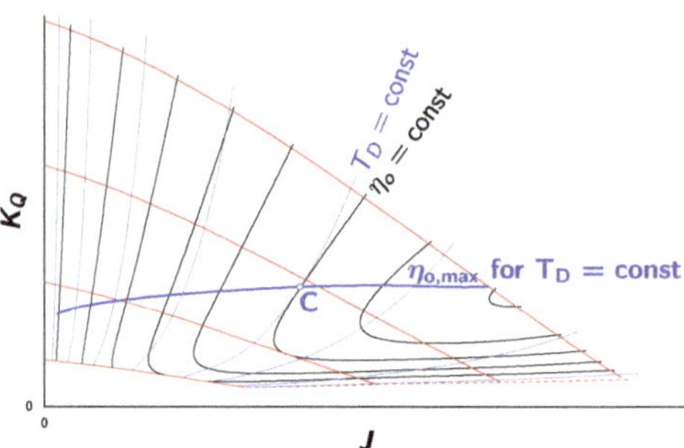

(**b**) Extended efficiency map additionally showing T_D = const (thin blue lines) and the construction of "$\eta_{o,max}$ for T_D = const" (point C, bold blue line).

Figure 3. Composition of the K_Q–J efficiency map. For its construction, see Appendix A.1.

2.2. Enhanced Efficiency Maps

To enhance the basic efficiency map, these families of curves can be added:

$$T_D = \frac{K_T}{J^2} = \text{const}, \tag{7}$$

$$P_D = \frac{K_Q}{J^3} = \text{const}, \tag{8}$$

$$T_n = \frac{K_T}{J^4} = \text{const, and} \tag{9}$$

$$P_n = \frac{K_Q}{J^5} = \text{const}. \tag{10}$$

These four curves have a very practical significance, because each of them eliminates one of the two unknowns in propeller optimisation: D or n, the propeller diameter and the shaft speed, respectively. (Examples of how these diagrams can be used for propeller selection are presented in Appendix A.4).

Note that in the K_Q–J efficiency map, the lines for P_D and P_n = const are ordinary curves to the single power of three and five, whereas the curves for T_D and T_n = const are truly parametric curves. Having established these families of curves, we can now draw the four lines of maximum propeller efficiency for each of them: "$\eta_{o,max}$ for T_D = const", "$\eta_{o,max}$ for P_D = const", "$\eta_{o,max}$ for T_n = const", and "$\eta_{o,max}$ for P_n = const", all connecting the points of maximum efficiency (see Appendix A.1.3 for how they are constructed). Figure 3b shows the family of curves for T_D = const (thin blue lines) and for "$\eta_{o,max}$ for T_D = const" (bold blue line).

As mentioned earlier, there exists a second diagram: the K_T–J efficiency map with K_T as the ordinate. It is composed of the families of $K_T(J)$ curves for the set of constant $^P/_D$-values and the contour lines $K_Q\,(J, K_T(J))$ and $\eta_o\,(J, K_T(J))$ = const for the set of selected K_Q- and η_o-values. The lines for maximum efficiency are called "$\eta_{o,max}$ for J = const" and "$\eta_{o,max}$ for $^P/_D$ = const" as before. This efficiency map can also be enhanced with the families of curves "T_D, P_D, T_n, and P_n = const", where the curves for T_D and T_n = const are curves to the single power of three and five.

There are certain advantages of efficiency maps over open-water diagrams for finding the optimum propeller. These are described in Appendixes A.3 and A.4. For the purpose of this article, we will concentrate on the shape of the lines for maximum efficiencies. To recap, these lines are the solutions of the optimisation problems under different constraints. We will use the K_Q–J efficiency map enhanced by the addition of the T_D and T_n = const curves and the K_T–J map enhanced with the P_D and P_n = const curves, as proposed by Danckwardt [6]. We will call them the T–J and P–J Danckwardt diagrams (see Appendix A.2 for a short historical overview).

2.3. Remake

The author presented efficiency maps for the Wageningen B-screw Series during the smp'19 conference [7]. For the purpose of this paper, the two Danckwardt diagrams for all propellers of the Wageningen B-screw Series, as outlined in Table 1, were recreated with the help of a purpose-made computer program. This program employs the polynomials (5) and (6), as described in Section 1 and published by Oosterveld and van Oossanen in 1975 [2]. For these newly generated diagrams, the symbols were updated to the ITTC nomenclature [8] (see also Table A1 for the differences to the nomenclature used by Danckwardt). In addition to the curves presented in the original diagrams, the line for "$\eta_{o,max}$ for P/D = const" and the contour lines for $K_T(J, K_Q)$ or $K_Q(J, K_T)$ = const were added. All these recalculated Danckwardt diagrams are presented in Appendix B, and Table A4 shows an overview of the composition of these diagrams. Examples of how these diagrams are used are explained in Appendix A.4.

Table 1. Summary of the propeller models of the Wageningen B-screw Series.

Z	A_e/A_0																
2	0.30		0.38														
3		0.35				0.50			0.65			0.80					
4				0.40			0.55			0.70			0.85			1.00	
5					0.45			0.60			0.75			0.90			1.05
6						0.50			0.65			0.80			0.95		
7							0.55			0.70			0.85				

The computer program calculates the lines of maximum efficiency by finding all points, where the tangents to the T_D, T_n, P_D, and P_n curves are tangential to the efficiency contour lines. As a free variable, the pitch to diameter ratio P/D was used. This choice was found to be necessary to be able to remove possible multiple solutions for the $\eta_{o,max}$ lines. It must be mentioned that, as a possible consequence, the calculation of the $\eta_{o,max}$ lines for T_D, T_n, P_D, and P_n = const sometimes did not succeed at one or the other boundary due to numerical difficulties. The artefacts on the left boundary at low J- and P/D-values were manually deleted and extrapolated by hand whenever possible. The right border, where the J- and P/D-values are high, posed a different numerical challenge. In all cases however, it was possible to reconstruct the valid line manually, but sometimes not right up to the maximum value of P/D. It should be mentioned that these difficulties never arose at high P/D-values when the $\eta_{o,max}$ line doubles back, as discussed in the following section.

3. Ambiguity of the $\eta_{o,max}$ Lines

3.1. Introductory Example

We have seen that the Danckwardt diagrams lend themselves to finding the optimum propeller under certain constraints. For the sake of argument, let us assume that we have already decided on the blade number ($Z = 5$) and blade area ratio ($A_e/A_0 = 0.9$), and we know the torque Q (from the given available power P_P), the inflow velocity v_a, and propeller diameter D but not the shaft speed n, which should be optimised together with the pitch to diameter ratio P/D. Using these known values, we can calculate P_D from Equation (A4), which we assume to be 0.15. On the Danckwardt P–J diagram for the Wageningen B5-90 propeller (see Appendix B), we find the intersection of the curve for P_D = 0.15 with the "$\eta_{o,max}$ for P_D = const" line (blue lines). We can read off the values for J, K_T, K_Q, P/D, and η_o (approximately 0.65, 0.24, 0.041, 1.04, and 0.60, respectively).

Let us now assume that we want to investigate a three-bladed propeller with the blade area ratio A_e/A_0 of 0.80 for the same operating condition. In the diagram for the Wageningen B3-80 propeller, we find the intersection for our assumed value of $P_D = 0.15$ with the "$\eta_{o,max}$ for $P_D = $ const" line (blue lines) at J, K_T, K_Q, P/D, and η_o equal to 0.62, 0.20, 0.035, 1.00, and 0.57, respectively. However, there also exists another intersection between the $P_D = 0.15$ and the "$\eta_{o,max}$ for $P_D = $ const" lines at a higher P/D-value: 0.74, 0.295, 0.062, 1.30, and 0.56. It looks to be that there are two optimum propellers for this condition, since the $\eta_{o,max}$ lines are solutions to the optimisation problem under the given constraints!

The reason for this somewhat puzzling behaviour is obviously the doubling back of the $\eta_{o,max}$ lines. Two solutions to the optimisation problem exist in the region of this overlap, whereas no solution to the optimisation problem exists right of this region of overlap.

3.2. Classification

To classify this overlap, the value of $T_D|_{P/D|_{\max}}$, where the "$\eta_{o,max}$ for $T_D = $ const" line intersects the maximum $P/D|_{\max}$ curve, was calculated for every propeller in the Wageningen B-screw Series (please refer to Figure 4) as proposed in [7]. The intersection of this $T_D = $ const curve with the "$\eta_{o,max}$ for $T_D = $ const" line was found. The pitch to diameter ratio at this intersection is denoted as $\widehat{P/D}|_{T_D}$. The minimum value of $T_D|_{\min}$ is determined, where there is also no optimum propeller right of this curve! The difference between these two values for T_D is denoted as ΔT_D. For the T_n, P_D, and P_n curves, these values are calculated accordingly. Table 2 shows an overview of all these values for all propellers in the Wageningen B-screw Series. The table clearly shows that this overlap does not just sporadically occur, but that it is a widespread phenomena.

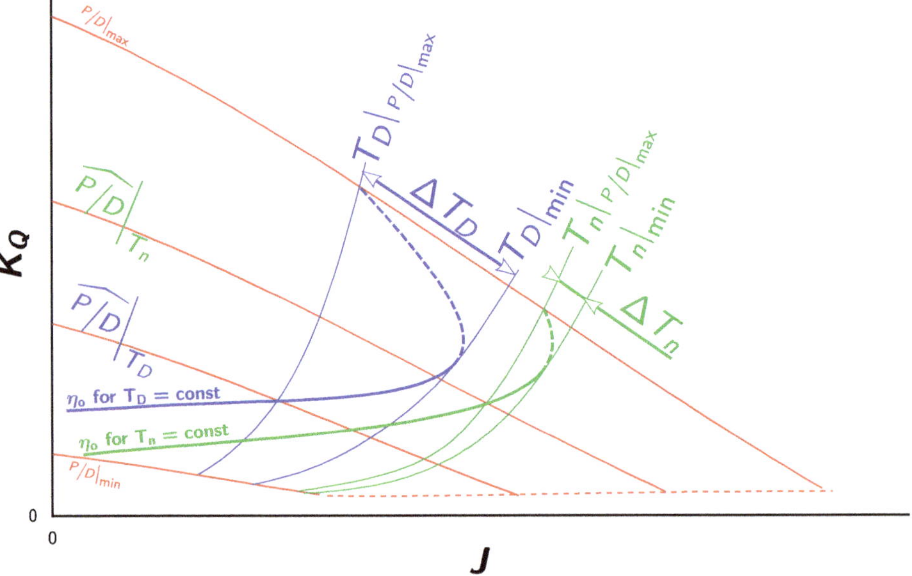

Figure 4. Symbols and definitions used to describe the overlap of the $\eta_{o,max}$ lines given in Table 2 using the example of the T_D and T_n curves in the Danckwardt T–J efficiency map.

Table 2. Main characteristics of all overlaps found in the recreated Danckwardt diagrams. (For symbols and definitions, see Section 3.2 and Figure 4).

Propeller	T_D			T_n			P_D			P_n		
	$\widehat{P/D}$	$T_D\vert_{\min}$	ΔT_D	$\widehat{P/D}$	$T_n\vert_{\min}$	ΔT_n	$\widehat{P/D}$	$P_D\vert_{\min}$	ΔP_D	$\widehat{P/D}$	$P_n\vert_{\min}$	ΔP_n
B2-30		—			—			—			—	
B2-38		—			—			—			—	
B3-35	0.94	0.155	0.182	1.14	0.102	0.043	0.93	0.033	0.049	1.14	0.021	0.010
B3-50	0.89	0.167	0.225	1.13	0.114	0.042	0.88	0.037	0.065	1.12	0.025	0.010
B3-65	0.86	0.228	0.385	1.12	0.162	0.056	0.86	0.054	0.127	1.11	0.037	0.014
B3-80	0.86	0.384	0.932	1.11	0.266	0.094	0.85	0.101	0.389	1.11	0.063	0.025
B4-40	1.02	0.272	0.134	1.18	0.177	0.042	1.01	0.063	0.039	1.17	0.039	0.010
B4-55	1.04	0.229	0.100	1.23	0.144	0.018	1.04	0.052	0.028	1.22	0.031	0.004
B4-70	1.10	0.229	0.064	1.31	0.138	0.004	1.10	0.053	0.018	1.31	0.030	0.001
B4-85	1.21	0.254	0.027		—		1.21	0.060	0.008		—	
B4-100	1.38	0.286	0.000		—		1.38	0.070	0.000		—	
B5-45	1.21	0.301	0.028	1.30	0.191	0.008	1.21	0.071	0.008	1.30	0.043	0.002
B5-60	1.25	0.231	0.015	1.35	0.141	0.001	1.25	0.052	0.004	1.35	0.031	0.000
B5-75	1.32	0.195	0.003		—		1.32	0.043	0.001		—	
B5-90		—			—			—			—	
B5-105		—			—			—			—	
B6-50	1.37	0.246	0.001	1.40	0.165	0.000	1.37	0.057	0.000	1.40	0.037	0.000
B6-65	1.37	0.210	0.001		—		1.37	0.047	0.000		—	
B6-80	1.40	0.197	0.000		—		1.40	0.044	0.000		—	
B6-95		—			—			—			—	
B7-55	1.39	0.204	0.000		—		1.39	0.047	0.000		—	
B7-70		—			—			—			—	
B7-85		—			—			—			—	

An overlap can only be observed for the "$\eta_{o,max}$ for T_D, T_n, P_D, and P_n = const" lines, but never for "$\eta_{o,max}$ for J = const" or "$\eta_{o,max}$ for P/D = const".

4. Discussion

4.1. Evaluation

For the following discussion, it should be kept in mind that the $\eta_{o,max}$ lines were calculated in a descriptive way by finding all points, where the tangents to the families of the T_D, T_n, P_D, and P_n = const curves coincide with the tangents to the contour lines of $\eta_{o,max}$ = const. Mathematically speaking, this is equivalent to solving an extrema problem.

To investigate the solutions, we take the propeller used in the example in Section 3.2 and plot the efficiencies along the $P_D = 0.15$ curve against the P/D-value, see the blue line in Figure 5. It can be seen, that the extremum at the lower P/D-value of about 1.0 is a maximum, whereas the extremum at the higher P/D-value of about 1.3 is a minimum. Additionally shown in this figure is the run of the efficiency curves for $P_D = 0.1$ and 0.07 (green and orange lines), the first represents $P_D|_{\min}$ and just touches the "$\eta_{o,max}$ for P_D = const" curve in the apex; the second comes to lay right of the apex. For these two cases, it is apparent that the propeller with the highest possible efficiency is situated beyond the boundary of $P/D = 1.4$.

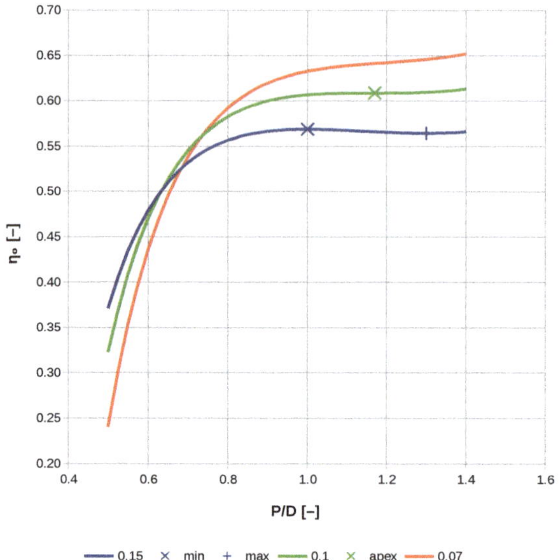

Figure 5. Run of the efficiencies versus pitch ratio P/D along the lines $P_D = 0.15$, 0.1, and 0.07, corresponding to the region of the overlap $P_D|_{\text{max}}$ and the region right of the overlap, respectively. Additionally shown are the extrema for $P_D = 0.15$ and the apex for $P_D|_{\text{max}}$.

The author of this paper believes that these overlaps of the $\eta_{o,max}$ lines causing the ambiguities described are physically not explainable.

There are the obvious reasons: Firstly, the open-water efficiency drops and starts to climb again in the region of the overlap. Secondly, there is no optimum except at the boundary of the available data, right of the overlap.

Generally, we can argue that we can extend the propeller series to even higher P/D-values. Eventually we will arrive at a pitch setting, where the open-water efficiency will become zero, since such a propeller would have blades perpendicular to the section inflow and hence would not be able to accelerate water in the axial direction. Thus, it is not unreasonable to assume that a pitch to diameter ratio must exist, where the open-water efficiency is globally at its highest. Indeed, this can be seen in both Danckwardt diagrams for the Wageningen B2-30 and B2-38 propellers (see Appendix B): a peak of the open-water efficiency can be noticed at a J-value of about 1 and a P/D ratio of about 1.1. It can be observed that all four lines for $\eta_{o,max}$ pass (and must pass) through this absolute maximum of η_o. Even if this point of the absolute maximum of η_o comes to lie right of and above the set of the K_T and K_Q curves—and thus is not displayed in the diagram—the four lines for $\eta_{o,max}$ must still converge towards this single point of the global absolute maximum of η_o. Following this thought, it is evident that the lines of $\eta_{o,max}$ can not bend back, as can be seen with certain propellers, and this behaviour is deemed as physically inexplicable.

4.2. Implications

Admittedly, paper charts are seldom used nowadays in propeller design work, but depending on the computer algorithm used for automatically searching the propeller with the highest achievable efficiency, the following problems can be encountered: Firstly, in the region of the overlap, the computer program could pick the solution on the upper branch of the $\eta_{o,max}$ curve, if no appropriate checks are implemented. It could also jump between the two extrema. Equipped with the knowledge of the described behaviour, the algorithm can be tweaked to find the correct solution. Secondly, in the region right of the overlap, where there exists no optimum, the algorithm could calculate the optimum propeller to have a pitch ratio of 1.4, which is right on the boundary of the available data. As can be clearly seen in Figure 5, the line for $P_D = 0.07$ still climbs in the vicinity of the boundary, indicating that there exist propellers with even higher efficiencies beyond the boundary of the tested propellers. It has to be emphasised that, based on the polynomials, there simply does not exist an optimum propeller in this region, where such an optimum propeller must exist. Thirdly, it can be argued that there exists an optimum propeller up to the point where the $\eta_{o,max}$ curve doubles back. Nevertheless, special care must be taken in the region of the apex, because the line of optimum efficiency already starts to swerve away.

4.3. Accuracy of the Polynomials

As the issue of the doubling back of the $\eta_{o,max}$ lines cast some doubts on the accuracy of the underlying polynomials, the question of the overall accuracy of the polynomials also arises. Whereas Figure 6a,b show a good agreement of the $\eta_{o,max}$ lines between the original and the recreated diagrams for the whole range but the area of doubling back, Figure 6c,d show a big discrepancy between the polynomials published in 1969 and 1975. These deviations can be of varying significance, depending on the subsequently employed and more detailed optimisation procedures. When the Wageningen B-screw Series is used to find the optimum propeller and the thus obtained dimensions will be kept fixed and only small corrections are applied to them—as is common practice at the smaller end of the market—it is of paramount importance that this optimisation routine based on the polynomials gives consistent and accurate results.

For large propellers, the outcome of the optimisation based on the Wageningen B-screw Series polynomials is used as the starting point for further and more detailed optimisation. An accurate starting point would speed up the subsequent full optimisation processes, but should not change the outcome. In reality, the optimised variable D or n found in the first step is very often kept fixed and will not be optimised further in the final optimisation. In this case, it is again of highest importance to get accurate results from the polynomials.

4.4. Provenance and Causes of Overlaps

It must be emphasised, that the overlaps observed are not a feature of the presentation in the form of efficiency maps, but of the underlying data, i.e., the polynomials (5) and (6) published by Oosterveld and van Oossanen in 1975 [2]. It should also be noted that at the time when Danckwardt published his diagrams—which show no overlaps at all, see Figure A1a,b—the and polynomials were not known yet. The design charts published by Yosifov et al. are already based on the polynomials [9]. On all their efficiency maps, the lines for $\eta_{o,max}$ stop before they reach the maximum P/D-value. Yosifov et al. do not mention or explain this behaviour. Those diagrams, where the lines stop far from the maximum P/D ratio, are for the same propellers, where we have identified an overlap.

Nonetheless, it is not clear where these ambiguities were introduced during the process of manufacturing, measuring, fairing, scaling to uniform Reynolds number, and calculating the regression polynomials. Without further investigation into all of these steps, the source of this behaviour is not known, but some possibilities spring to mind: Between the testing of the first and the last propeller, a time span of more than 30 years passed. During this time span, it can be assumed that the manufacturing of the model propellers and the testing technology improved. The propellers were tested at different basins and also at different Reynolds numbers, and were only later corrected to a uniform Reynolds number of $2 \cdot 10^6$. Even the numerical regression used to calculate the polynomials could have introduced this behaviour. Helma shows in [3], for selected propellers, that the lines of maximum efficiency in the recreated Danckwardt diagrams follow the published lines by Danckwardt at low P/D-values (see Figure 6a,b). However, the regression curves given by van Lammeren et al. for propellers with four blades [5] already exhibit a troublesome behaviour at higher values of the pitch to diameter ratio (see Figure 6c,d).

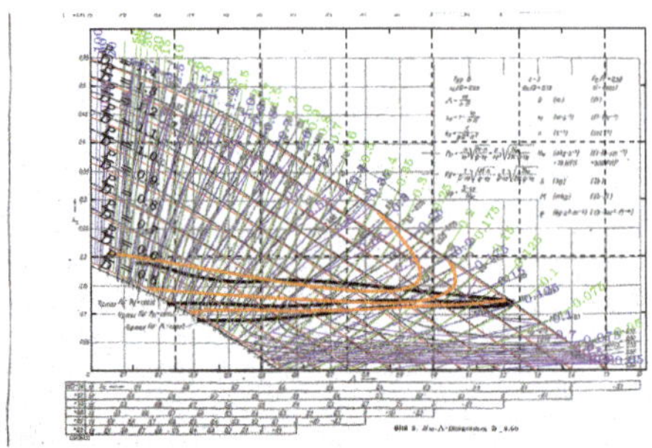

(**a**) P–J (K_T–J) diagram of propeller B3-50, 1956 (original black and white print) and 1975 (coloured).

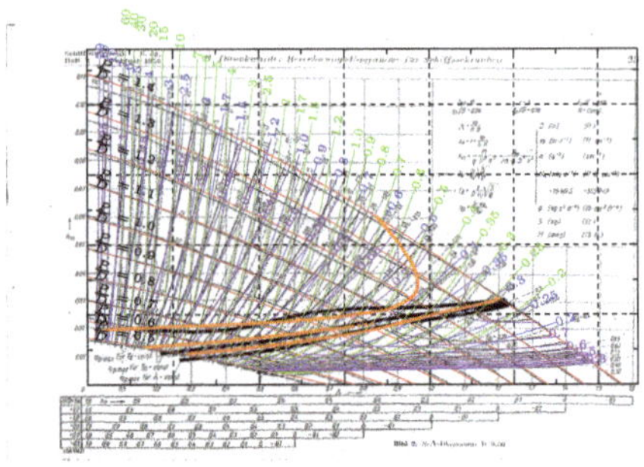

(**b**) T–J (K_Q–J) diagram of propeller B3-50, 1956 (original black and white print) and 1975 (coloured).

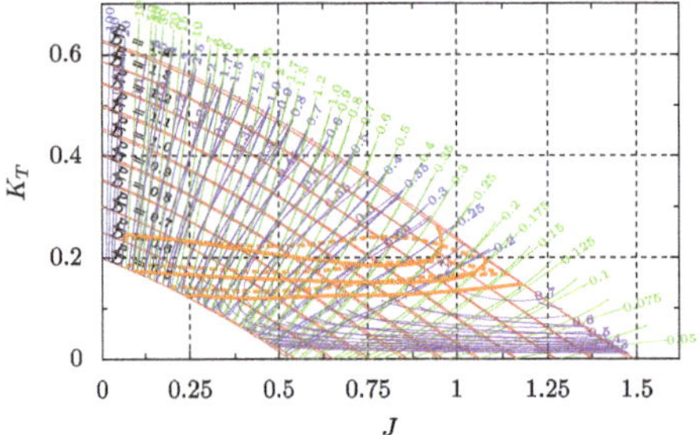

(**c**) T–J (K_Q–J) diagram of propeller B4-70, based on the polynomials from 1969 (broken lines) and 1975 (solid lines).

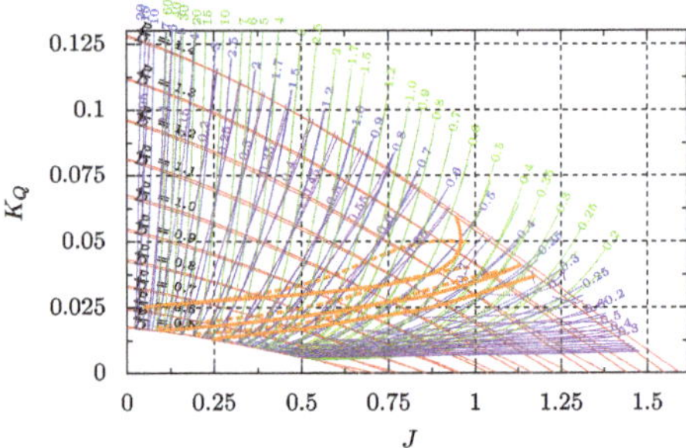

(**d**) P–J (K_T–J) diagram of propeller B4-70, based on the polynomials from 1969 (broken lines) and 1975 (solid lines).

Figure 6. Comparison between the original Danckwardt diagrams, 1956 [6] (probably based on the original data from 1949 [4]); diagrams calculated with polynomials for the scaled and faired open-water characteristics, 1969 [5]; and the most recent polynomials, 1975 [2]. Reproduced from [3], with permission from AMT'17, 2017.

5. Conclusions

With the help of the alternative presentation of open-water characteristics as efficiency maps, it was shown that the current set of polynomials for the Wageningen B-screw Series, as published by Oosterveld and van Oossanen in 1975 [2], shows some troublesome behaviours for higher pitch to diameter ratios for many propellers of the series.

Considering the widespread use of these polynomials, it is suggested to revisit the originally tested data and check all steps involved in the processing of the data sets for the deduction of the polynomials.

The propeller designers would be very well advised to take caution when designing propellers whenever any line of $\eta_{o,max}$ doubles back.

Funding: This research received no external funding.

Conflicts of Interest: The author declares no conflict of interest.

Abbreviations

The following abbreviations are used in this manuscript:

AMT	International Conference on Advanced Model Measurement Technology for the Maritime Industry
ITTC	International Towing Tank Conference
MARIN	Maritime Research Institute Netherlands
smp	International Symposium on Marine Propulsors

Appendix A. Efficiency Maps

Appendix A.1. Lines of Maximum Efficiency

Section 2 introduces the efficiency map and describes the elements found therein. It also introduces the lines "$\eta_{o,max}$ for J, P/D, T_D, P_D, T_n, and P_n = const", where the efficiency is maximum under certain constraints. Efficiency maps lend themselves to construct these lines of maximum efficiency graphically, but they can also be calculated with the help of any optimisation procedures solving for maximum efficiency with the appropriate constraints. For the following discussion on how these lines are derived and their significance, please consult Figure 3.

Appendix A.1.1. Line "$\eta_{o,max}$ for J = const"

If the J-value is known, the propeller with the maximum efficiency can be found by drawing a vertical line at this given value of J (dotted black line). The propeller with the maximum efficiency can be found at the point where this vertical line just touches the efficiency contour line (point A). The K_Q-value can be read off the ordinate and the K_T- and P/D-values can be interpolated between the curves for $K_Q|_{P/D=\text{const}}$ and K_T = const (thin red and dotted red lines). Connecting all points of these maximum efficiencies for every J-value gives us the line "$\eta_{o,max}$ for J = const", which can be drawn into the efficiency map (bold black line). Note that the equivalent to this line on the conventional open-water diagram is the envelope to all $\eta_o(J)$ lines.

Appendix A.1.2. Line "$\eta_{o,max}$ for P/D = const"

Another line of maximum efficiencies is the line called "$\eta_{o,max}$ for P/D = const" (bold red line). This line can be used to find the propeller with the maximum efficiency, if the P/D ratio is known. It connects all points where the tangents of the K_Q curve (thin red line) and the η_o contour line (thin black line) coincide (point B). On the conventional open-water diagram, this line corresponds to the line connecting the maxima of all $\eta_o(J)$ (which is situated below the envelope to the efficiency curves, resulting in a lower efficiency for the same advance coefficient).

Appendix A.1.3. Line "$\eta_{o,max}$ for T_D = const"

If the propeller diameter, the speed of advance, and the (required) thrust are known, we can plot the curve

$$T_D = \frac{1}{D^2 v_a^2}\frac{T}{\rho} = \text{const}, \tag{A1}$$

which is equal to

$$T_D(J) = \frac{K_T(J)}{J^2} = \text{const}, \tag{A2}$$

into the efficiency map (thin blue lines). This formulation uses the old trick of eliminating the unknown shaft speed n, which now becomes part of the solution, when optimising for the highest possible efficiency.

This curve can be drawn either into the K_T–J or the K_Q–J efficiency map. In the K_T–J diagram, the curve is a simple quadratic curve in the form of cJ^2, where c = suitable constant; whereas in the K_Q–J diagram, the line becomes the truly parametric curve with J as parameter

$$T_D\left(J, K_Q(J)\right) = \frac{K_T\left(J, K_Q(J)\right)}{J^2}. \tag{A3}$$

In case of a given set of constant T_D-values, a family of curves results.

Once again, a line for propellers with the highest possible efficiency can be constructed by connecting all points, where the tangent to the "T_D = const" curve coincides with the tangent to the "η_o = const" contour line (point C); this is called the "$\eta_{o,max}$ for T_D = const" line (thick blue line).

Please refer to Appendix A.4 to see how this can be used to find the optimum propeller quickly if the thrust T, the propeller diameter D, and the speed of advance v_a is known.

Appendix A.1.4. Lines "$\eta_{o,max}$ for P_D, T_n, and P_n = const"

Finally, other families of curves can be drawn:

$$P_D = \frac{K_Q}{J^3} = \frac{1}{D^2 v_a^2}\frac{Qn}{\rho v_a} = \frac{1}{D^2 v_a^2}\frac{P_P}{2\pi\rho v_a}, \tag{A4}$$

$$T_n = \frac{K_T}{J^4} = \frac{n^2}{v_a^4}\frac{T}{\rho}, \text{ and} \tag{A5}$$

$$P_n = \frac{K_Q}{J^5} = \frac{n^2}{v_a^4}\frac{Qn}{\rho v_a} = \frac{n}{v_a^4}\frac{P_P}{2\pi\rho v_a}. \tag{A6}$$

These formulations eliminate the shaft speed n, Equation (A4), and the propeller diameter D, Equations (A5) and (A6). In Appendix A.4, it is shown how these formulations help in solving each of the six possible optimisation problems a propeller designer can encounter.

The corresponding lines of maximum efficiency are called "$\eta_{o,max}$ for P_D = const", "$\eta_{o,max}$ for T_n = const", and "$\eta_{o,max}$ for P_n = const", respectively.

Appendix A.2. Origins

The first diagrams using the presentation discussed were published in 1917 by Bendemann and Madelung [10] and in 1923 by von der Steinen (Von der Steinen argues in his paper that he had finished it earlier, but could not publish it for 6 years because of a exceptionally high work load, thus claiming the intellectual of these diagrams.) [11]. Bendeman and Madelung based their idea and how the data should be presented on the polar diagram of aerofoils. In 1936, Papmel published design charts [12] using the same setup. Schoenherr included all four lines of maximum efficiency in 1949 [13]. All authors mentioned so far suggested to plot the T_D and T_n curves into the K_T–J and P_D and to plot P_n into the K_Q–J efficiency map. Bendemann and Madelung pointed out that the usage of a double logarithmic scale results in the T_D and T_n curves becoming straight lines, making it easier for the designer to use these maps.

Danckwardt calculated design charts for the Wageningen B-screw Series in 1956 [6], but instead of drawing the T_D and T_n curves into the K_T–J efficiency map, he plotted the P_D and P_n curves (and vice versa). This deliberate decision makes life easier for the propeller designer (but not for the draftsman plotting these efficiency maps!), since now only one single chart is required to get the missing torque or thrust coefficient, which are not available in the efficiency maps suggested by previous authors. To honour the inventor, these came to be known as Danckwardt diagrams. We will refer to them as P–J and T–J diagrams to distinguish them from the general K_T–J and K_Q–J efficiency maps. As a mnemonic, remember that you use the T–J diagram whenever the thrust T is known and P–J whenever the power P_P (or the torque Q) is known.

In 1983, Yosifov et al. calculated the design charts according to Papmel for the Wageningen B-screw Series with the aid of a computer using the polynomials [14], which previously became available in 1975 [2]. Finally, Yosifov et al. published polynomials for the $\eta_{o,max}$ lines in 1986 [9], removing the need to resort to paper and pencil.

All these diagrams might use different symbols or alternative definitions of the variables (mostly multiplied by constant values or using inverse values). They also show different degrees of details, but they all build on the same idea of the efficiency map. Examples for original Danckwardt diagrams and design charts calculated by Yosifov et al. are shown in Figures A1 and A2, respectively.

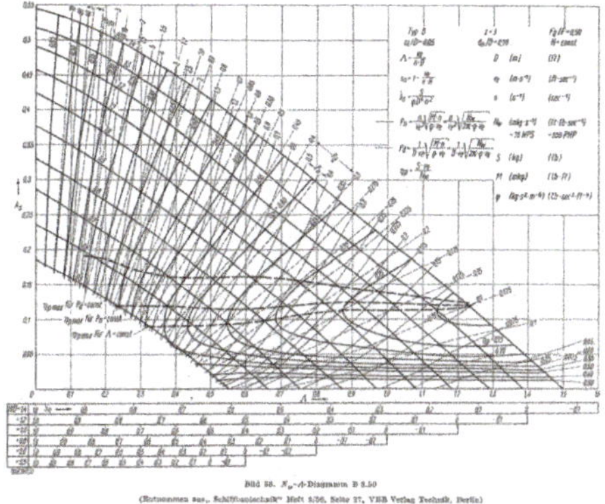
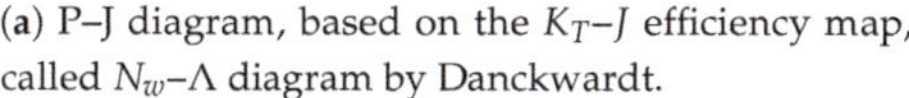

(**a**) P–J diagram, based on the K_T–J efficiency map, called N_w–Λ diagram by Danckwardt.

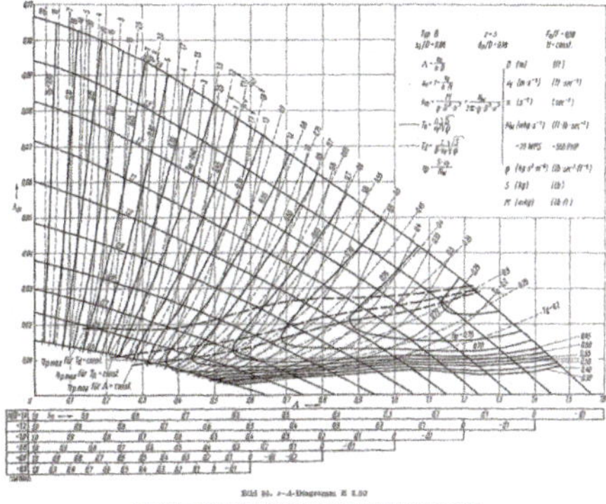

(**b**) T–J diagram, based on the K_Q–J efficiency map, called s–Λ diagram by Danckwardt.

Figure A1. Example of the original Danckwardt diagrams for the propeller B3-50. See Table A1 for the differences between Danckwardt's and ITTC's nomenclature. Reproduced from [6].

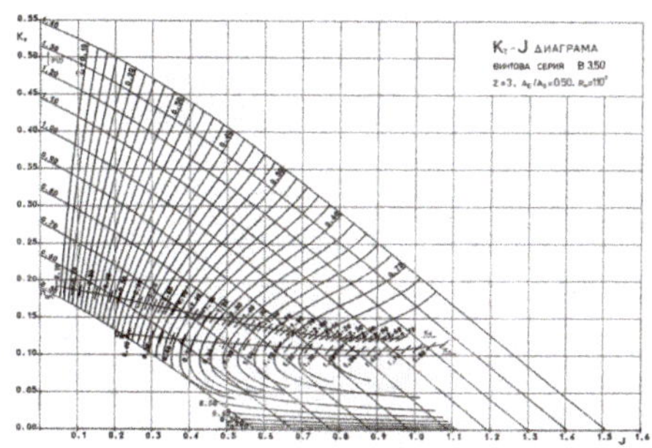
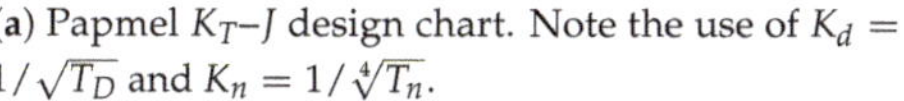

(**a**) Papmel K_T–J design chart. Note the use of $K_d = 1/\sqrt{T_D}$ and $K_n = 1/\sqrt[4]{T_n}$.

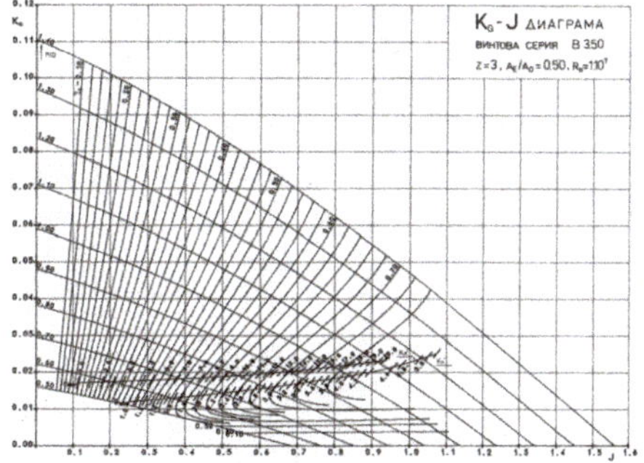

(**b**) Papmel K_Q–J design chart. Note the use of $K'_d = 3.455/\sqrt{P_D}$ and $K'_n = 3.455/\sqrt[4]{P_n}$.

Figure A2. Example of the original Papmel design charts for the propeller B3-50, as recreated by Yosifov et al. Reproduced from [14], with permission from Bulgarian Ship Hydrodynamics Centre, 1983.

Table A1. Differences between the nomenclature used by Danckwardt and ITTC.

Name	Danckwardt	ITTC
Pitch	H	P
Blade area ratio	F_a/F	A_e/A_0
Speed of advance	v_e	v_a
Thrust	S	T
Torque	M	Q
Delivered power	N_W	P_D
Advance coefficient	Λ	J
Thrust coefficient	k_s	K_T
Torque coefficient	k_m	K_Q
Open-water efficiency	η_p	η_o
Slip	s_n	S_R

Appendix A.3. Advantages of Efficiency Maps

It was certainly noticed by the reader that the efficiency maps introduced in Section 2 and above are comparable, but not identical, to the B_p–δ diagram. One of the benefits of the efficiency maps is that they include the bollard pull condition, whereas this condition disappears into infinity on the B_p–δ diagram, because δ—as the inverse of the J-value— becomes infinity for $J = 0$.

The efficiency maps can also include all six solutions lines for the optimum propeller in one diagram, but such diagrams were never published because they would be even more confusing than the published diagrams with three or four solution lines. In comparison, the B_p–δ diagrams can only be used to solve one single problem. This is certainly another big advantage for the propeller designer.

Compared to the conventional open-water diagrams, the efficiency maps can contain the solutions for real-world optimisation problems, whereas the open-water diagrams can only show the "$\eta_{o,max}$ for J = const" and "$\eta_{o,max}$ for P/D = const" lines (but very seldom do).

Appendix A.4. How Are Efficiency Maps Used to Optimise a Propeller for Given Conditions?

The main purpose of using efficiency maps in the scope of this paper is to show the ambiguity of polynomials, as explained in Section 3. Nevertheless, we want to show in this section the practical significance of these diagrams for optimising a propeller.

The optimisation challenges a propeller designer can encounter can be categorised into six basic problems. These are tabulated in Table A2 together with the corresponding solution path using the Danckwardt diagrams. Efficiency maps can be used to solve any of these problems in a direct way without the need to resort to an iterative process.

Table A2. The six basic optimisation problems encountered by propeller designers and their solution paths. All unknown values can finally be calculated from the solution.

N°	Problem Definition		Solution Path			Solution
	Known Values	Unknown Values	Calculate ...	... Using Equation	To Find Optimum Use Line "$\eta_{o,max}$ for ... = const"	
1	v_a, n, D	P/D	J	(1)	J	$P/D, K_T, K_Q$
2	D, P	v_a, n	P/D	—	P/D	J, K_T, K_Q
3	T, v_a, D	$n, P/D$	T_D	(A1)	T_D	$J, P/D, K_Q$
4	Q, v_a, D	$n, P/D$	P_D	(A4)	P_D	$J, P/D, K_T$
5	T, v_a, n	$D, P/D$	T_n	(A5)	T_n	$J, P/D, K_Q$
6	Q, v_a, n	$D, P/D$	P_n	(A6)	P_n	$J, P/D, K_T$

We will explain the solution path using the "Example 1: Optimum rotation rate for a given diameter" from Kuiper's book "The Wageningen Propeller Series" [15]. Kuiper presents the problem, where the propeller thrust T (1393 kN), the propeller diameter D (7 m), and the advance speed v_a (16.8 kn) are given. The density of water ρ is assumed to be 1025 kg m^{-3}. These values were typical for a container vessel of that time with a speed of 21 kn. He also assumes that a four-bladed propeller has been chosen. The task at hand is to find the optimum propeller, the required power P_P, and especially the shaft speed n. Kuiper calculates the required blade area ratio A_e/A_0 to be 0.48. For the optimisation process, he selects a blade area ratio of 0.55 to agree with the diagrams published.

Having established the basic conditions of the optimisation problem, he explains that "the thrust and the diameter are known, but the rotation rate is not. This means that the parameters K_T and J cannot be calculated yet. However, the parameter K_T/J^2 can be calculated because it does not contain the rotation rate (as we already have seen in Appendix A.1.3, where we called this parameter T_D). Using the figures above in Equation (A1), he gets a value of 0.3707. Kuiper now starts the program supplied with the book and searches the P/D value for the fixed K_T/J^2 value of 0.3707, where the open-water efficiency η_o becomes a maximum. Stopping at an accuracy of 5⁄100 for P⁄D, he obtains the values given at iteration 2 of Table A3.

Table A3. The solutions for "Example 1" in Kuiper's book [15], as obtained by Kuiper using the program supplied with the book by using Danckwardt diagrams and the exact solution.

	Iteration	P/D	J	K_T	K_Q	η_o	T_D
	1	0.95	0.674	0.168	0.0278	0.650	0.371
Kuiper	2	1.00	0.699	0.181	0.0310	0.651	0.371
	3	1.05	0.723	0.194	0.0344	0.650	0.371
Danckwardt diagram		1.00	0.70	0.18	0.031	0.65	0.371
Exact solution		1.004	0.7007	0.1823	0.031 24	0.6509	0.371 310

Kuiper states that "the conclusion is that the optimum efficiency can be reached with a pitch ratio of 1.0". From the advance ratio J, the required shaft speed is derived as $1.767\,\mathrm{s}^{-1}$ (Equation (1)) and the required power from the torque coefficient K_T as 18 513 kW (Equation (3)).

Using the Danckwardt diagrams to solve this optimisation problem follows the same path, but instead of finding the optimum by manually searching for the maximum efficiency, we use the T–J diagram for the propeller B4-55 (see Appendix B). We pencil the line for $T_D = 0.371$ between the thin blue lines for $T_D = 0.25$ and $T_D = 0.5$ and find its intersection with the thick blue line "$\eta_{o,max}$ for $T_D =$ const". The values are given in line 4 of Table A3. The unknown values for n and P_P are calculated as before and we get the same figures.

For comparison, the exact solution is given in the last line of Table A3 with an accuracy of 1⁄1000 for P⁄D.

The other design challenges from Table A2 are solved accordingly.

Appendix B. Efficiency Maps for Wageningen B-Screw Series

The following pages contain the recreated Danckwardt diagrams of all propellers of the Wageningen B-screw Series, according to Table 1. They are valid for a sectional Reynolds number of $2 \cdot 10^6$. The diagrams are based on the polynomials published in 1975 by Oosterveld and van Oossanen [2]. Table A4 explains the composition of the diagrams and the line colours and types used.

Table A4. Composition of the Danckwardt P–J (K_T–J) and T–J (K_Q–J) efficiency maps. Additionally shown is the significations of line colour and type. K_T = thrust coefficient; K_Q = torque coefficient; η_o = open-water efficiency; J = advance coefficient; T = thrust; Q = torque; P_P = propeller power; D = propeller diameter; P = propeller pitch; n = shaft speed (in s^{-1}); v_a = speed of advance; and ρ = water density. If not stated otherwise, all values are in SI base units.

Diagram:	P–J	T–J
Abscissa:	$J = \frac{v_a}{nD}$	$J = \frac{v_a}{nD}$
Ordinate:	$K_T = \frac{T}{\rho n^2 D^4}$	$K_Q = \frac{Q}{\rho n^2 D^5} = \frac{P_P}{2\pi\rho n^3 D^5}$
One set of curves:	— $K_T(J)$ for $P/D = \text{const}$	— $K_Q(J)$ for $P/D = \text{const}$
Two sets of contour lines:	— $\eta_o = \frac{v_a}{nD} = \text{const}$	— $\eta_o = \frac{v_a}{nD} = \text{const}$
	- - - $K_Q = \frac{Q}{\rho n^2 D^5} = \frac{P_P}{2\pi\rho n^3 D^5} = \text{const}$	- - - $K_T = \frac{T}{\rho n^2 D^4} = \text{const}$
Two families of (parametric) curves for sets of constant values:	— $P_D = \frac{K_Q}{J^3} = \frac{1}{D^2 v_a^2}\frac{Qn}{\rho v_a} = \frac{1}{D^2 v_a^2}\frac{P_P}{2\pi\rho v_a} = \text{const}$	— $T_D = \frac{K_T}{J^2} = \frac{1}{D^2 v_a^2}\frac{T}{\rho} = \text{const}$
	— $P_n = \frac{K_Q}{J^5} = \frac{n}{v_a^4}\frac{Qn}{\rho v_a} = \frac{n}{v_a^4}\frac{P_P}{2\pi\rho v_a} = \text{const}$	— $T_n = \frac{K_T}{J^4} = \frac{n}{v_a^4}\frac{T}{\rho} = \text{const}$
Four lines of $\eta_{o,max}$:	"for P_D= const" (for known Q, v_a, D)	"for T_D= const" (for known T, v_a, D)
	"for P_n= const" (for known Q, v_a, n)	"for T_n= const" (for known T, v_a, n)
	"for **J= const**" (for known v_a, n, D)	"for **J= const**" (for known v_a, n, D)
	"for P/D= const" (for known P/D)	"for P/D= const" (for known P/D)

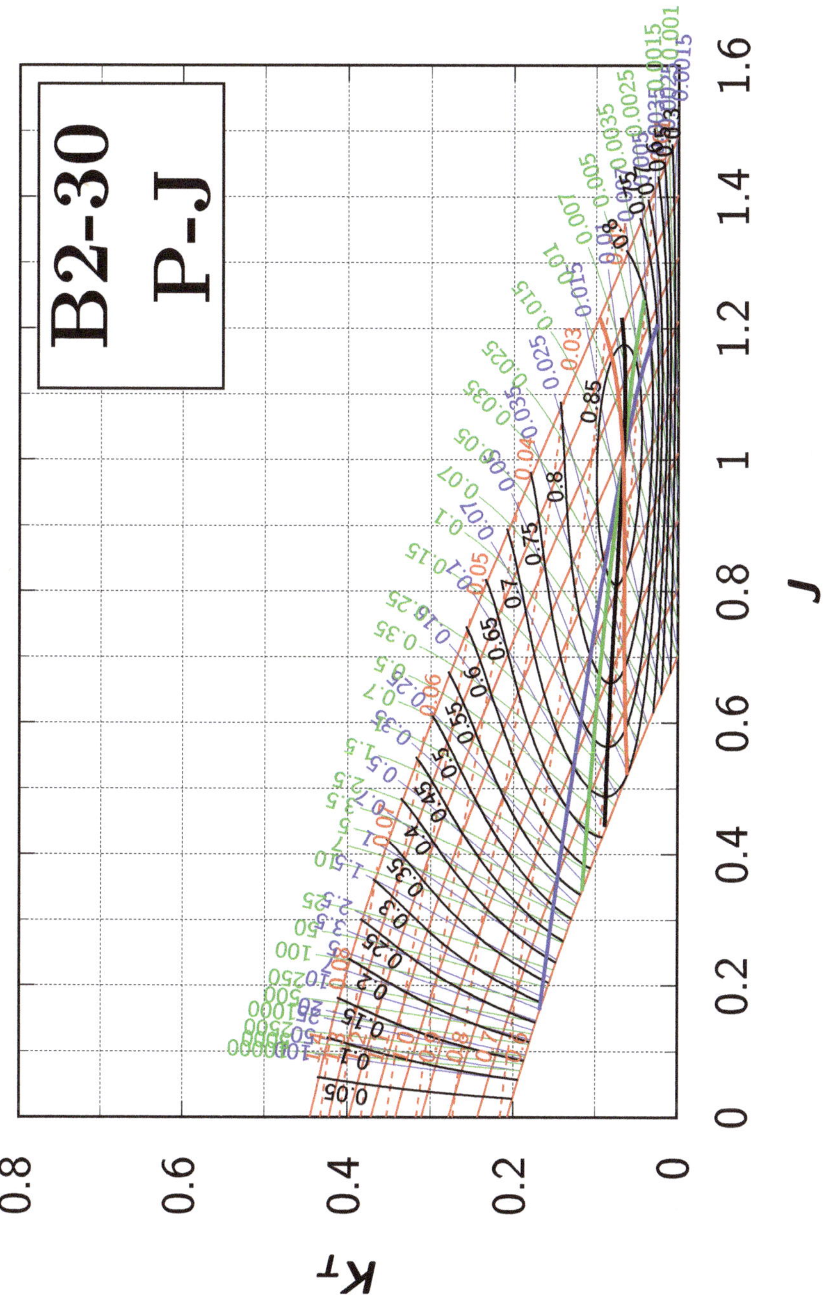
B2-30
P-J
J
K_T
0
0.2
0.4
0.6
0.8
1
1.2
1.4
1.6

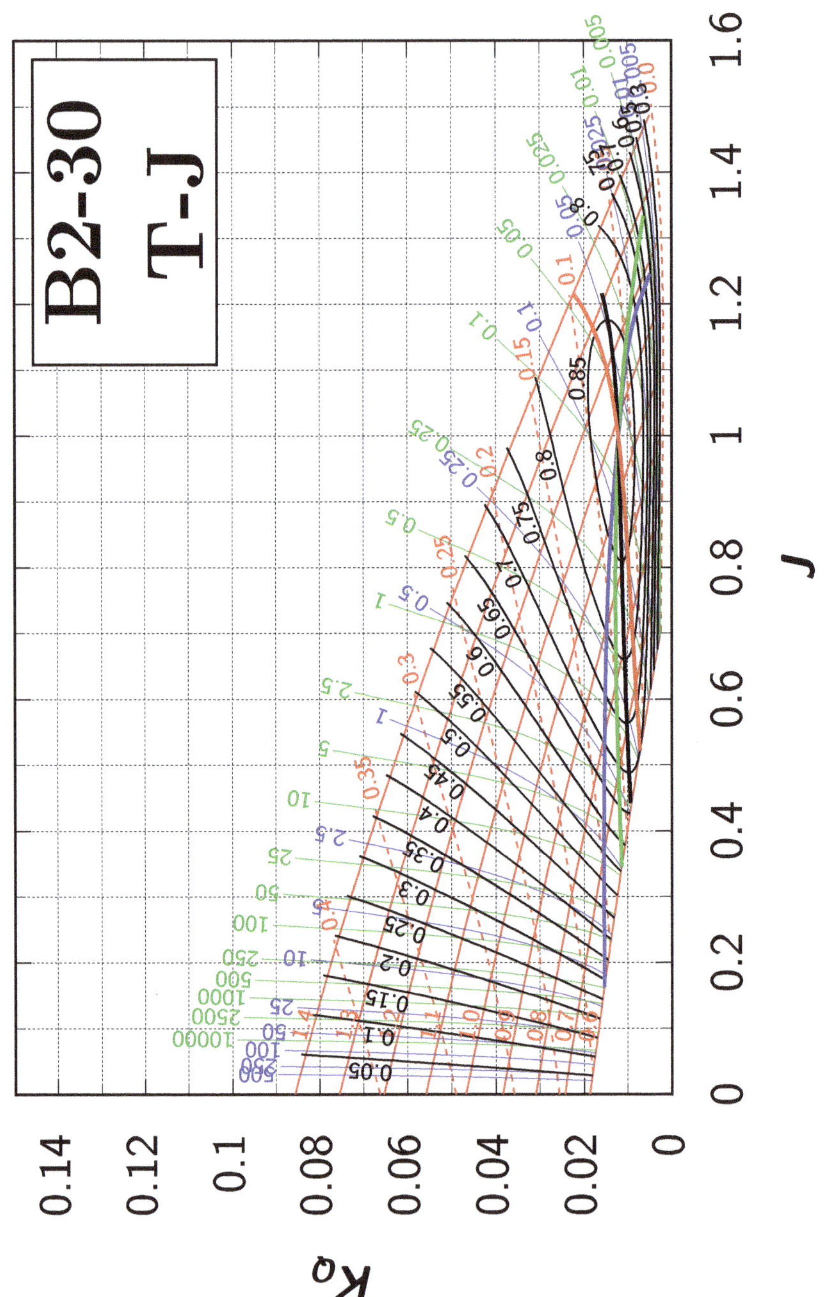
B2-30
T-J
K_Q
J

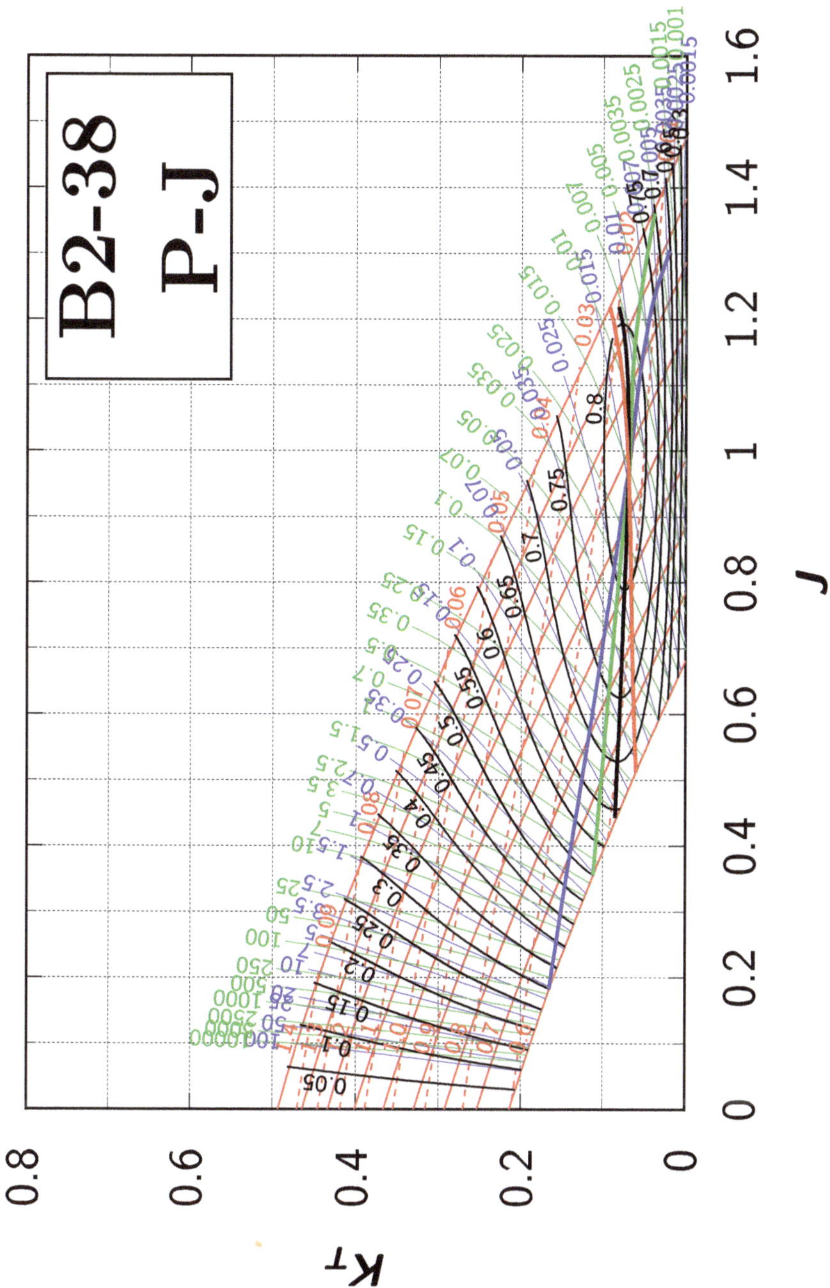
B2-38
P-J
K_T
J
0.8
0.6
0.4
0.2
0
0
0.2
0.4
0.6
0.8
1
1.2
1.4
1.6

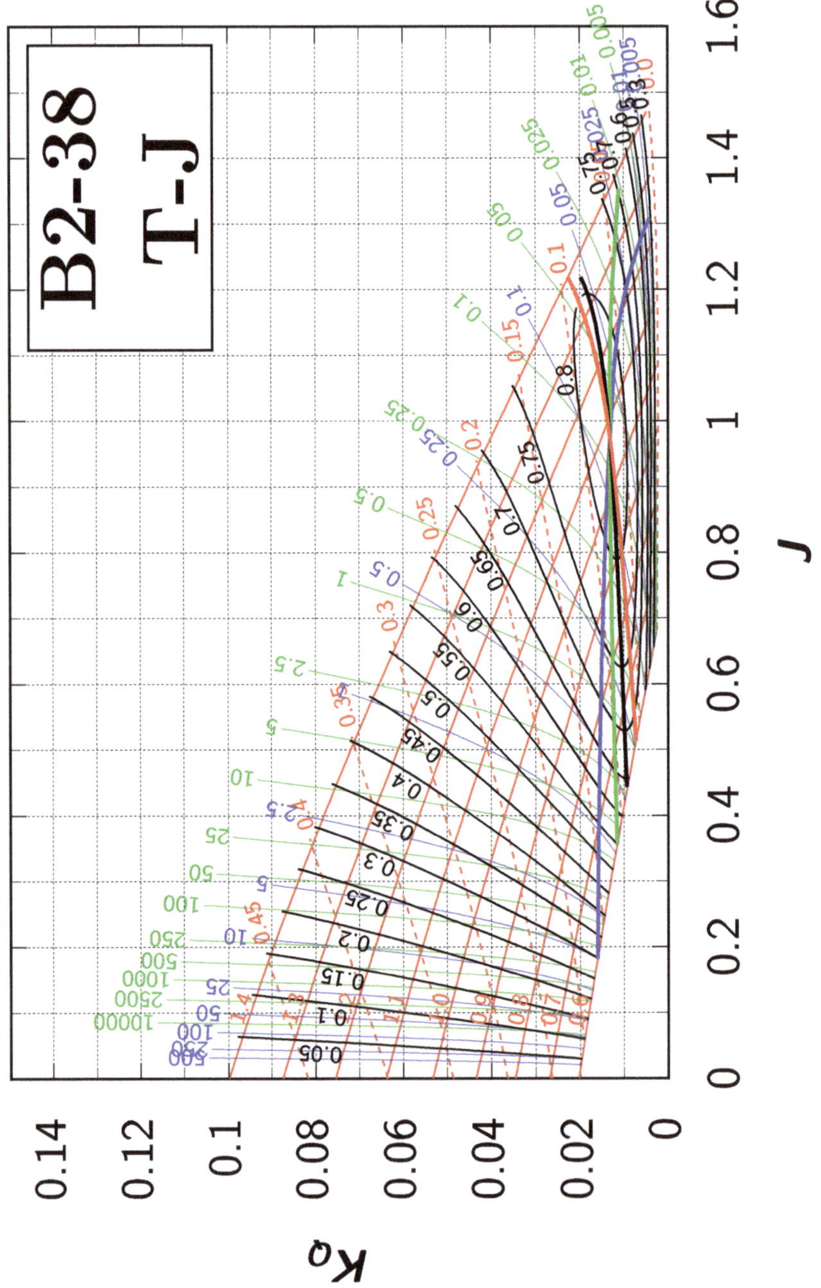
B2-38
T-J
J
K_Q
0
0.2
0.4
0.6
0.8
1
1.2
1.4
1.6
0
0.02
0.04
0.06
0.08
0.1
0.12
0.14

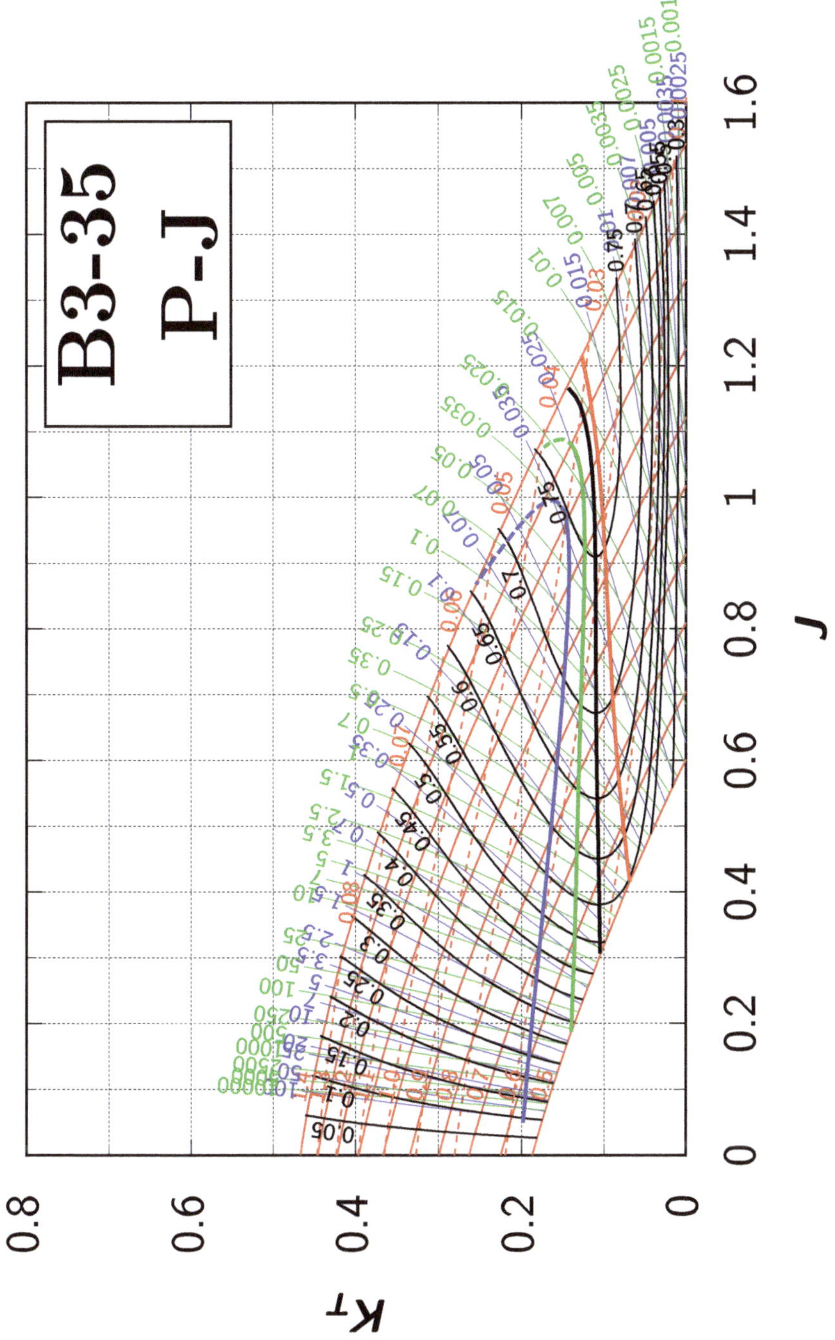
B3-35
P-J
K_T
J
0.8
0.6
0.4
0.2
0
0.2
0.4
0.6
0.8
1
1.2
1.4
1.6

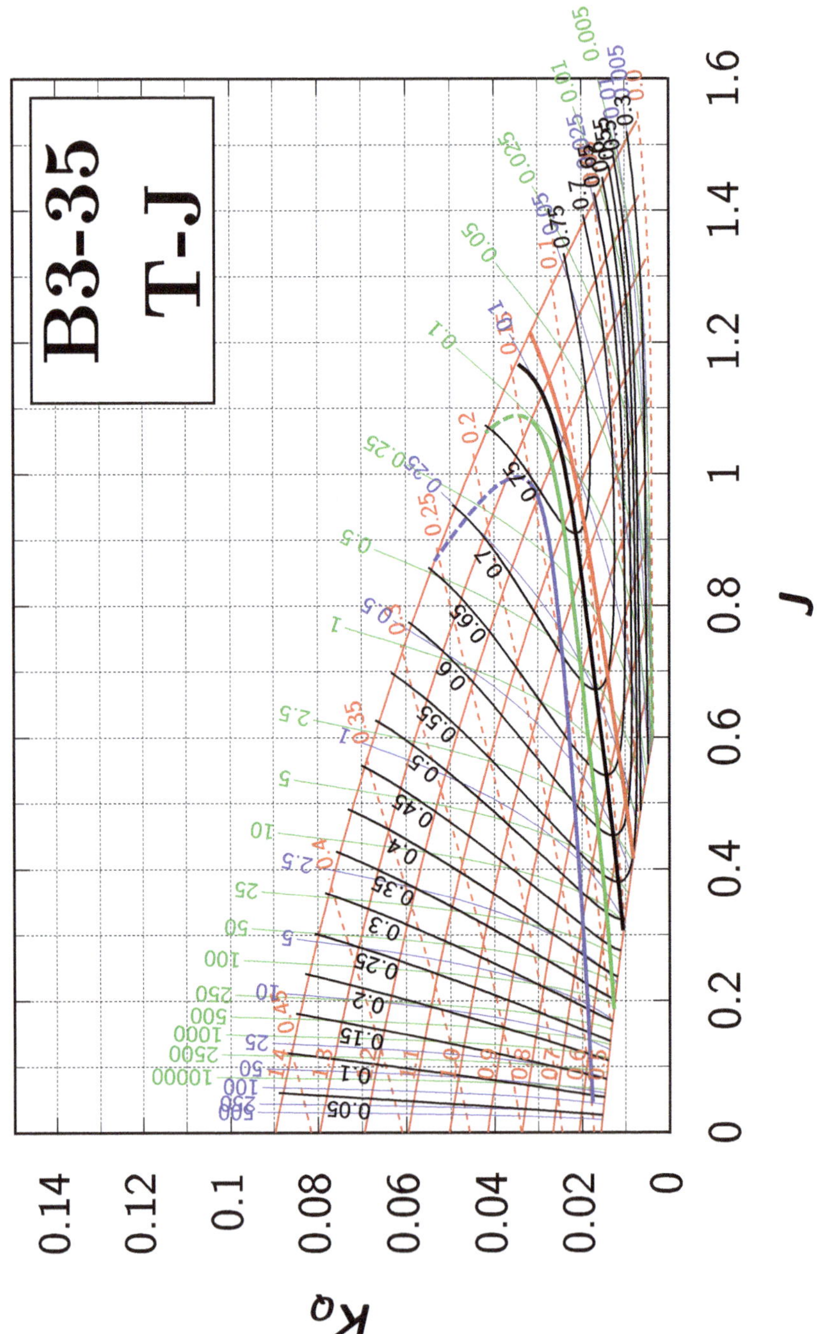
B3-35
T-J
K_Q
J
0.14
0.12
0.1
0.08
0.06
0.04
0.02
0
0
0.2
0.4
0.6
0.8
1
1.2
1.4
1.6

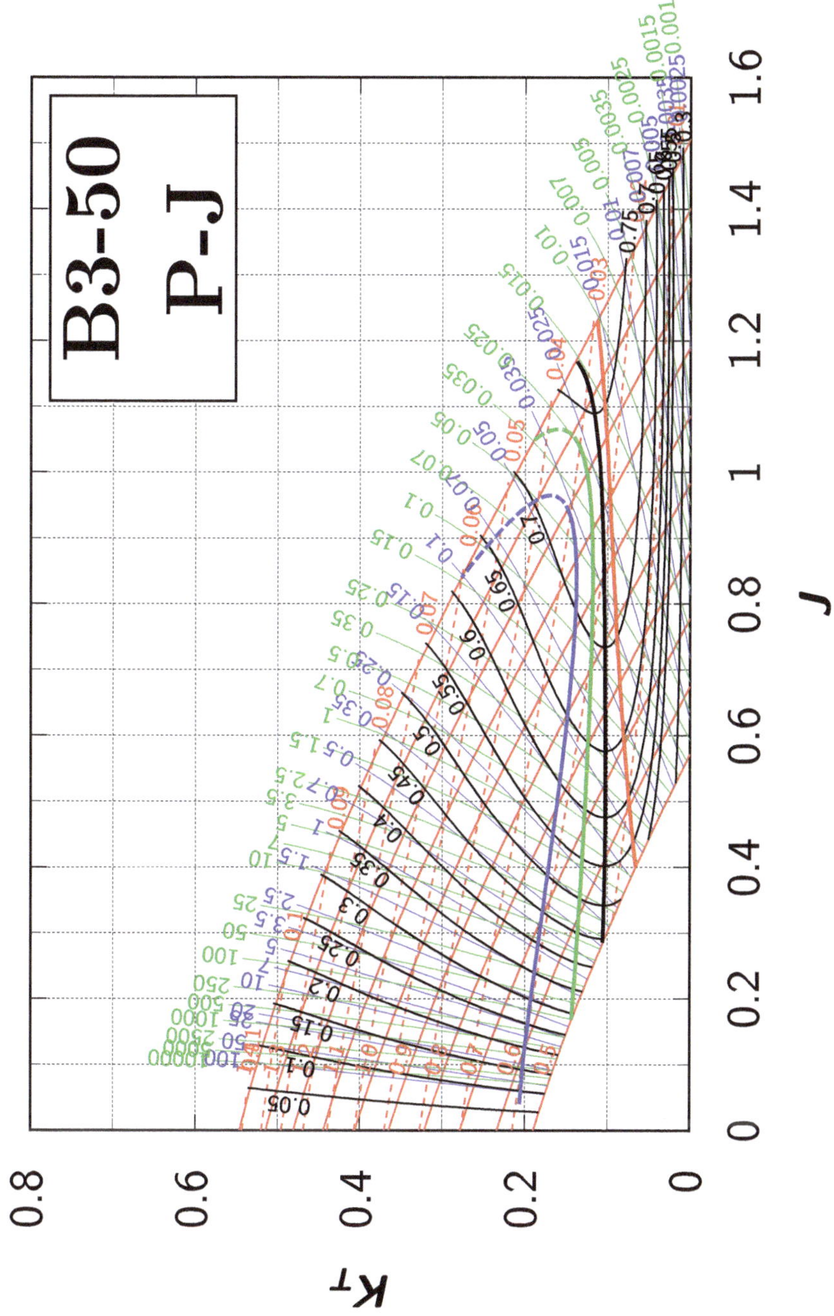
B3-50
P-J
J
K_T
0
0.2
0.4
0.6
0.8
1
1.2
1.4
1.6

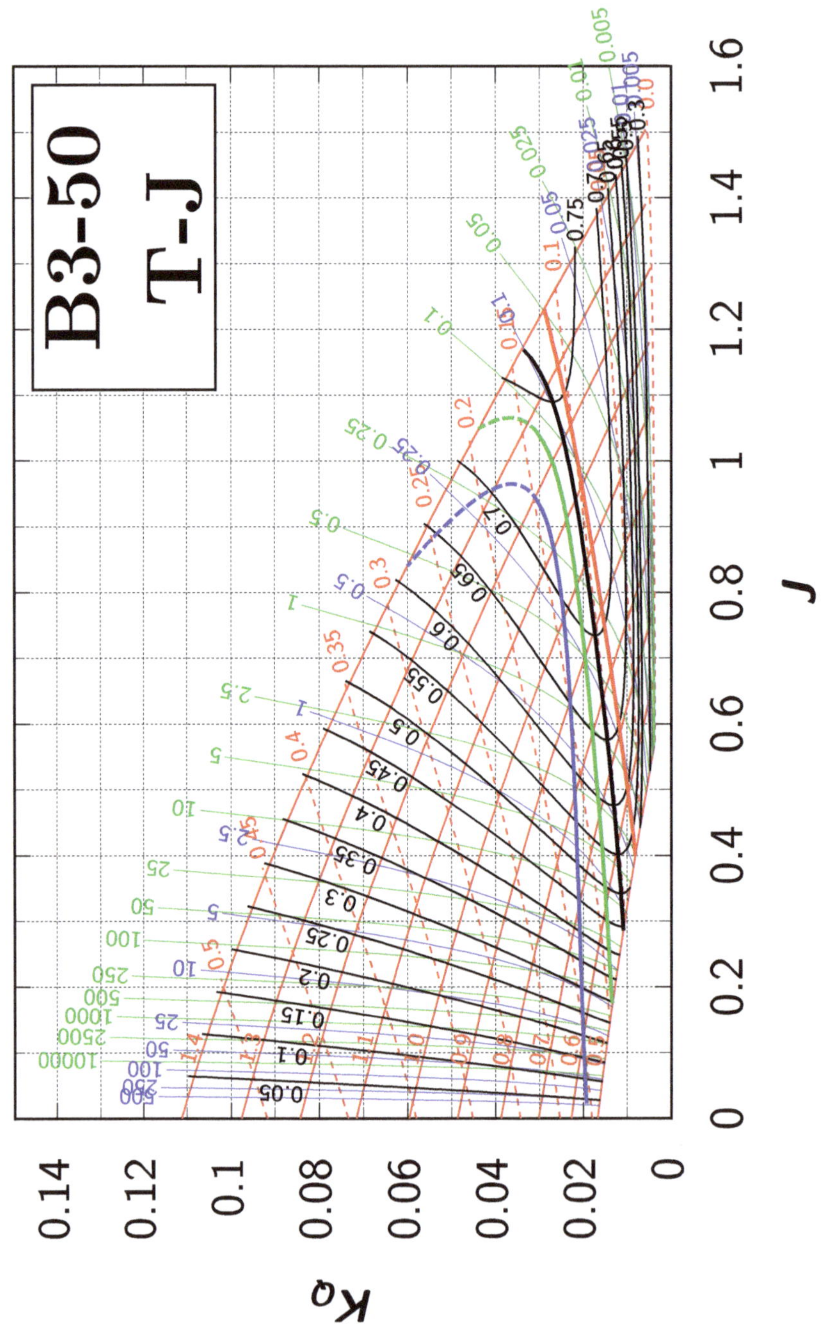
B3-50
T-J
J
K_Q

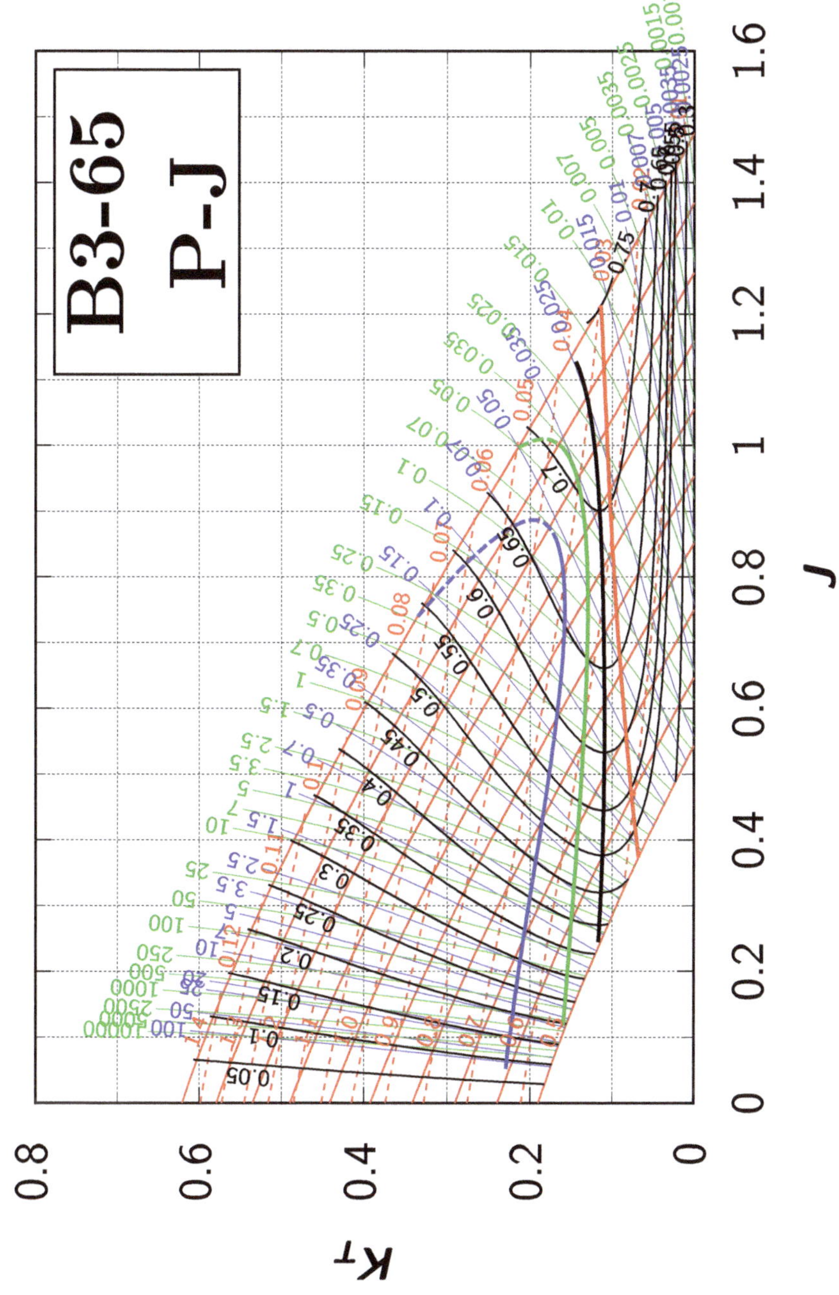
B3-65
P-J
J
K_T

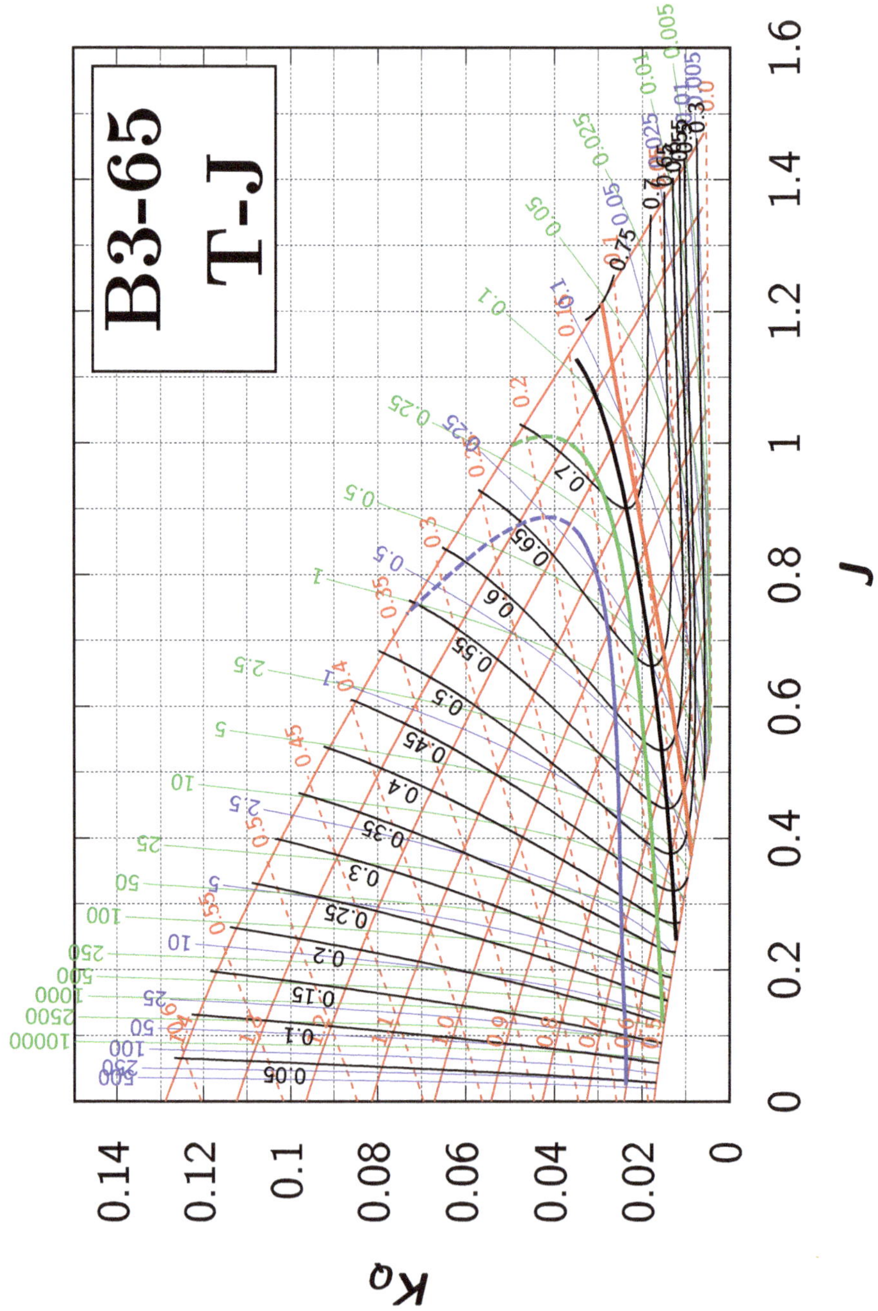
B3-65
T-J
J
K_Q

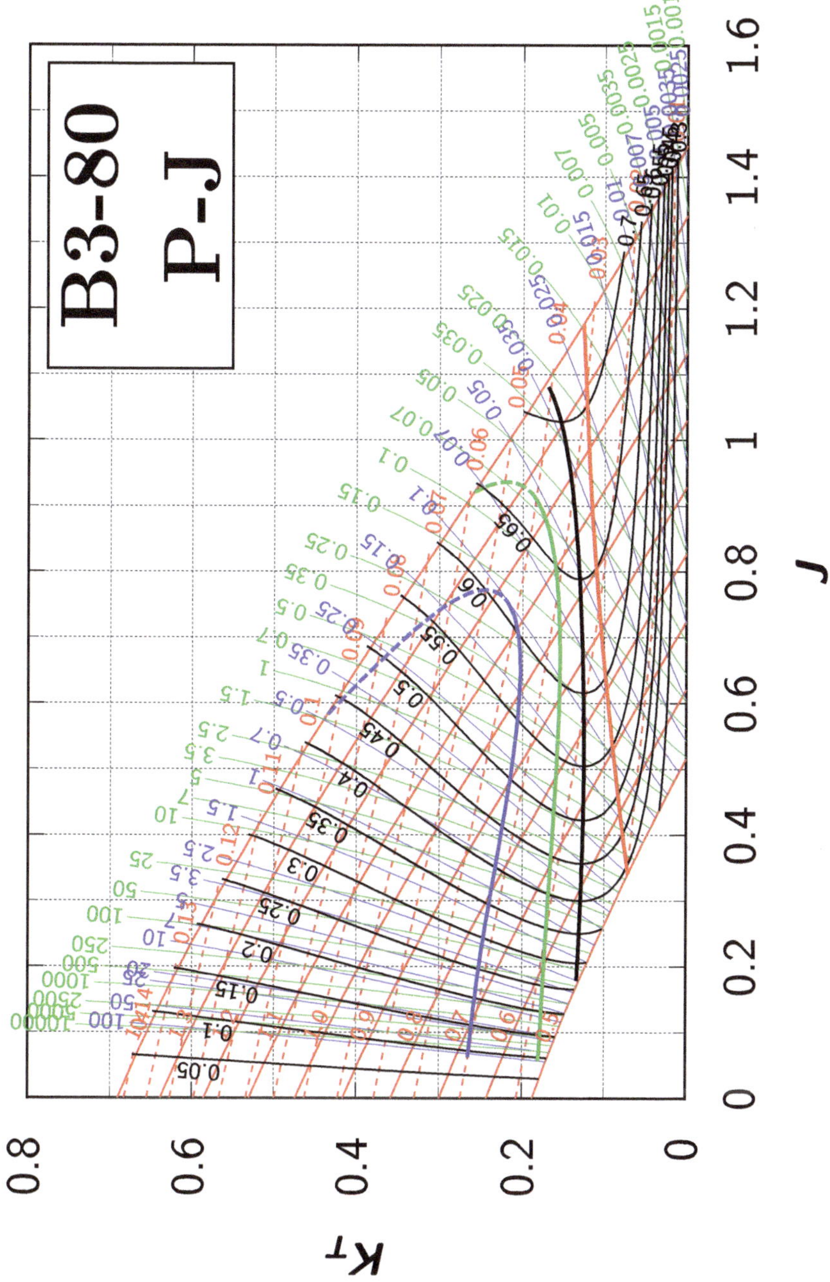
B3-80
P-J
J
K_T

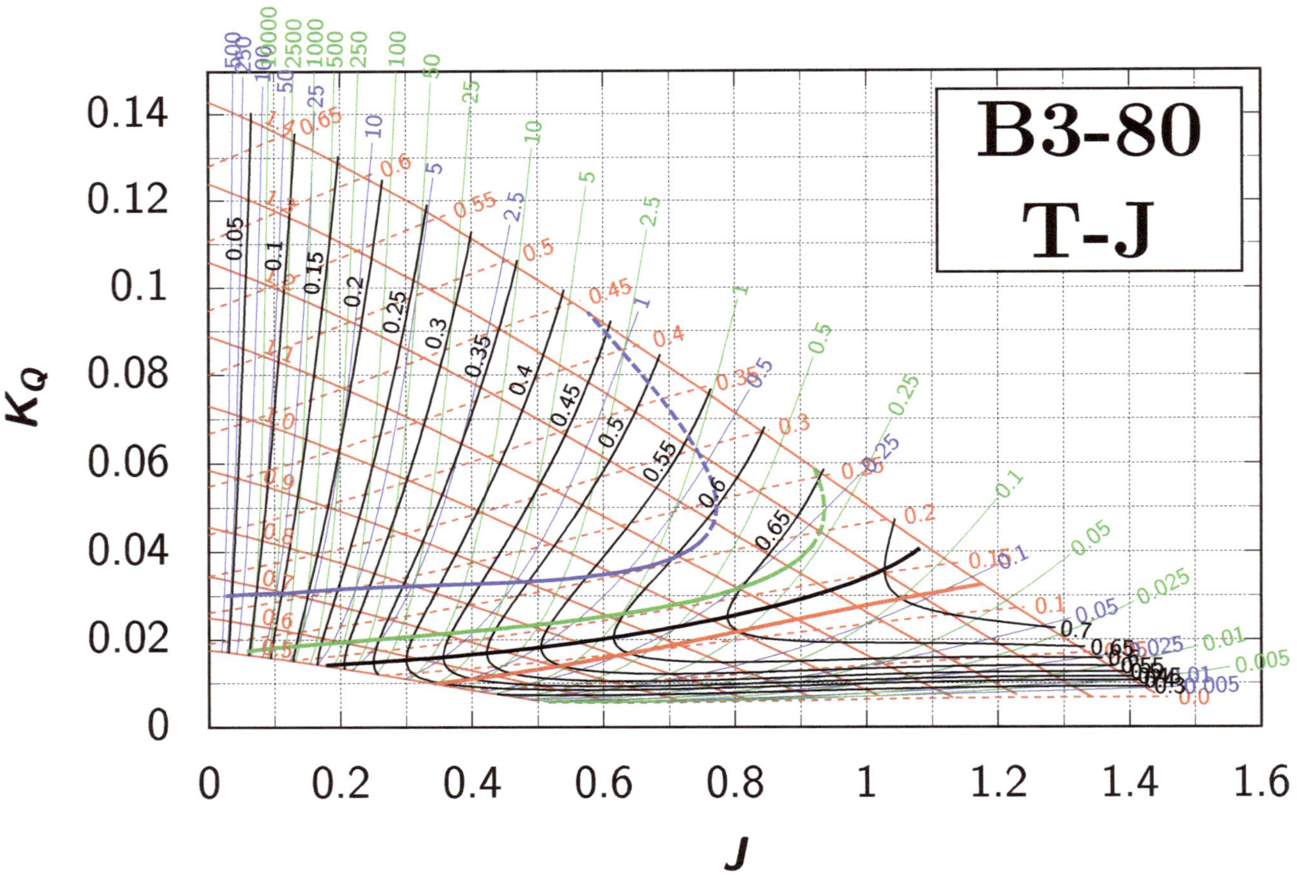
B3-80
T-J
J
K_Q

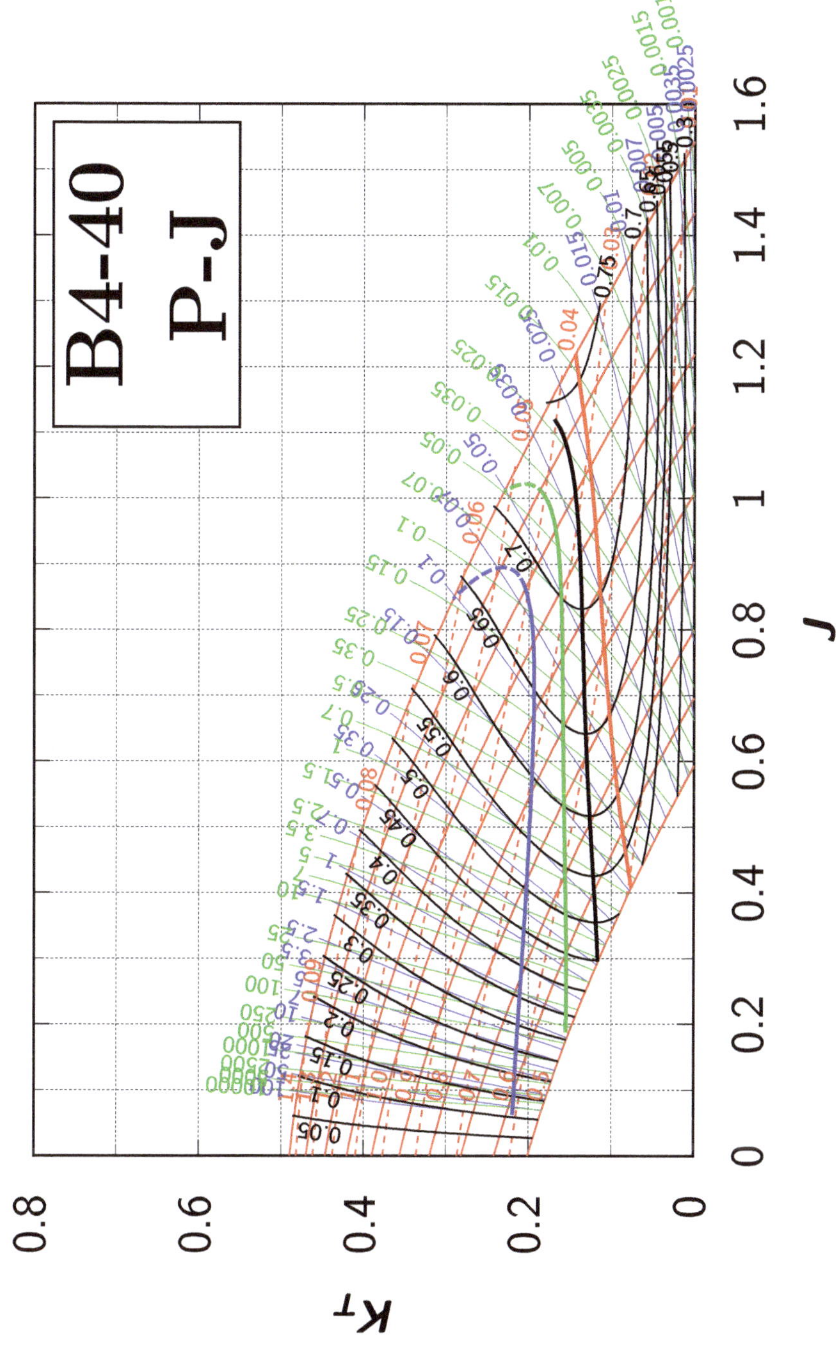
B4-40
P-J
K_T
J
0
0.2
0.4
0.6
0.8
1
1.2
1.4
1.6

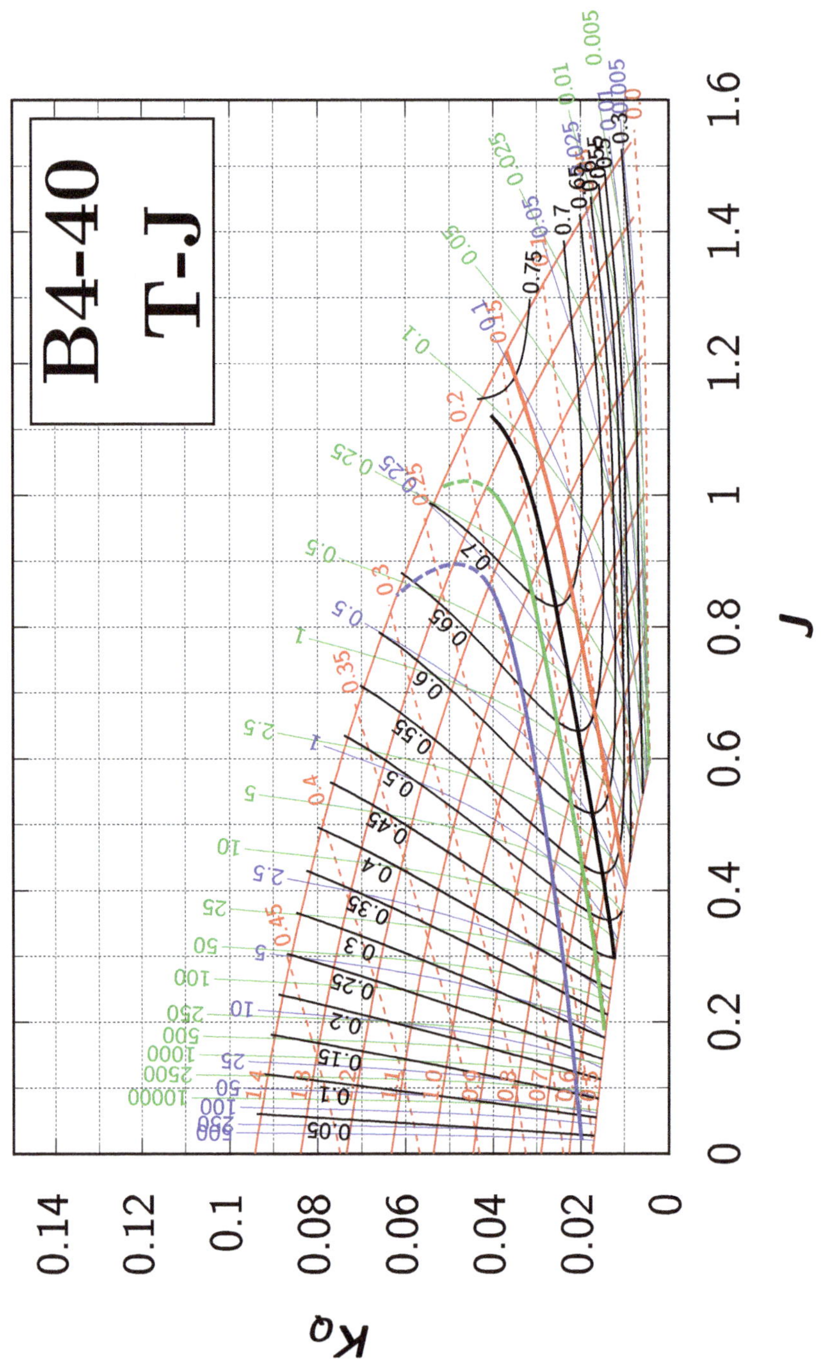
B4-40
T-J
J
K_Q

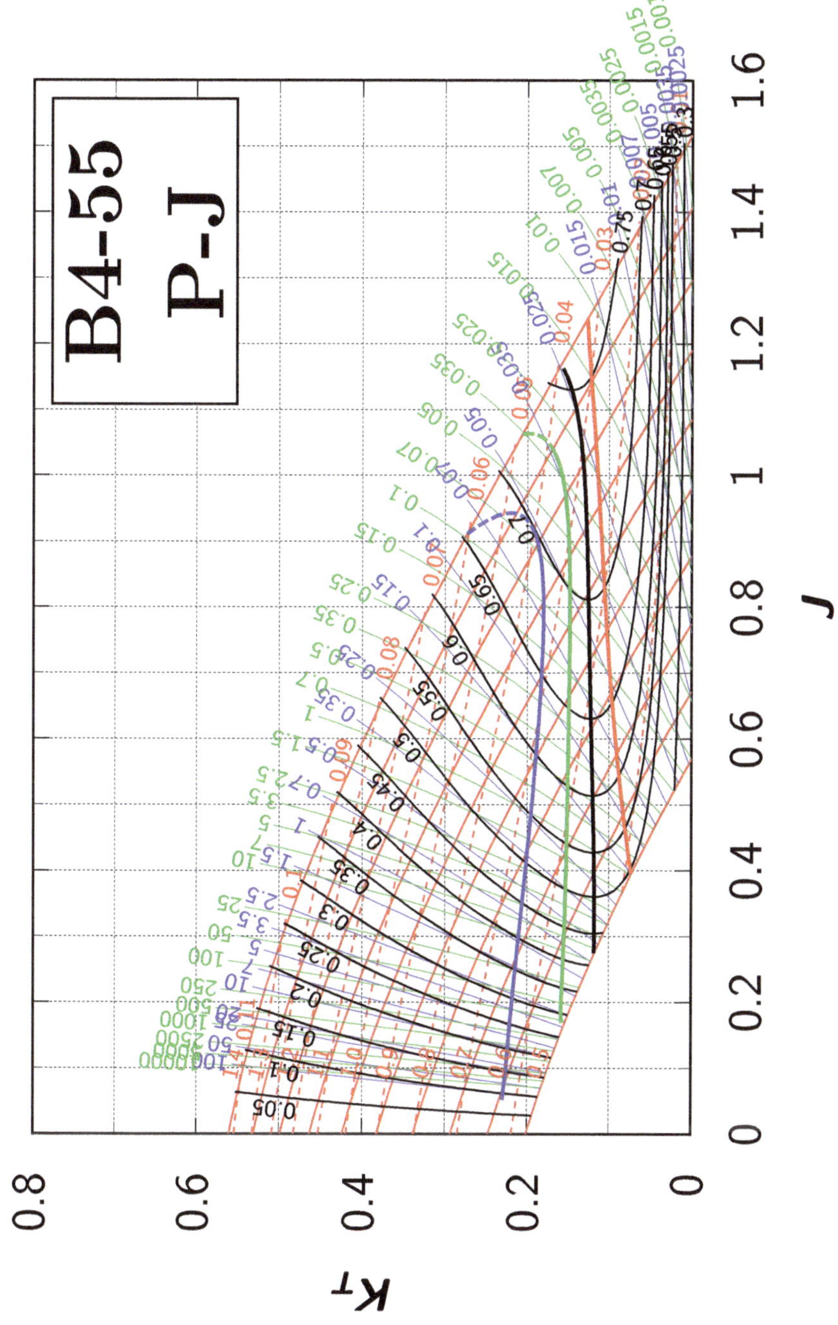
B4-55
P-J
J
K_T
0
0.2
0.4
0.6
0.8
1
1.2
1.4
1.6

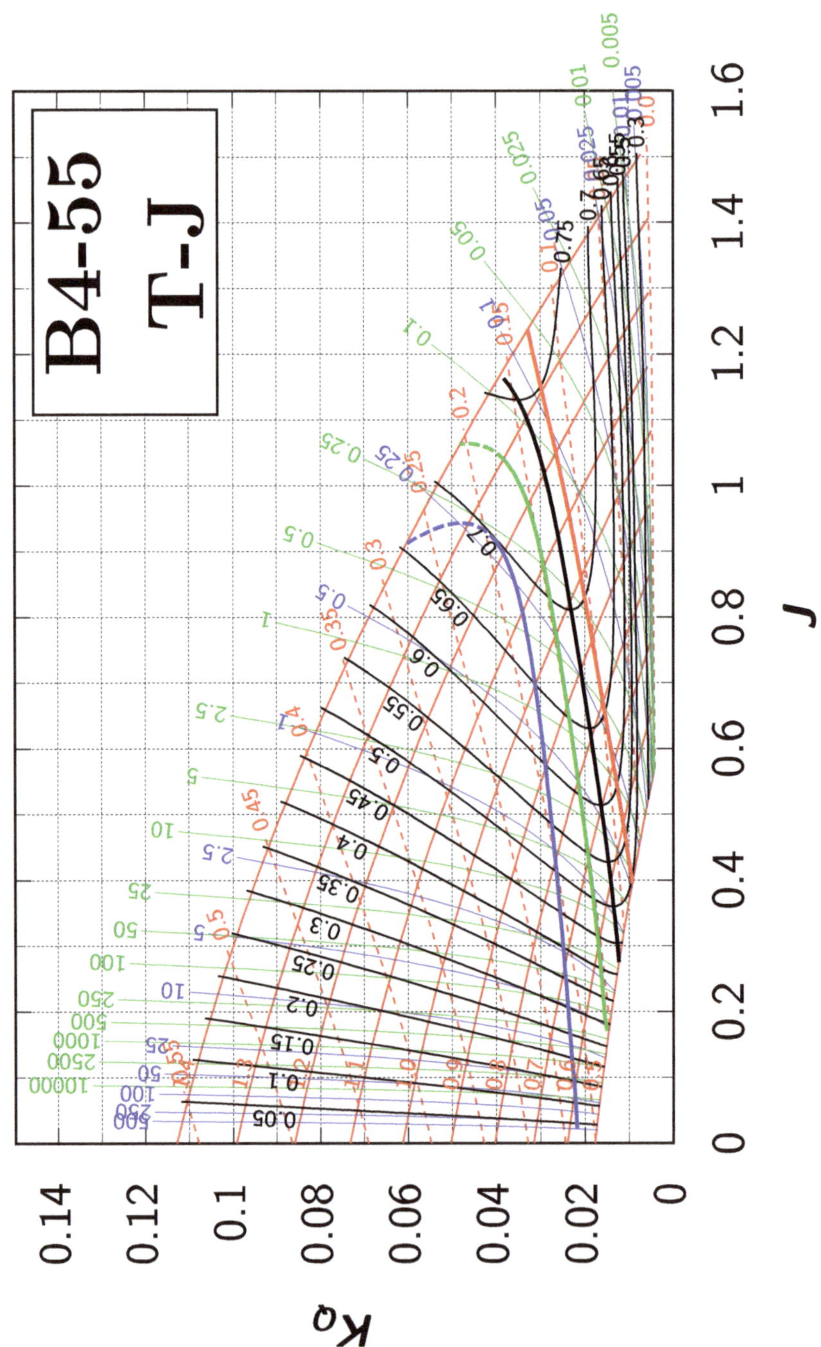
B4-55
T-J
J
K_Q

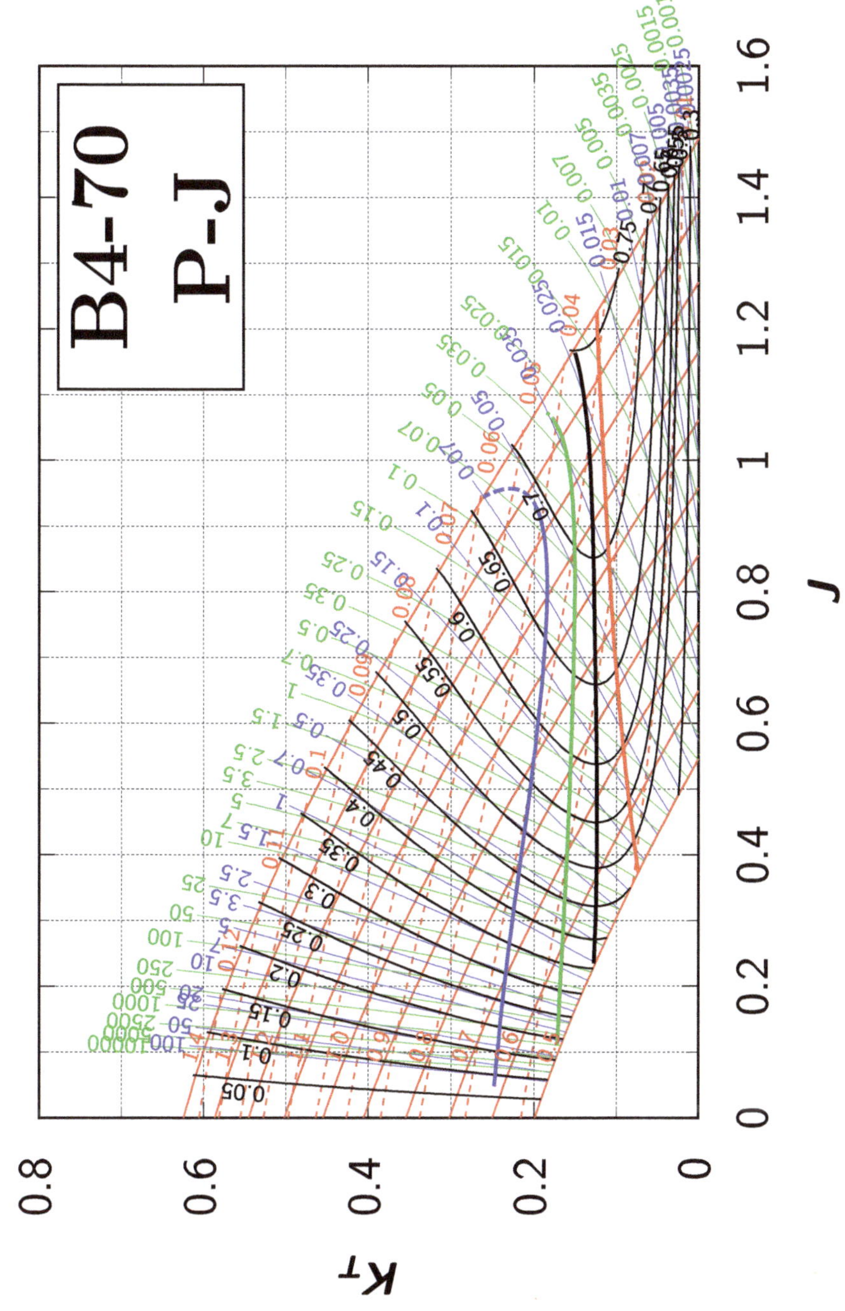

B4-70
P-J
J
K_T
0
0.2
0.4
0.6
0.8
1
1.2
1.4
1.6

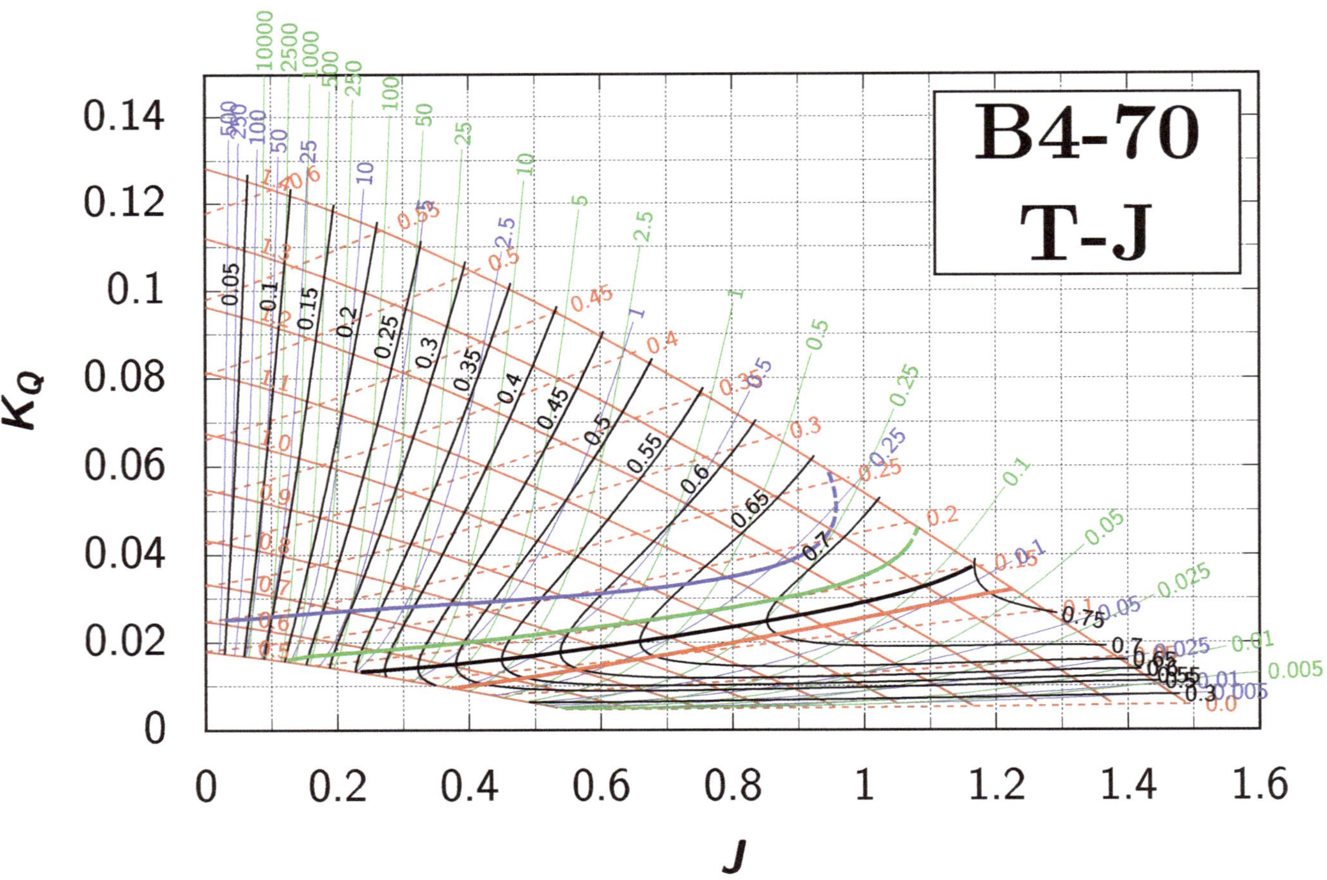

B4-70
T-J
K_Q
J
0.14
0.12
0.1
0.08
0.06
0.04
0.02
0
0.2
0.4
0.6
0.8
1
1.2
1.4
1.6

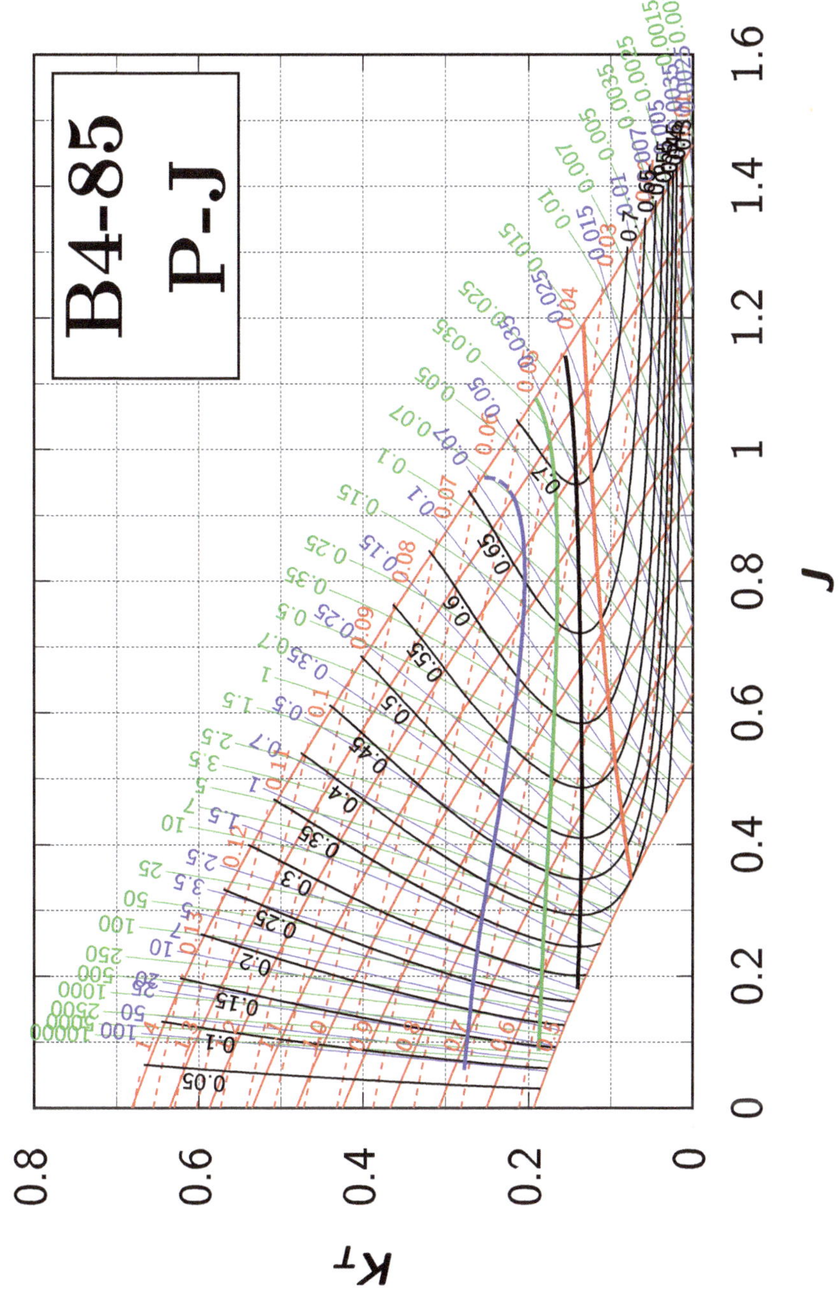
B4-85
P-J
K_T
J

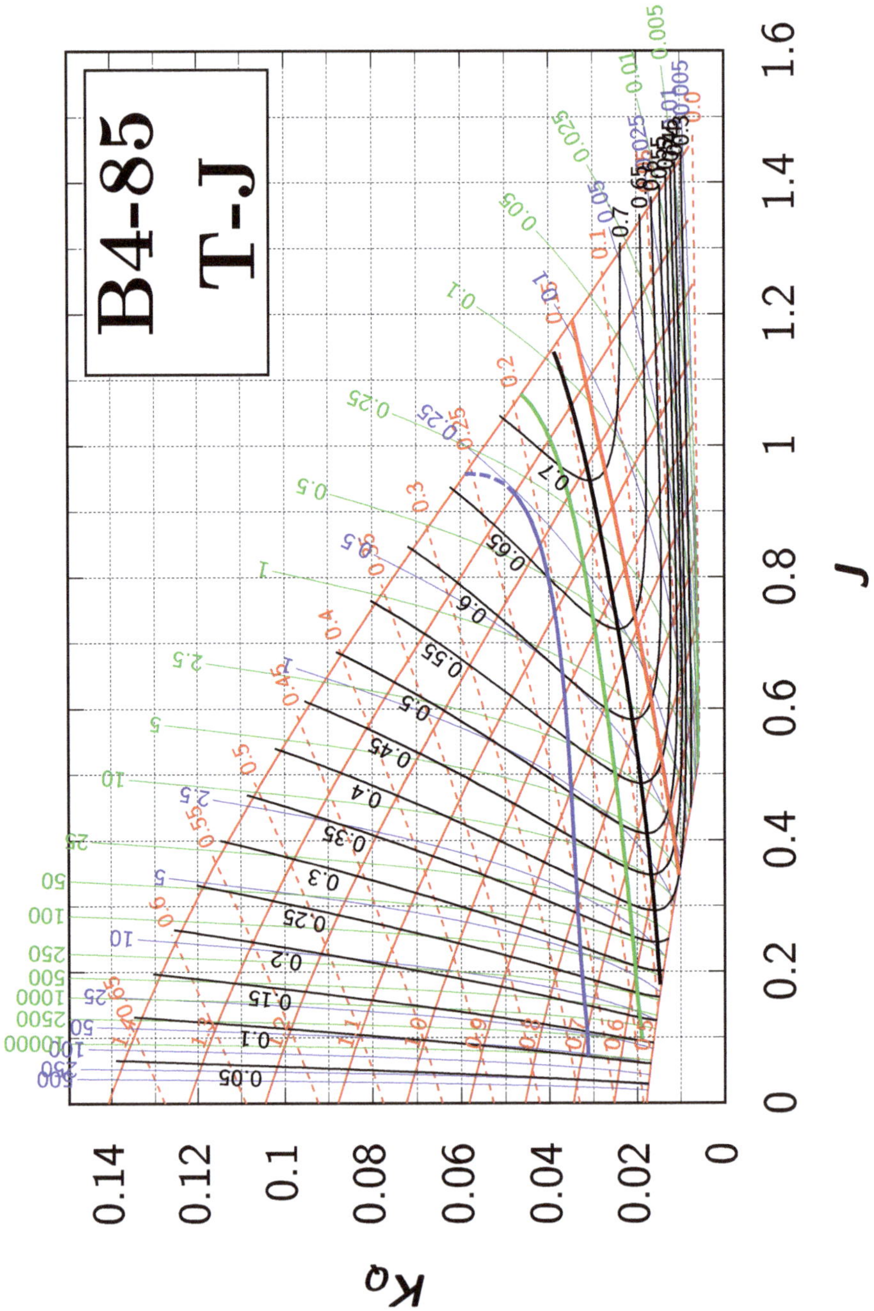
B4-85
T-J
J
K_Q

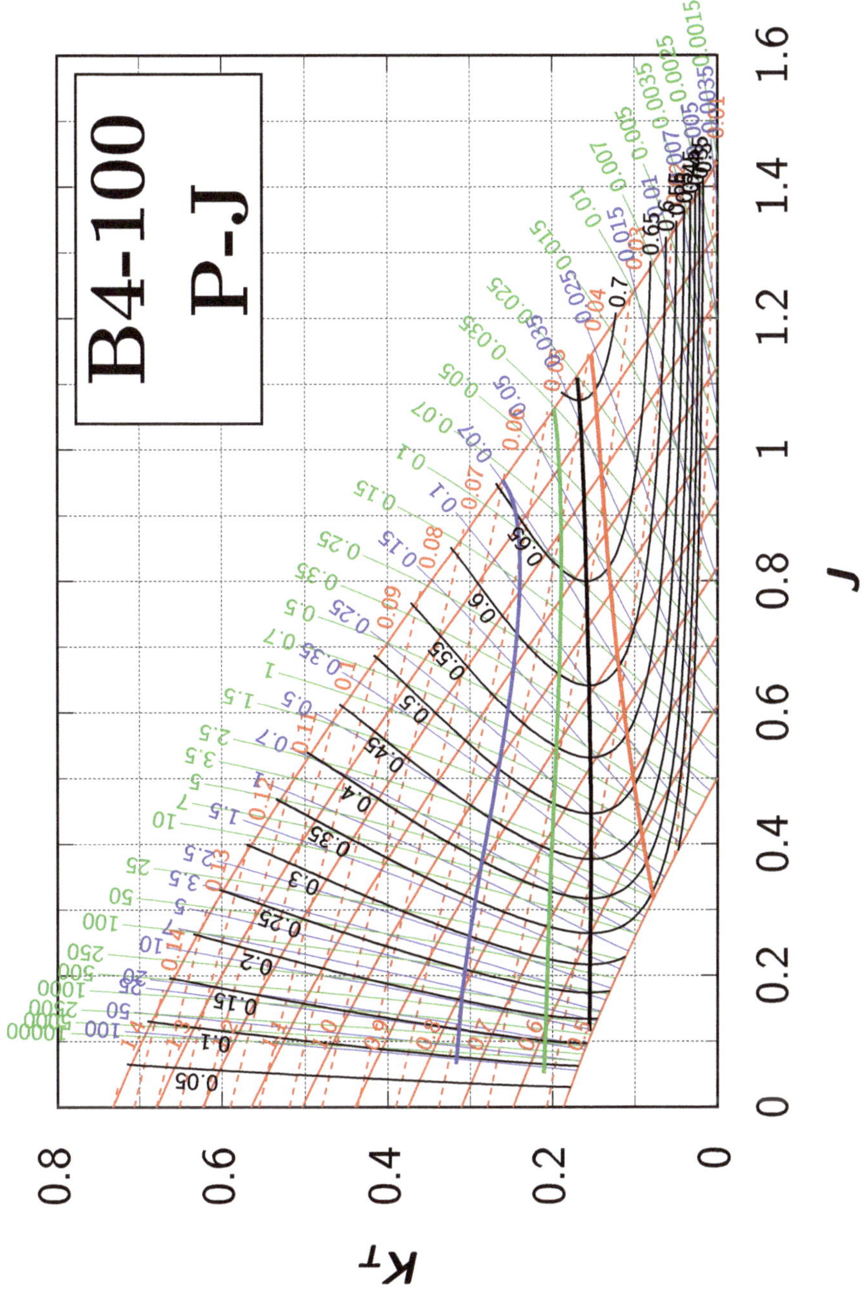
B4-100
P-J
J
K_T
0
0.2
0.4
0.6
0.8
1
1.2
1.4
1.6

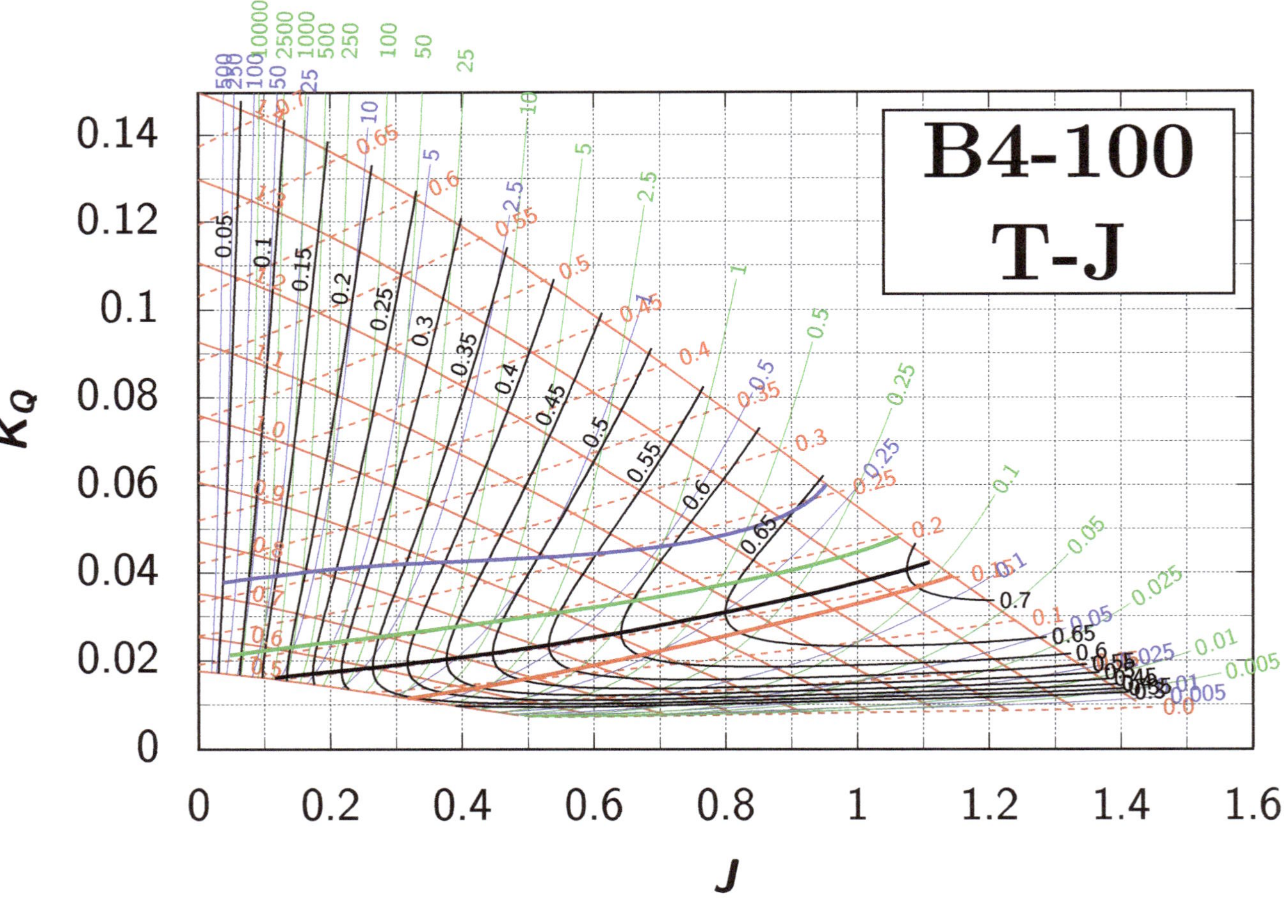
B4-100
T-J
J
K_Q

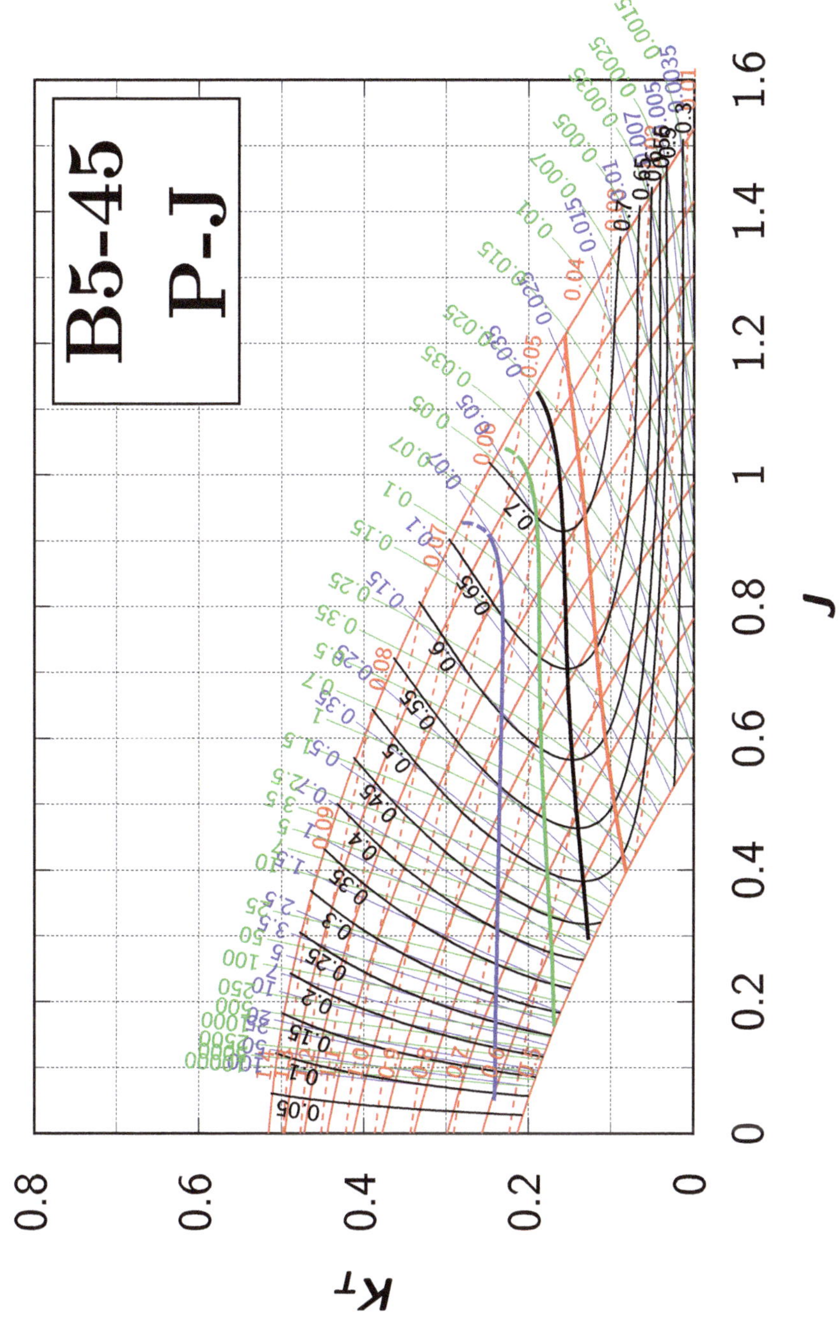
B5-45
P-J
J
K_T

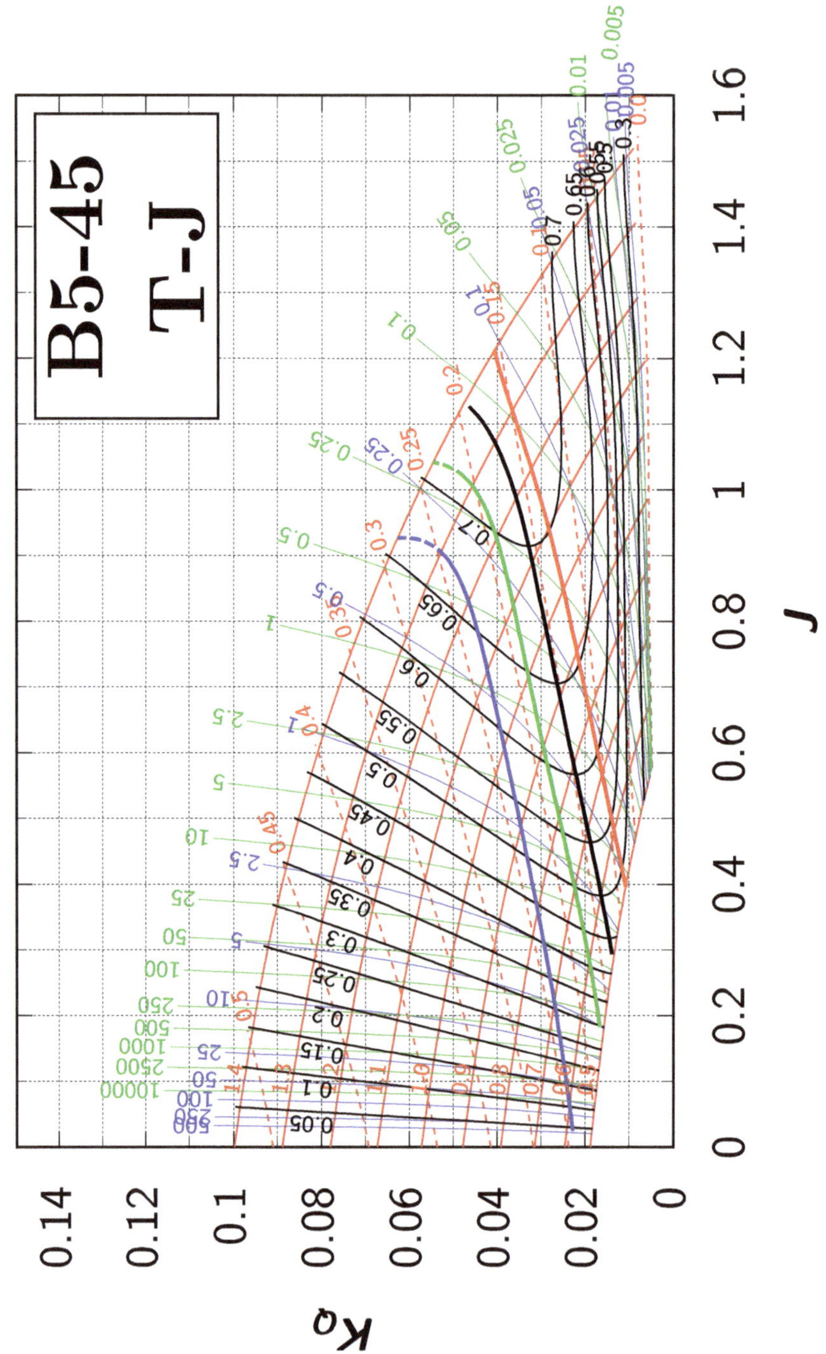
B5-45
T-J
K_Q
J

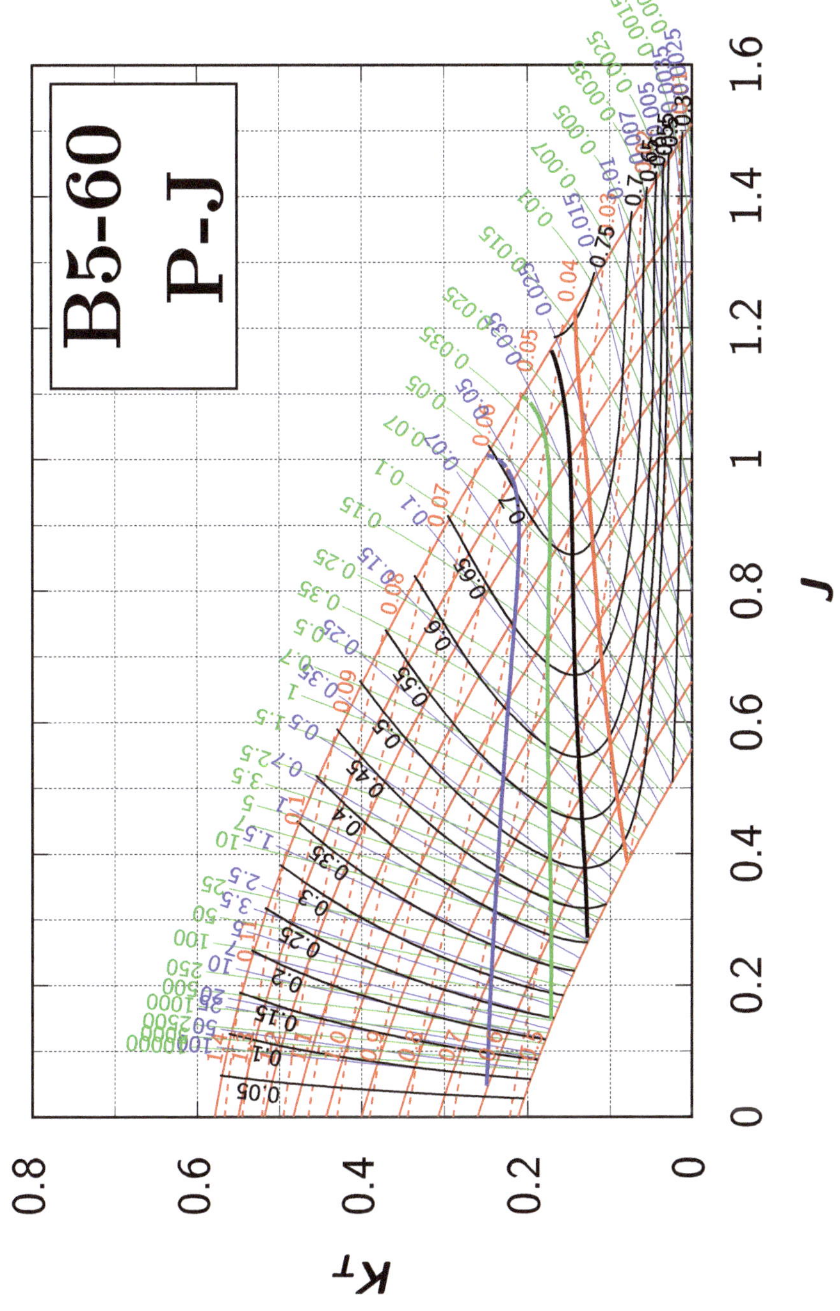
B5-60
P-J
J
K_T

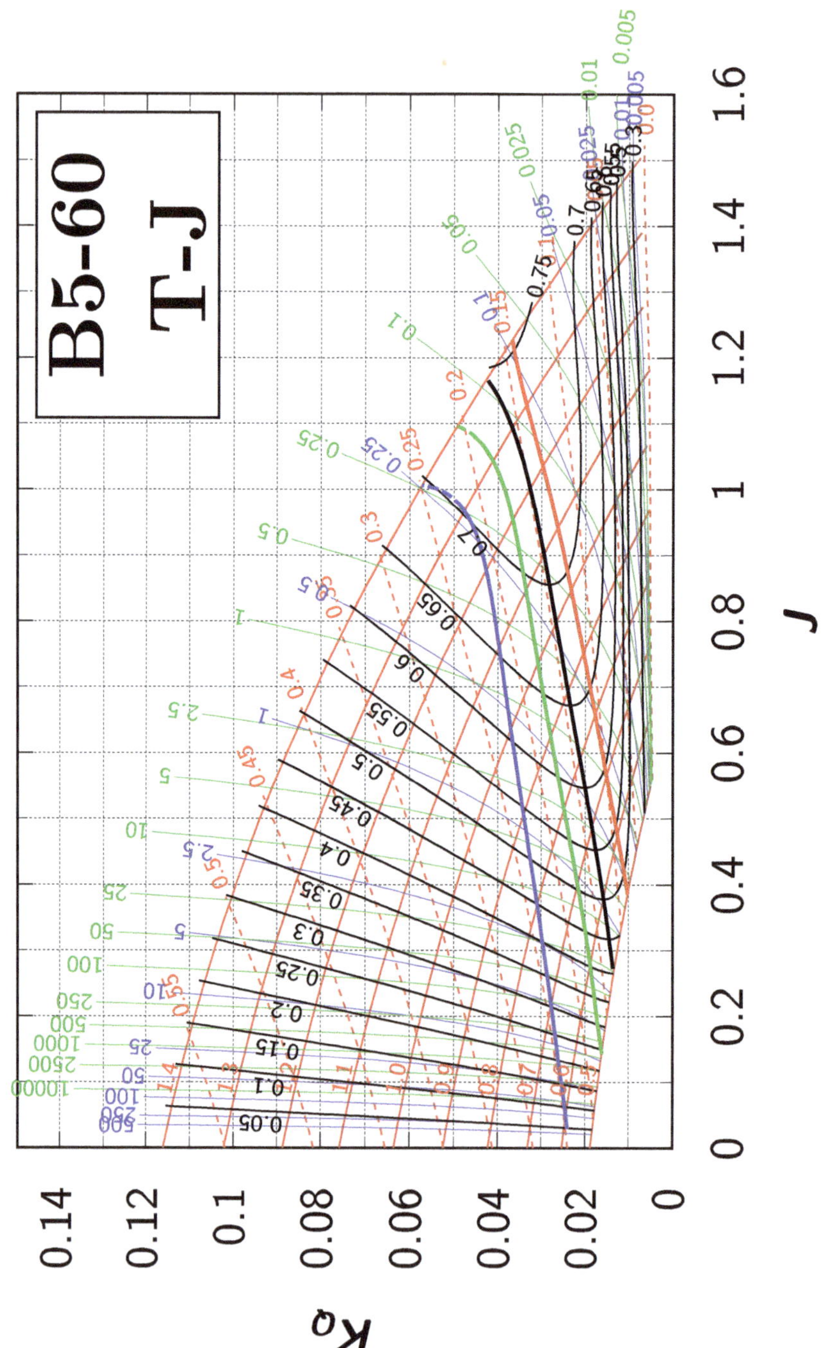
B5-60
T-J
J
K_Q
0 0.2 0.4 0.6 0.8 1 1.2 1.4 1.6
0 0.02 0.04 0.06 0.08 0.1 0.12 0.14

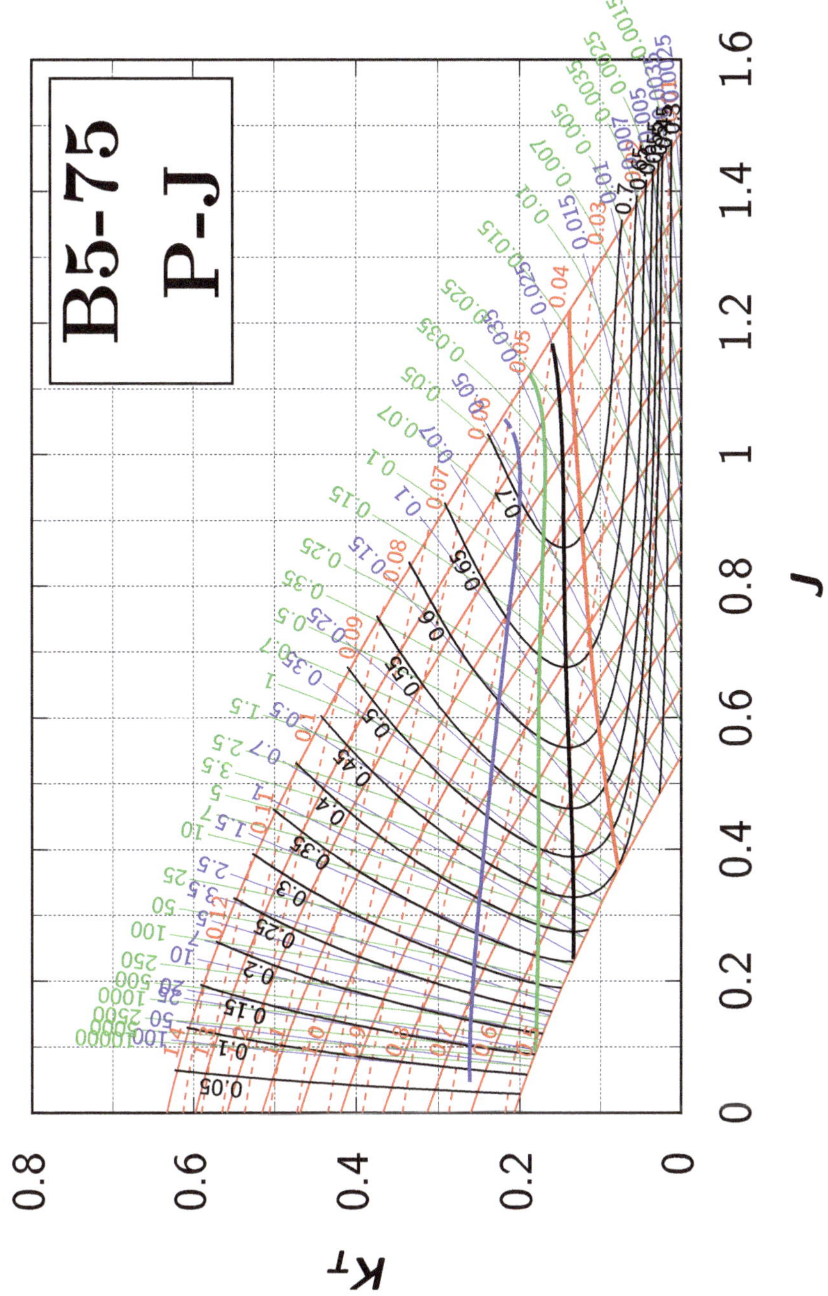
B5-75
P-J
J
K_T

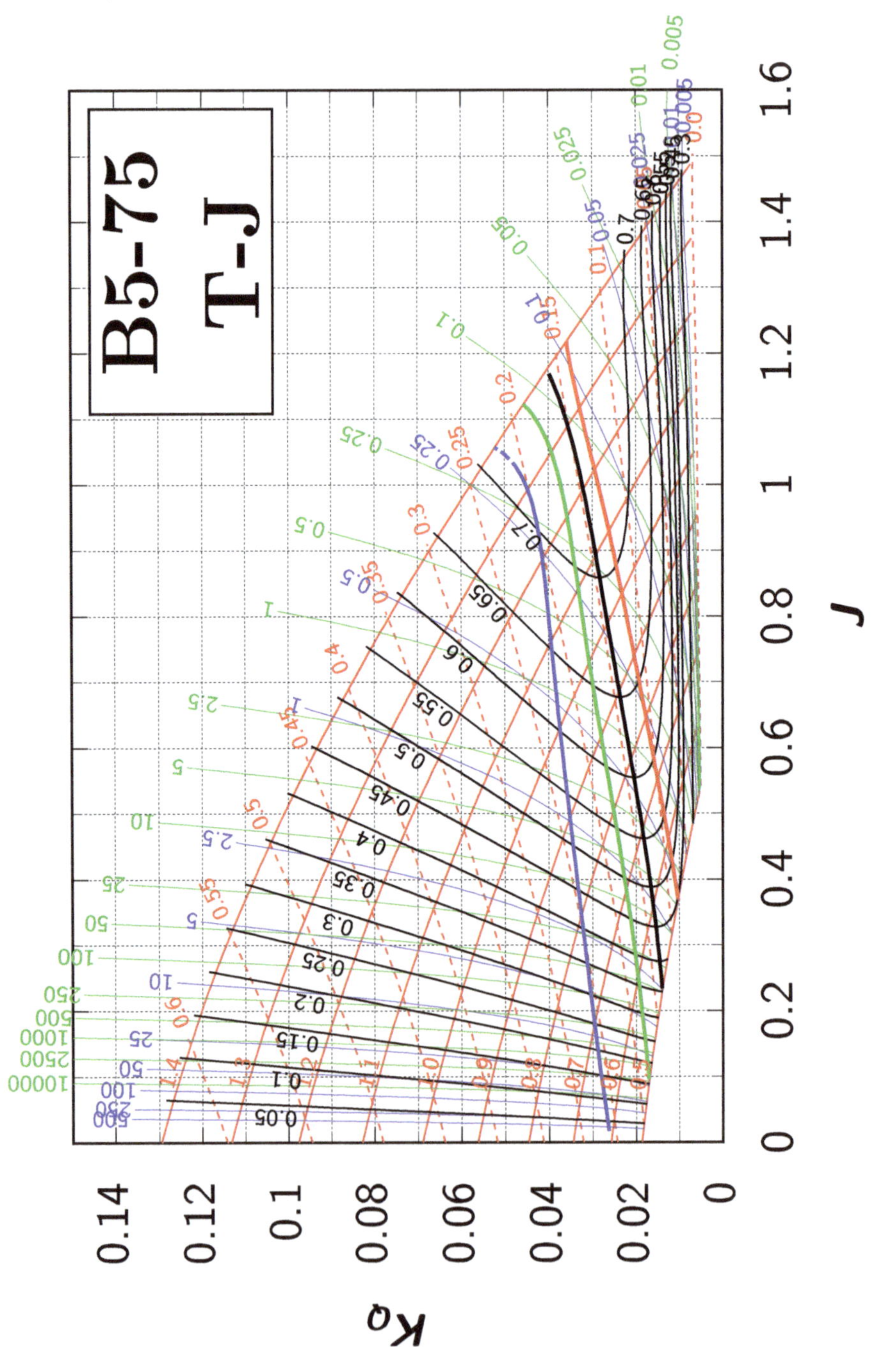

B5-75
T-J
J
K_Q

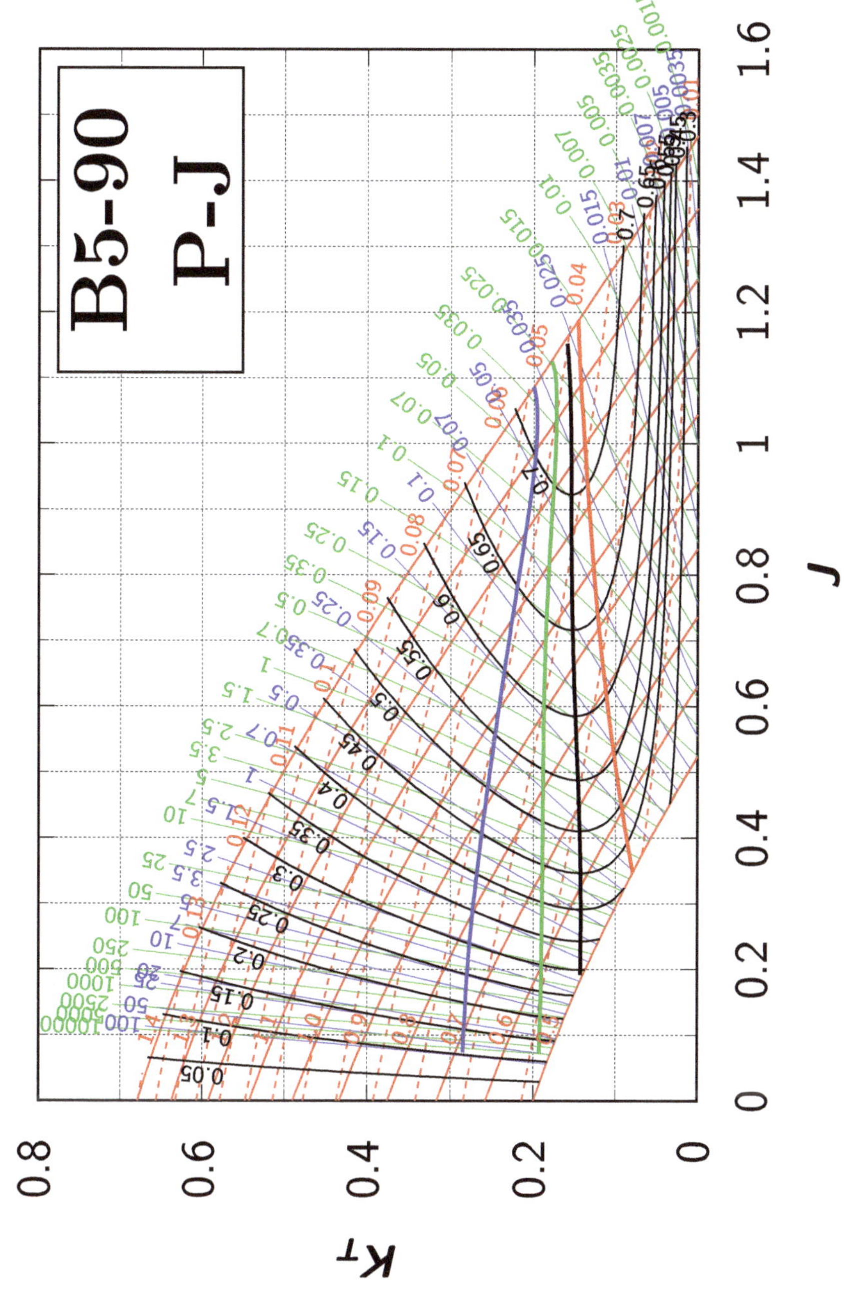
B5-90
P-J
K_T
J
0
0.2
0.4
0.6
0.8
1
1.2
1.4
1.6

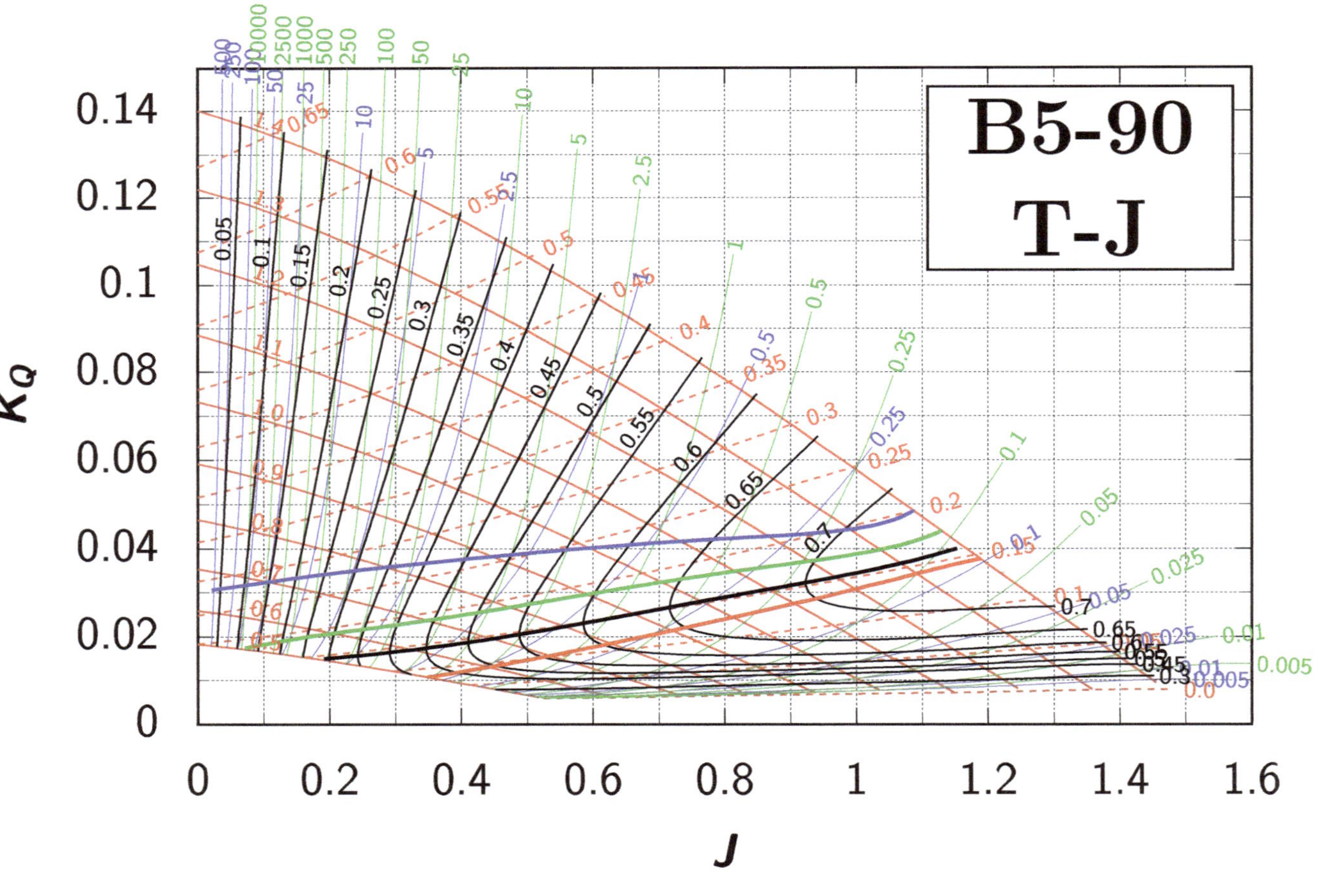

B5-90
T-J
K_Q
J
0.14
0.12
0.1
0.08
0.06
0.04
0.02
0
0
0.2
0.4
0.6
0.8
1
1.2
1.4
1.6

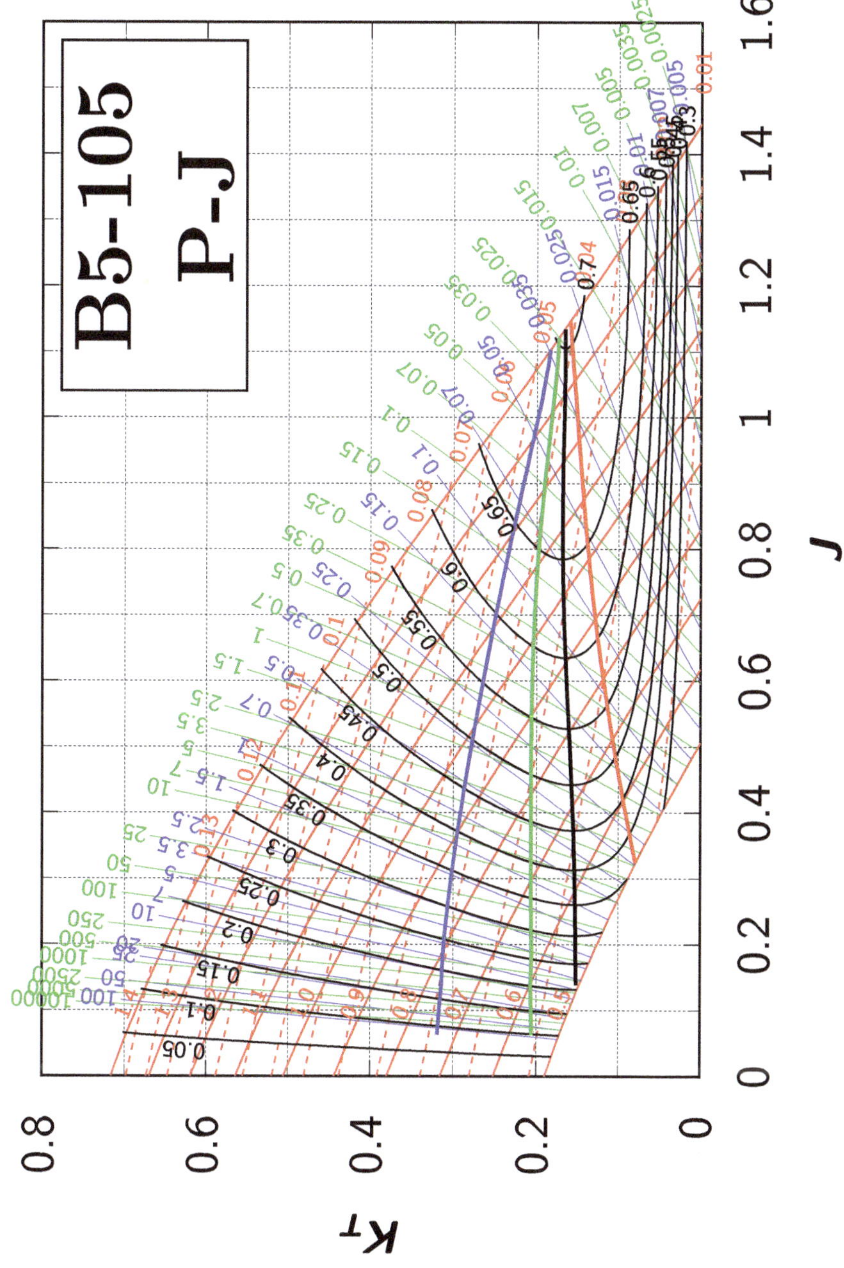
B5-105
P-J
J
K_T

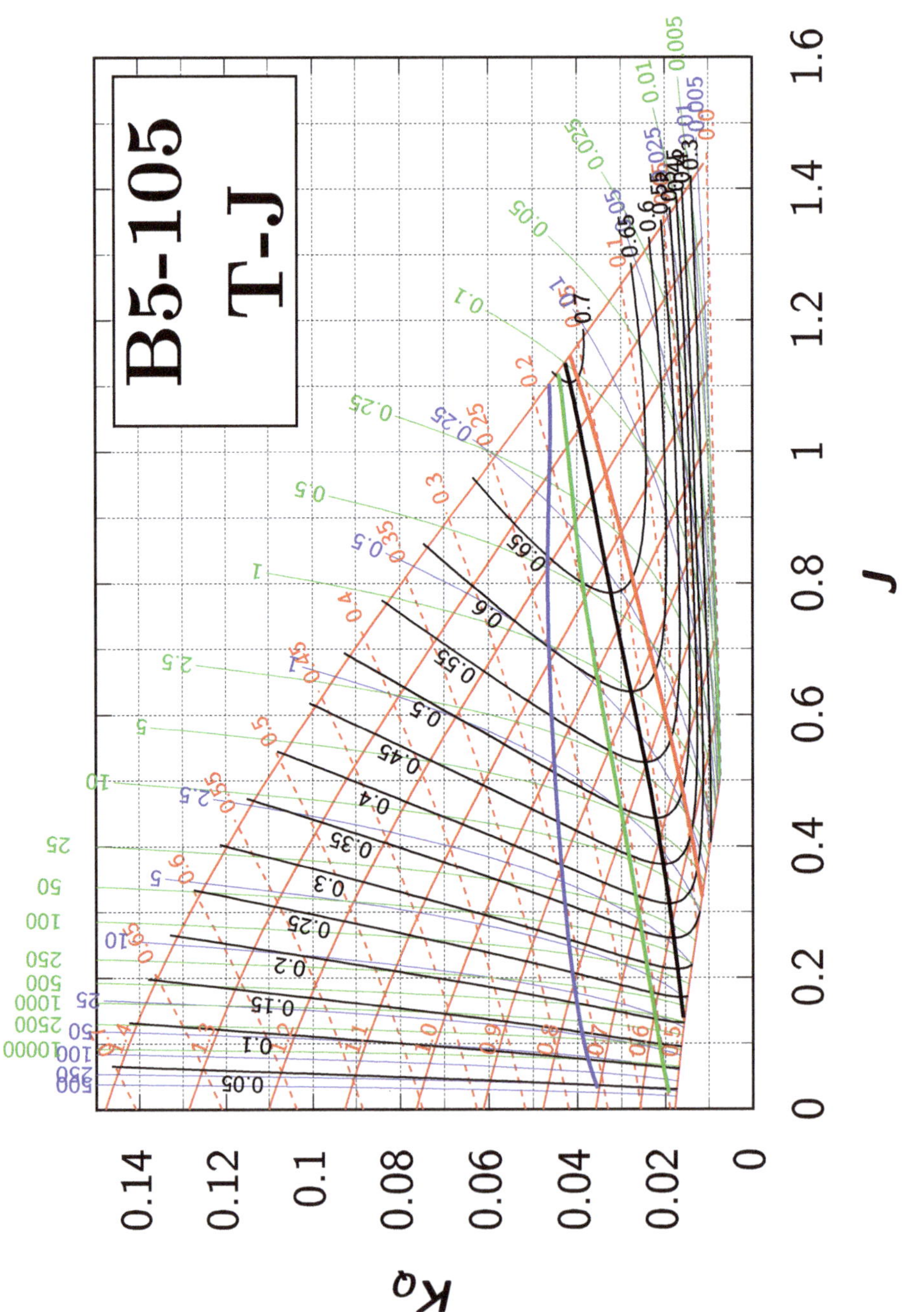

B5-105
T-J
J
K_Q

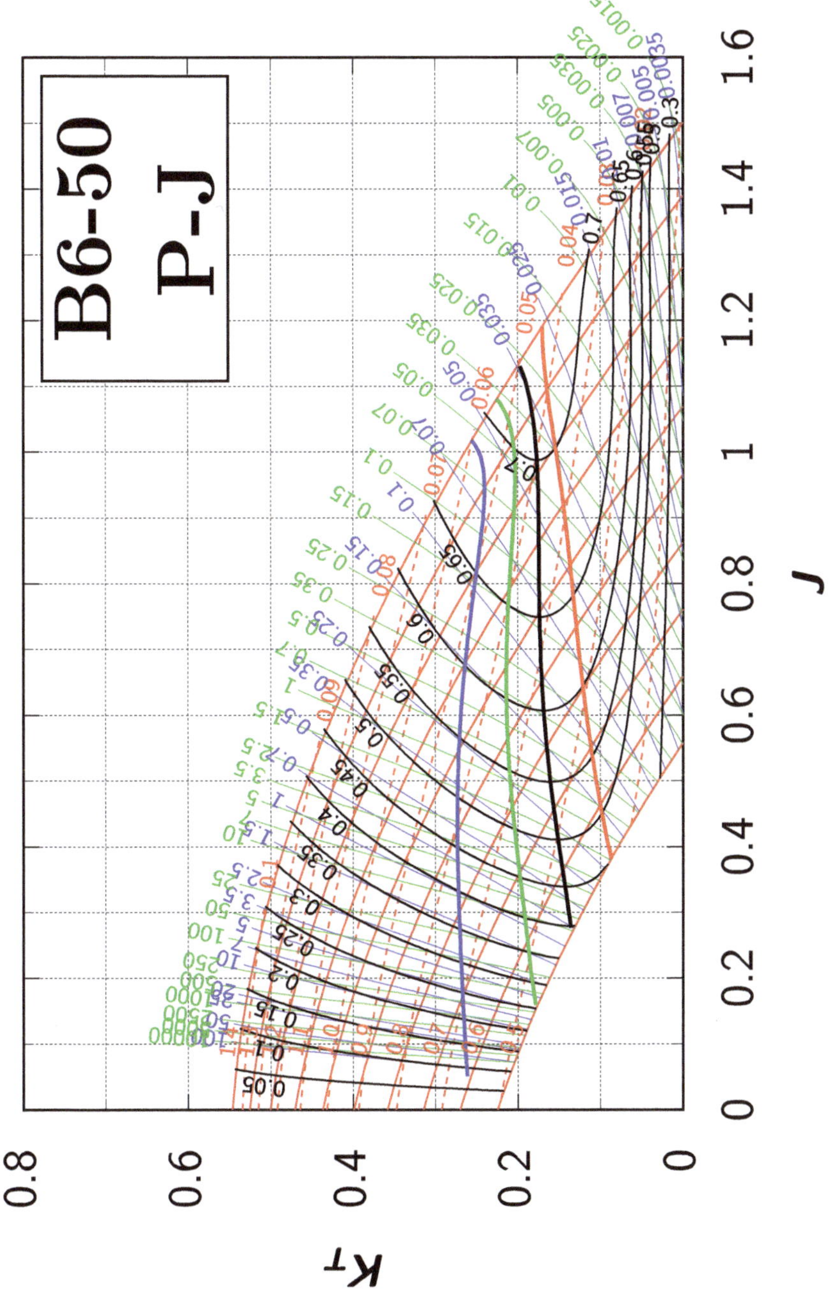
B6-50
P-J
J
K_T
0
0.2
0.4
0.6
0.8
1
1.2
1.4
1.6
0
0.2
0.4
0.6
0.8

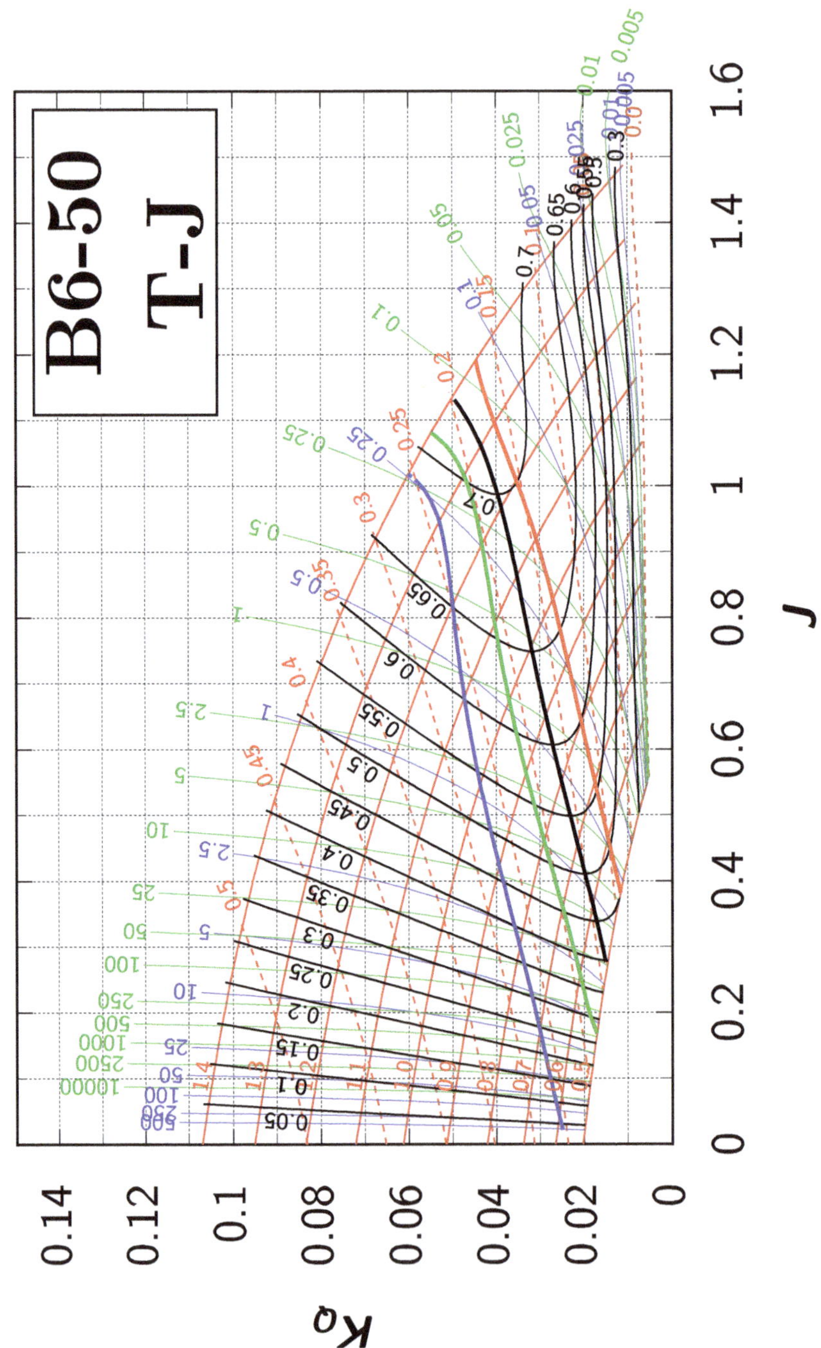
B6-50
T-J
J
K_Q

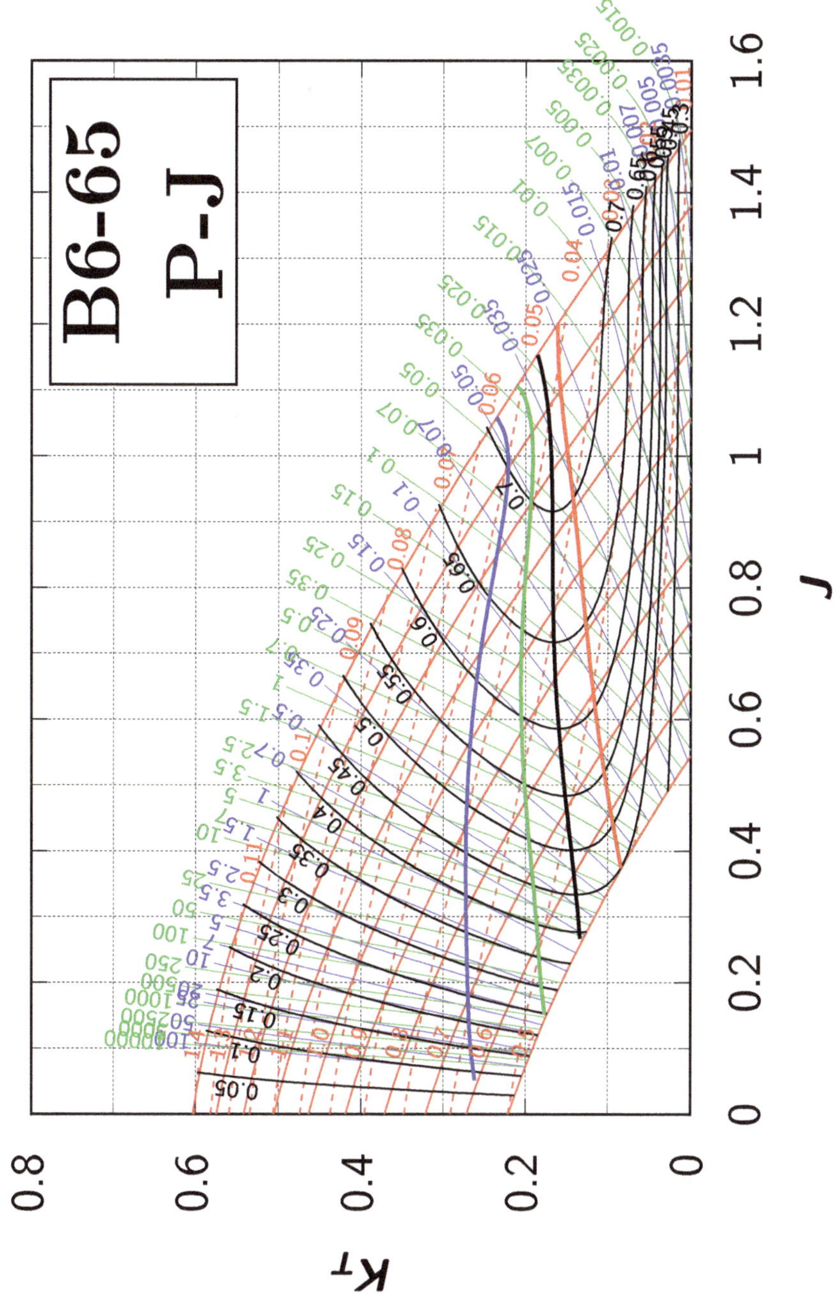
B6-65
P-J
J
K_T

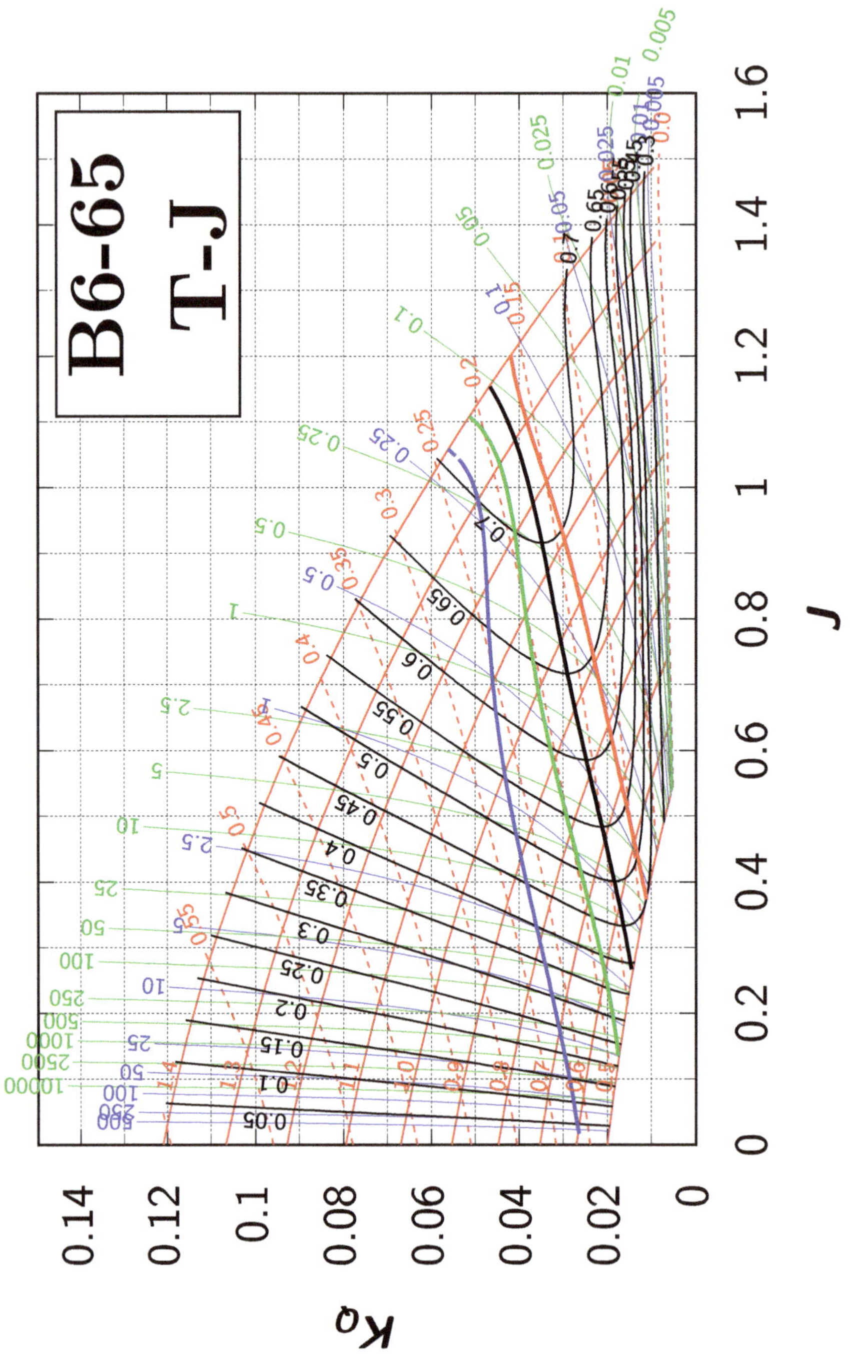
B6-65
T-J
K_Q
J

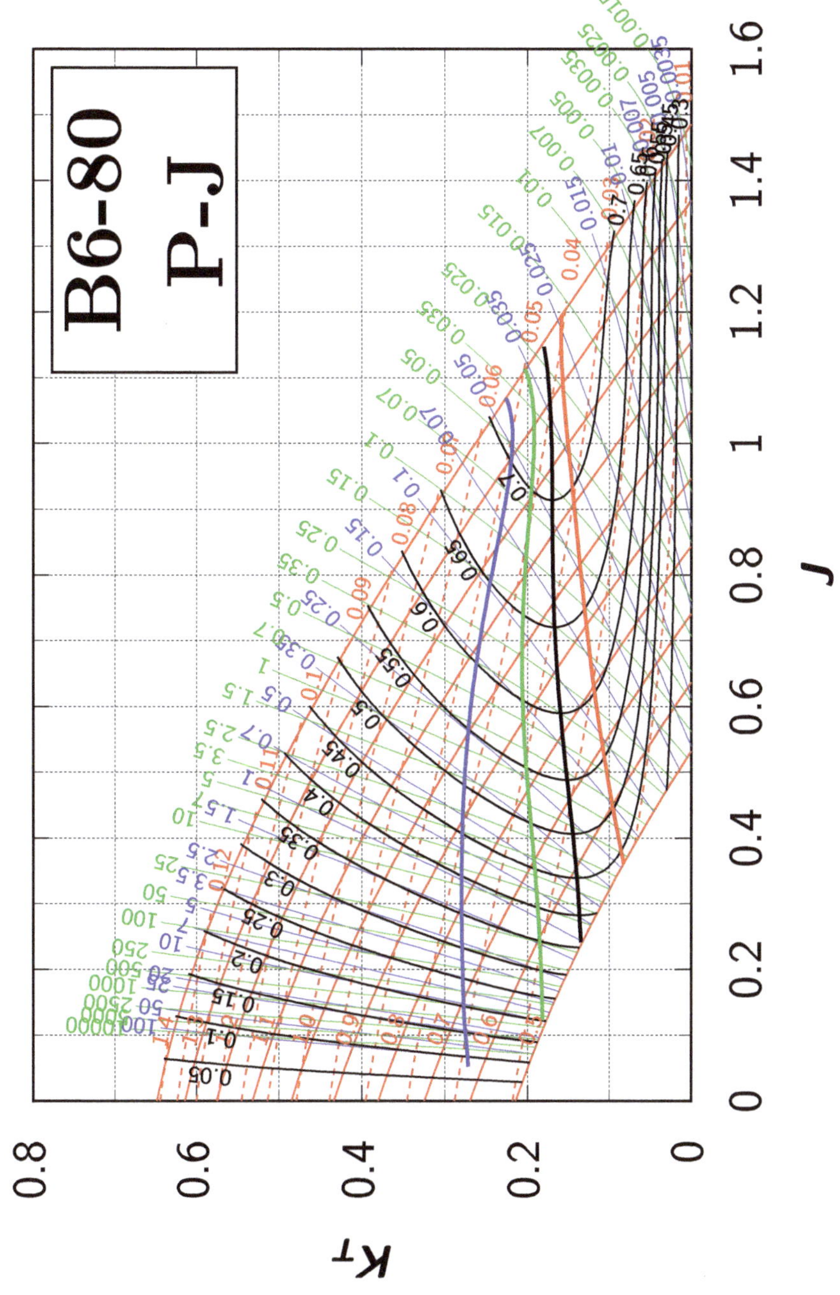

B6-80
P-J
J
K_T

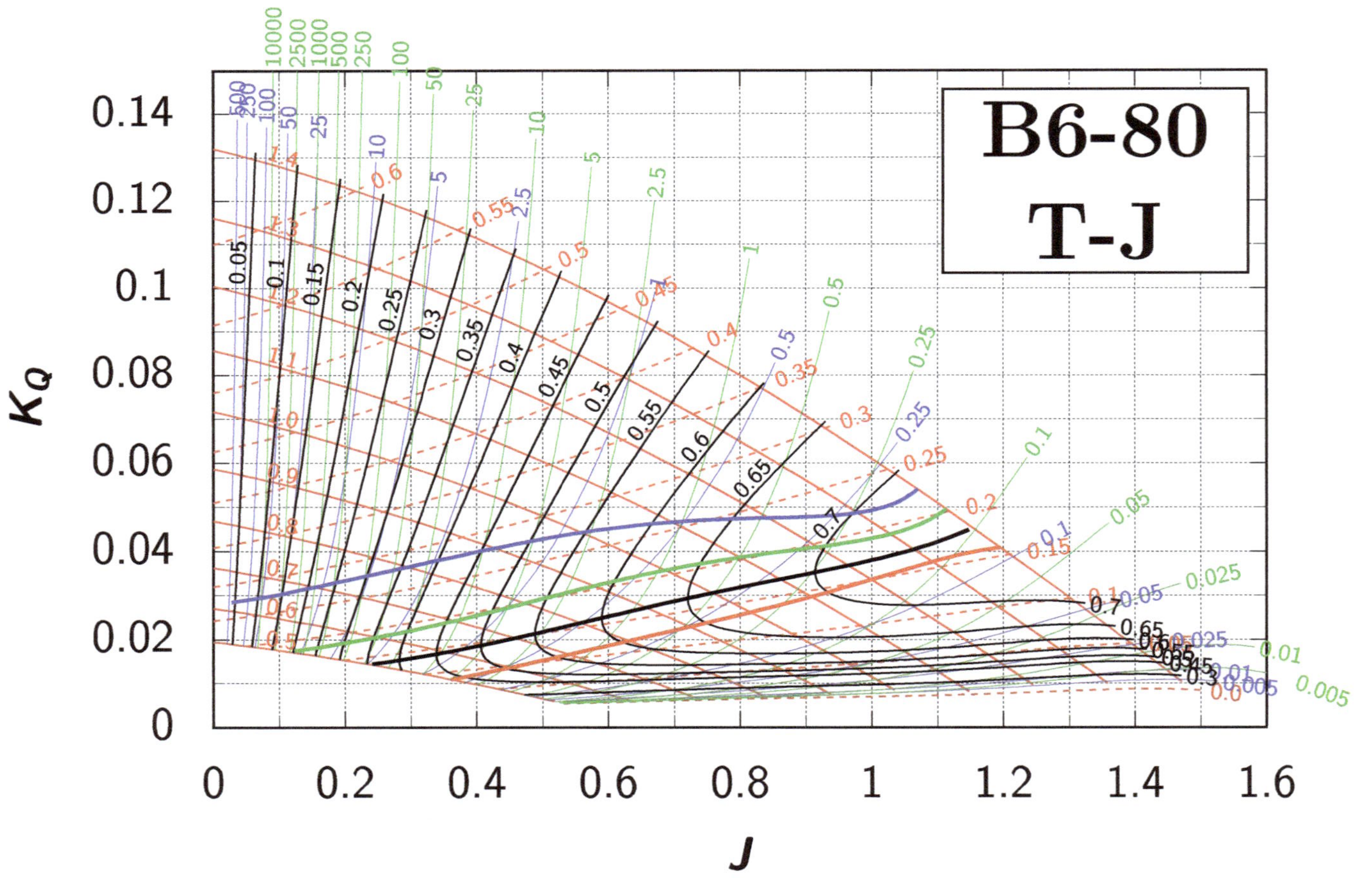
B6-80
T-J
J
K_Q
0
0.2
0.4
0.6
0.8
1
1.2
1.4
1.6
0
0.02
0.04
0.06
0.08
0.1
0.12
0.14

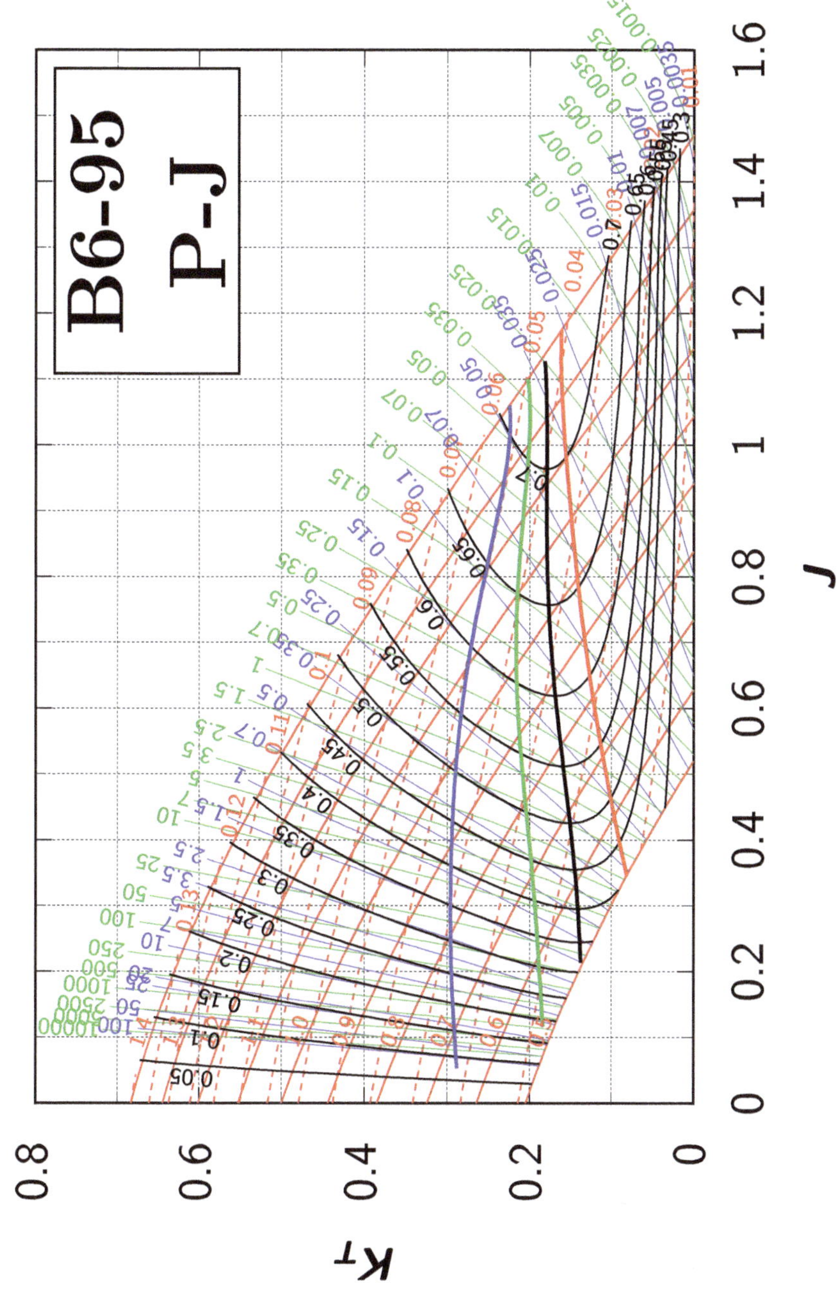

B6-95
P-J
J
K_T

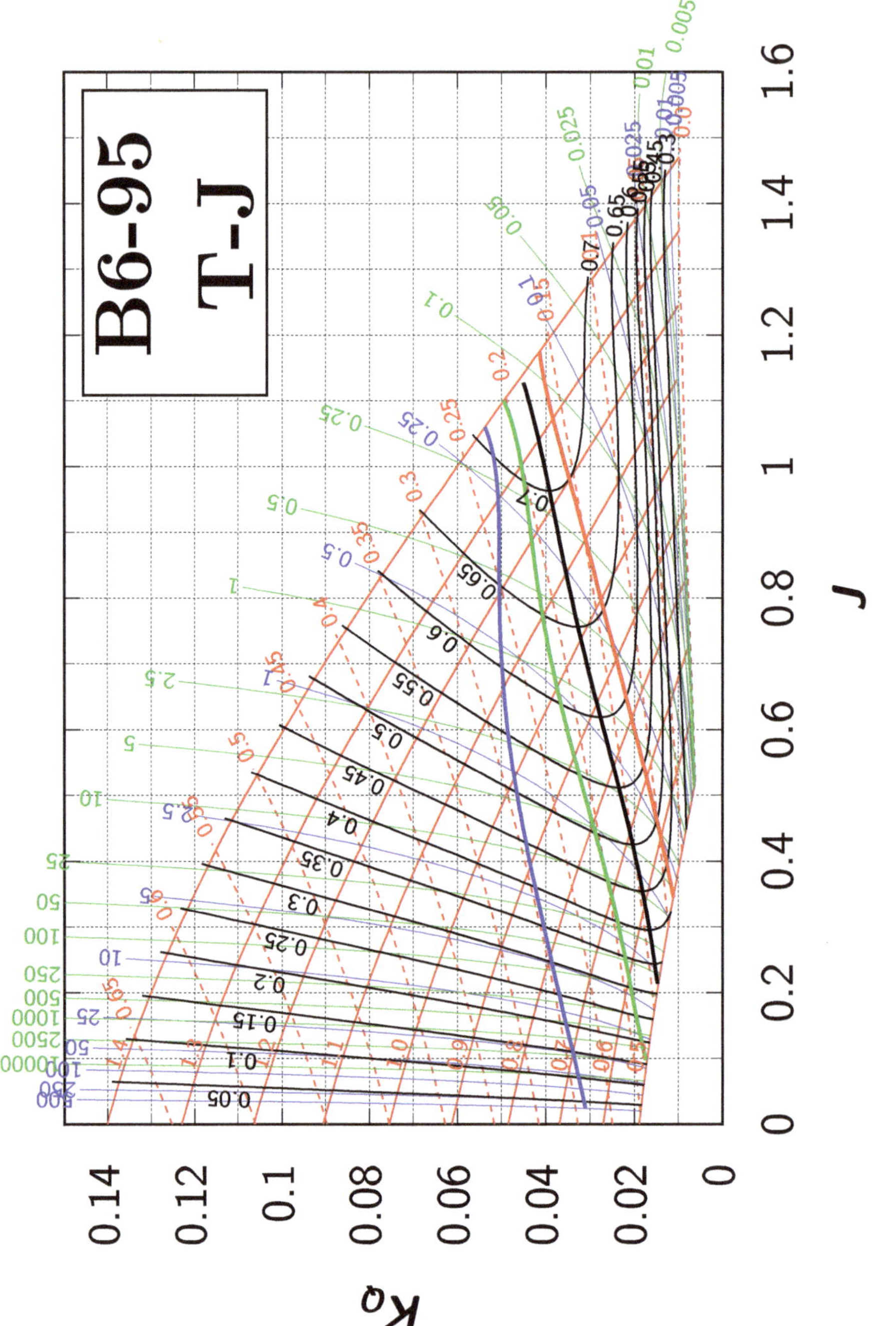
B6-95
T-J
J
K_Q

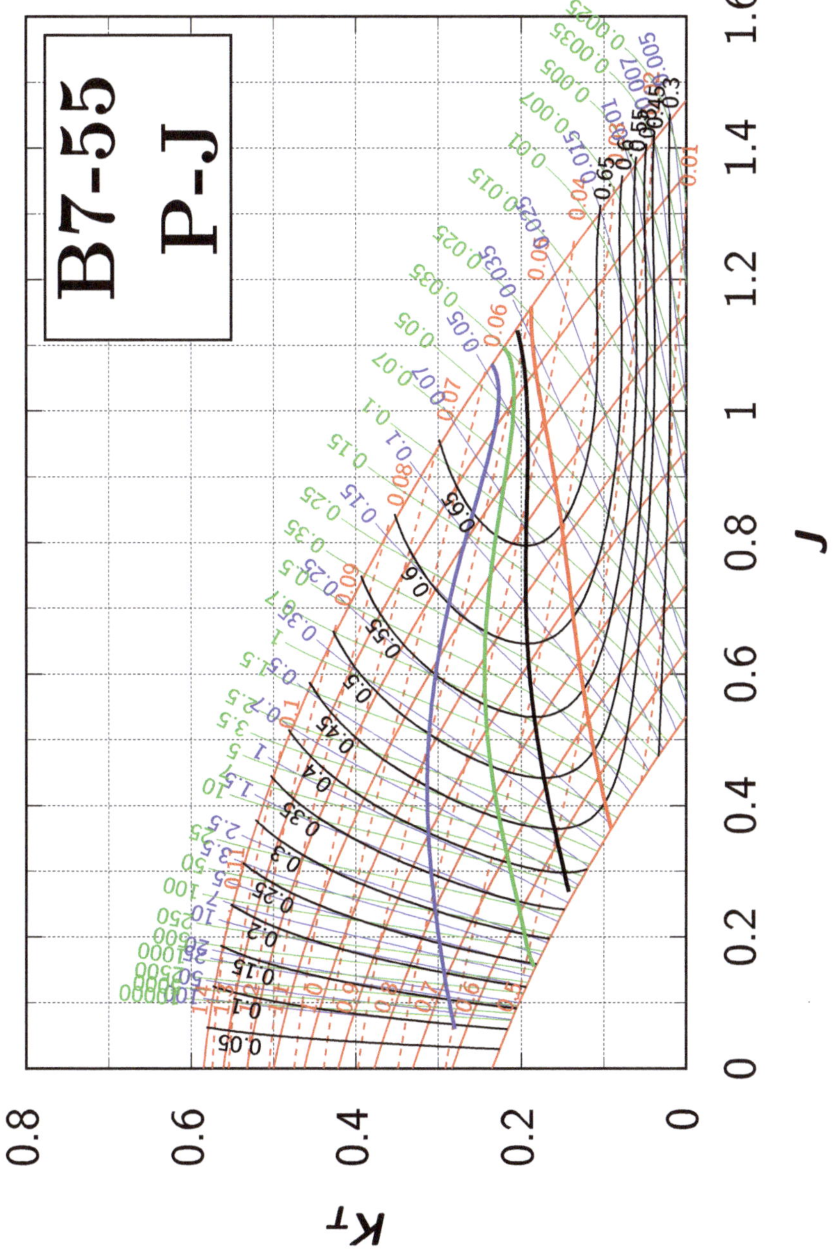
B7-55
P-J
K_T
J
0.8
0.6
0.4
0.2
0
0
0.2
0.4
0.6
0.8
1
1.2
1.4
1.6

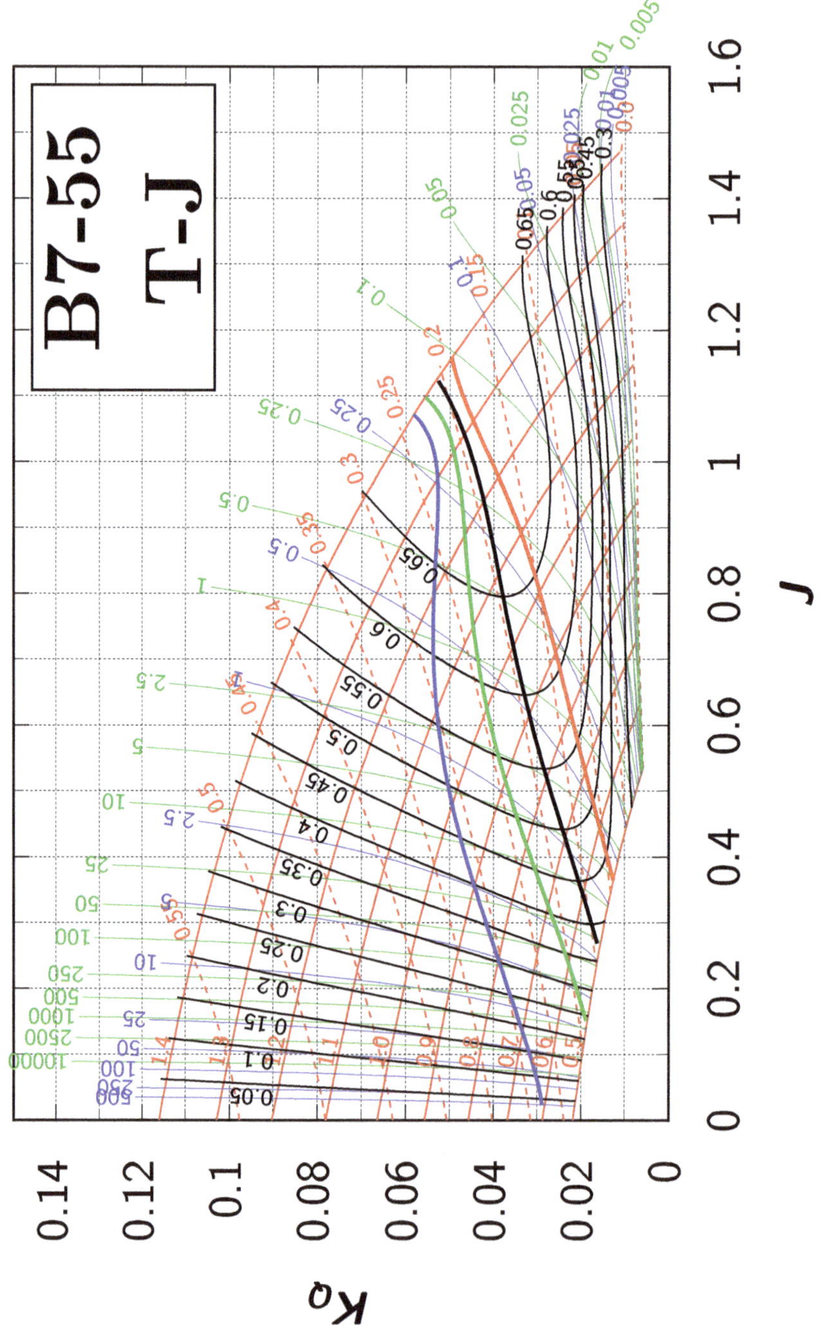
B7-55
T-J
J
K_Q
0
0.2
0.4
0.6
0.8
1
1.2
1.4
1.6
0.02
0.04
0.06
0.08
0.1
0.12
0.14

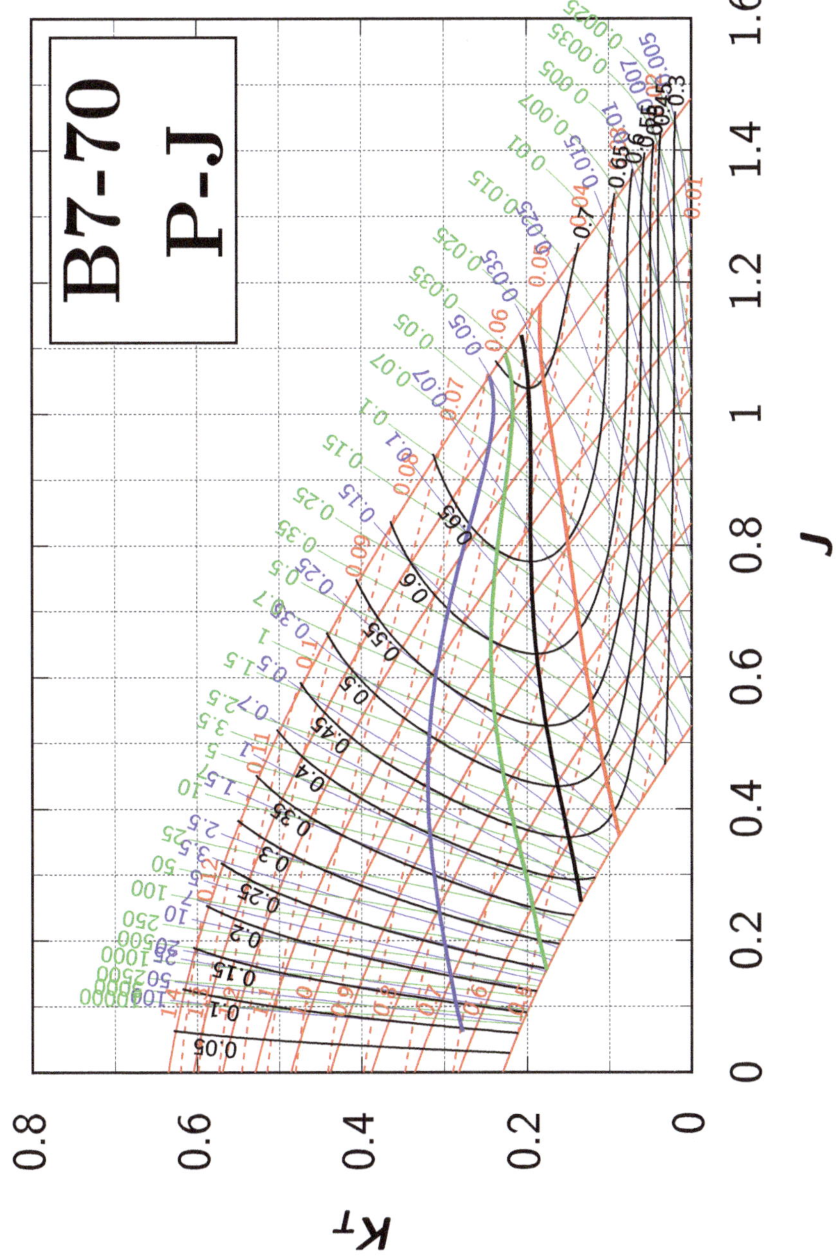
B7-70
P-J
J
K_T

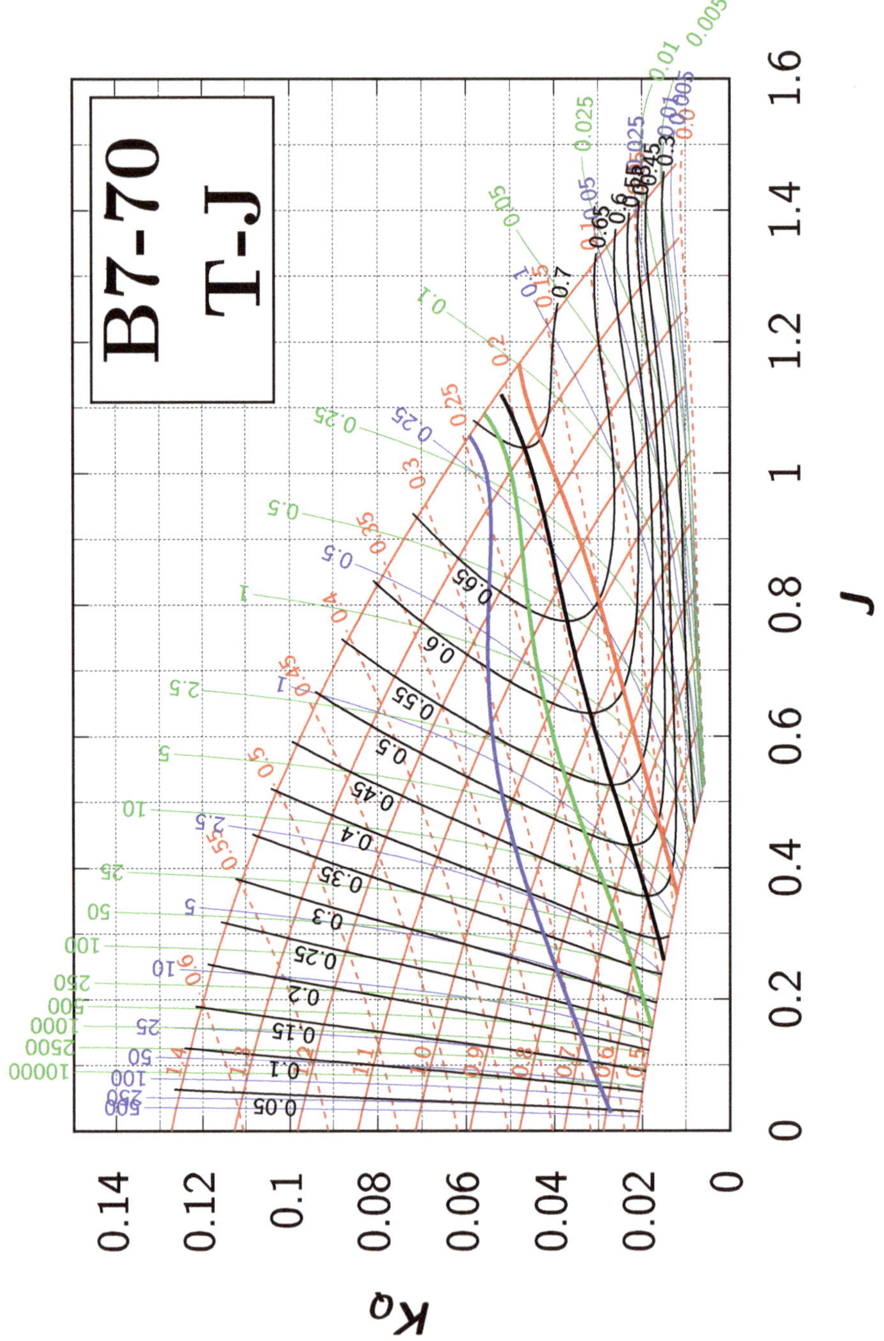
B7-70
T-J
J
K_Q

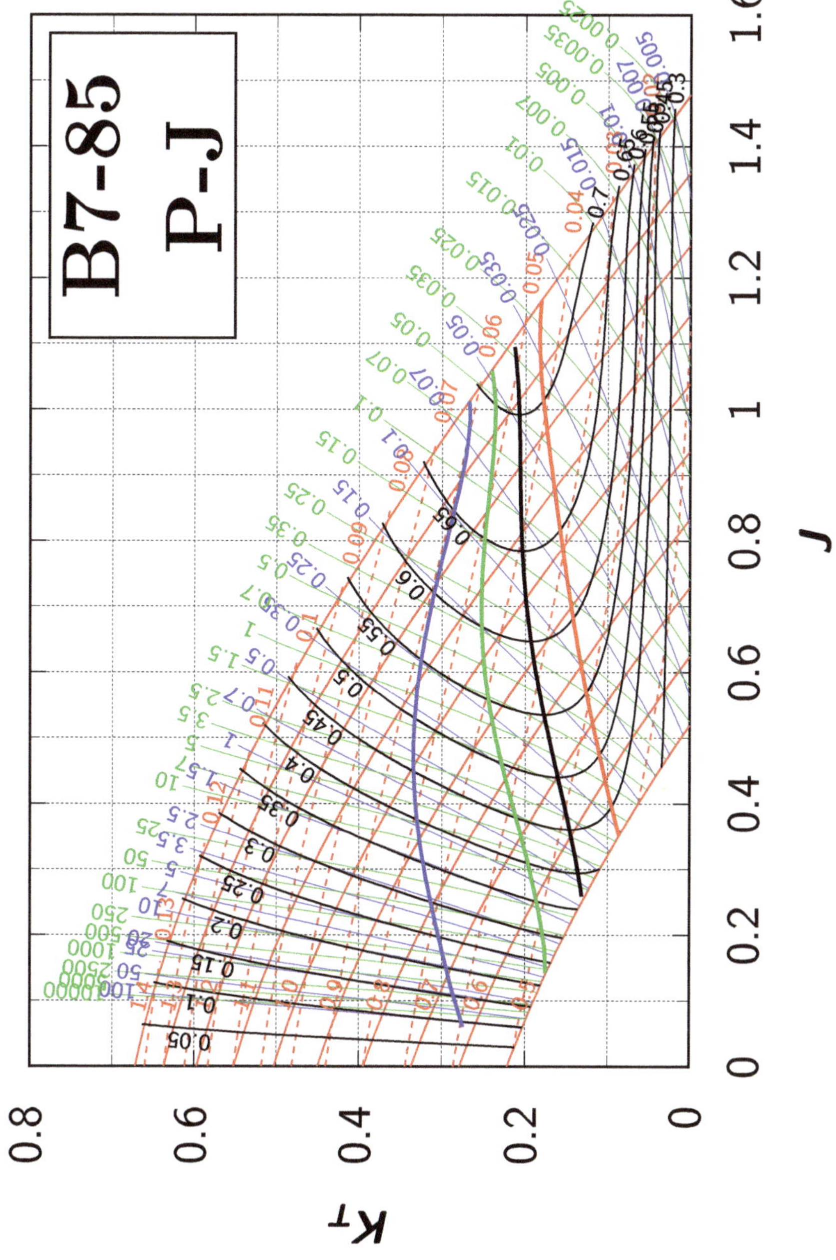
B7-85
P-J
J
K_T

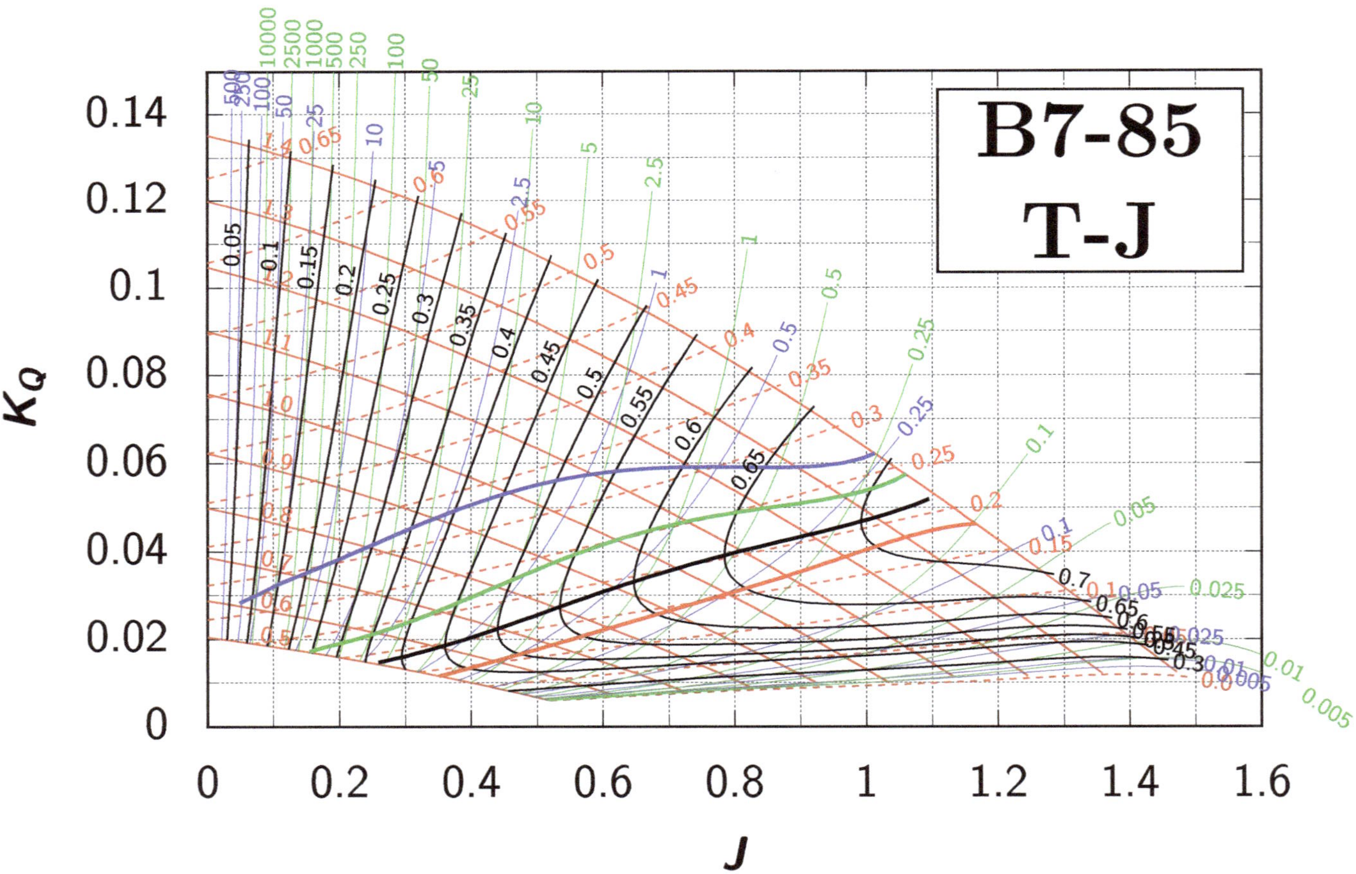

B7-85
T-J
K_Q
J
0.14
0.12
0.1
0.08
0.06
0.04
0.02
0
0.2
0.4
0.6
0.8
1
1.2
1.4
1.6

References

1. Resultaten van systematische proeven met vrijvarenden 4-bladige schroeven, type A.4-40. *Het Ship* **1936**, *140*. (cited in [4]).
2. Oosterveld, M.W.C.; van Oossanen, P. Further Computer-Analysed Data of the Wageningen B-screw Series. *Int. Shipbuild. Prog.* **1975**, *22*, 251–262. [CrossRef]
3. Helma, S. Wageningen B-Series revisited. In Proceedings of the 5th International Conference on Advanced Model Measurement Technology for The Maritime Industry (AMT'17), Glasgow, UK, 11–13 October 2017; Strathclyde University: Glasgow, UK, 2017.
4. van Lammeren, W.P.A.; van Aken, J.A. Een Uitbreiding van de Systematische 3- en 4-bladige Schroefseries van het Nederlandsch Scheepsbouwkundig Proefstation. *Schip en Werf* **1949**, *13*, 264–281.
5. van Lammeren, W.P.A.; van Manen, J.D.; Oosterveld, M.W.C. The Wageningen B-screw Series. *Trans. SNAME* **1969**, *77*, 269–317.
6. Danckwardt, E. Berechnungsdiagramme für Schiffsschrauben. *Schiffbautechnik* **1956**, *2*, 22–30.
7. Helma, S. Surprising behaviour of the Wageningen B-screw Series polynomials. In Proceedings of the Sixth International Symposium on Marine Propulsors (smp'17), Rome, Italy, 27–30 May 2019.
8. 28th ITTC Quality Systems Group. ITTC Symbols and Terminology List. 2017. Available online: https://ittc.info/media/7937/structured-list_2017_a.pdf (accessed on 6 February 2010).
9. Yosifov, K.; Zlatev, Z.; Staneva, A. Optimum Characteristics Equations for the 'K-J' Propeller Design Charts, based on the Wageningen B-screw Series. *Int. Shipbuild. Prog.* **1986**, *33*, 101–111. [CrossRef]
10. Bendemann, F.; Madelung, G. Praktische Schraubenberechnung, 15. Bericht der Deutschen Versuchsanstalt für Luftfahrt. *Tech. Ber. Flugz.* **1917**, *2*, 53–80.
11. von der Steinen, C. Praktische Schraubendiagramme. *Werft-Reederei-Hafen* **1923**, *4*, 61–72.
12. Papmel, E.E. *Practical Design of the Screw Propeller*; cited from [9]; Leningrad, Russia, 1936. (In Russian)
13. Schoenherr, K. Propulsion and propellers. In *Principles of Naval Architecture*; Rossell, H.E., Chapman, L.B., Eds.; Society of Naval Architects and Marine Engineers (SNAME): New York, NY, USA, 1949; pp. 158–168.
14. Yosifov, K.Y.; Staneva, A.D. *Propeller Design Charts*; Bulgarian Ship Hydrodynamics Centre (BSHC): Varna, Bulgaria, 1983.
15. Kuiper, G. *The Wageningen Propeller Series*; MARIN Publication: Wageningen, The Netherlands, 1992.

Journal of
Marine Science and Engineering

Article

Time Domain Modeling of Propeller Forces due to Ventilation in Static and Dynamic Conditions

Anna Maria Kozlowska [1,2,*], Øyvind Øksnes Dalheim [2,3], Luca Savio [1,2] and Sverre Steen [2,3]

1 SINTEF Ocean, Otto Nielsens vei 10, 7052 Trondheim, Norway; Luca.Savio@sintef.no
2 Kongsberg University Technology Centre "Performance in a Seaway" at NTNU, N-7491 Trondheim, Norway; oyvind.dalheim@ntnu.no (Ø.Ø.D.); sverre.steen@ntnu.no (S.S.)
3 Department of Marine Technology, Faculty of Engineering Science and Technology, Norwegian University of Science and Technology, N-7491 Trondheim, Norway
* Correspondence: anna.kozlowska@sintef.no; Tel.: +47-453-92-065

Received: 5 November 2019; Accepted: 23 December 2019; Published: 9 January 2020

Abstract: This paper presents experimental and theoretical studies on the dynamic effect on the propeller loading due to ventilation by using a simulation model that generates a time domain solution for propeller forces in varying operational conditions. For ventilation modeling, the simulation model applies a formula based on the idea that the change in lift coefficient due to ventilation computes the change in the thrust coefficient. It is discussed how dynamic effects, like hysteresis effects and blade frequency dynamics, can be included in the simulation model. Simulation model validation was completed by comparison with CFD (computational fluid dynamics) calculations and model experiments. Experiments were performed for static and dynamic (heave motion) conditions in the large towing tank at the SINTEF Ocean in Trondheim and in the Marine Cybernetics Laboratories at NTNU (Norwegian University of Science and Technology). The main focus of this paper is to explain and validate the prediction model for thrust loss due to ventilation and out of water effects in static and dynamic heave conditions.

Keywords: intermittent ventilation; vortex ventilation; thrust loss; lift coefficient; propeller simulation model

---

1. Introduction

Ventilation is the phenomenon by which air is drawn on structures operating below the free surface, such as hydrofoils, rudders, and propellers. Propeller ventilation might happen when the propeller is coming close to the free surface and air is drawn to it, or when the blades are piercing the free surface and the air is sucked down to the below-water parts of the propeller. Propeller ventilation leads to a sudden and large loss of propeller thrust and torque, which might lead to propeller racing and possibly damaging dynamic loads, as well as noise and vibrations. Ventilation typically occurs when the propeller loading is high and the propeller submergence is limited, which is most likely to happen in heavy seas when the relative motions between the free surface and the propeller are large. Propeller ventilation inception depends on different parameters, i.e., propeller loading, forward speed, and the distance from the propeller to the free surface, see for instance Califano [1], Koushan [2], Kozlowska et al. [3], Kozlowska and Steen [4] and Smogeli [5].

Koushan performed extensive model tests on an azimuth thruster with 6 DoF measurements of forces on one of the four blades, as reported in three papers by Koushan [2,6,7]. Koushan [2] described the dynamics of the ventilated propeller blade axial force on the pulling thruster at bollard condition running at several constant immersion ratios and a constant propeller rate of revolution. Koushan [6] presented the dynamics of ventilated propeller blade axial force on a pulling thruster at bollard condition and constant propeller rate of revolution in forced sinusoidal heave motion. Koushan [7] presented the dynamics of ventilated propeller blade and duct loadings at bollard condition and constant propeller rate of revolution.

A difficulty when creating a calculation model to study ventilation is covering all the ventilation regimes and submergences. Kozlowska et al. [8] showed two different ventilation inception mechanisms depending on the level of submergence of the propeller. Either ventilation can start by forming an air-filled vortex from the free surface, and the free surface can be sucked down to the propeller, or it becomes surface piercing, such that air can enter the suction side of the blade directly from the atmosphere.

Free surface vortex ventilation is characterized by severe thrust losses occurring when a vortex appears on the blade surface, funneling a considerable quantity of air from the free surface down to the suction side of the blade. Vortex ventilation happens only for high thrust loadings at low forward speeds. Surface piercing ventilation is characterized by uniform thrust losses during the complete revolution of the propeller. The propeller might be non-ventilated, partially, or fully-ventilated depending on several factors, where submergence and advance number are clearly important. The typical thrust losses are not only a function of ventilation. Even bigger thrust loss is caused by the propeller coming partly out of the water so that the effective propeller disk area is significantly reduced. When the propeller is partly out of the water, thrust loss can be computed from the fraction of the propeller disc area that is above the water. This is typically complemented by adding the loss caused by the so-called Wagner effect [9], which accounts for the dynamic lift effect of the recently immersed propeller blades. Thrust losses due to the out-of-water effect have been studied by Gutsche [10], Faltinsen et al. [11], and Minsaas et al. [12].

The propeller simulation model is a further development of Dalheim's model, see Steen et al. [13] and Dalheim [14], which was updated by including a physical model for estimating the ventilated blade area based on propeller loading for the vortex ventilation regime. The ventilated blade area ratio is computed using the steady-state vortex ventilation model based on the vortex model by Rott [15] and the propeller momentum theory. It is also discussed in this paper how the dynamic effects, i.e., hysteresis effect and blade frequency dynamics, are included in the existing simulation model PropSim (2018).

2. Calculation Model for Thrust Loss due to Ventilation and Out-of-Water Effects

2.1. Calculation Model

The calculation model predicts the total thrust loss factor $\beta_T = K_T/K_{T0}$ where K_T is the actual thrust coefficient and K_{T0} is the time-averaged value of the thrust coefficient at the relevant advance number J obtained from the calm water, deeply submerged non-ventilated propeller. The calculation model also predicts the ventilated blade area ratio A_V/A_0 since it is required for the calculation of thrust loss in partial ventilation regimes. The formulation of the model depends on the submergence. When the propeller is deeply submerged, it is considered not to experience any ventilation at any advance number. Therefore, $A_v/A_0 = 0$ and the total thrust loss factor $\beta_T = 1.0$, meaning that there is no thrust reduction due to ventilation or out-of-water effects. In the calculation model, the limit for "deeply submerged" is set to $h/R > 3.4$; h is the propeller submergence from the shaft centre to the free surface, and R is the propeller radius.

2.2. Submerged, Vortex Ventilation

In this regime of submergence, the propeller is prone to ventilation due to the impact of a free surface vortex. Thrust loss for ventilating fully submerged propellers might be calculated using the idea presented by Kozlowska and Steen [4], where the change in propeller blade lift coefficient due to ventilation is used to calculate the change of K_T.

Minsaas et al. [12] developed an expression for reduced thrust due to ventilation for fully ventilated propellers assuming that the suction side of the propeller blade is fully ventilated and the pressure on the pressure side of the propeller blade section is equal to the static pressure. The thrust loss factor due to vortex ventilation can be expressed as follows:

$$\beta_{VC} = \frac{1.5EAR}{K_{Tn}} \cdot \left(\frac{\pi}{2}\alpha + \frac{2gh}{V_\infty^2} \right) \tag{1}$$

where g is acceleration of gravity, V_∞ is the relative velocity at the 70% radius propeller blade section, EAR is the expanded blade area ratio of the propeller, and $\propto$ is the angle of attack of the 70% radius propeller blade section.

Kozlowska and Steen [4] concluded that this formula overestimates the thrust loss for deeply submerged propellers and underestimates the thrust loss for propellers working near the free surface. Kozlowska and Steen [4] proposed a correction to Equation (1) based on the assumption that the thrust loss depends also on how much the blade area is covered by air. Thus, the lift coefficient for a partially ventilated propeller might be approximated from the formulas for a lift coefficient of a non-ventilated flat plate and a fully ventilated flat plate, weighted by the ratios of ventilated and non-ventilated areas.

The resulting formula for the thrust loss due to ventilation is:

$$\beta_{VC} = \left(\frac{1.5EAR}{K_{Tn}} \cdot c_{LV} \cdot \frac{A_V}{A_0} \right) + \left(1 - \frac{A_V}{A_0} \right) \tag{2}$$

where A_0 is the propeller disk area ($A_0 = \pi R^2$), A_V is ventilated propeller disc area. $A_V = A_0$ means that the propeller is fully ventilated. c_{LV} is the lift coefficient of the ventilated propeller, which can be calculated as:

$$c_{LV} = c_L(\sigma_V = 0) + \frac{2gh}{V_\infty^2} = \frac{\pi}{2}\alpha + \frac{2gh}{V_\infty^2} \tag{3}$$

The main problem with using Equation (2) is to estimate the blade area that is covered by air A_V/A_0. For bollard and low advance ratio conditions ($J < 0.1$), a polynomial relation between the ventilated blade area ratio and submergences developed by Dalheim and presented by Steen et al. [13] is used.

By using the steady state vortex ventilation model based on the vortex model from Rott [15] and the propeller momentum theory, it is possible to estimate A_V/A_0 due to vortex formation for advance numbers $J \geq 0.1$. The vortex model depends on two parameters: a source strength, which is related to the propeller loading, and the ambient vorticity; the sink gathers vorticity to form the vortex. The ambient vorticity in the towing tank is partly generated by the propeller wake and, therefore, again related to the propeller load through the circulation. A tuning constant η_Γ is added to the calculation model in order to account for the effect that not all the propeller blade circulation is converted into ambient vorticity ($0.6 \leq \eta_\Gamma \leq 0.8$). The application of the vortex model from Rott [15] is described in Kozlowska et al. [8].

2.3. Thrust Loss due to Ventilation and Out-of-Water Effects

Free surface ventilation occurs for propeller submergences $-1 < h/R < 1.2$. The dominating thrust losses are, at least in most cases with significant forward speed, not due to ventilation but due to loss of submerged propeller disk area. We can separate the thrust losses as follows: thrust loss due to loss of propeller disc area, thrust loss due to wave making, thrust loss due to ventilation and due to the Wagner effect. For propeller submergence $h/R < 1$, the thrust has to be corrected for loss of propeller disc area.

The total thrust losses can be divided in loss of propeller disc area (β_0), Wagner effect (β_W), steady wave motion (β_1), and ventilation (β_{VC}) as follows:

$$\beta_T = \beta_{VC} \cdot \beta_W \cdot \beta_0 \cdot \beta_1 \tag{4}$$

Loss of propeller disc area for $h/R < 1$ can be estimated using purely geometrical considerations, as in Gutsche [10], as follows:

$$\beta_0 = 1 - \frac{\arccos\left(\frac{h}{R}\right)}{\pi} + \frac{h}{\pi R}\sqrt{1 - \left(\frac{h}{R}\right)^2}. \quad (5)$$

Dalheim [14], reproduced in Steen et al. [12], gave a more elaborate formula, where the influence of the hub is included, but the difference is not very significant, and the use of the above, much simpler formula is acceptable in most cases. The Wagner effect (Wagner, [9]) accounts for the dynamic lift effect. If the lift of a foil is changed suddenly by a sudden change in geometric angle of attack, the corresponding change in lift is first only half of the final steady value due to the induced angle of attack from the shed vortex formed by the time rate of change of circulation. This effect diminishes gradually, and a curve-fit formula is used in the equation below. It shows that the foil must travel about 20 chord lengths to almost recover full lift. The idea is that a similar effect occurs when a propeller blade is suddenly passing through the surface and into the water. The thrust loss factor β_W is calculated from the average value during the submerged part of the blade rotation. Thus, it will, in general, depend on the propeller radius as well as propeller submergence. For the simplified formula, β_W is modeled at the characteristic section $\frac{r}{R} = 0.7$. Therefore, the thrust loss factor due to the dynamic lift effect, which is relevant only for $h/R < 1$, is calculated as:

$$\beta_W = 0.5 + 0.5\sqrt{1 - \left(\frac{155 - V_\infty \cdot t/c}{155}\right)^{27.59}} \quad (6)$$

where V_∞ is the local relative velocity at the blade section, which, when ignoring induced velocities, can be calculated as $V_\infty = \sqrt{V_A^2 + (0.7\pi nD)^2}$, t is propeller blade thickness, and c is the chord length.

Minsaas et al. [11] proposed an empirical expression for the combined effect of loss of disk area, wave making, and Wagner effect:

$$\beta = \beta_0 \cdot \beta_1 \cdot \beta_W, \quad (7)$$

$$\beta = \begin{cases} 1 - 0.675\left(1 - 0.769 \cdot \frac{h}{R}\right)^{1.258} & for\ h/R < 1.3 \\ 1 & for\ h/R \geq 1.3 \end{cases} \quad (8)$$

2.4. *Torque Loss due to Ventilation and Out-of-Water Effects*

Figure 1 shows a relation between the torque loss factor and the thrust loss factor. As it is observed from Figure 1, the propeller torque has similar behavior as propeller thrust and shows good agreement with experimental results by Minsaas et al. [12],

$$K_{Tt} = \beta_T \cdot K_{Tn}, \quad (9)$$

$$K_{Qt} = \beta_T^m \cdot K_{Qn}, \quad (10)$$

where m is constant between 0.8 and 0.85 [13].

This means that if the torque is measured, one can also accurately know the thrust in ventilated conditions, and it means that when thrust loss is predicted or simulated, for instance, using the methods in this paper, the corresponding torque loss can easily be determined using Equations (9) and (10) with $m = 0.8$.

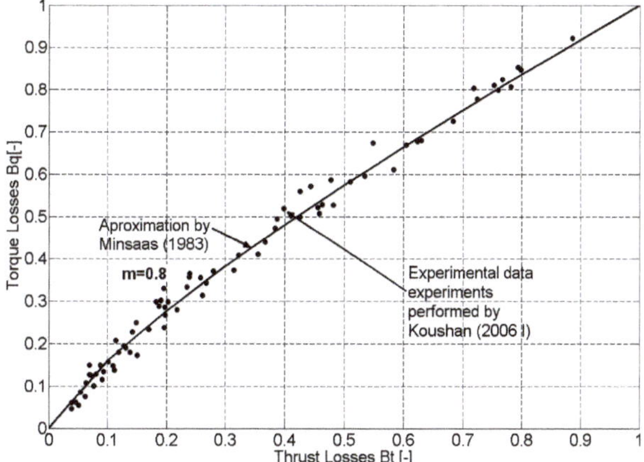

Figure 1. Relation between thrust and torque loss factors, based on experiments Kou2006_I.

2.5. Time Domain Simulation Model: PropSim (2016) and PropSim (2018)

The propeller simulation model PropSim (2016) was developed by Dalheim [14] and implemented in Simulink. The simulation model generates a time domain solution to the six degrees of freedom propeller forces in varying operating conditions, including change of operating point, unsteady axial and tangential flow field, effect of oblique inflow in manoeuvring condition, Wagner effect, reduced propeller submergence, and ventilation. The time domain simulation model in PropSim (2016) is quasi-static. For ventilation modelling, PropSim (2016) used a formula based on an idea presented in Kozlowska and Steen [4], where the change in lift coefficient due to ventilation is used to compute the change in K_T resulting in the formula for the thrust loss calculations due to ventilation presented in Equation (2).

For the PropSim (2016) simulation model, Dalheim used model tests from Kozlowska and Steen [4] to construct a polynomial relation for the value of the ventilated blade area ratio A_V/A_0 Figure 2 contains two different curves, one for increasing and one for decreasing propeller submergence, which indicate the propeller hysteresis behavior of the propeller ventilation.

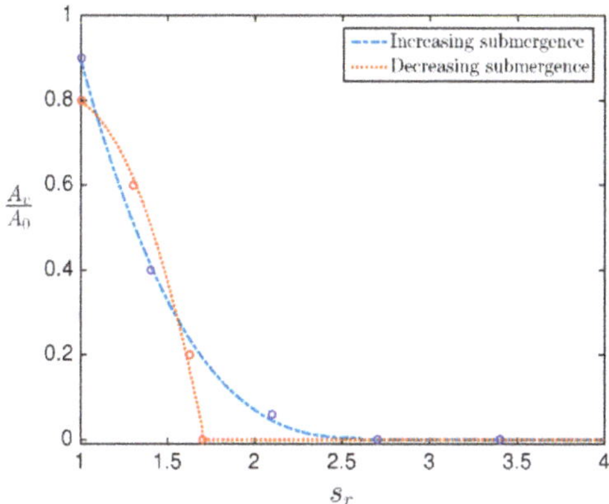

Figure 2. Ratio of ventilated propeller disk area to nominal disk area as function of propeller submergences ratios S_r, Dalheim [14]. Submergence ratio is denoted by the symbol S_r.

The empirical relation presented in Figure 1 is used for the simulation model PropSim (2016) and can hardly be viewed as generally valid since other factors like forward speed and propeller loading must be expected to matter, Steen et al. [13]. Therefore, the simulation model (PropSim(2018)) has been updated by adding a physics-based model for estimating the ventilated area of the propeller disc based on propeller loading and submergence, outlined in Section 2.3 above. Like PropSim (2016), the simulation model PropSim (2018) is quasi static, since it is assuming that the response is quasi steady and based on the calculation model presented in Kozlowska et al. [8].

3. Results

3.1. Simulation Model Validation

Validation of the simulation model denoted PropSim (2018) is carried out using model experiments performed by Koushan [2] and Kozlowska et al. [8]. These two experimental campaigns are referred to in this article by using acronyms: experiments reported by Koushan [2]—*Kou2006_I*, and experiments reported by Kozlowska et al. [8]—*Koz17*.

Kou2006_I experiments were performed on an open pulling propeller exposed to forced sinusoidal heave motion. Carriage speed U and the propeller shaft frequency n were varied in order to obtain different low advance numbers J (around 0.1). In order to obtain more data for validation purposes in higher advance numbers, *Koz17* experiments were performed. The testing conditions are listed in Table 1.

The same propeller model (P1374) was used for these experiments: the propeller has a diameter of 250 mm, design pitch ratio of 1.1, and expanded area ratio of 0.595. The propeller hub diameter is 65 mm.

The *Kou2006_I* experiments were conducted in the Marine Cybernetics Laboratories at the Norwegian University of Science and Technology. The tank is 40 m long, 6.45 m wide, and 1.5 m deep. Ventilation is generated by sinusoidal vertical motion of the propeller with different amplitudes. Blade axial, radial, tangential forces, and moments about all three axes were measured during the experiments. A pulse meter indicating the angular position of the reference blade.

The *Koz17* experiments took place in the large towing tank at SINTEF Ocean in Trondheim, with dimensions (length × breadth × depth) of 260 m × 10.5 m × 5.6 m. The tests were performed at different submergence ratios and propeller rotational speeds. For each draught and propeller rotational speed, the propeller was tested at different advance numbers ranging from $J = 0$ to $J = 1.0$. Different advance numbers were obtained at various propeller rotational speeds so that for the same advance numbers, different propeller thrusts were obtained, thus varying the Weber number W_e. A conventional two components propeller open water dynamometer was used to measure propeller thrust and torque. The main purpose for these experiments was to obtain more data at higher advance numbers. Also, the different advance numbers were obtained at a range of propeller rotational speeds that, for the same advance number, different thrust coefficients were tested.

Figure 3 shows the comparison between a simulation performed by using PropSim (2016) and PropSim (2018) simulation models and experimental results, *Kou2006_I*. PropSim (2016) model relates thrust loss to the estimated ventilated blade area using an empirical relation that is based on the same model experiments, as shown in Figure 3 (*Kou2006_I*). This is believed to be the reason why, for this particular case, the agreement between experimental results and calculation fitted better to the 2016 version of the simulation model.

Table 1. Summary of performed test campaigns presented in this paper.

Experiments	Publication	Measurements	Instruments:
Acronym: Kou2006_I	Kozlowska et al. [3], Kozlowska and Steen [4] Koushan [2] Koushan [6]	- ventilation in static condition (different propeller submergence h/R) - ventilation in dynamic condition (sinusoidal heave motion with different amplitudes) Draughts: $h/R = 2.4, 1.6, 1.2, 1.0, 0.5, 0$ Propeller speeds: $n = 11$ rps, 14 rps Advance number: $0 \leq J \leq 0.143$	- 5 DoF blade dynamometer - 6 DoF dynamometer for the whole thruster - pulse meter (blade position) - two underwater lamps - high speed camera
Acronym: Koz17	Kozlowska et al. [8]	- ventilation in static condition (different propeller submergence h/R) Draughts: $h/R = 2.4, 2.0, 1.6, 1.5, 1.4, 1.2, 1.0, 0.5, 0, -0.5$ Propeller speeds: $n = 9$ rps, 12 rps, 14 rps, 16 rps Advance number: $0 \leq J \leq 1.0$	- conventional two components propeller open water dynamometer - two high speed cameras - two lamps (one underwater and one above the free surface)

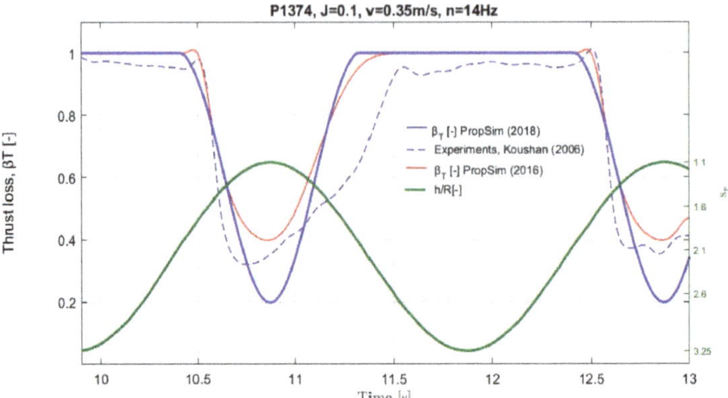

Figure 3. Comparison of thrust loss between simulation performed by using the PropSim (2016) simulation model, PropSim (2018) simulation model, and experimental results of *Kou 2006_I*, the sampling rate is above four times the propeller revolutions, and it is equal to 60 Hz.

Figure 4 shows the comparison of thrust loss calculations using two different versions of the simulation model: PropSim (2016) and PropSim (2018). Propeller thrust loss is calculated for a propeller working at constant propeller revolutions $n = 12$ Hz under dynamic heave motion conditions $\left(1.2 \leq \frac{h}{R} \leq 2.2\right)$ for high advance number $J = 0.4$. There are no experimental conditions testing propeller ventilation during dynamic heave motion for high advance numbers, thus for validation purposes, the simulation model results have to be compared with static conditions based on the *Koz17* experiments, implicitly assuming that the behavior is quasi-static. It can be observed from Figure 4 that the thrust loss prediction agrees better with the calculations performed by using the simulation model PropSim (2018). The reason for this result is that the simulation model PropSim(2016) does not include the forward speed and propeller loading as a factor for calculating the blade area ratio A_V/A_0. Simulation model PropSim (2016) overestimated thrust losses due to ventilation. For example, for $h/R = 1.2$, the thrust loss due to ventilation is 0.84 based on the PropSim (2018) simulation model and in the range of 0.45–0.5 for the PropSim (2016) simulation model. The experimental measurements are equal to 0.78, which is closer to the updated simulation model PropSim (2018). The same behavior was observed for other submergence ratios $h/R = 1.4$, 1.6, and 2.0.

Figure 5 shows the comparison of thrust loss calculation using the PropSim (2016) and PropSim (2018) simulation model. Propeller thrust loss is calculated for the propeller working with constant propeller revolutions $n = 12$ Hz under dynamic heave motion conditions $\left(1.2 \leq \frac{h}{R} \leq 2.2\right)$ for the high advance number $J = 0.6$. It can be observed from Figure 5 that the thrust loss prediction agrees better with the calculation for the simulation model PropSim (2018). Simulation model PropSim (2016) overestimates thrust losses due to ventilation. For example, for $h/R = 1.2$, the thrust loss due to ventilation is 1.0 based on the PropSim (2018) simulation model and is in the range of 0.45–0.5 for the PropSim (2016) simulation model. The experimental measurements are equal to 0.92, which is closer to the updated simulation model PropSim (2018). Also, experimental data based on experiments *Koz17* shows no thrust loss for $J = 0.6$ for the submergence ratio $\left(1.4 \leq \frac{h}{R} \leq 2.2\right)$ the same as predicted by using the PropSim (2018) simulation model.

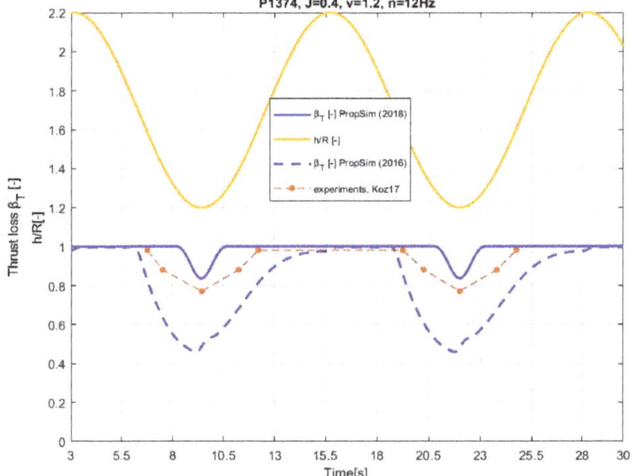

Figure 4. Comparison of thrust loss between the simulation performed by Dalheim (PropSim(2016)) and Kozlowska (PropSim (2018)) and experimental results of *Koz17*.

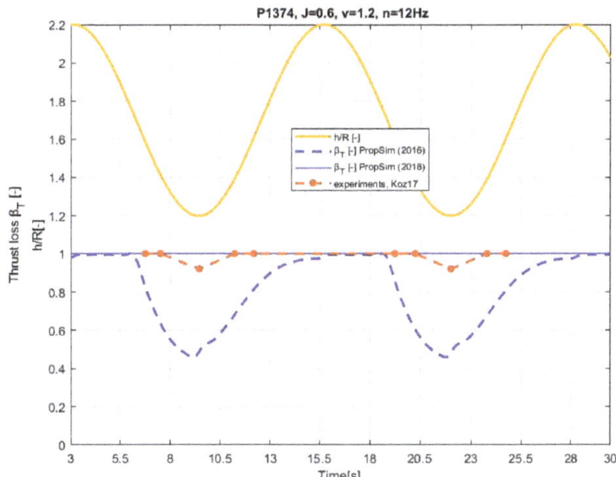

Figure 5. Comparison of thrust loss between simulation performed by Dalheim (PropSim(2016)) and Kozlowska (PropSim (2018)) and experimental results of *Koz17*.

3.2. Dynamic Effect Causing Hysteresis of the Thrust Loss during Heave Motion of the Propeller Calculated by PropSim (2018_hysteresis)

A significant dynamic effect of propeller ventilation is connected with the thrust and torque hysteresis effect, appearing mostly in connection with intermittent vortex ventilation. The hysteresis effect is caused by the fact that it takes a while for ventilation of a submerged propeller to be established, so in a situation with decreasing submergence or increasing propeller loading, there is less thrust loss than for the same condition in static operation, while when ventilation disappears, it takes time for the thrust to build up, due to the Wagner effect, so then thrust loss is larger than the corresponding static operation.

In order to account for this effect, the PropSim (2018) simulation model was updated. The dynamic effect was added by making propeller circulation, described in Equation (11) as a time dependent function.

$$\Gamma = \frac{V_c \cdot C_{0.7} \cdot K_{Tn}}{3 \cdot EAR} \quad (11)$$

The time dependent function was divided into two different cases. One, which corresponds with time, which is required to establish ventilation, and the other, which corresponds with time, which is desired for ventilation to disappear. Symbol n_s, which is used in Figure 6, is the minimum number of propeller revolutions needed to establish ventilation, thus forming a ventilating vortex from the free surface, and symbol n_v used in Figure 6 is the minimum number of propeller revolutions needed for a vortex, and thereby ventilation to disappear. n_s and n_v are functions of propeller submergence for low advance numbers $0.08 < J < 0.15$. Higher advance numbers are not covered since, then, vortex ventilation is found to be very unlikely, and for free surface ventilation, little or no hysteresis effect is observed in experiments. Propeller circulation as a function of time for the creation and disappearance of vortex ventilation is presented in Figure 7. Since in principle vortex ventilation is dependent on the surface tension, while the proposed model does not include any effect from surface tension, the model presented here is only valid for high Weber numbers, taken to be $W_e > 180$. Polynomial approximations of n_s and n_v are presented in the equations below.

$$n_s = \begin{cases} 20\left(\frac{h}{R}\right) - 24 & for\ 1.2 \leq \frac{h}{R} \leq 1.5 \\ 15664\left(\frac{h}{R}\right)^4 - 106470\left(\frac{h}{R}\right)^3 + 270657\left(\frac{h}{R}\right)^2 - 304831\left(\frac{h}{R}\right) + 128309 & for\ 1.5 \leq \frac{h}{R} \leq 2.0 \end{cases} \quad (12)$$

$$n_v = -1619.3\left(\frac{h}{R}\right)^3 + 907.1\left(\frac{h}{R}\right)^2 - 16843\left(\frac{h}{R}\right) + 10496 \qquad for\ 1.2 \leq \frac{h}{R} \leq 2.0 \quad (13)$$

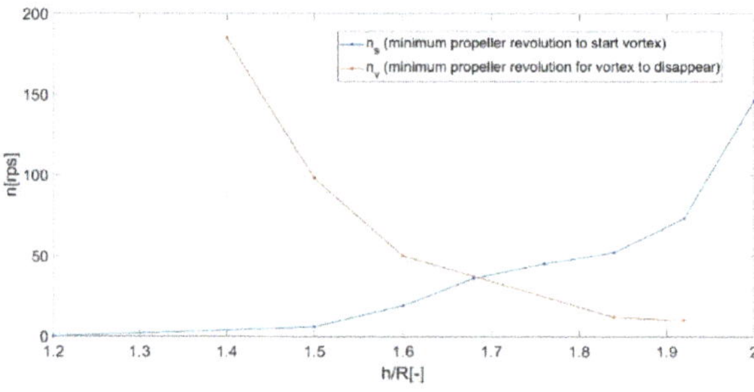

Figure 6. Minimum number of propeller revolution to establish a ventilating vortex (n_s) and minimum number of propeller revolution for the vortex to disappear (n_v) as a function of propeller submergence, valid only for high Weber number and low advance numbers $0.08 < J < 0.15$.

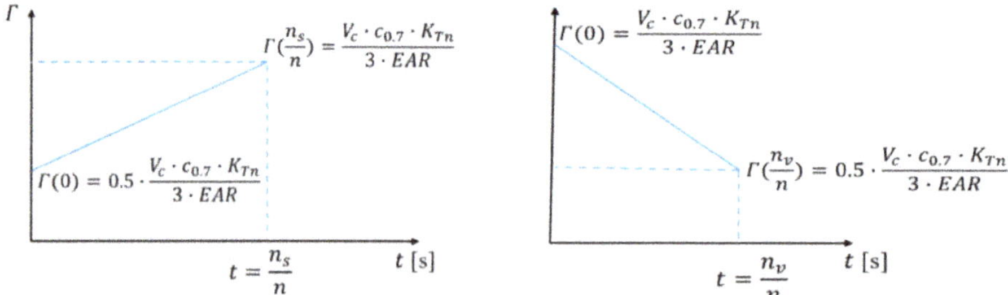

Figure 7. Propeller circulation as a function of time to establish the vortex ventilation (top side) and as a function of time for ventilation to dissapear (bottom side), K_{Tn} is the thrust coefficient for non ventilating condition and K_T is the thrust coefficient for ventilating condition for given and constant propeller submergence *h/R*.

Table 2 shows five different experimental conditions that were used for the comparison between calculations using the PropSim (2018 hysteresis) simulation model and experimental results. A is the amplitude of the harmonic heave motion, while T is the period of the harmonic heave motion.

Table 2. Different experimental condition used for comparison between the simulation model and experimental results.

Case Number	h/R [-]		A/R [-]	A/T [m/s]	T [s]	J [-]	n [rps]
Dynamic condition (heave motion), *Kou2006_I*, low advance numbers							
	min	max					
1	1.15	2.15	2.4	0.075	4	0.143	14
2	1.15	2.15	1.2	0.1	1.5	0.143	14
3	1.05	3.25	2.4	0.15	2	0.143	14
4	1.05	3.25	2.4	0.075	4	0.1	14
5	1.15	2.15	1.2	0.1	1.5	0.1	14

Figures 8 and 9 show the comparison of thrust loss in the experiments under dynamic heave motion for different heave amplitudes and calculations by using the simulation model PropSim (2018_hysteresis), which includes the hysteresis effect in the simulation. As can be observed in the figures, calculations performed by the PropSim (2018_hysteresis) simulation model agree quite well with experiments. This means that the simulation model correctly accounts for the hysteresis effect on ventilation due to the propeller working with periodically varying submersions.

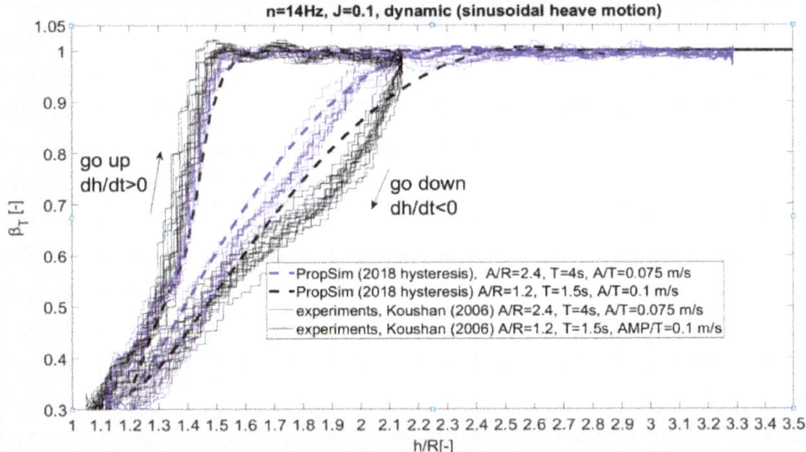

Figure 8. Comparison for thrust loss between experiments (dynamic) and simulation (PropSim2018_hysteresis) for different amplitudes of the propeller (heave), see Case 4 and 5 presented in Table 2; the simulation includes the hysteresis effect for two different motions of the propeller (upwards and downwards), PropSim2018_hysteresis, the blue line for the up behavior is similar and lies under the black line.

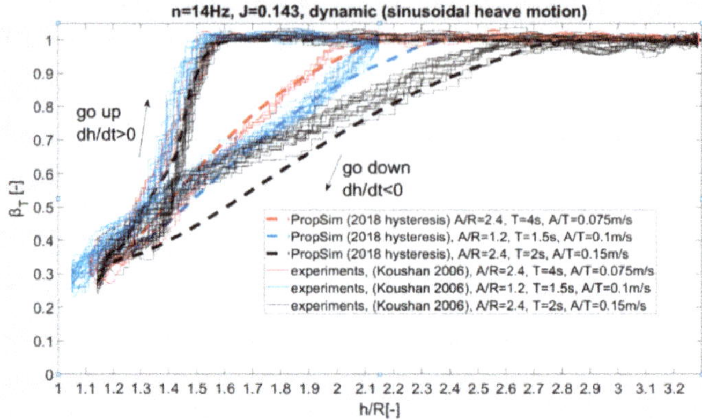

Figure 9. Comparison for thrust loss between experiments (dynamic) and simulations (PropSim2018_hysteresis) for different amplitudes of the propeller (heave), see Case 1, 2, and 3 presented in Table 2, simulation includes hysteresis the effect for two different motions of the propeller (upwards and downwards), PropSim2018_hysteresis.

3.3. Dynamic Effect, Thrust Loss due to Ventilation as a Function of Blade Position Calculated by PropSim (2018_blade_dynamics)

From the experiments, it can be observed that the thrust varies with the position of the blade during one cycle of rotation when the propeller is ventilating and/or coming partly out of water. For deep and constant propeller submergence and low advance numbers (i.e., h/R = 1.5 and J = 0.15, no out of water effect), the biggest thrust loss is found when the blade is close to the free surface (between 315° and 90°). For the propeller blade position between 90° and 315°, the thrust is rebuilt and achieves values close to the nominal thrust. The different thrust losses correspond to the different ventilation extent. It is clear from the experiments that in this condition the propeller blade can be both fully, partially, or non-ventilating depending on the blade position. When the propeller is coming out of the water (i.e., h/R = 0), the thrust loss also varies due to the blade position. The three reasons for this variation are ventilation, loss of propeller disk area, and Wagner effect. The previous versions of the simulation model denoted PropSim (2016) and PropSim (2018) both include the loss of propeller disk area and the Wagner effect as a function of propeller position. In the previously described versions of PropSim (2018), the thrust averaged over a propeller revolution is calculated, meaning that propeller blade frequency dynamics are not represented in the simulation, see Figure 10a. In order to include blade frequency dynamics, the blade thrust is computed as a function of the blade position, see Figure 10b.

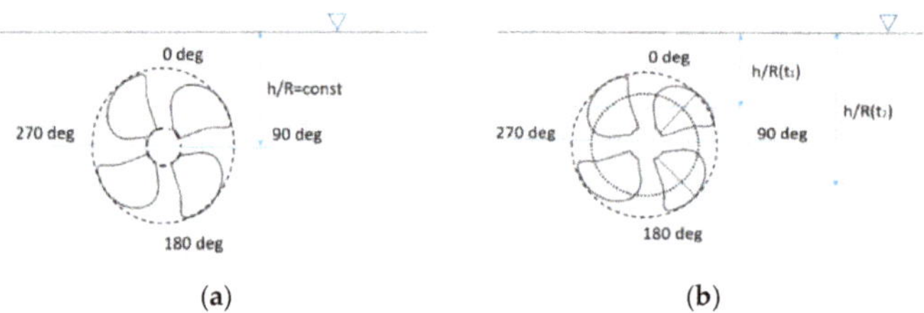

Figure 10. Modification of PropSim (2018) in order to account for the effect that the blade thrust loss due to ventilation varies during one cycle of revolution PropSim (2018_blade_dynamics). (**a**) PropSim (2018); (**b**) PropSim (2018_blade_dynamics).

Figure 11 shows the comparison between calculations (CFD) and experiments *Koz10* of the thrust loss due to ventilation as a function of blade position. The CFD calculations were performed by TUHH (Hamburg University of Technology) and based on their in-house code FreSco+. The method used in the calculation was a RANS (Reynolds Averaged Navier Stokes) code based on a finite volume discretisation of the computational domain. For the investigation presented below, the propeller is modelled in the cylindrical domain, which rotates with the number of revolutions of the propeller; see Wockner-Kluwe [16]. It can be observed from the figure that the prediction of thrust loss is more repeatable between revolutions for CFD calculations than for experiments. Figure 12 shows the calculation of the thrust loss due to ventilation, which varies due to different blade positions (PropSim (2018_blade_dynamics)).

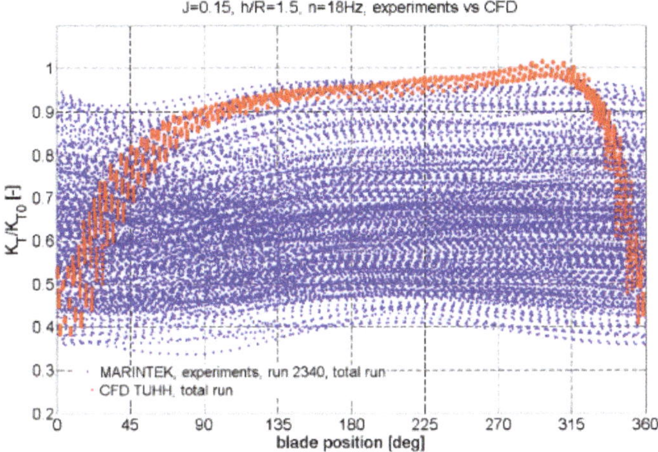

Figure 11. Thrust loss factor as a function of blade position, red markers correspond to CFD calculations, and blue markers correspond to experimental results (*Koz10*).

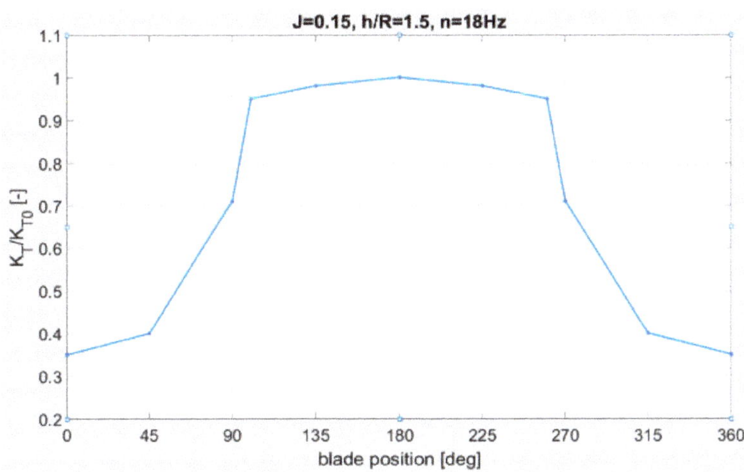

Figure 12. Calculation of thrust loss factor as a function of blade position using the simulation model PropSim (2018_blade_dynamics).

It can be observed from the comparison of Figures 11 and 12 that the calculations made by simulation using PropSim (2018_blade_dynamics) are closer to the CFD computational results than the experimental results. For both CFD and simulations, similar thrust loss is observed for every propeller revolution. During measurements, different thrust losses appear depending on the time in the experiments, see Figure 11. The reason for the large variation of thrust loss between the propeller revolutions is not known. Measurement error is unlikely, given the repeatability of measurements in fully-submerged conditions in the same experimental campaign. It is believed that the variation might be caused by the ventilatilon in this condition being unstable and on the limit between the ventilated and non-ventilated condition. When the propeller ventilates, the thrust is reduced, so the thrust loading is reduced, and thereby reducing the ventilation probability, which, in turn, causes the ventilation to disappear after some time. The fact that it takes some time for the ventilation to disappear is what causes the hysteresis effect discussed earlier. When the ventilation disappears, the thrust and, therefore, thrust loading increases again, leading eventually to new ventilation. The reason for the perfectly symmetric appearance of the thrust loss factor versus the blade angle in Figure 12 is that the calculation method is quasi-static—it does not capture the "memory effect" of the flow, which is present in both reality and CFD.

Figures 13 and 14 show the thrust loss variations at different blade positions for n = 18 Hz and higher advance numbers. Figure 13 shows the comparison between calculations (CFD) and experiments *Koz10* of the thrust loss due to ventilation as a function of blade position. Figure 14 shows the calculation of the thrust loss by using the PropSim (2018_blade_dynamics) simulation model. By comparing the two figures, it is seen that both experiments and PropSim-calculations show hardly any thrust loss. For the experiments, there is a very slight variation of thrust, which is not due to ventilation but might be caused by the free surface (wave making) effect. For PropSim, this type of free surface effect is not modeled. The CFD shows a thrust that varies with position, although showing a partly higher thrust than the nominal. The reason for this variation is not known but might also be due to the free surface (wave making) effect.

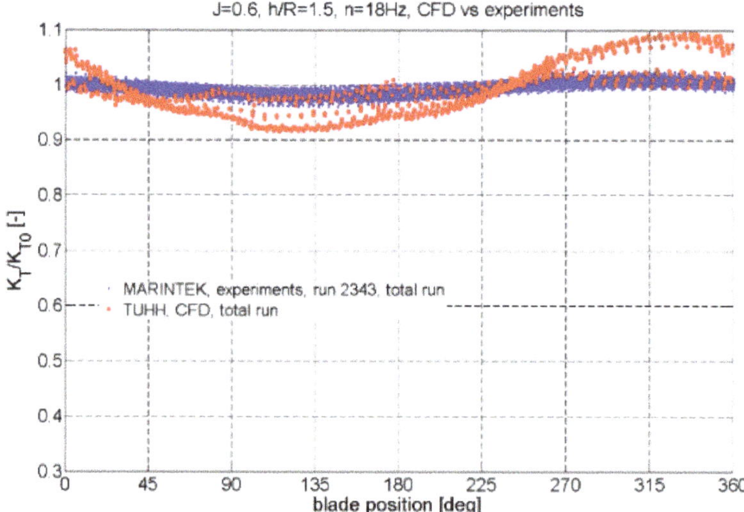

Figure 13. Thrust loss factor as a function of blade position red markers correspond to CFD calculations, and blue markers correspond to experimental results (*Koz10*).

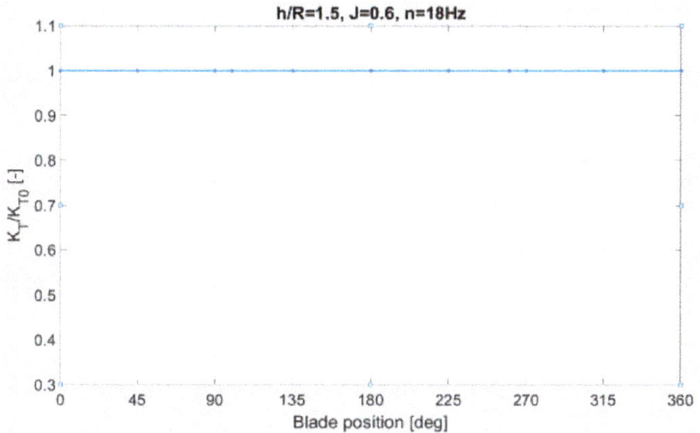

Figure 14. Calculation of thrust loss factor as a function of blade position.

- $h/R = 1.5$, $n = 18$ Hz, $J = 0.6$ (high advance number)

Figures 15 and 16 show the thrust variations at different blade positions in one cycle of revolution for $n = 18$ Hz and a low advance number. For this case, the propeller is half submerged, so different thrust loss is a consequence of a combination of the out-of-water effect, Wagner effect, and ventilation. Figures 15 and 16 show relatively good agreement between the experimental results and calculations, although the experimental results are more gradually changing compared to the simulation. The reason for this is because in the proposed method, the propeller blade forces are considered at the lifting line. This means that a half-submerged propeller will get zero propeller blade forces in the upper half of one revolution, as seen in Figure 16, which is inconsistent with the gradual variation seen from the experimental results in Figure 15. In the experimental results, the thrust has a peak of around 170°–180°; however, the thrust continues to increase until the lifting line leaves the water. This is due to the included Wagner effect, which gives a gradual increase in the thrust loss factor towards unity as the blade section travels up to around six chord lengths.

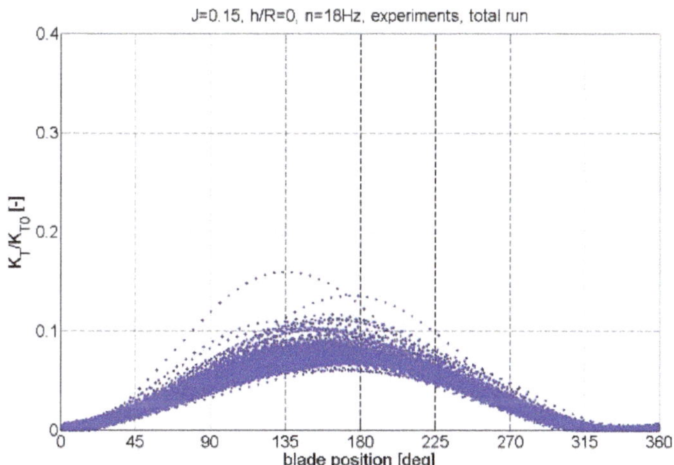

Figure 15. Thrust loss factor as a function of blade position based on experimental results (Koz10).

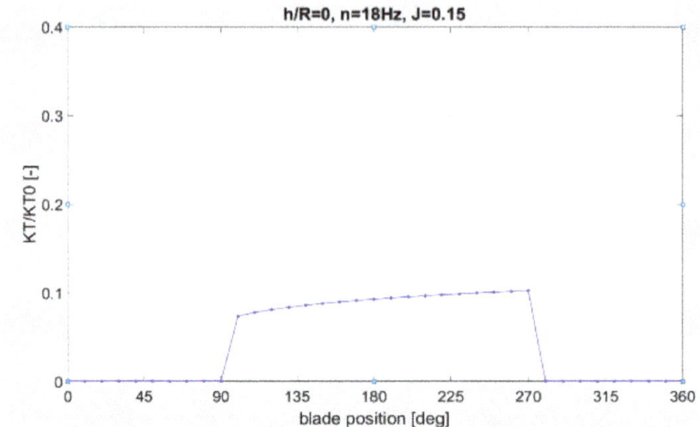

Figure 16. Calculation of thrust loss factor as a function of blade position.

- $h/R = 0$ (half submerged), $n = 18$ Hz, $J = 0.15$ (low advance number)

Figures 17 and 18 show the thrust variations at different blade positions in one cycle of revolution for $n = 18$ Hz and a high advance number $J = 0.9$. For these cases, the propeller is half submerged, so a different thrust loss is a consequence of a combination of the out-of-water effect, Wagner effect, and ventilation. Figures 17 and 18 show similar agreement for a high advance number as for a low advance number. In the experimental results, the thrust has a peak around 200°, i.e., after the blade has passed its lowest position. This can be explained by air dragged down by the propeller blade, causing loss of effective propeller blade area, which means that it will take a longer time for the thrust to build. In the proposed method, however, the thrust continues to increase until the lifting line leaves the water. Similar to a low advance number, this is due to the included Wagner effect. The thrust loss caused by air dragged down by the propeller blade does not vary along one revolution in the proposed method because it is calculated based on propeller submergence, not blade submergence. This is why a similar displacement of the thrust (relative to the bottom blade position) is not visible in the proposed method. The integrated value of thrust will, however, have its center around 200° in the proposed method, similar to the experimental results.

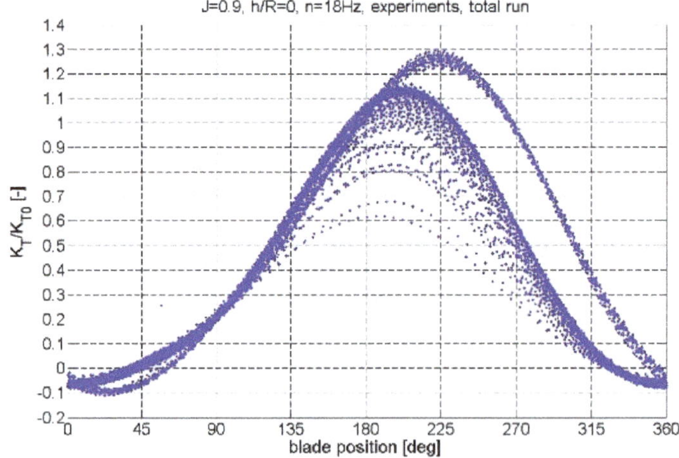

Figure 17. Thrust loss factor as a function of blade position, based on experimental results (*Koz10*).

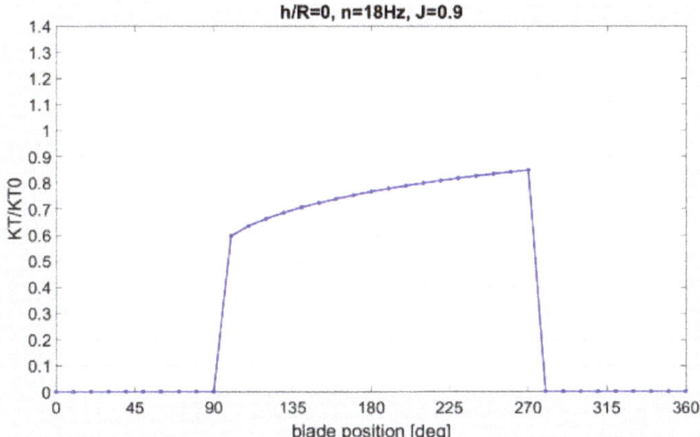

Figure 18. Calculation of thrust loss factor as a function of blade position.

- $h/R = 0$ (half submerged), $n = 18$ Hz, $J = 0.9$ (high advance number)

Figure 19 shows a time series of the computed thrust coefficient for a single blade and the propeller for $J = 0.15$, $h/R = 0$, $n = 18$ Hz made using the PropSim (2018_blade_dynamics) simulation model.

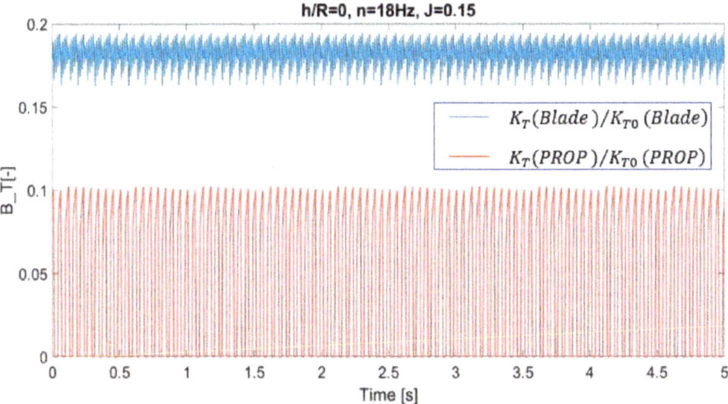

Figure 19. Time series of the computed thrust coefficient for a single blade and the propeller for $J = 0.15$, $h/R = 0$, $n = 18$ Hz, $K_T(Blade)/K_{T0}(Blade)$ is the thrust loss coefficient for a single blade and $K_T(PROP)/K_{T0}(PROP)$ is the thrust loss coefficient for the whole propeller.

Figure 20 shows a time series of the computed thrust coefficient for a single blade and the propeller for $J = 0.9$, $h/R = 0$, $n = 18$ Hz made using the PropSim (2018_blade_dynamics) simulation model. Figure 19 is presenting results from the same computation as Figure 16, while Figure 20 is presenting results from the same computation as Figure 18. However, in Figures 19 and 20, thrust loss is presented as function of time instead of angular position. Note also that while Figures 16 and 18 use a thrust loss factor dividing the actual blade thrust with the nominal (non-ventilated) blade thrust, Figures 19 and 20 are using a factor where the actual thrust is divided with the total propeller thrust.

Both Figures 19 and 20 show how the amplitude of variation decreases, and how the dominating frequency of the variation is increasing for the entire propeller compared to a single blade.

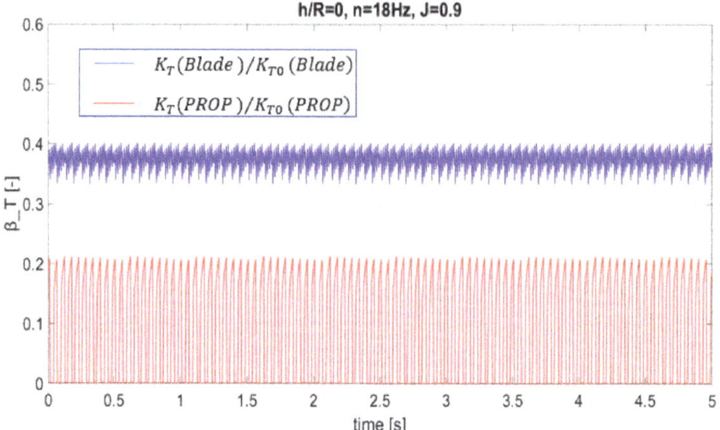

Figure 20. Time series of the computed thrust coefficient for a single blade and the propeller for $J = 0.9$, $h/R = 0$, $n = 18$ Hz, $K_T(Blade)/K_{T0}(Blade)$ is the thrust loss coefficient for a single blade, and $K_T(PROP)/K_{T0}(PROP)$ is the thrust loss coefficient for the whole propeller.

4. Discussion and Conclusions

This paper presents a simple time-domain simulation model for thrust loss due to ventilation and the out-of-water effect. The thrust loss model can be used also for predicting torque loss, through a simple, empirical relation between thrust and torque loss. The presented simulation model is an extension of the model previously presented by Kozlowska et al. [8] and Dalheim [14].

Two different dynamic effects of the ventilating vortex have been added. One effect is connected with the dynamic effect causing hysteresis and connected with the propeller loading. The other dynamic effect is connected to the thrust loss variation with the position of the blade during one propeller revolution.

A significant dynamic effect of the propeller ventilation is connected with the thrust and torque hysteresis effect, appearing mostly in connection with intermittent vortex ventilation. The hysteresis effect is caused by the fact that it takes a while for the ventilation of a submerged propeller to be established, so in a situation with decreasing submergence or increasing propeller loading, there is less thrust loss than for the same condition in static operation. Also, when ventilation starts to disappear, it takes time for thrust to build up again, so that the thrust loss is larger than in the corresponding static operation. In order to account for this effect, the PropSim (2018) simulation model was further developed. The dynamic effect was added by including the time dependency of the propeller circulation in the calculation of propeller ventilation. The comparison between the simulation model PropSim (2018_hysteresis) and model experiments shows good agreement, which means that the simulation model correctly accounts for the hysteresis effect on ventilation due to propeller working with periodically varying submersion.

The other dynamic effect, which is connected to the blade position during one cycle of rotation, has been added to the simulation model denoted PropSim (2018_blade_dynamics). The comparison between the results obtained by the simulation model, experiments and CFD calculations shows that the simulation model is closer to the CFD computational results than the experimental results. For both CFD and simulation results, similar thrust loss was observed for every propeller revolution. During the experiments, different thrust losses depending on the time in the experiments were observed.

Author Contributions: Conceptualization, L.S., A.M.K. and Ø.Ø.D.; Data curation, A.M.K.; Formal analysis, A.M.K.; Investigation, A.M.K.; Methodology, A.M.K. and S.S.; Supervision, S.S.; Writing–original draft, A.M.K. All authors have read and agreed to the published version of the manuscript.

Funding: This research was founded by Kongsberg Maritime AS, NTNU and SINTEF Ocean.

Acknowledgments: This work has been carried out at the University Technology Centre at NTNU (Trondheim) sponsored by Rolls Royce Marine (now Kongsberg Maritime AS) and SINTEF Ocean.

Conflicts of Interest: The authors declare no conflict of interest.

References

1. Califano, A. Dynamics Loads on Marine Propellers Due to Intermittent Ventilation. Ph.D. Thesis, Norwegian University of Science and Technology, Trondheim, Norway, 2010.
2. Koushan, K. Dynamics of ventilated propeller blade loading on thrusters. In Proceedings of the World Maritime Technology Conference (WMTC), Rome, Italy, 17–22 September 2006.
3. Kozlowska, A.; Steen, S.; Koushan, K. Classification of different type of propeller ventilation and ventilation inception mechanisms. In Proceedings of the First International Symposium on Marine Propulsors, smp09, Trondheim, Norway, 9 June 2009.
4. Kozlowska, A.; Steen, S. Ducted and Open Propeller Subjected to Intermittent Ventilation. In Proceedings of the Eighteen International Conference on Hydrodynamics in Ship Design, Safety and Operation, Gdansk, Poland, 12–13 May 2010.
5. Smogeli, Ø.N. Control of Marine Propellers: From Normal to Extreme Conditions. Ph.D. Thesis, Norwegian University of Science and Technology, Faculty of Engineering Service and Technology, Department of Marine Technology, Trondheim, Norway, 2006.
6. Koushan, K. Dynamics of Ventilated Propeller Blade Loading on Thruster due to Forced Sinusoidal Heave Motion. In Proceedings of the 26th Symposium on Naval Hydrodynamics, Rome, Italy, 17–22 September 2006.
7. Koushan, K. Dynamics of Propeller Blade and Duct Loading on Ventilated Ducted Thruster Operating at Zero Speed. In Proceedings of the International Conference on Technological Advances in Podded Propulsion, Brest, France, 3–5 October 2006.
8. Kozlowska, A.; Savio, L.; Steen, S. Predicting Thrust loss of ship propellers due to ventilation and out of water effect. *J. Ship Res.* **2017**, *61*, 198–213. [CrossRef]
9. Wagner, H. Über Stoß- und Gleitvorgänge an der Oberfläche von Flüssigkeiten. *Z. Angew. Math. Mech.* **1932**, *12*, 192–235. [CrossRef]
10. Gutsche, F. *Der Einfluss der Kavitation auf die Profileigenschaften von Propellerblattschnitten, Schiffbauforschung, Heft 1*; Wasserbau und Schiffbau: Berlin, Germany, 1962.
11. Faltinsen, O.; Minsaas, K.; Liapas, N.; Skjørdal, S.O. *Prediction of Resistance and Propulsion of a Ship in a Seaway*; The Shipbuilding Research Association Japan: Tokyo, Japan, 1981.
12. Minsaas, K.; Faltinsen, O.; Person, B. On the importance of added resistance, propeller immersion and propeller ventilation for large ships in seaway. In Proceedings of the International Symposium on Practical Design of Ships and other Floating Structures PRADS 1983, Tokyo, Japan, 17–22 October 1983.
13. Steen, S.; Dalheim, Ø.Ø.; Savio, L.; Koushan, K. Time Domain modelling of propeller forces. In Proceedings of the PRADS2016, Copenhagen, Denmark, 4–8 September 2016.
14. Dalheim, Ø.Ø. Development of a Simulation Model for Propeller Performance. Master's Thesis, Norwegian University of Science and Technology, Trondheim, Norway, 2015.
15. Rott, N. On the viscous core of a line vortex. *Z. Angew. Math. Phys.* **1958**, *9*, 543–553. [CrossRef]
16. Wockner-Kluwe, K. Evaluation of the Unsteady Propeller Performance behind Ship in Waves. Ph.D. Thesis, TUHH, Hamburg, Germany, 2013.

Journal of
Marine Science and Engineering

Article

VIScous Vorticity Equation (VISVE) for Turbulent 2-D Flows with Variable Density and Viscosity

Spyros A. Kinnas

Ocean Engineering Group, Department of Civil, Architectural, and Environmental Engineering, The University of Texas at Austin, Austin, TX 78712, USA; kinnas@mail.utexas.edu; Tel.: +1-512-475-7969

Received: 23 January 2020; Accepted: 4 March 2020; Published: 11 March 2020

Abstract: The general vorticity equation for turbulent compressible 2-D flows with variable viscosity is derived, based on the Reynolds-Averaged Navier-Stokes (RANS) equations, and simplified versions of it are presented in the case of turbulent or cavitating flows around 2-D hydrofoils.

Keywords: vorticity; viscous flow; turbulent flow; mixture model; cavitating flow

1. Introduction

The vorticity equation has been utilized by several authors in the past to analyze the viscous flow around bodies. Vortex element (or particle, or blub) and vortex-in-cell methods have been used for several decades for the analysis of 2-D or 3-D flows, as described in Chorin [1], Christiansen [2], Leonard [3], Koumoutsakos et al. [4], Ould-Salihi et al. [5], Ploumhans et al. [6], Cottet and Poncet [7], and Cottet and Poncet [8]. Those methods essentially decouple the vortex dynamics (convection and stretching) from the effects of viscosity.

In recent years, the VIScous Vorticity Equation (VISVE), has been solved by using a finite volume method, *without* decoupling the vorticity dynamics from the effects of viscosity. This method has been applied to 2-D and 3-D hydrofoils, cylinders, as well as propeller blades as described in Tian [9], Tian and Kinnas [10], Wu et al. [11], Wu and Kinnas [12], Wu et al. [13], Wu and Kinnas [14]. The major advantage of this approach is that it requires a significantly smaller computational domain than RANS over which VISVE must be solved, as shown in Figure 1, due to the fact that the vorticity vector vanishes much closer to the body surface (in the order of 1–3 maximum body thickness), as opposed to the velocity vector which vanishes much farther (in the order of 5–10 body lengths) from the body surface. Due to the significantly smaller domain, VISVE requires a much smaller number of cells (5–50 times smaller) than RANS for the same accuracy of the predictions, and subsequently, a significantly smaller CPU time, as reported in Wu et al. [11].

All applications of VISVE mentioned above have addressed laminar flow of an incompressible fluid. However, in the case of incompressible turbulent flow the vorticity equation must be modified since the viscosity (molecular + turbulent) varies within the flow. In addition, in the case of cavitating flow, treated via a mixture model, both the density and the viscosity vary within the flow. In this work the general equation in terms of the mean vorticity is derived in the case of turbulent flows with variable density and viscosity, and then simplified in the case of turbulent or cavitating flows around 2-D hydrofoils.

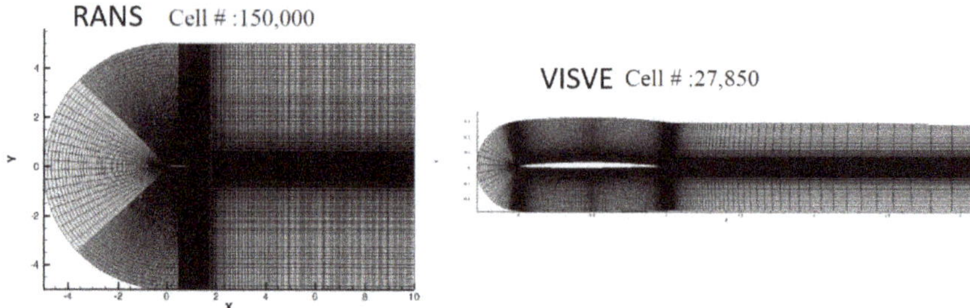

Figure 1. Typical domain and grids for flow around hydrofoil, from RANS with 150K cells (**left**) and from VISVE with 28K cells (**right**). Note the image on the right is drawn at a larger scale than that on the left.

2. Review of Reynolds-Averaged Navier-Stokes Equations

We consider flow of a fluid, where $\vec{q} = (u_1, u_2, u_3)$ is the *mean* velocity in a x_1, x_2, x_3 coordinate system. The Navier-Stokes equations are as follows:

$$\rho\frac{\partial\vec{q}}{\partial t} + \rho(\vec{q}\cdot\vec{\nabla})\vec{q} = -\vec{\nabla}p + \vec{\nabla}\cdot\mathbf{T} - \vec{\nabla}\Phi \tag{1}$$

where ρ is the density of the fluid, p is the *mean* pressure, Φ is the potential of a conservative body force per unit volume ($\Phi = \rho g z$ in the case the body force is the gravity, with the gravity acceleration g pointing in the -z direction.), and $\mathbf{T}$ is the (symmetric) tensor of viscous stresses

$$\mathbf{T} = \begin{bmatrix} \tau_{11} & \tau_{12} & \tau_{13} \\ \tau_{21} & \tau_{22} & \tau_{23} \\ \tau_{31} & \tau_{32} & \tau_{33} \end{bmatrix} \quad \text{and} \quad \vec{\nabla}\cdot\mathbf{T} = \begin{bmatrix} \frac{\partial\tau_{11}}{\partial x_1} + \frac{\partial\tau_{21}}{\partial x_2} + \frac{\partial\tau_{31}}{\partial x_3} \\ \frac{\partial\tau_{12}}{\partial x_1} + \frac{\partial\tau_{22}}{\partial x_2} + \frac{\partial\tau_{32}}{\partial x_3} \\ \frac{\partial\tau_{13}}{\partial x_1} + \frac{\partial\tau_{22}}{\partial x_2} + \frac{\partial\tau_{33}}{\partial x_3} \end{bmatrix} = \begin{bmatrix} \frac{\partial\tau_{11}}{\partial x_1} + \frac{\partial\tau_{12}}{\partial x_2} + \frac{\partial\tau_{13}}{\partial x_3} \\ \frac{\partial\tau_{21}}{\partial x_1} + \frac{\partial\tau_{22}}{\partial x_2} + \frac{\partial\tau_{23}}{\partial x_3} \\ \frac{\partial\tau_{31}}{\partial x_1} + \frac{\partial\tau_{32}}{\partial x_2} + \frac{\partial\tau_{33}}{\partial x_3} \end{bmatrix}$$

with

$$\tau_{ij} = \mu\left[\frac{\partial u_i}{\partial x_j} + \frac{\partial u_j}{\partial x_i}\right] - \delta_{ij}\frac{2}{N}\left[\mu\vec{\nabla}\cdot\vec{q} + \rho k\right] \tag{2}$$

where $\mu = \mu_m + \mu_t$ is the total dynamic viscosity, with μ_m being the molecular viscosity and μ_t being the turbulent viscosity of the fluid, under the Boussinesq approximation for the turbulent stresses; δ_{ij} is the Kronecker delta; k is the turbulent kinetic energy; $N = 2$ for two-dimensional flows, and $N = 3$ for three-dimensional flows. The second term in Equation (2) guarantees that $\tau_{ii} = -\rho\overline{u_i'u_i'} = -2\rho k$, where u_i' is the turbulent velocity in the ith direction.

Equation (1) can also be written as follows, as shown in the Appendix A:

$$\begin{aligned} \rho\frac{\partial\vec{q}}{\partial t} + \rho(\vec{q}\cdot\vec{\nabla})\vec{q} &= -\vec{\nabla}p + 2\vec{\nabla}\cdot(\mu\vec{\nabla}\vec{q}) + \vec{\nabla}\times(\mu\vec{\omega}) \\ &\quad -\frac{2}{N}\vec{\nabla}[\mu\vec{\nabla}\cdot\vec{q}] - \frac{2}{N}\vec{\nabla}[\rho k] - \vec{\nabla}\Phi \end{aligned} \tag{3}$$

where $\vec{\omega} = (\omega_1, \omega_2, \omega_3) = \vec{\nabla}\times\vec{q}$ is the vorticity of the flow, and the tensor $\vec{\nabla}\vec{q}$ is the *gradient* of $\vec{q}$, defined as follows

$$\vec{\nabla}\vec{q} = \begin{bmatrix} \frac{\partial u_1}{\partial x_1} & \frac{\partial u_2}{\partial x_1} & \frac{\partial u_3}{\partial x_1} \\ \frac{\partial u_1}{\partial x_2} & \frac{\partial u_2}{\partial x_2} & \frac{\partial u_3}{\partial x_2} \\ \frac{\partial u_1}{\partial x_3} & \frac{\partial u_2}{\partial x_3} & \frac{\partial u_3}{\partial x_3} \end{bmatrix}$$

With the help of some identities of vector calculus, as shown in the Appendix A, Equation (3) becomes:

$$\begin{aligned}\rho\frac{\partial\vec{q}}{\partial t}+\rho(\vec{q}\cdot\vec{\nabla})\vec{q} &= -\vec{\nabla}p+\mu\nabla^2\vec{q}+2(\vec{\nabla}\mu\cdot\vec{\nabla})\vec{q}+\mu\vec{\nabla}(\vec{\nabla}\cdot\vec{q})+\vec{\nabla}\mu\times\vec{\omega}\\ &\quad -\frac{2}{N}\vec{\nabla}[\mu\vec{\nabla}\cdot\vec{q}]-\frac{2}{N}\vec{\nabla}[\rho k]-\vec{\nabla}\Phi\end{aligned} \tag{4}$$

Dividing Equation (4) by ρ and rearranging will give us:

$$\begin{aligned}\frac{\partial\vec{q}}{\partial t}+(\vec{q}\cdot\vec{\nabla})\vec{q} &= -\frac{\vec{\nabla}p}{\rho}+\nu\nabla^2\vec{q}+\nu\vec{\nabla}(\vec{\nabla}\cdot\vec{q})+\frac{2}{\rho}(\vec{\nabla}\mu\cdot\vec{\nabla})\vec{q}+\frac{1}{\rho}\vec{\nabla}\mu\times\vec{\omega}\\ &\quad -\frac{2}{N\rho}\vec{\nabla}[\mu\vec{\nabla}\cdot\vec{q}]-\frac{2}{N\rho}\vec{\nabla}[\rho k]-\frac{\vec{\nabla}\Phi}{\rho}\end{aligned} \tag{5}$$

where $\nu=\frac{\mu}{\rho}$ is the kinematic (total) viscosity

2.1. Flow with Constant Density

In that case, since $\vec{\nabla}\cdot\vec{q}=0$ and $\rho = constant$ (In this work we consider incompressible flow when $\rho = constant$, as opposed to the more general definition of incompressible flow when $\frac{D\rho}{Dt}=0$), Equation (5) becomes:

$$\frac{\partial\vec{q}}{\partial t}+(\vec{q}\cdot\vec{\nabla})\vec{q}=-\frac{\vec{\nabla}p}{\rho}+\nu\nabla^2\vec{q}+2(\vec{\nabla}\nu\cdot\vec{\nabla})\vec{q}+\vec{\nabla}\nu\times\vec{\omega}-\frac{2}{N}\vec{\nabla}k-\frac{\vec{\nabla}\Phi}{\rho} \tag{6}$$

In the case of Newtonian fluid in laminar flow ($\mu=\mu_m$ =constant and $k=0$) and incompressible flow, in which case $\vec{\nabla}\cdot\vec{q}=0$, Equation (4) reduces to its most common form:

$$\rho\frac{\partial\vec{q}}{\partial t}+\rho(\vec{q}\cdot\vec{\nabla})\vec{q}=-\vec{\nabla}p+\mu_m\nabla^2\vec{q}-\vec{\nabla}\Phi \tag{7}$$

3. Vorticity Equation in 2D

We consider 2D flow, in which case $x_1=x$, $x_2=y$, $x_3=z$; $u_1=u$, $u_2=v$, $u_3=w$; $\frac{\partial}{\partial y}=0$ and $v=0$. The velocity has two components: $\vec{q}=(u,w)$, and the vorticity has only one component in the y direction $\omega_2=\omega=\frac{\partial u}{\partial z}-\frac{\partial w}{\partial x}$. As shown in the Appendix A the vorticity equation will become as follows:

$$\begin{aligned}\frac{\partial\omega}{\partial t}+\vec{\nabla}\cdot(\omega\vec{q}) &= -\vec{\nabla}\left(\frac{1}{\rho}\right)\times\vec{\nabla}p+\vec{\nabla}\nu\times\nabla^2\vec{q}+\nu\nabla^2\omega+\vec{\nabla}\nu\times\vec{\nabla}(\vec{\nabla}\cdot\vec{q})\\ &\quad -\omega\nabla^2\nu-\omega\vec{\nabla}\cdot\left[\nu\frac{\vec{\nabla}\rho}{\rho}\right]+\vec{\nabla}\nu\cdot\vec{\nabla}\omega+\frac{\nu}{\rho}\vec{\nabla}\rho\cdot\vec{\nabla}\omega\\ &\quad +2\left[\frac{\partial}{\partial z}\left(\frac{\vec{\nabla}\mu}{\rho}\right)\cdot\vec{\nabla}u-\frac{\partial}{\partial x}\left(\frac{\vec{\nabla}\mu}{\rho}\right)\cdot\vec{\nabla}w\right]\\ &\quad -\vec{\nabla}\left(\frac{1}{\rho}\right)\times\vec{\nabla}(\mu\vec{\nabla}\cdot\vec{q})-\vec{\nabla}\left(\frac{1}{\rho}\right)\times\vec{\nabla}(\rho k)-\vec{\nabla}\left(\frac{1}{\rho}\right)\times\vec{\nabla}\Phi\end{aligned} \tag{8}$$

As shown in the Appendix A Equation (8) may also be written as follows:

$$\begin{aligned}\frac{\partial\omega}{\partial t}+\vec{\nabla}\cdot(\omega\vec{q}) &= -\vec{\nabla}\left(\frac{1}{\rho}\right)\times\vec{\nabla}p+\nabla^2(\nu\omega)-2\omega\nabla^2\nu \\ &+2\vec{\nabla}\nu\times\vec{\nabla}(\vec{\nabla}\cdot\vec{q})-\omega\vec{\nabla}\cdot\left[\nu\frac{\vec{\nabla}\rho}{\rho}\right]+\frac{\nu}{\rho}\vec{\nabla}\rho\cdot\vec{\nabla}\omega \\ &+2\left[\frac{\partial}{\partial z}\left(\frac{\vec{\nabla}\mu}{\rho}\right)\cdot\vec{\nabla}u-\frac{\partial}{\partial x}\left(\frac{\vec{\nabla}\mu}{\rho}\right)\cdot\vec{\nabla}w\right] \\ &-\vec{\nabla}\left(\frac{1}{\rho}\right)\times\vec{\nabla}(\mu\vec{\nabla}\cdot\vec{q})-\vec{\nabla}\left(\frac{1}{\rho}\right)\times\vec{\nabla}(\rho k)-\vec{\nabla}\left(\frac{1}{\rho}\right)\times\vec{\nabla}\Phi \end{aligned} \quad (9)$$

3.1. 2D Flow of Fluid with Constant Density

In that case $\rho = constant$, $\vec{\nabla}\rho = 0$, $\vec{\nabla}\left(\frac{1}{\rho}\right) = 0$, and $\vec{\nabla}\cdot\vec{q} = 0$. Then Equation (9) reduces to:

$$\frac{\partial\omega}{\partial t}+\vec{\nabla}\cdot(\omega\vec{q}) = \nabla^2(\nu\omega)-2\omega\nabla^2\nu-2\frac{\partial\vec{\nabla}\nu}{\partial x}\cdot\vec{\nabla}w+2\frac{\partial\vec{\nabla}\nu}{\partial z}\cdot\vec{\nabla}u \quad (10)$$

or, by expressing $\omega = \frac{\partial u}{\partial z}-\frac{\partial w}{\partial x}$ in the second term on the RHS of Equation (10), and by making use of the continuity equation, $\frac{\partial u}{\partial x}+\frac{\partial w}{\partial z} = 0$, we get:

$$\frac{\partial\omega}{\partial t}+\vec{\nabla}\cdot(\omega\vec{q}) = \nabla^2(\nu\omega)+2\frac{\partial^2\nu}{\partial z^2}\frac{\partial w}{\partial x}-2\frac{\partial^2\nu}{\partial x^2}\frac{\partial u}{\partial z}+4\frac{\partial^2\nu}{\partial x\partial z}\frac{\partial u}{\partial x} \quad (11)$$

It is worth noting that Equation (10) is valid for turbulent flows (with the vorticity ω being the *mean* vorticity), but the turbulent kinetic energy k is *not* involved in the equation.

In the case of laminar flow of Newtonian fluid $\nu = \frac{\mu_m}{\rho} = constant$, Equation (10) becomes the vorticity equation in its most common diffusion equation form:

$$\frac{\partial\omega}{\partial t}+\vec{\nabla}\cdot(\omega\vec{q}) = \nu\nabla^2\omega \quad (12)$$

or, since $\vec{\nabla}\cdot\vec{q} = 0$:

$$\frac{D\omega}{Dt}\equiv\frac{\partial\omega}{\partial t}+\vec{q}\cdot\vec{\nabla}\omega = \nu\nabla^2\omega \quad (13)$$

3.2. 2D Flow around Hydrofoil

In that case we assume that the hydrofoil is placed along axis x, with the inflow also along x axis, and that $\frac{\partial}{\partial x} << \frac{\partial}{\partial z}$, especially within the narrow region close to the hydrofoil and its wake where $\omega \neq 0$. Then, as shown in the Appendix A, Equation (9) reduces to:

$$\frac{\partial\omega}{\partial t}+\vec{\nabla}\cdot(\omega\vec{q}) = \nabla^2(\nu\omega)+\vec{\nabla}\cdot\left[\nu\frac{\vec{\nabla}\rho}{\rho}\omega\right] \quad (14)$$

4. Conclusions and Future Work

The vorticity equation (in terms of the mean vorticity) is derived in the case of turbulent flows with variable density and viscosity, and its simplified version in the case of flows around 2-D hydrofoils has been provided. Equation (14) has already been implemented by Yao and Kinnas [15] to address the turbulent non-cavitating flow around 2-D hydrofoils and cylinders, as well as the cavitating laminar flow around 2-D hydrofoils, using the mixture model, by Xing et al. [16].

Representative results from the work of Yao and Kinnas [15] are shown in Figures 2–4 together with results from RANS by using the commercial code ANSYS/Fluent (https://www.ansys.com/

products/fluids/ansys-fluent). In this case Equation (14) was used where the turbulent viscosity was evaluated by *synchronous* coupling of VISVE with the open source RANS code OpenFOAM (https://www.openfoam.com/). As described in more detail in Yao and Kinnas [15], at every time step the velocities determined by VISVE were passed into the part of OpenFOAM which solved the k,ϵ equations and then returned the turbulent kinematic viscosity back to VISVE, in order to solve Equation (14) with the updated values of the kinematic viscosity.

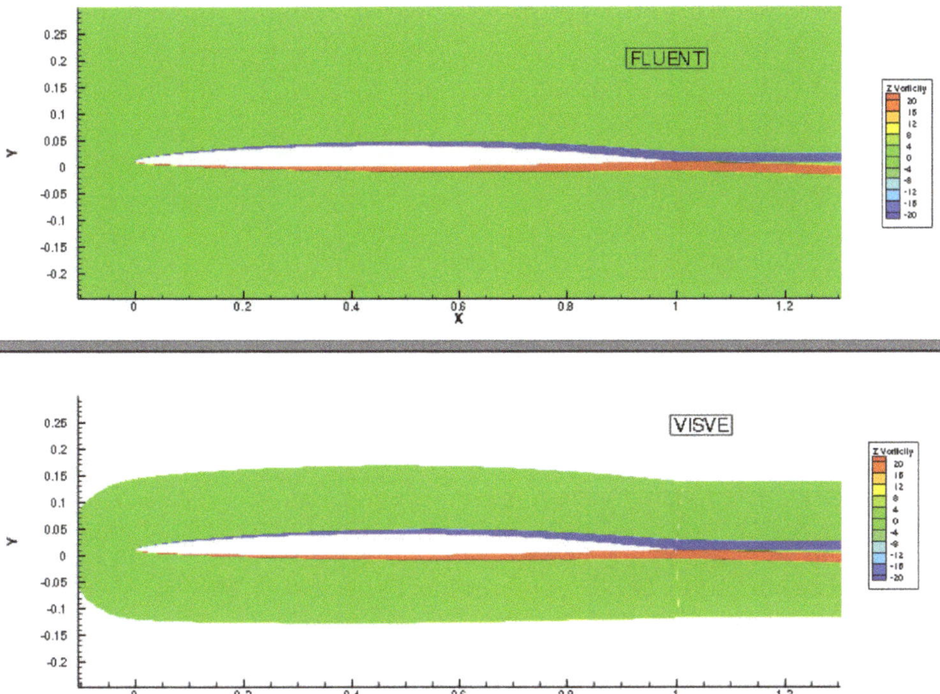

Figure 2. Vorticity contour plots from RANS (ANSYS/Fluent, **top**) and VISVE (**bottom**) for turbulent flow around hydrofoil. $Re = 2 \times 10^6$.

In the case of the hydrofoil the velocity profiles predicted by VISVE, shown in Figure 3, are in good agreement to those predicted by RANS. It should be noted that the hydrofoil assumption has also been used in the case of the cylinder, for which results are shown in Figure 4. The results from VISVE compare well with those from RANS, even though with somewhat larger differences downstream of the cylinder, up to the time shown in the figure, but deteriorate for later times as shown in Yao and Kinnas [15], once asymmetry appears between the top and bottom flow (not shown in this paper).

In the future, the author and his students intend to assess numerically the effect of the last three terms in the right-hand side of Equation (10), on the prediction. These terms are currently ignored, but they might affect the accuracy of predictions, especially in the case of a hydrofoil at high angles of attack where the hydrofoil assumptions, made in this paper, would not hold. The ultimate objective of this research is to extend VISVE in the case of turbulent flows around 3-D hydrofoils, and eventually propeller blades.

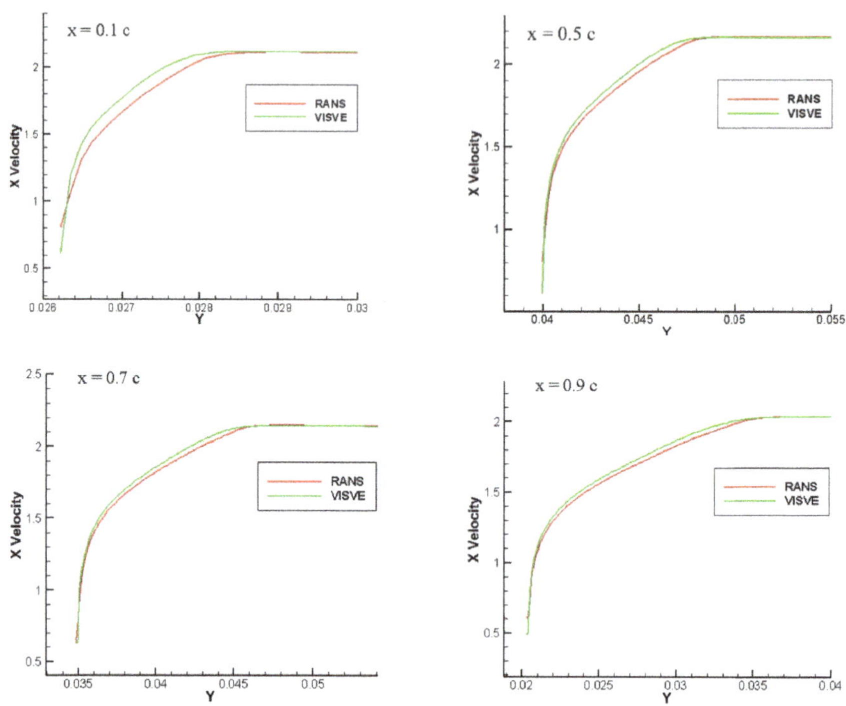

Figure 3. Velocity profiles of turbulent flow around hydrofoil, from RANS (ANSYS/Fluent) and from VISVE, at different locations along the chord, *c*, on the suction side. $Re = 2 \times 10^6$.

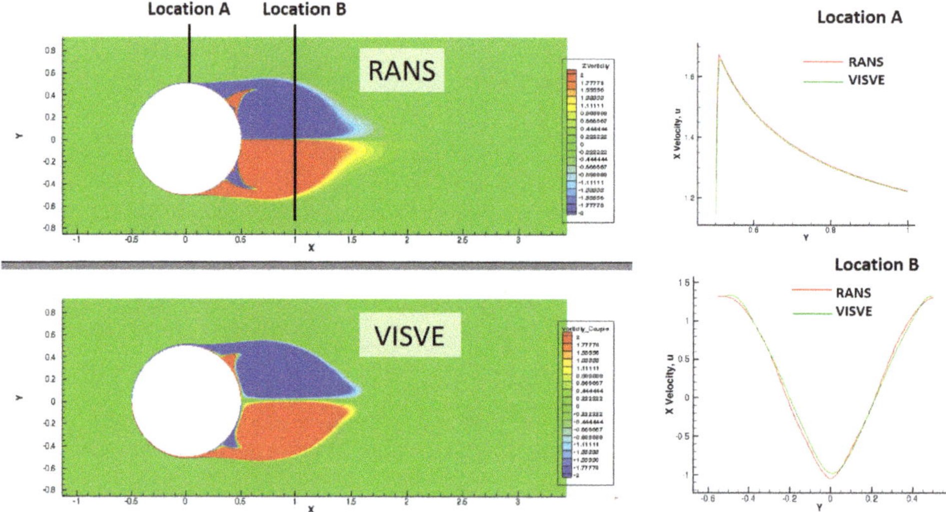

Figure 4. Vorticity contour plots and velocity profiles for turbulent flow over cylinder from RANS (ANSYS/Fluent) and from VISVE, at t = 4 s and $Re = 10^6$.

Acknowledgments: Support for this research was provided by the U.S. Office of Naval Research (Grant Number N00014-18-1-2276; Ki-Han Kim) and by Phase VIII of the "Consortium on Cavitation Performance of High Speed Propulsors".

Conflicts of Interest: The author declares no conflict of interest.

Appendix A. Proofs

Appendix A.1. How to Get from Equation (1) to Equation (3)

We consider the ith component of Equation (1):

$$\rho\frac{\partial u_i}{\partial t} + \rho\vec{q}\cdot\vec{\nabla}u_i = -\frac{\partial p}{\partial x_i} + \frac{\partial \tau_{ji}}{\partial x_j} - \frac{\partial \Phi}{\partial x_i} = -\frac{\partial p}{\partial x_i} + \frac{\partial \tau_{ij}}{\partial x_j} - \frac{\partial \Phi}{\partial x_i} \tag{A1}$$

with the $\frac{\partial \tau_{ji}}{\partial x_j}$ term meant in the Einstein notation with $j = 1,2,3$

We then consider the rate of strain term of the expression for τ_{ij}, as given by Equation (2):

$$\mu\left[\frac{\partial u_i}{\partial x_j} + \frac{\partial u_j}{\partial x_i}\right] = 2\mu\frac{\partial u_i}{\partial x_j} + \mu\left[\frac{\partial u_j}{\partial x_i} - \frac{\partial u_i}{\partial x_j}\right] = 2\mu\frac{\partial u_i}{\partial x_j} + \epsilon_{kij}\mu\omega_k \tag{A2}$$

where ϵ_{kij} is the Levi-Civita symbol.

Then for given i and with $j = 1,2,3$, using the Einstein notation we have:

$$\begin{aligned}
\frac{\partial}{\partial x_j}\left[\mu\left(\frac{\partial u_i}{\partial x_j} + \frac{\partial u_j}{\partial x_i}\right)\right] &= 2\frac{\partial}{\partial x_j}\left(\mu\frac{\partial u_i}{\partial x_j}\right) + \epsilon_{kij}\frac{\partial(\mu\omega_k)}{\partial x_j} \\
&= 2\vec{\nabla}\cdot(\mu\vec{\nabla}u_i) + \epsilon_{ijk}\frac{\partial(\mu\omega_k)}{\partial x_j} \\
&= 2\vec{\nabla}\cdot(\mu\vec{\nabla}u_i) + \epsilon_{ij'k'}\frac{\partial(\mu\omega_{k'})}{\partial x_{j'}} + \epsilon_{ik'j'}\frac{\partial(\mu\omega_{j'})}{\partial x_{k'}} \quad (i \neq k' \neq j' \neq i) \\
&= 2\vec{\nabla}\cdot(\mu\vec{\nabla}u_i) + \epsilon_{ij'k'}\frac{\partial(\mu\omega_{k'})}{\partial x_{j'}} - \epsilon_{ij'k'}\frac{\partial(\mu\omega_{j'})}{\partial x_{k'}} \quad (i \neq k' \neq j' \neq i) \\
&= 2\vec{\nabla}\cdot(\mu\vec{\nabla}u_i) + [\vec{\nabla}\times(\mu\vec{\omega})]_i
\end{aligned} \tag{A3}$$

where $[\vec{\nabla}\times(\mu\vec{\omega})]_i$ means the ith component of $\vec{\nabla}\times(\mu\vec{\omega})$

Based on Equations (2) and (A3) we get:

$$\frac{\partial \tau_{ij}}{\partial x_j} = 2\vec{\nabla}\cdot(\mu\vec{\nabla}u_i) + [\vec{\nabla}\times(\mu\vec{\omega})]_i - \frac{2}{N}\frac{\partial}{\partial x_i}\left[\mu\vec{\nabla}\cdot\vec{q} + \rho k\right] \tag{A4}$$

which eventually leads to Equation (3).

Appendix A.2. How to Get from Equation (3) to Equation (4)

$$\vec{\nabla}\cdot(\mu\vec{\nabla}\vec{q}) = \begin{bmatrix}\vec{\nabla}\cdot(\mu\vec{\nabla}u_1)\\ \vec{\nabla}\cdot(\mu\vec{\nabla}u_2)\\ \vec{\nabla}\cdot(\mu\vec{\nabla}u_3)\end{bmatrix} \tag{A5}$$

Using vector identities we get:

$$\vec{\nabla}\cdot(\mu\vec{\nabla}u_i) = \vec{\nabla}\mu\cdot\vec{\nabla}u_i + \mu\nabla^2 u_i \tag{A6}$$

or

$$\vec{\nabla}\cdot(\mu\vec{\nabla}\vec{q}) = (\vec{\nabla}\mu\cdot\vec{\nabla})\vec{q} + \mu\nabla^2\vec{q} \tag{A7}$$

Then

$$\vec{\nabla} \times (\mu\vec{\omega}) = \mu\vec{\nabla} \times \vec{\omega} + \vec{\nabla}\mu \times \vec{\omega} \tag{A8}$$

$$\vec{\nabla} \times \vec{\omega} = \vec{\nabla} \times (\vec{\nabla} \times \vec{q}) = \vec{\nabla}(\vec{\nabla} \cdot \vec{q}) - \nabla^2\vec{q} \tag{A9}$$

Using the equations above we get to Equation (4).

Appendix A.3. How to Get from Equation (5) to Equation (8)

We consider 2D flow, in which case $x_1 = x, x_2 = y, x_3 = z; u_1 = u, u_2 = v, u_3 = w; \frac{\partial}{\partial y} = 0$ and $v = 0$. The velocity has two components: $\vec{q} = (u, w)$, and the vorticity has only one component in the y direction $\omega_2 = \omega = \frac{\partial u}{\partial z} - \frac{\partial w}{\partial x}$.

First:

$$\vec{\nabla} \times \left(\frac{\partial\vec{q}}{\partial t}\right) = \frac{\partial(\vec{\nabla} \times \vec{q})}{\partial t} = \frac{\partial\vec{\omega}}{\partial t} \tag{A10}$$

Then by using the identity:

$$(\vec{q} \cdot \vec{\nabla})\vec{q} = (\vec{\nabla} \times \vec{q}) \times \vec{q} + \vec{\nabla}\left(\frac{q^2}{2}\right) = \vec{\omega} \times \vec{q} + \vec{\nabla}\left(\frac{q^2}{2}\right) \tag{A11}$$

we get

$$\vec{\nabla} \times [(\vec{q} \cdot \vec{\nabla})\vec{q}] = \vec{\nabla} \times (\vec{\omega} \times \vec{q}) \tag{A12}$$

due to the identity:$\vec{\nabla} \times \vec{\nabla} f = 0$

Then using the following identities:

$$\vec{\nabla} \times (\vec{\omega} \times \vec{q}) = (\vec{q} \cdot \vec{\nabla})\vec{\omega} - (\vec{\omega} \cdot \vec{\nabla})\vec{q} + \vec{\omega}(\vec{\nabla} \cdot \vec{q}) - \vec{q}(\vec{\nabla} \cdot \vec{\omega}) \tag{A13}$$

$$\vec{\nabla} \cdot \vec{\omega} = \vec{\nabla} \cdot (\vec{\nabla} \times \vec{q}) = 0 \tag{A14}$$

We get:

$$\vec{\nabla} \times [\text{LHS of equation (5)}] = \frac{\partial\vec{\omega}}{\partial t} + (\vec{q} \cdot \vec{\nabla})\vec{\omega} - (\vec{\omega} \cdot \vec{\nabla})\vec{q} + \vec{\omega}(\vec{\nabla} \cdot \vec{q}) \tag{A15}$$

Now, in 2-D, since the vortex stretching term $(\vec{\omega} \cdot \vec{\nabla})\vec{q} = 0$, the above equation becomes:

$$\vec{\nabla} \times [\text{LHS of equation (5)}]_{2D} = \frac{\partial\omega}{\partial t} + (\vec{q} \cdot \vec{\nabla})\omega + \omega(\vec{\nabla} \cdot \vec{q}) = \frac{\partial\omega}{\partial t} + \vec{\nabla} \cdot (\omega\vec{q}) \tag{A16}$$

We now take the $\vec{\nabla}\times$ of each term on the RHS of Equation (5):

- The $\vec{\nabla}\times$ of the 1st term on the RHS of (5):

$$\vec{\nabla} \times \left(\frac{\vec{\nabla}p}{\rho}\right) = \vec{\nabla}\left(\frac{1}{\rho}\right) \times \vec{\nabla}p + \left(\frac{1}{\rho}\right)\vec{\nabla} \times \vec{\nabla}p = \vec{\nabla}\left(\frac{1}{\rho}\right) \times \vec{\nabla}p \tag{A17}$$

  The $\vec{\nabla}\times$ of the last 3 terms of Equation (5) are treated in a similar manner as that of the first term, and the corresponding resulting terms are the last 3 terms of Equation (8). Please note the first and the last 3 terms in Equation (8) vanish in the case $\rho = constant$.
- The $\vec{\nabla}\times$ of the 2nd term on the RHS of (5):

$$\vec{\nabla} \times (\nu\nabla^2\vec{q}) = \vec{\nabla}\nu \times \nabla^2\vec{q} + \nu\nabla^2(\vec{\nabla} \times \vec{q}) = \vec{\nabla}\nu \times \nabla^2\vec{q} + \nu\nabla^2\omega \tag{A18}$$

- The $\vec{\nabla}\times$ of the 3rd term on the RHS of (5):

$$\vec{\nabla} \times [\nu \vec{\nabla}(\vec{\nabla} \cdot \vec{q})] = \vec{\nabla}\nu \times \vec{\nabla}(\vec{\nabla} \cdot \vec{q}) + \nu \vec{\nabla} \times [\vec{\nabla}(\vec{\nabla} \cdot \vec{q})] = \vec{\nabla}\nu \times \vec{\nabla}(\vec{\nabla} \cdot \vec{q}) \tag{A19}$$

- The $\vec{\nabla}\times$ of the 4th term (divided by 2) on the RHS of (5):

  We only handle this term in 2D. In that case:

$$\left\{\vec{\nabla} \times \left[\left(\frac{\vec{\nabla}\mu}{\rho} \cdot \vec{\nabla}\right)\vec{q}\right]\right\}_{2D} = \frac{\partial}{\partial z}\left(\frac{\vec{\nabla}\mu}{\rho} \cdot \vec{\nabla}u\right) - \frac{\partial}{\partial x}\left(\frac{\vec{\nabla}\mu}{\rho} \cdot \vec{\nabla}w\right) = \tag{A20}$$

$$= \frac{\partial}{\partial z}\left(\frac{\vec{\nabla}\mu}{\rho}\right) \cdot \vec{\nabla}u + \frac{\vec{\nabla}\mu}{\rho} \cdot \frac{\partial \vec{\nabla}u}{\partial z} - \frac{\partial}{\partial x}\left(\frac{\vec{\nabla}\mu}{\rho}\right) \cdot \vec{\nabla}w - \frac{\vec{\nabla}\mu}{\rho} \cdot \frac{\partial \vec{\nabla}w}{\partial x} = \tag{A21}$$

$$= \frac{\partial}{\partial z}\left(\frac{\vec{\nabla}\mu}{\rho}\right) \cdot \vec{\nabla}u - \frac{\partial}{\partial x}\left(\frac{\vec{\nabla}\mu}{\rho}\right) \cdot \vec{\nabla}w + \frac{\vec{\nabla}\mu}{\rho} \cdot \vec{\nabla}\left(\frac{\partial u}{\partial z} - \frac{\partial w}{\partial x}\right) = \tag{A22}$$

$$= \frac{\partial}{\partial z}\left(\frac{\vec{\nabla}\mu}{\rho}\right) \cdot \vec{\nabla}u - \frac{\partial}{\partial x}\left(\frac{\vec{\nabla}\mu}{\rho}\right) \cdot \vec{\nabla}w + \frac{\vec{\nabla}\mu}{\rho} \cdot \vec{\nabla}\omega \tag{A23}$$

- The $\vec{\nabla}\times$ of the 5th term on the RHS of (5):

$$\vec{\nabla} \times \left(\frac{1}{\rho}\vec{\nabla}\mu \times \vec{\omega}\right) = \vec{\nabla}\left(\frac{1}{\rho}\right) \times (\vec{\nabla}\mu \times \vec{\omega}) + \frac{1}{\rho}\vec{\nabla} \times (\vec{\nabla}\mu \times \vec{\omega}) \tag{A24}$$

  then

$$\vec{\nabla} \times (\vec{\nabla}\mu \times \vec{\omega}) = (\vec{\omega} \cdot \vec{\nabla})\vec{\nabla}\mu - (\vec{\nabla}\mu \cdot \vec{\nabla})\vec{\omega} + \vec{\nabla}\mu(\vec{\nabla} \cdot \vec{\omega}) - \vec{\omega}(\vec{\nabla} \cdot \vec{\nabla}\mu) \tag{A25}$$

  with $\vec{\nabla} \cdot \vec{\omega} = 0$, and since $\vec{\omega} \cdot \vec{\nabla} = 0$ in 2D we get:

$$[\vec{\nabla} \times (\vec{\nabla}\mu \times \vec{\omega})]_{2D} = -\vec{\nabla}\mu \cdot \vec{\nabla}\omega - \omega\nabla^2\mu \tag{A26}$$

  In addition, in 2D:

$$(\vec{\nabla}\mu \times \vec{\omega})_{2D} = \left(\frac{\partial\mu}{\partial x}, 0, \frac{\partial\mu}{\partial z}\right) \times (0, \omega, 0) = \left(-\omega\frac{\partial\mu}{\partial z}, 0, \omega\frac{\partial\mu}{\partial x}\right) \tag{A27}$$

  and

$$\left[\vec{\nabla}\left(\frac{1}{\rho}\right) \times (\vec{\nabla}\mu \times \vec{\omega})\right]_{2D} = \left(\frac{\partial(1/\rho)}{\partial x}, 0, \frac{\partial(1/\rho)}{\partial z}\right) \times \left(-\omega\frac{\partial\mu}{\partial z}, 0, \omega\frac{\partial\mu}{\partial x}\right) = \tag{A28}$$

$$= -\omega\left[\frac{\partial(1/\rho)}{\partial x}\frac{\partial\mu}{\partial x} + \frac{\partial(1/\rho)}{\partial z}\frac{\partial\mu}{\partial z}\right] = -\omega\vec{\nabla}\left(\frac{1}{\rho}\right) \cdot \vec{\nabla}\mu \tag{A29}$$

  Finally

$$\left[\vec{\nabla} \times \left(\frac{1}{\rho}\vec{\nabla}\mu \times \vec{\omega}\right)\right]_{2D} = -\omega\vec{\nabla}\left(\frac{1}{\rho}\right) \cdot \vec{\nabla}\mu - \frac{\vec{\nabla}\mu \cdot \vec{\nabla}\omega}{\rho} - \frac{\omega\nabla^2\mu}{\rho} = \tag{A30}$$

$$= -\omega\vec{\nabla}\left(\frac{\vec{\nabla}\mu}{\rho}\right) - \frac{\vec{\nabla}\mu \cdot \vec{\nabla}\omega}{\rho} \tag{A31}$$

The combination of the $\vec{\nabla}\times$ of the 4th and the 5th terms will give us:

$$2\frac{\partial}{\partial z}\left(\frac{\vec{\nabla}\mu}{\rho}\right) \cdot \vec{\nabla}u - 2\frac{\partial}{\partial x}\left(\frac{\vec{\nabla}\mu}{\rho}\right) \cdot \vec{\nabla}w + \frac{\vec{\nabla}\mu \cdot \vec{\nabla}\omega}{\rho} - \omega\vec{\nabla}\left(\frac{\vec{\nabla}\mu}{\rho}\right) \tag{A32}$$

The last two terms above can also be rewritten by using $\mu = \nu\rho$ and $\vec{\nabla}(\nu\rho) = \nu\vec{\nabla}\rho + \rho\vec{\nabla}\nu$, which renders the RHS of Equation (8)

Appendix A.4. How to Get from Equation (8) to Equation (9)

In 2D $v = 0$ and $\frac{\partial}{\partial y} = 0$, and $\omega = \frac{\partial u}{\partial z} - \frac{\partial w}{\partial x}$

$$\begin{aligned}
[\vec{\nabla}\nu \cdot \vec{\nabla}\omega]_{2D} = {} & \frac{\partial \nu}{\partial x}\frac{\partial \omega}{\partial x} + \frac{\partial \nu}{\partial z}\frac{\partial \omega}{\partial z} = \frac{\partial \nu}{\partial x}\frac{\partial}{\partial x}\left(\frac{\partial u}{\partial z} - \frac{\partial w}{\partial x}\right) + \frac{\partial \nu}{\partial z}\frac{\partial}{\partial z}\left(\frac{\partial u}{\partial z} - \frac{\partial w}{\partial x}\right) = \\
& \frac{\partial \nu}{\partial x}\frac{\partial^2 u}{\partial x \partial z} - \frac{\partial \nu}{\partial x}\frac{\partial^2 w}{\partial x^2} + \frac{\partial \nu}{\partial z}\frac{\partial^2 u}{\partial z^2} - \frac{\partial \nu}{\partial z}\frac{\partial^2 w}{\partial x \partial z} = \\
& \frac{\partial \nu}{\partial x}\frac{\partial}{\partial z}\left(-\frac{\partial w}{\partial z} + \vec{\nabla}\cdot\vec{q}\right) - \frac{\partial \nu}{\partial x}\frac{\partial^2 w}{\partial x^2} + \frac{\partial \nu}{\partial z}\frac{\partial^2 u}{\partial z^2} - \frac{\partial \nu}{\partial z}\frac{\partial}{\partial z}\left(-\frac{\partial u}{\partial x} + \vec{\nabla}\cdot\vec{q}\right) = \\
& -\frac{\partial \nu}{\partial x}\nabla^2 w + \frac{\partial \nu}{\partial z}\nabla^2 u + \frac{\partial \nu}{\partial x}\frac{\partial(\vec{\nabla}\cdot\vec{q})}{\partial z} - \frac{\partial \nu}{\partial z}\frac{\partial(\vec{\nabla}\cdot\vec{q})}{\partial x} = \\
& \left[\vec{\nabla}\nu \times \nabla^2\vec{q}\right]_{2D} - [\vec{\nabla}\nu \times \vec{\nabla}(\vec{\nabla}\cdot\vec{q})]_{2D} \qquad \text{(A33)}
\end{aligned}$$

or

$$\left[\vec{\nabla}\nu \times \nabla^2\vec{q}\right]_{2D} = [\vec{\nabla}\nu \cdot \vec{\nabla}\omega]_{2D} + [\vec{\nabla}\nu \times \vec{\nabla}(\vec{\nabla}\cdot\vec{q})]_{2D} \qquad \text{(A34)}$$

where we have made use of $\frac{\partial u}{\partial x} + \frac{\partial w}{\partial z} = \vec{\nabla}\cdot\vec{q}$

We finally get to Equation (9) by considering the identity:

$$\nabla^2(\nu\omega) = \nu\nabla^2\omega + \omega\nabla^2\nu + 2\vec{\nabla}\nu \cdot \vec{\nabla}\omega \qquad \text{(A35)}$$

Appendix A.5. How to Get from Equation (9) to Equation (14)

In that case we assume that the hydrofoil is placed along axis x, with the inflow also along x axis, and that $\frac{\partial}{\partial x} << \frac{\partial}{\partial z}$, especially within the narrow region close to the hydrofoil and its wake where $\omega \neq 0$.

Then, since we ignore $\frac{\partial}{\partial x}$ then all $\vec{\nabla}F$ point in the z direction and thus $\vec{\nabla}F \times \vec{\nabla}G \approx 0$. In addition the following approximations can be made:

$$-2\omega\nabla^2\nu \approx -2\omega\frac{\partial^2 \nu}{\partial z^2} \qquad \text{(A36)}$$

$$-\omega\vec{\nabla}\left(\nu\frac{\partial\vec{\nabla}\rho}{\rho}\right) \approx -\omega\frac{\partial}{\partial z}\left(\frac{\nu}{\rho}\frac{\partial \rho}{\partial z}\right) \qquad \text{(A37)}$$

$$\frac{\nu}{\rho}\vec{\nabla}\rho \cdot \vec{\nabla}\omega \approx \frac{\nu}{\rho}\frac{\partial \rho}{\partial z}\frac{\partial \omega}{\partial z} \qquad \text{(A38)}$$

$$2\frac{\partial}{\partial z}\left(\frac{\vec{\nabla}\mu}{\rho}\right)\cdot\vec{\nabla}u \approx 2\frac{\partial}{\partial z}\left(\frac{1}{\rho}\frac{\partial \mu}{\partial z}\right)\frac{\partial u}{\partial z} \approx 2\frac{\partial}{\partial z}\left(\frac{1}{\rho}\frac{\partial \mu}{\partial z}\right)\omega = \qquad \text{(A39)}$$

$$2\frac{\partial}{\partial z}\left(\frac{\partial \nu}{\partial z} + \frac{\nu}{\rho}\frac{\partial \rho}{\partial z}\right)\omega = 2\omega\frac{\partial^2 \nu}{\partial z^2} + 2\omega\frac{\partial}{\partial z}\left(\frac{\nu}{\rho}\frac{\partial \rho}{\partial z}\right) \qquad \text{(A40)}$$

By adding the approximate expressions for all the terms shown above we get

$$\frac{\nu}{\rho}\frac{\partial \rho}{\partial z}\frac{\partial \omega}{\partial z} + \omega\frac{\partial}{\partial z}\left(\frac{\nu}{\rho}\frac{\partial \rho}{\partial z}\right) = \frac{\partial}{\partial z}\left(\omega\frac{\nu}{\rho}\frac{\partial \rho}{\partial z}\right) \qquad \text{(A41)}$$

Finally we get to Equation (14) by making the following approximation:

$$\frac{\partial}{\partial z}\left(\omega\frac{\nu}{\rho}\frac{\partial\rho}{\partial z}\right)\approx\vec{\nabla}\cdot\left[\nu\frac{\vec{\nabla}\rho}{\rho}\omega\right] \tag{A42}$$

References

1. Chorin, A.J. Numerical study of slightly viscous flow. *J. Fluid Mech.* **1973**, *57*, 785–796. [CrossRef]
2. Christiansen, I. Numerical simulation of hydrodynamics by the method of point vortices. *J. Comput. Phys.* **1973**, *13*, 363–379. [CrossRef]
3. Leonard, A. Vortex methods for flow simulation. *J. Comput. Phys.* **1980**, *37*, 289–335. [CrossRef]
4. Koumoutsakos, P.; Leonard, A.; Pepin, F. Boundary conditions for viscous vortex methods. *J. Comput. Phys.* **1994**, *113*, 52–61. [CrossRef]
5. Ould-Salihi, M.L.; Cottet, G.H.; El Hamraoui, M. Blending finite-difference and vortex methods for incompressible flow computations. *SIAM J. Sci. Comput.* **2001**, *22*, 1655–1674. [CrossRef]
6. Ploumhans, P.; Winckelmans, G.; Salmon, J.K.; Leonard, A.; Warren, M. Vortex methods for direct numerical simulation of three-dimensional bluff body flows: Application to the sphere at Re= 300, 500, and 1000. *J. Comput. Phys.* **2002**, *178*, 427–463. [CrossRef]
7. Cottet, G.H.; Poncet, P. Particle methods for direct numerical simulations of three-dimensional wakes. *J. Turbul.* **2002**, *3*, 1–9. [CrossRef]
8. Cottet, G.H.; Poncet, P. Advances in direct numerical simulations of 3D wall-bounded flows by Vortex-in-Cell methods. *J. Comput. Phys.* **2004**, *193*, 136–158. [CrossRef]
9. Tian, Y. Leading Edge Vortex Modeling and Its Effect on Propulsor Performance. Ph.D. Thesis, Ocean Engineering Group, CAEE, UT Austin, Austin, TX, USA, 2014.
10. Tian, Y.; Kinnas, S.A. A Viscous Vorticity Method for Propeller Tip Flows and Leading Edge Vortex. In Proceedings of the 4th International Symposium on Marine Propulsors, SMP15, Austin, TX, USA, 31 May–4 June 2015.
11. Wu, C.; Xing, L.; Kinnas, S.A. A Viscous Vorticity Equation (VISVE) Method applied to 2-D and 3-D hydrofoils in both forward and backing conditions. In Proceedings of the 23rd SNAME Offshore Symposium, Houston, TX, USA, 14 February 2018.
12. Wu, C.; Kinnas, S.A. A 3-D VIScous Vorticity Equation (VISVE) Method Applied to Flow Past a Hydrofoil of Elliptical Planform and a Propeller. In Proceedings of the 6th International Symposium on Marine Propulsors, SMP19, Rome, Italy, 26–30 May 2019.
13. Wu, C.; Kinnas, S.A.; Li, Z.; Wu, Y. A conservative viscous vorticity method for unsteady unidirectional and oscillatory flow past a circular cylinder. *Ocean Eng.* **2019**, *191*, 106504. [CrossRef]
14. Wu, C.; Kinnas, S.A. Parallel Implementation of a VIScous Vorticity Equation (VISVE) Method in 3-D Incompressible Flow. *J. Comput. Phys.* **2019**. under review.
15. Yao, H.; Kinnas, S.A. Coupling Viscous Vorticity Equation (VISVE) Method with OpenFOAM to Predict Turbulent Flow Around 2-D Hydrofoils and Cylinders. In Proceedings of the 29th International Ocean and Polar Engineering Conference, Honolulu, HI, USA, 16–21 June 2019.
16. Xing, L.; Wu, C.; Kinnas, S.A. VISVE, a Vorticity Based Model Applied to Cavitating Flow around a 2-D Hydrofoil. In Proceedings of the 10th International Symposium on Cavitation (CAV2018), Baltimore, MD, USA, 14–16 May 2018; ASME Press: New York, NY, USA, 2018.

MDPI
St. Alban-Anlage 66
4052 Basel
Switzerland
Tel. +41 61 683 77 34
Fax +41 61 302 89 18
www.mdpi.com

Journal of Marine Science and Engineering Editorial Office
E-mail: jmse@mdpi.com
www.mdpi.com/journal/jmse

www.ingramcontent.com/pod-product-compliance
Lightning Source LLC
LaVergne TN
LVHW070931160826
845679LV00018B/1771